BIOSURFACTANTS

SURFACTANT SCIENCE SERIES

1. Nonionic Surfactants, *edited by Martin J. Schick* (see also Volumes 19 and 23)
2. Solvent Properties of Surfactant Solutions, *edited by Kozo Shinoda* (out of print)
3. Surfactant Biodegradation, *by R. D. Swisher* (see Volume 18)
4. Cationic Surfactants, *edited by Eric Jungermann* (see also Volumes 34 and 37)
5. Detergency: Theory and Test Methods (in three parts), *edited by W. G. Cutler and R. C. Davis* (see also Volume 20)
6. Emulsions and Emulsion Technology (in three parts), *edited by Kenneth J. Lissant*
7. Anionic Surfactants (in two parts), *edited by Warner M. Linfield* (out of print)
8. Anionic Surfactants: Chemical Analysis, *edited by John Cross* (out of print)
9. Stabilization of Colloidal Dispersions by Polymer Adsorption, *by Tatsuo Sato and Richard Ruch*
10. Anionic Surfactants: Biochemistry, Toxicology, Dermatology, *edited by Christian Gloxhuber* (see Volume 43)
11. Anionic Surfactants: Physical Chemistry of Surfactant Action, *edited by E. H. Lucassen-Reynders* (out of print)
12. Amphoteric Surfactants, *edited by B. R. Bluestein and Clifford L. Hilton* (out of print)
13. Demulsification: Industrial Applications, *by Kenneth J. Lissant*
14. Surfactants in Textile Processing, *by Arved Datyner*
15. Electrical Phenomena at Interfaces: Fundamentals, Measurements, and Applications, *edited by Ayao Kitahara and Akira Watanabe*
16. Surfactants in Cosmetics, *edited by Martin M. Rieger*
17. Interfacial Phenomena: Equilibrium and Dynamic Effects, *by Clarence A. Miller and P. Neogi*
18. Surfactant Biodegradation, Second Edition, Revised and Expanded, *by R. D. Swisher*

19. Nonionic Surfactants: Chemical Analysis, *edited by John Cross*
20. Detergency: Theory and Technology, *edited by W. Gale Cutler and Erik Kissa*
21. Interfacial Phenomena in Apolar Media, *edited by Hans-Friedrich Eicke and Geoffrey D. Parfitt*
22. Surfactant Solutions: New Methods of Investigation, *edited by Raoul Zana*
23. Nonionic Surfactants: Physical Chemistry, *edited by Martin J. Schick*
24. Microemulsion Systems, *edited by Henri L. Rosano and Marc Clausse*
25. Biosurfactants and Biotechnology, *edited by Naim Kosaric, W. L. Cairns, and Neil C. C. Gray*
26. Surfactants in Emerging Technologies, *edited by Milton J. Rosen*
27. Reagents in Mineral Technology, *edited by P. Somasundaran and Brij M. Moudgil*
28. Surfactants in Chemical/Process Engineering, *edited by Darsh T. Wasan, Martin E. Ginn, and Dinesh O. Shah*
29. Thin Liquid Films, *edited by I. B. Ivanov*
30. Microemulsions and Related Systems: Formulation, Solvency, and Physical Properties, *edited by Maurice Bourrel and Robert S. Schecter*
31. Crystallization and Polymorphism of Fats and Fatty Acids, *edited by Nissim Garti and Kiyotaka Sato*
32. Interfacial Phenomena in Coal Technology, *edited by Gregory D. Botsaris and Yuli M. Glazman*
33. Surfactant-Based Separation Processes, *edited by John F. Scamehorn and Jeffrey H. Harwell*
34. Cationic Surfactants: Organic Chemistry, *edited by James M. Richmond*
35. Alkylene Oxides and Their Polymers, *by F. E. Bailey, Jr., and Joseph V. Koleske*
36. Interfacial Phenomena in Petroleum Recovery, *edited by Norman R. Morrow*
37. Cationic Surfactants: Physical Chemistry, *edited by Donn N. Rubingh and Paul M. Holland*
38. Kinetics and Catalysis in Microheterogeneous Systems, *edited by M. Grätzel and K. Kalyanasundaram*
39. Interfacial Phenomena in Biological Systems, *edited by Max Bender*
40. Analysis of Surfactants, *by Thomas M. Schmitt*
41. Light Scattering by Liquid Surfaces and Complementary Techniques, *edited by Dominique Langevin*
42. Polymeric Surfactants, *by Irja Piirma*
43. Anionic Surfactants: Biochemistry, Toxicology, Dermatology. Second Edition, Revised and Expanded, *edited by Christian Gloxhuber and Klaus Künstler*
44. Organized Solutions: Surfactants in Science and Technology, *edited by Stig E. Friberg and Björn Lindman*
45. Defoaming: Theory and Industrial Applications, *edited by P. R. Garrett*

46. Mixed Surfactant Systems, *edited by Keizo Ogino and Masahiko Abe*
47. Coagulation and Flocculation: Theory and Applications, *edited by Bohuslav Dobiáš*
48. Biosurfactants: Production · Properties · Applications, *edited by Naim Kosaric*
49. Wettability, *edited by John C. Berg*

ADDITIONAL VOLUMES IN PREPARATION

Fluorinated Surfactants, *Erik Kissa*

BIOSURFACTANTS

Production • Properties • Applications

edited by
Naim Kosaric
University of Western Ontario
London, Ontario, Canada

CRC Press is an imprint of the
Taylor & Francis Group, an **informa** business

CRC Press
Taylor & Francis Group
6000 Broken Sound Parkway NW, Suite 300
Boca Raton, FL 33487-2742

First issued in paperback 2019

ISBN-13: 978-0-8247-8811-7 (hbk)
ISBN-13: 978-0-367-40245-7 (pbk)

Library of Congress Cataloging-in-Publication Data

Biosurfactants : production · properties · applications / edited by Naim Kosaric.
p. cm. -- (Surfactant science series ; v. 48)
Includes bibliographical references and index.
ISBN 0-8247-8811-7 (alk. paper)
1. Biosurfactants. I. Kosaric, Naim. II. Series.
TP248.B57B58 1993
668'.14--dc20 92-40312
CIP

Visit the Taylor & Francis Web site at
http://www.taylorandfrancis.com

and the CRC Press Web site at
http://www.crcpress.com

Preface

Following our first book on biosurfactants (*Biosurfactants and Biotechnology*, Marcel Dekker, 1987), this book further expands this exciting and relatively new field of biotechnology. In the last five years, interest in and active research on, as well as applications of, biosurfactants, have increased. Although initial interest and applications were primarily in the area of petroleum engineering and enhanced oil recovery, new applications in medicine and industry have evolved. Because of their advantages over synthetic surfactants, biosurfactants are also of increasing interest in cosmetics, foods, environmental control and abatement, and in any industry where surface-active phenomena play a role in processing and product formulation.

The 17 chapters in this book have been written by leading international authorities on these topics, representing the state of the art. Further potential applications of biosurfactants are shown in a variety of processes and products. In this respect, this book is an excellent resource for both research and development and industrial use.

Biosurfactants have also found wide and excellent application in environmental management, as they have been demonstrated to help and enhance biodegradation of toxic pollutants in water and soil. At the present time not much information is available in the literature, but there is a solid indication that many private companies are actively pursuing research in this direction, as well as in applications of biosurfactants in cosmetics and foods. One can predict with confidence that the application of biosurfactants in these areas will be further enhanced in the future.

At this stage, biosurfactants as high-value biotechnological products have been fully recognized worldwide. It took about 25 years to reach this state, after the initial pioneering research on this subject. A distinguished scientist, researcher, personal friend, and colleague, Professor J. E. Zajic was one of the leading pioneers in this field, and he deserves full recognition for his enthusiasm and vision. In his early work on petroleum microbiology and accompanied biosurfactant production, Professor Zajic not only personally contributed, but enthusiastically disseminated interest in biosurfactants to his numerous students and colleagues, who have contributed considerably to the world of biosurfactants. It is a privilege and honor to dedicate this book to Professor J. E. Zajic (1928–1987) in memory of this great pioneer, scientist, and applied microbiologist.

Naim Kosaric

Contents

Contributors

Roman Blaszczyk Department of Chemical and Biochemical Engineering, University of Western Ontario, London, Ontario, Canada

Horst Chmiel* Fraunhofer-Institut für Grenzflächen- und Bioverfahrenstechnik, Stuttgart, Germany

Lina Cloutier Department of Chemical and Biochemical Engineering, University of Western Ontario, London, Ontario, Canada

V. Deltrieu Institute of Biotechnology, Swiss Federal Institute of Technology, Zurich, Switzerland

Anjana J. Desai Department of Microbiology and Biotechnology Centre, The M. S. University of Baroda, Baroda, India

Jitendra D. Desai Research Centre, Indian Petrochemicals Corporation Ltd., Baroda, India

Armin Fiechter† Institute of Biotechnology, Swiss Federal Institute of Technology, Zurich, Switzerland

Peter Gehr Department of Anatomy, University of Berne, Berne, Switzerland

Current affiliations:
*Institut für Industrielle Reststoff und Abfallwirtschaft GmbH, Saarbrücken, Germany
†Fraunhofer-Institut für Grenzflächen- und Bioverfahrenstechnik, Stuttgart, Germany

Marianne Geiser Department of Anatomy, University of Berne, Berne, Switzerland

Donald F. Gerson* Process Development, Connaught Laboratories Ltd., Willowdale, Ontario, Canada

Bernard F. Gibbs Department of Protein Engineering, National Research Council of Canada, Montreal, Quebec, Canada

Thomas Gruber[†] Fraunhofer-Institut für Grenzflächen- und Bioverfahrenstechnik, Stuttgart, Germany

Dietmar Haltrich Institute for Biotechnology, Technical University of Graz, Graz, Austria

Rolf K. Hommel Institute of Biochemistry, University of Leipzig, Leipzig, Germany

Katharina Jenny Institute of Biotechnology, Swiss Federal Institute of Technology, Zurich, Switzerland

O. Käppeli Biosystems Division, BIDECO AG, Dübendorf, Switzerland

Václav Klekner Institute of Microbiology, Czechoslovak Academy of Sciences, Prague, Czechoslovakia

Andreas K. Koch Institute of Biotechnology, Swiss Federal Institute of Technology, Zurich, Switzerland

Naim Kosaric Department of Chemical and Biochemical Engineering, University of Western Ontario, London, Ontario, Canada

Robert M. Lafferty Institute for Biotechnology, Technical University of Graz, Graz, Austria

Siegmund Lang Institute of Biochemistry and Biotechnology, Technical University of Braunschweig, Braunschweig, Germany

---

Current affiliations:
*Research and Development, Apotex Fermentation, Inc., Winnipeg, Manitoba, Canada
[†]Bayer AG, Krefeld, Germany

Reinhard Müller-Hurtig Institute of Biochemistry and Biotechnology, Technical University of Braunschweig, Braunschweig, Germany

Catherine N. Mulligan* Department of Biochemical Engineering, National Research Council of Canada, Montreal, Quebec, Canada

Urs A. Ochsner Institute of Biotechnology, Swiss Federal Institute of Technology, Zurich, Switzerland

Colin Ratledge Department of Applied Biology, University of Hull, Hull, England

Jakob Reiser Institute of Biotechnology, Swiss Federal Institute of Technology, Zurich, Switzerland

Samuel Schürch Respiratory Research Group, The University of Calgary, Calgary, Alberta, Canada

Martin Siemann Institute of Biochemistry and Biotechnology, Technical University of Braunschweig, Braunschweig, Germany

Walter Steiner Institute for Biotechnology, Technical University of Graz, Graz, Austria

Patrick Sticher† Institute of Biotechnology, Swiss Federal Institute of Technology, Zurich, Switzerland

Joran Velikonja Department of Chemical and Biochemical Engineering, University of Western Ontario, London, Ontario, Canada

Fritz Wagner Institute of Biochemistry and Biotechnology, Technical University of Braunschweig, Braunschweig, Germany

---

Current affiliations:
*SNC Research Corporation, Montreal, Quebec, Canada
†EAW AG, Dübendorf, Switzerland

BIOSURFACTANTS

I

Production

1

Biosynthetic Mechanisms of Low Molecular Weight Surfactants and Their Precursor Molecules

ROLF K. HOMMEL Institute of Biochemistry, University of Leipzig, Leipzig, Germany

COLIN RATLEDGE Department of Applied Biology, University of Hull, Hull, England

I. INTRODUCTION

Biosurfactants display a range of structures but have the common ability to cause emulsification of oil–water mixtures. Accordingly, biosurfactants must be able to dissolve, at least partially, in both water and a water-immiscible liquid, thereby effecting a decreased surface tension enabling mixing and microsolubilization (i.e., emulsification) to occur. The range of biological components that are available as building blocks for any biosurfactant is limited. They may vary from components that are wholly water-soluble but have no solubility in oil to those that are virtually water-insoluble but can, and do, dissolve in any oil or lipid material.

Biological molecules of the first category include the carbohydrates, especially the mono- and disaccharides. Also in this category are the polyols that are derived from the carbohydrates, as well as a number of the hydrophilic amino acids, for example, glutamate, aspartate, lysine, ornithine, and arginine. One should also include the principal nucleotide bases—guanine, cytosine, adenine, and thymine—but, although these are water-soluble, they do not appear to be used in any of the biosurfactants whose structure has been established. Peptides consisting primarily of hydrophilic amino acids are also water soluble and the acids associated with the tricarboxylic acid cycle are strongly water-soluble biological molecules. The tricarboxylic acid, citric acid, is of course produced from *Aspergillus niger* on a commercial scale and processes for the production of the dicarboxylic acids, fumaric acid, and malic acid also exist but have limited commercial productions [1]. Biosurfactants based on them or related acids have yet to be reported through chemically produced compounds, such as tributylcitrate [1] are clearly useful in the surfactant industry.

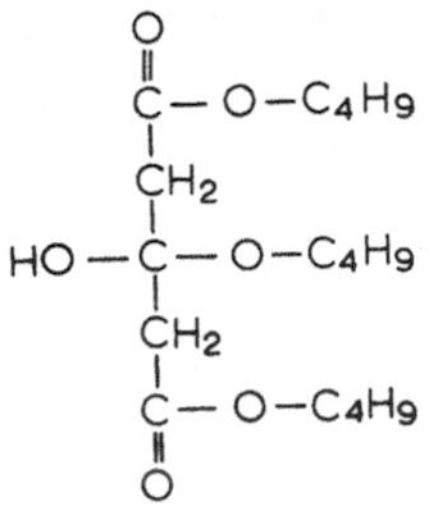

[1]

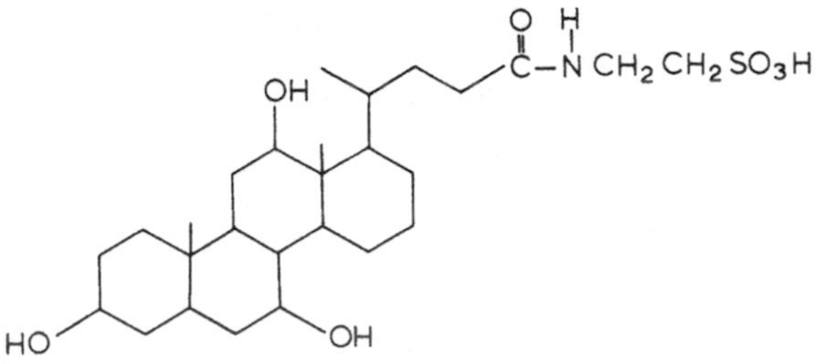

[2]

For water-insoluble biomolecules, the range naturally includes all of the microbial lipids whose molecular complexities go from those based on fatty acids to those based on isoprenoid structures, for example, sterols, terpenes, carotenes, and polyprenols [2]. Although terpenoid compounds are not known to be involved in any of the molecules usually considered as biosurfactants, it should be pointed out that taurocholic acid [2] which is derived from the triterpene, cholesterol, and the amine, taurine is the lipoidal emulsificant used in mammalian lipid metabolism to effect the dispersion of fatty materials in aqueous environments. Such biosurfactants, although adequately produced in animals as a bile acid, do not appear to have any microbial counterpart that has so far been identified as a lipid-emulsifying agent. However, also included here are the hydrophobic amino acids, such as phenylalanine, leucine, isoleucine, valine, and alanine, which have only a very limited solubility in water. Polypeptides that have a predominance of such residues also have limited solubilities in water with a concomitant increase in their ability to associate with lipids especially at lipid–water interfaces, where biosurfactants exert their greater influence.

A biosurfactant is usually found to be a combination of water-soluble and water-insoluble components, thus enabling it to associate at any water–oil interface so that mutual solubilization or emulsification may begin. Although at first inspection one may be surprised at the apparently large range of biosurfactants, the number in fact is somewhat limited as the range of water-soluble to water-insoluble components found in nature is far greater than those actually used by microorganisms. Perhaps as progress continues to be made, we may see new biosurfactants emerging that do incorporate some of the molecules mentioned above that have not yet been implicated as components of microbial surfactants.

In this chapter, we give the pathways of biosynthesis for the principal components of microbial surfactants: fatty acids and related components including the various long-chain acylated components, the carbohydrates and polyol moieties of the glycolipids, and the amino acids used in a number of different ways both as water-soluble entities and as water-insoluble moieties. Greater detail concerning the range of microbial lipids, which are constituents of most biosurfactants, are found in the book on this subject edited by Ratledge and Wilkinson [2], although briefer accounts by Weete [3], Harwood and Russell [4], and Gurr and Harwood [5] may be equally useful.

II. BIOSYNTHESIS OF FATTY ACIDS FROM GLUCOSE

Microbial fatty acids are usually of a narrow range of chain lengths, C_{16} and C_{18}, with relatively small amounts of shorter (C_{12} and C_{14}) and longer (C_{20}) acids. Although the biosynthesis of fatty acids is similar in all biological systems, there are important differences between some bacterial systems and those of eukaryotic microorganisms (yeasts and molds). It is this difference that then accounts for the difference in how unsaturated fatty acids are synthesized in the two systems and why the resulting unsaturated C_{18} fatty acids are not the same. Bacteria usually produce 18:1 (*c*11),* *cis*-vaccenic acid, whereas yeasts and molds and all other living cells, produce oleic acid, 18:1 (*c*9). These differences occur because of the organization of the enzymes making up the individual fatty acid synthetase complexes. In all cases, however, fatty acid biosynthesis begins with acetyl-coenzyme A (acetyl-CoA). As this is the key intermediate for fatty acid biosynthesis and also, via formation of mevalonic acid, for the biosynthesis of all the terpenoid lipids, it is important to consider the metabolic origins of this molecule.

A. Formation of Acetyl-CoA

In bacteria, there is no mitochondrion and consequently acetyl-CoA is formed directly in the cell compartment (the cytoplasm) from pyruvate (the end-product of glucose metabolism) by pyruvate dehydrogenase:

$$\text{Pyruvate} + \text{NAD}^+ + \text{CoA} \rightarrow \text{acetyl-CoA} + \text{CO}_2 + \text{NADH} \qquad (1)$$

In eukaryotic microorganisms, that is yeasts and molds, pyruvate dehydrogenase is a mitochondrial enzyme and thus, although pyruvate can enter the mitochondrion, its product, acetyl-CoA, cannot leave because its molecular size is too large. As fatty acid biosynthesis takes place in the cytoplasm (as in bacteria) there has to be some mechanism for acetyl units to be translocated from inside the mitochondrian into the main cell compartment.

Two principal routes for translocation of acetyl units occur in yeasts and molds:

1. The carnitine acetyl transferase (CAT) route. This system appears to occur in all yeasts and molds. No exceptions have yet been reported. CAT catalyzes the reversible formation of acetyl-carnitine:

$$\text{Acetyl-CoA} + \text{carnitine} \rightarrow \text{acetyl carnitine} + \text{CoA} \qquad (2)$$

Acetyl-carnitine, being smaller than acetyl-CoA, is readily transported across the mitochondrial membrane. The reverse of reaction 2 then occurs in the cytoplasm to regenerate acetyl-CoA.

* The standard nomenclature for lipids is used throughout this chapter. Thus, for example, 18:1 (*c*11) refers to a fatty acid having 18 carbon atoms and one double bond; the position of the double bond is then denoted by enumerating the first C atom, starting at the carboxyl end, of the bond and, where needed, whether this is *cis* (*c*) or *trans* (*t*). Thus 18:1 (*c*11) refers to *cis*-vaccenic acid. Branched-chain fatty acids are designated by br, although if the branching is at the ω-1 or ω-2 C atoms these acids are referred to as *iso* and *anteiso* acids, respectively. Cyclopropane or cyclopropene rings are designated by *cyc*. [A fuller discussion and description of microbial fatty acids may be found in Refs. 2–4.]

2. The citrate translocation and cleavage route. This system operates only in a selected number of yeasts and molds, because it is associated with the phenomenon of oleaginicity whereby lipid contents over 20% of the biomass can be created in yeasts [6,7] and in molds [7]. As implied above, this system operates in conjunction with the acetyl-carnitine route and not in place of it. The key reactions of this route depend upon citrate being synthesized in the mitochondrion faster than it is metabolized by the tricarboxylic acid cycle. (This is accomplished usually by isocitrate dehydrogenase ceasing to function at its full capacity because of a diminution in the supply of adenosine monophosphate (AMP) on which it depends. A full discussion of the biochemical sequence of oleaginicity is given elsewhere [6–8].) When citrate begins to accumulate in the mitochondrion, its concentration is kept constant by the surplus being translocated across the mitochondrial membrane into the cytosol, there to be cleaved by ATP: citrate lyase (ACL):

$$\text{Citrate} + \text{CoA} + \text{ATP} \rightarrow \text{acetyl-CoA} + \text{oxaloacetate} + \text{ADP} + \text{P}_i \tag{3}$$

The oxaloacetate arising as coproduct in the ACL reaction is converted to malate (eq. 4), which then acts as the counterion for citrate translocation. The overall scheme is given in Fig. 1.

$$\text{Oxaloacetate} + \text{NADH} \rightarrow \text{L-malate} + \text{NAD}^+ \tag{4}$$

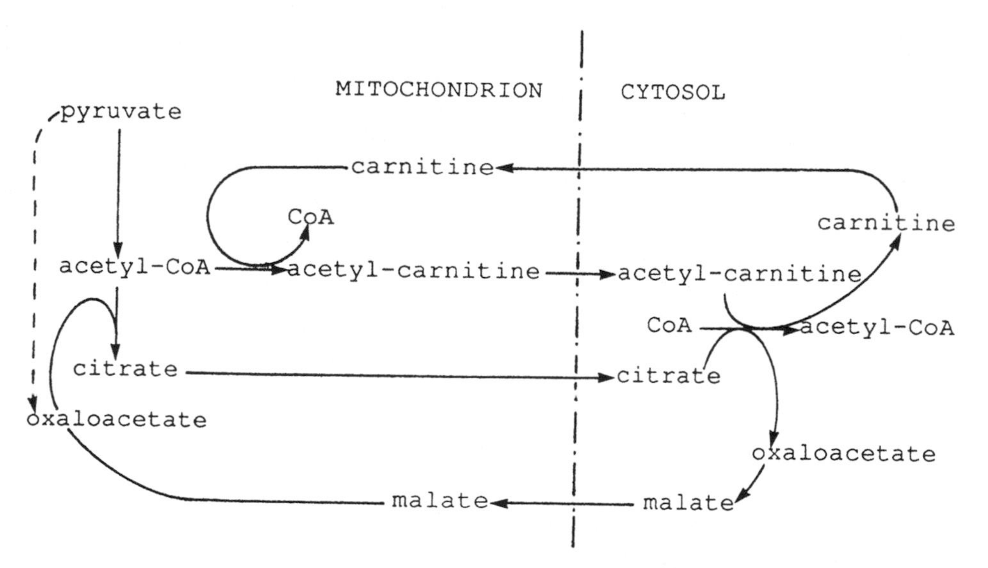

FIG. 1 Routes to the information of acetyl-CoA in eukaryotic microorganisms. The carnitine route is thought to operate in all cells; the citrate route occurs also in oleaginous microorganisms. The citrate-malate exchanged is a linked reaction.

Other routes for the creation of acetyl-CoA in the cytoplasm of eukaryotic microorganisms have been considered [9], but the above routes are the two principal ones. Acetyl-CoA is synthesized directly from acetate in those yeasts that produce ethanol and then use the ethanol when the supply of glucose becomes exhausted. This route involves acetyl-CoA synthetase:

$$(\text{Pyruvate} \rightarrow \text{ethanol} \rightarrow)\ \text{acetate} + \text{CoA} + \text{ATP} \rightarrow \text{acetyl-CoA} + \text{ADP} + P_i \quad (5)$$

B. Acetyl-CoA Carboxylation

In all cells—bacterial, yeast, or fungal—acetyl-CoA is carboxylated to malonyl-CoA by acetyl-CoA carboxylase (ACC); the sequence that occurs is the sum of two partial reactions:

$$\text{Enzyme-biotin} + \text{ATP} + HCO_3 \rightarrow \text{enzyme-biotin-}CO_2 + \text{ADP} + P_i \quad (6a)$$

$$\text{Enzyme-biotin-}CO_2 + \text{acetyl-CoA} \rightarrow \text{enzyme-biotin} + \text{malonyl CoA} \quad (6b)$$

$$\text{Net: acetyl-CoA} + HCO_3 + \text{ATP} \rightarrow \text{malonyl-CoA} + \text{ADP} + P_i \quad (6c)$$

The sequence is irreversible. The enzyme has been investigated in some detail in *Escherichia coli* when the ACC enzyme is composed of three functional units: biotinyl carboxyl carrier protein (BCCP), biotin carboxylase (which catalyzes 6a) and carboxyl transferase (which catalyzes 6b). There are variations in the constitution of this complex [10] both in other bacteria and in yeasts. In the yeasts *Saccharomyces cerevisiae* and *Yarrowia (Candida) lipolytica* the three components are combined into a single, trifunctional protein [11,12]. A more complex ACC, with additional enzyme activities of phosphoenolpyruvate carboxylase and malate dehydrogenase, occurs in the alga, *Euglena gracilis* [13].

C. Fatty Acid Synthetase

The elongation of acetyl-CoA to long-chain (C_{16}) fatty acids is mechanistically the same in both prokaryotes and eukaryotes: the essential steps are condensation of an acetyl group (7) with a malonyl group (8) to yield a C_4 unit + CO_2 (9) followed by reduction (10), dehydration (11), and further reduction of the C_4 unit (12) until a saturated C_4 (butyryl) group is formed. The cycle is then repeated by condensation of the butyryl group with a further malonyl group leading to a C_6 moiety + CO_2 (9a). This continues until a long-chain fatty acyl group is formed. This cycle may be summarized as follows:

Acetyl transacylase (priming reaction):

$$CH_3CO\text{-SCoA} + \text{protein} \rightarrow CH_3CO\text{-S-protein} + \text{CoA} \quad (7)$$

Malonyl transacylase:
$COOHCH_2CO\text{-}SCoA + protein \rightarrow COOHCH_2CO\text{-}S\text{-}protein + CoA$ (8)

Condensing enzyme (β-ketoacyl synthase):
$^{\blacktriangle}CH_3{}^{\blacktriangle}CO\text{-}S\text{-}protein + {}^{*}COOH{}^{*}CH_2\text{-}{}^{*}CO\text{-}S\text{-}protein \rightarrow$
$^{\blacktriangle}CH_3\ {}^{\blacktriangle}CO.{}^{*}CH_2{}^{*}CO\text{-}S\text{-}protein + {}^{*}CO_2 + protein$ (9)

The fate of the three *C atoms of the malonyl group and of the two ▲C atoms of the acetyl group are indicated.

β-Ketoacyl reductase
$CH_3COCH_2CO\text{-}S\text{-}protein + NADPH \rightarrow CH_3CHOHCH_2CO\text{-}S\text{-}protein + NADP^+$ (10)

The hydroxy fatty acid which is formed is the D-isomer.

β-Hydroxyacyl dehydratase:
$CH_3CHOHCH_2CO\text{-}S\text{-}protein \rightarrow CH_3CH{:}CHCO\text{-}S\text{-}protein + H_2O$ (11)

Enoyl reductase:
$CH_3CH{:}CHCO\text{-}S\text{-}protein + NADPH \rightarrow CH_3CH_2CH_2CO\text{-}S\text{-}protein + NADP^+$ (12)

The cycle now repeats with a new malonyl group (see reaction 8 above) condensing with the butyryl group (12) using the same condensing enzyme (9) as before:

$CH_3CH_2CH_2CO\text{-}S\text{-}protein + {}^{*}COOH{}^{*}CH_2{}^{*}CO\text{-}S\text{-}protein \rightarrow$
$CH_3CH_2CH_2CO{}^{*}CH_2{}^{*}CO\text{-}S\text{-}protein + {}^{*}CO_2 + protein$ (9a)

The new keto acid is then reduced, dehydrated, and reduced again as before with the reaction cycle continuing until the acyl chain reaches 16 carbons in length. At this point the palmitoyl group is then transferred from the protein back to the coenzyme A:

Palmitoyl transacylase:
$CH_3(CH_2)_{14}CO\text{-}S\text{-}protein + CoA \rightarrow CH_3(CH_2)_{14}CO\text{-}S\text{-}CoA + protein$ (13)

The overall reaction may therefore be written as:

$$7Malonyl\text{-}CoA + acetyl\text{-}CoA + 14NADPH \rightarrow palmitoyl\text{-}CoA + 7CO_2 + 7CoA + 14NADP^+ + 7H_2O$$

In bacteria, as typified by *E. coli*, the protein involved in carrying the growing acyl chain is a small separable polypeptide termed the acyl carrier protein (ACP). ACP has a molecular weight of 8847, is very acidic (pI = 4.1) and its complete amino acid sequence is now established [15]. The acyl group is attached to a

4-phosphopantetheine group, which is the same moiety as occurs at the terminus of CoA, and thus the acyl groups continue to be thioesters throughout the sequence of reactions. All the enzymes (7-13) in *E. coli* have been separated in a catalytically active form. However, recent work [15] on the 3-ketoacyl (acetoacetyl)-ACP synthase (9) "condensing enzyme" suggests that this enzyme activity may not require the synthesis of acetyl-ACP (7), possibly inferring that acetyl-CoA could be used instead. Clearly, there is still much to be unravelled about the intricacies of this multistep process.

Plant fatty acid synthetases [FASs] likewise dissociate into their component enzymes and are thus closely allied to the bacterial synthetases. Such separable synthetases are termed type II synthetases and are distinct from the type I synthetases that occur in a few bacteria (possibly confined to the *Mycobacterium–Nocardia–Corynebacterium* group), yeasts and animals, including birds. Eukaryotic algae, as exemplified by *Euglena gracilis*, appear to possess both types of FAS enzymes [4].

The type I synthetases are large molecular weight, multifunctional, and non-separable proteins. In yeasts, the functional organization of the FAS enzyme complex from *Saccharomyces cerevisiae* has been extensively studied [10,16,17]. The yeast synthetase is a complex of two nonidentical subunits (α and β) formed into a complex of molecular weight 2.4×10^6, having an $\alpha_6\beta_6$ structure. Fungal FASs are of similar organization [18,19].

Thus in yeasts and fungi the individual reactions (7-13) are carried out by just two proteins. One protein carries out the condensation reaction (9) and the β-ketoacyl reductase step (10), as well as possessing the acyl carrier protein function; the other protein carries out the remaining functions of acetyl and malonyl transfer (7 and 8), the dehydratase (11) and enoyl reductase (12), as well as the terminal transfer of the final palmitoyl chain to CoA (13).

The efficiency of the final long-chain acyl transacylase reaction determines whether a C_{16} acid (palmitate) is produced or whether the chain undergoes a further cycle to emerge as the C_{18} acid (stearate). There is some evidence to suggest that in yeast this may be sensitive to temperature, so that at increased temperatures a higher proportion of palmitate to stearate results [20]. When organisms contain higher than normal amounts of C_{12} fatty acids (lauric acids) or C_{14} fatty acids (myristic acids)—as occurs in species of *Entomophthora* [2,21,22]—it is probable that this is due to an increased affinity or activity of the final transacylase toward C_{12} and C_{14} fatty acyl chains.

In the *Mycobacterium–Nocardia–Corynebacterium* (MNC) group of bacteria, elongation of fatty acids beyond C_{18} is more complex and is due only partly to FAS (type I) continuing to operate. Elongation is primarily due to a second system, which is exactly the same as the bacterial ACP (type II) system, except that with these MNC organisms, the primary substrate is not acetyl-ACP but palmitoyl-ACP. The type I FAS produces C_{24}- and C_{26}-CoA esters; the type II FAS produces from C_{18}- to C_{30}-ACP derivatives. This is summarized in Fig. 2 (also see ref. 23 for review). These long-chain fatty acids are then used in the biosynthesis of mycolic acids (see below).

D. Biosynthesis of Unsaturated Fatty Acids

As mentioned in the introductory paragraph to Sec. II, there is a distinction between the unsaturated fatty acids of bacteria and those of yeasts and other eukaryotic microorganisms that is directly attributable to the two different types of FAS complex.

In those bacteria, which have the ACP system of fatty acid synthesis (i.e., all those outside the MNC complex), a branch point occurs after the synthesis of D-β-hydroxydecanoyl-ACP which, for the biosynthesis of unsaturated fatty acids, is acted on by a 2,3-dehydratase (14) to give 10:1 (*c*3) fatty acyl-ACP:

$$CH_3(CH_2)_5CH_2CHOHCH_2CO\text{-}S\text{-}ACP \rightarrow$$
$$CH_3(CH_2)_5CH{:}CHCH_2CO\text{-}S\text{-}ACP \quad (14)$$

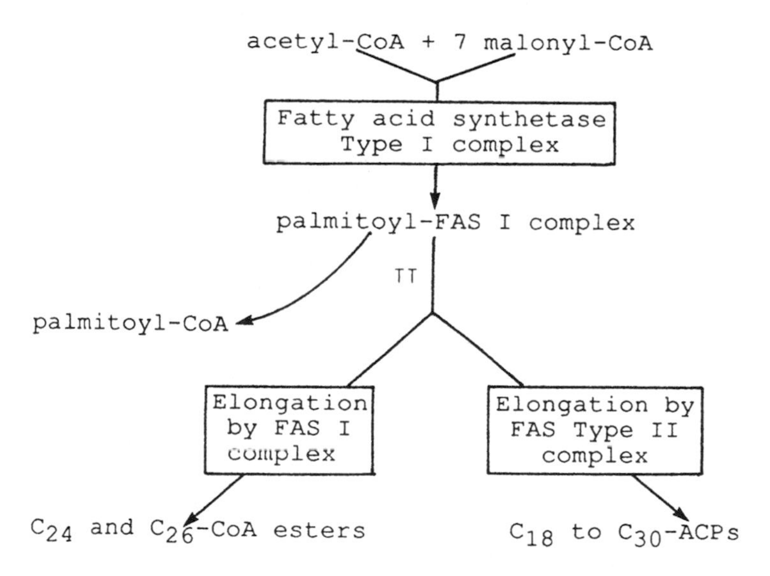

FIG. 2 Summary of fatty acid synthesizing and elongation systems in *M. smegmatis* as a typical representative of the CMN group of bacteria. TT, transacyl transferase enzyme; FAS, fatty acid synthetase; ACP, acyl carrier protein. (See Ref. 23 for further details and review.)

Subsequent elongation of the *cis*-3-decenoyl-ACP is by a different set of FAS enzymes than are used for the biosynthesis of the saturated acids. The sequence of reactions is shown in Fig. 3: the double bond, initially between C3 and C4 atoms, progressively moves away from the carboxyl group as the chain is successively lengthened by the addition of further C2 units between the double bond and the carboxyl group. As this route to the biosynthesis of unsaturated fatty acids does not involve molecular oxygen, it is termed the anaerobic route.

Without the means to desaturate fatty acids directly, bacteria, with a few exceptions, are unable to effect further unsaturations of the monounsaturated acids and consequently polyunsaturated fatty acids are not normally found in these organisms.

In yeasts and other organisms possessing the nonseparable FAS type I complex, which therefore includes the MNC group of bacteria, unsaturated fatty acids, including the polyunsaturated fatty acids, are formed by specific desaturase enzyme systems. (In plants, acyl-ACP derivatives are the substrates for these reactions and in animals CoA esters of fatty acids are used [4,24].) The reaction is a complex one involving cytochrome b_5 as electron carrier (Fig. 4).

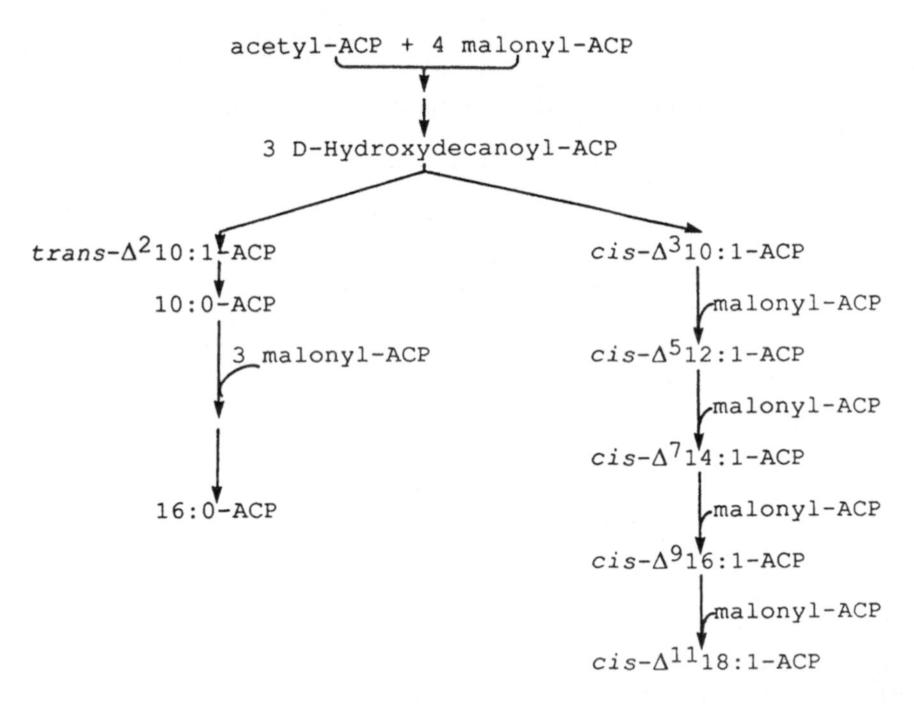

FIG. 3 Biosynthetic pathway to unsaturated fatty acid biosynthesis in bacteria (the anaerobic route). ACP, acyl carrier protein.

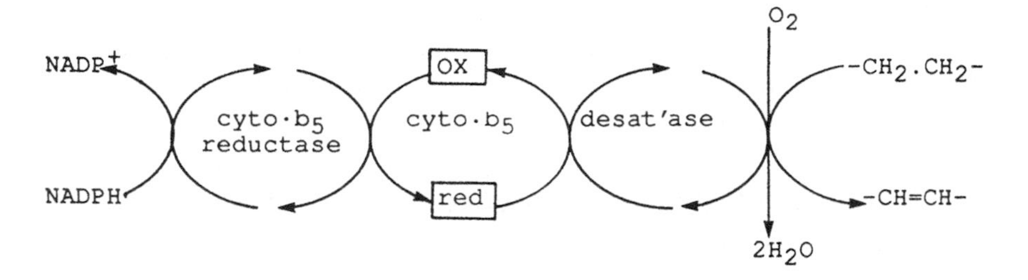

FIG. 4 Desaturation mechanism for fatty acyl groups. Fatty acyl groups being desaturated are attached to phospholipids in eukaryotic microorganisms and, when present, also in bacteria.

The product from stearoyl-CoA is the 18:1(*c*9)-CoA ester—oleoyl-CoA—which is then clearly distinct from the Δ^{11}-monounsaturated acid of bacteria. Further desaturation of oleoyl-CoA requires transfer of the oleoyl group onto a phospholipid [25], so that these reactions now must occur within the internal membranes of the cell—the endoplasmic reticulum (also known as the microsomal fraction when cells are disrupted and fractionated). The first reaction occurs via a Δ^{12}-desaturase to give linoleoyl-phospholipid [18:2 (*c*9, *c*12)] which may then be followed by either a Δ^{15}-desaturase or a Δ^{6}-desaturase. In the former case, α-linolenoyl-phospholipid [18:3 (*c*9, *c*12, *c*15)] results, which is found in small amounts (usually <10%) in all yeasts and molds except for the Mucorales. In the Mucorales, the Δ^{6}-desaturase occurs and yields γ-linolenoyl-phospholipid, 18:3 (*c*6, *c*9, *c*12). In *Mucor circinelloides* recent work has shown that the NADPH needed to drive the desaturation reactions in the microsomal membranes is generated by a specific membrane-bound malic enzyme:

$$\text{Malate} + \text{NADP}^+ \rightarrow \text{pyruvate} + CO_2 + \text{NADPH} \quad (15)$$

The phospholipid that acts as the carrier for the fatty acyl group undergoing desaturation in the mold is phosphatidylinositol, and it is specifically the fatty acyl group on the *sn*-2 position that is desaturated [26].

Control of fatty acid desaturation is regarded as a temperature-regulated event, such that at lower growth temperatures, a higher proportion of unsaturated fatty acyl groups occurs than at higher temperatures [27]. Such a switch to unsaturated fatty acids maintains membrane fluidity and therefore functionality. The control appears to be exerted both at the level of the fatty acyl desaturases and also at the level of the fatty acyl synthetase generating shorter chain fatty acids, which decrease the proportion of C_{18} fatty acids being produced. For example, in *Candida utilis* grown at 10°C the content of palmitoleic acid (16:1) is increased to approximately 10% of the total fatty acids as opposed to only 2% in cells grown at 35°C [28]. Full discussions of factors, including temperature, affecting the fatty acyl composition of microorganisms are given by Neidleman [27] and Rose [29].

E. Formation of Hydroxy, Branched-Chain, and Other Fatty Acids

Hydroxy fatty acids may arise by a number of different routes.

1. They may be formed as intermediates of fatty acid biosynthesis. Such fatty acids are characterized by being the D-isomer and this is the probable origin of D-3-hydroxypalmitic acid found in *Saccharomycopsis* (=*Hansenula*) *malanga* [30]. Similarly, 3-hydroxydecanoate, 3-hydroxylaurate, and 3-hydroxymyristrate (the C_{10}, C_{12}, and C_{14} acids, respectively), which are found in the lipid A component of Gram-negative bacteria, probably arise in this way as they are also of the D isomeric form [31].

2. They may arise by oxidative degradation of fatty acids. Although the β-oxidation cycle, like biosynthesis, leads to the formation of 3-hydroxy fatty acids, these are the L-isomers. A specific and different hydroxylating system is used to produce the 2-hydroxy fatty acids and is considered part of the α-oxidation route of fatty acid degradation [4,10]. Such acids also occur in many bacteria [4,10,31].

3. They may arise by terminal or penultimate oxidation of fatty acids, thus giving ω-hydroxy or ω-1-hydroxy fatty acids, respectively. Such acids are found in both bacteria and yeasts particularly, although not exclusively, when alkanes or fatty acids have been used as growth substrate (see Sec. III.C).

4. They may arise by direct hydration of an unsaturated fatty acid so that linoleic acid [18:2 (*c*9, *c*12)] would produce 12-hydroxyoleic acid and oleic acid [18:1 (*c*9)] would produce 9-hydroxystearic acid:

$$-CH{=}CH- + H_2O \rightarrow -CH_2-CHOH- \qquad (16)$$

The biosynthetic origin of 15-hydroxyoctadece-9,12-enoic acid, which is found as an extracellular lipid of a *Candida* sp. when grown on ethanol [32], is thus probably linolenic acid [18:3 (*c*9, *c*12, *c*15)], rather than linoleic acid [18:2 (*c*9, *c*12)], being directly hydroxylated. However, Soda [33] recently reported the biotransformation of oleic acid [18:1 (*c*9)] to ricinoleic acid (12-hydroxyoleic acid) by various microorganisms, although the veracity of this claim has not apparently been upheld by other workers. If Soda is correct, then this reaction would be classed as a midchain hydroxylation; such a reaction does occur for the formation of ricinoleic acid in plants [4] but not apparently in *Claviceps purpurea*, which hydroxylates linoleic acid to give ricinoleic acid.

Dihydroxy fatty acids are not unknown, although if the hydroxyl groups are on adjacent carbon atoms, as with 9,10-dihydroxypalmitate in *Bacillus megaterium*, this arises via formation of epoxypalmitate by a soluble epoxide hydrolase acting on oleic acid [31]:

```
                          O
                         / \
–CH=CH– + H2O → –CH–CH– → –CHOH–CHOH–        (17)
```

However, the origin of 15,16-dihydroxypalmitic acid (ustilagic acid A) in *Ustilago maydis* is uncertain, although it is not likely to be via epoxide formation across a double bond which would then have to be in the terminal position. *Ustilago maydis* also produces 2,15,16-trihydroxypalmitic acid (ustilagic acid B) [3], whose immediate precursor is probably ustilagic acid A undergoing partial α-oxidation. Both the ustilagic acids occur as extracellular glycolipids attached to the disaccharide cellobiose and thus are structurally allied to the sophorolipids produced by *Torulopsis bombicola* and related species of yeasts. Such surfactants are discussed elsewhere in this volume.

Branched chain fatty acids are found in a number of bacteria as well as some molds. The side chains are usually only methyl groups, which may be at the ω-1 or ω-2 carbon atom, giving iso or anteiso fatty acids: $R\text{-}CH(CH_3)_2$ or $R\text{-}CH(CH)_3{\cdot}CH_2{\cdot}CH_3$, respectively, where R is the alkanoic acid chain. Such acids are produced from the C-skeletons of leucine or isoleucine, respectively [34]. Alternatively, side chains may be centrally located as exemplified by tuberculostearic acid (10-methylstearic acid) found in *Mycobacterium tuberculosis* and related species as well as related genera and some *Streptomyces* species. These acids are formed [35] by addition of a C_1 unit from *S*-adenosylmethionine (SAM) across the double bond of oleic acid:

$$-CH{=}CH- \rightarrow -CH(CH_3)-CH_2- \qquad (18)$$

Multiple branched fatty acids such as mycocerosic acid (2,4,5,7-tetramethyloctacosanoic acid) also occur in mycobacteria [23].

Cyclic fatty acids are found in bacteria and are usually of the cyclopropane type with a methylene group being introduced, via *S*-adenosylmethionine as methyl donor, across a double bond. Thus lactobacillic acid (*cis*-11,12-methyleneoctadecanoic acid) is produced from *cis*-vaccenic acid [18:1 (*c*11)]. Although this acid was originally discovered in the lactobacilli, it is of widespread occurrence in bacteria [4].

F. Biosynthesis of Mycolic Acids

Mycolic acids [3] are long-chain, β-hydroxy fatty acids substituted at the α-carbon atom with a moderately long aliphatic chain (Table 1). The total number

$$R-\underset{H}{\overset{OH}{C}}-\underset{\underset{CH_3}{(CH_2)_n}}{CH}-COOH$$

n up to 23

[3]

of carbon atoms varies from 30 to 86. They are produced by species in the genera of *Mycobacterium, Nocardia, Rhodococcus, Corynebacterium* and some minor genera such as *Gordona, Bacterionema, Micropolyspora,* and *Brevibacterium* [36,37]. They usually form part of the cell wall complex being linked to the arabinogalactan-peptidoglycan matrix of the cell wall [38,39]. However, mycolic acids are also found as the lipid component of a small number of extracellular glycolipids produced by these bacteria, particularly following growth on alkanes or related materials (see elsewhere in this book). The structural aspects of these complex fatty acids are shown in summary in Table 1 but have been described elsewhere in considerable detail [36–43]. There are a considerable number of variations in structures along the long fatty acid chain: there may be keto and methoxygroups, methyl groups and cyclopropane rings, and both *cis* and *trans* double bonds occurring within the chain itself. The side-chain fatty acid may also contain one double bond.

TABLE 1 Outline Structures of Mycolic Acids[a] of the CMN[b] Group and Related Taxa

Genus	Total no. of C atoms	*n* (see 3)
Corynebacterium	22–36	5–15
Bacterionema	30–36	11–15
Brevibacterium	30–48	7–15
Rhodococcus	34–64	9–15
Nocardia	44–60	7–15
Gordona	56–74	13–19
Mycobacterium		
M. smegmatis	62	21
M. kansasii	80	21
M. tuberculosis	86–90	21, 23

Source: Refs. 36, 39, and 43.
[a]These acids are considerably diverse and complex. Interested readers should consult one or more of the above mentioned reviews for full details.
[b]Corynebacteria, mycobacteria, and nocardia.

The synthesis of the mycolic acids is subject to considerable speculation as it is only within the last two years that the first reactions have been characterized in cell-free systems [44]. The formation of the mycolate is considered [38,39] to be a condensation of a Claisen-type reaction between two fatty acids, probably both activated as some thiol ester:

$$\text{R–CO–X} + \underset{\substack{|\\ \text{R}_1}}{\text{CH}_2}\text{–CO–Y} \rightarrow \text{R–CO–}\underset{\substack{|\\ \text{R}_1}}{\text{CH}}\text{CO–Y} \rightarrow \text{R–}\underset{\substack{|\\ \text{OH}}}{\text{CH}}\text{–}\underset{\substack{|\\ \text{R}_1}}{\text{CH}}\text{ COOH} \qquad (19)$$

where X and Y are considered as activating groups.

For the formation of the various mycolates [3], the long-chain fatty acid constituting the major part of the molecule is synthesized first before condensation with the second fatty acid. This part of the molecule is known as the meromycolate. The postulated pathway [38,39] is shown in Fig. 5. Methyl side chains or cyclopropane rings are introduced as with other fatty acids, presumably by action of *S*-adenosylmethionine.

III. BIOSYNTHESIS OF FATTY ACIDS FROM ALKANES

Microbial metabolism of aliphatic hydrocarbons has received continuing attention over the last two decades and has sought to answer the basic questions of the physiology, biochemistry, and genetics of the cell's adaptation to metabolize these strongly hydrophobic, water-immiscible substrates. Additionally, the biotechnological aspect with its commercial overtones has been focused mainly on the formation of derivatives of alkanes: fatty alcohols, long-chain diols, and fatty acids, including hydroxy acids and dicarboxylic acids [45–46]. Various reports

Elongation		Condensation		
meromycolate chain		side chain		mycolate
C_{14-20} acid	+	C_{8-18} acid	→	C_{22-38} corynomycolate
↓ C_2 units				
C_{34-42} acid	+	C_{10-18} acid	→	C_{44-60} nocardomycolate
↓ C_2				
C_{54-64} acid (meromycolate)	+	C_{24-26} acid	→	C_{78-90} mycolate

FIG. 5 Postulated pathway for the biosynthesis of mycolic acids. (From Refs. 38 and 39.)

demonstrate the ability of different microorganisms growing on alkanes or other water-insoluble substrates to synthesize large amounts of extracellular and cell-bound compounds in which different types of fatty acids represent the hydrophobic constituent of the molecule [48–50]. These biosurfactants are able to reduce the surface and interfacial tension and may serve to form micelles. Additionally, the appearance of bioemulsifiers originating in the cell envelope during microbial growth on alkanes has been reported [51, 52].

A. Uptake of *n*-Alkanes

The high degree of insolubility of hydrocarbons in aqueous media and their highly productive transformation to cell carbon material, reflected by appropriate growth rates and yields, are partly contradictory. The uptake mechanisms of alkanes are the most interesting and yet unclear processes in hydrocarbon assimilation. To explain substrate uptake three different models have been proposed, none of which, however, satisfactorily explain the passage of the alkane across the cell wall and cytoplasmic membrane into the cytoplasm. Uptake of, and growth on, alkanes have been strongly connected with the ability of the microbe to release surfactants to facilitate their uptake [53–55]; the ability to synthesize such surfactants may be regarded as a prerequisite for alkane metabolism.

The models developed will be discussed briefly to demonstrate the principle borderline cases. The *uptake of monodispersed-dissolved alkanes* proposed by Goma et al. [56] may not be valid for alkanes with chain lengths longer than C_9, because their solubility is in range of 10^{-5} to 10^{-10} g L^{-1} [48,57]. These concentrations do not explain the observed high growth rates on alkanes [43]. This model may be valid though for short-chain hydrocarbons, which, however, become increasingly toxic because of their increasing destruction of the cell membrane as the length of the alkane chain becomes less than C_8.

For different strains, growth on or in alkane drops, that is, *in direct contact with large oil drops*, has been reported for *Rhodococcus erythropolis* growing in alkane drops [58] and for *Rhodococcus aurantiacus* [59]. In both strains, the synthesis of the cell wall constituent, trehalose mycolate, is induced. In *R. erythropolis*, the increased synthesis of cell wall-bound trehalose mycolates coincides with growth in the droplets, but in *R. aurantiacus* the production accompanies the initial growth of the bacteria. These examples demonstrate different mechanisms of adaptation to achieve direct contact of the cell to large alkane drops. In general, an increased hydrophobicity of the cell surface is needed to establish direct contact of the cell to the alkane drop. Alterations in hydrophicity of the cell envelope are not restricted to alkane-utilizing microorganisms [60].

The model of *contact with fine oil droplets* (accommodated oil) is based on the formation of cellular metabolites with surface-active properties to lower the surface tension. Surfactants above their critical micelle concentration can form submicron droplets or micelles, which then accommodate or pseudosolubilize the alkane. By this process the interfacial area of the hydrocarbon is significantly enlarged, providing good conditions for achieving rapid uptake into the cells. This model is supported by observations that emulsified alkanes are assimilated faster by cells and by the stimulatory effect of surfactants on cell growth on lipophilic substrates [for review see Ref. 48]. Stimulatory effects of biosurfactants, such as rhamnolipids [61–63] and lactonic sophorosides [64,65], on hydrocarbon metabolism of the producing stains *Pseudomonas aeruginosa* and *Torulopsis bombicola*, respectively, suggest that this may be a general model. The sophoroside could not be replaced by other surfactants and was inactive or inhibitory toward other alkane-metabolizing yeasts [65]. Similar results have been reported for the sophorolipids from *Torulopsis* (*Candida*) *apicola* [66,67]. Otherwise, the degradation of oil pollutants by soil bacteria was stimulated by addition of biosurfactants [68,69], the adaptation phase was shortened, and the extent of hydrocarbon degradation was enhanced.

The role of biosurfactants in alkane degradation has been recently discussed by Hommel [67]. The bulk production of biosurfactants is, in general, not growth-associated (and mostly not restricted to growth on water-insoluble carbon sources). In the logarithmic phase of growth, small amounts of biosurfactants, above the critical micelle concentration, which were probably extracted from the cell wall by the hydrocarbon substrate [70–72] were detected, and may have been responsible for the pseudosolubilization of the alkane substrate. The reported critical micelle concentrations of biosurfactants in water are indeed very low [for reviews see Refs. 50 and 51].

The way by which the pseudosolubilized alkane, in conjunction with a surfactant, will be taken up into the cell cytoplasm is still obscure. The variety of cellular adaptations to facilitate the uptake of hydrocarbons was reviewed by Finnerty and Singer [73]. These include formation of a highly hydrophobic cell surface and extensive changes in membrane lipid composition during growth. Alterations in cell morphology are well documented in bacteria, yeasts, and molds [54]. The surface of the microbial cell thus plays an important function when considering the uptake of alkane. The formation of a hydrophobic complex of polysaccharide and fatty acids in the cell wall that would give the cells an amphipathic layer has been reported in *Candida tropicalis* [74]. Charged and neutral biosurfactants are assumed to have different kinds of action [55]. Nonionic biosurfactants like the trehalose lipids render the charged cell surface hydrophobic and then facilitate the attachment and subsequent (passive) uptake of alkanes into the cell [58,59]. Ionic biosurfactants like both the rhamnolipids and sophorolipids, which are negatively charged, may form micelles resulting in pseudosolubilization of the hydrocarbon and increase of its surface area. Microscopic studies of yeasts revealed pores/channels in the cell wall that permit penetration of hydrocarbons to the cell membrane [75,76]. The formation of fimbriae [77], extracellular vesicular components containing proteins, phospholipids, and lipopolysaccharides [76–80] have been reported in *Acinetobacter* strains.

Ratledge [47] proposed a hypothesis for the action of biosurfactants in alkane uptake (Fig. 6) which also includes the participation of a specific amphipathic receptor/channel. The alkane micelle or the directly attached alkane (cf. 58,59,80) would then be partitioned into the strongly hydrophobic channel. The biosurfactants would be excluded by size, by hydrophobicity, or by energetic considerations (interaction with other surfactant molecules to form micelles or reversed micelles) from the channel. Because of the alkane's hydrophobicity, it will coalesce inside the cell into microdroplets and into larger agglomerations, which become visible by electron microscopy [54]. The somewhat restricted activities of biosurfactants toward only acting with their producing microorganisms may be explained by specific interactions of the amphiphatic channel constituents and the biosurfactants to dissociate the biosurfactant monolayer of the emulsion droplet or those of the micelles. In *Cladosporium resinae*, the uptake of passively adsorbed hydrocarbon proceeds by an energy-requiring transfer of alkane into the cytosol [81,82], which is comparable to the energy-requiring alkane uptake systems in some yeasts [75,83]. The production of emulsifying agents by some marine, filamentous fungi, causing the formation of alkane droplets, which in turn were surrounded and then penetrated by hyphae, was reported by Kirk and Gordon [84].

Complexes of the alkane for transport through the plasma membrane have not been reported. There is no evidence for the participation of biosurfactants in this step. Deposits of alkanes in the plasma membrane and within the cell wall have been described for *Candida* sp. H [85]. Alkanes that have passed through the plasma membrane appear as alkane-inclusion bodies in the cytoplasm of *Acinetobacter* sp. [75,86–88]. Such bodies may be derived from the inner layer of the cytoplasmic membrane and apparently harbor degradative enzymes [89]. Similar results have been obtained from *Nocardia* sp. [90].

Microbial alkane uptake and the role of biosurfactants, including the extracellular high-molecular-weight polymeric materials, is a complex dynamic mechanism. The interplay of micellar or emulsified alkane droplets, the charged cell wall, and the subsequent transport of the alkane into the cytoplasm cannot be completely explained by any of the three models of alkane uptake. There is no direct evidence for the role of biosurfactants in this process. The question of whether the bulk production of one of those compounds is a prerequisite for or a result of alkane uptake must be answered in the individual microorganism [67]. Generalizations may not be possible. Recently, Casey et al. [91] reported an efficient technique for the enrichment of Alk$^-$ mutants of *Candida cloacae*: most

(a)

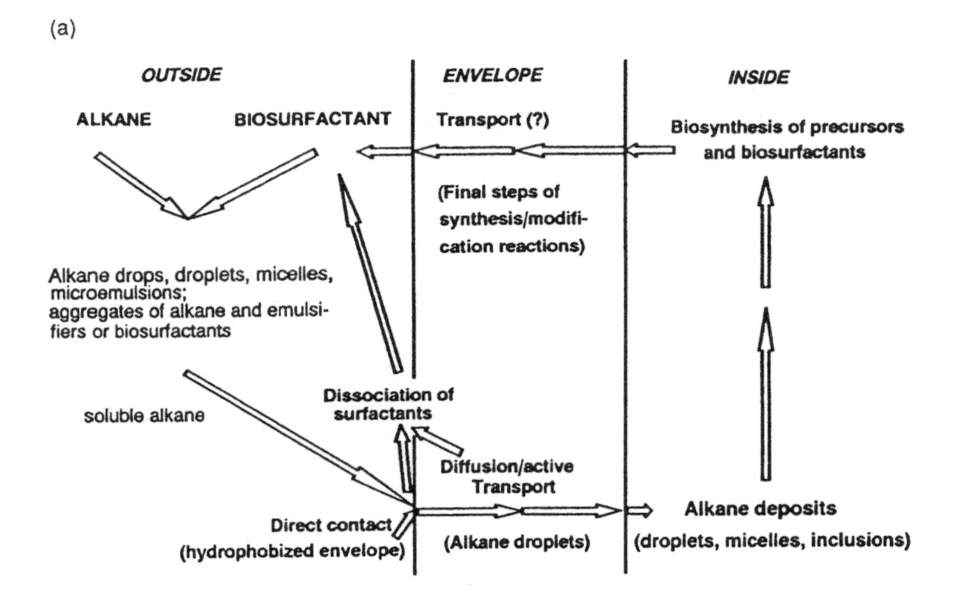

(b)

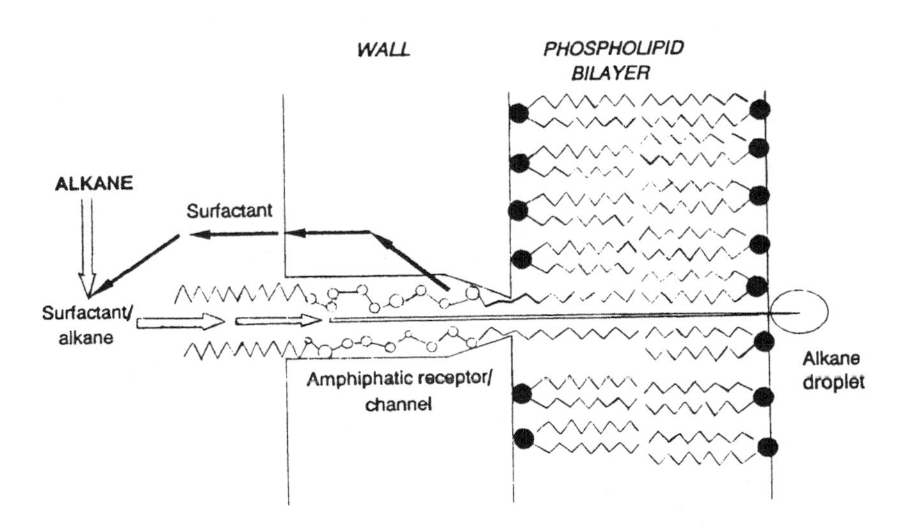

FIG. 6 Hypothetical scheme for surfactant-linked alkane uptake into a microbial cell (a) general scheme; (b) receptor channel concept. (From Ref. 47.)

of those generated were defective in the catabolic path but the addition of sophorolipid to cultures protected the cells from antibiotic attack, presumably by preventing hexadecane uptake. An Alk$^-$ PA$^+$ phenotype was assumed to be deficient at an early step of the pathway, either being unable to transport the alkane into the cells or having lost the alkane hydroxylase.

B. Hydrocarbon Oxidation

The main microbial oxidation pathways of hydrocarbons have been reviewed on many occasions [45,47,54,87,92–95,99] and are summarized in Fig. 7.

The oxidative attack of the alkane following the transport into the cell is catalyzed by a complex alkane hydroxylase linked to an electron carrier system. In

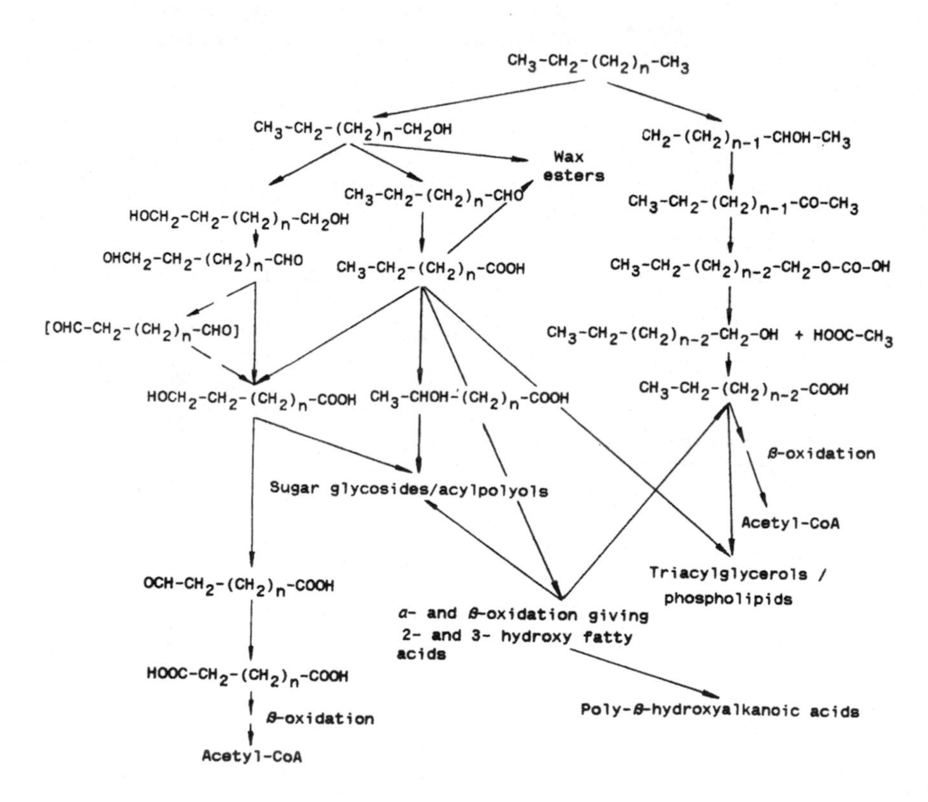

FIG. 7 Microbial oxidation products from alkanes and fatty acids. (Adapted from Ref. 95.)

the majority of microorganisms, the conversion to the corresponding alkane-1-ol is carried out by monooxygenase (hydroxylase) systems:

$$R\text{-}CH_3 + O_2 + NAD(P)H \rightarrow R\text{-}CH_2OH + NAD(P)^+ + H_2O \quad (20)$$

Rehm and Reiff [93] have also described subterminal oxidation pathways that lead to secondary alcohols being produced in some fungi.

The hydroxylation systems have been intensively studied. Two different systems are known. A rubredoxin-dependent system has been found *Pseudomonas* sp. [96] and in strains of *Acinetobacter calcoaceticus* [97,98] and other prokaryotes [99]. In yeasts and some prokaryotes, for example, *A. calcoaceticus* EB 104 [101,102], alkane hydroxylation is catalyzed by cytochrome P-450 systems. Yeast cytochrome P-450 systems involved in alkane metabolism have been reviewed by Käppeli [103] and more recently by Müller et al. [100]. These hydroxylation systems are specifically induced by long-chain alkanes and compounds with related structures [104–107]. The simultaneous addition of glucose to the culture medium (concentrations above 0.05 $g\ L^{-1}$) almost completely represses this induction [105,108]. Ilchenko et al. [108] reported the occurrence of cytochrome P-450 during cultivation of *Candida maltosa* on glucose upon transition to the stationary phase, which may indicate derepression of alkane-inducible cytochrome P-450 hydroxylase forms. The cytochrome P-450 alkane hydroxylase content of *C. maltosa* [109–111] and *C. tropicalis* [112,113] is strongly increased by transition of the cells to oxygen-limited growth during cultivation on long-chain *n*-alkanes. Recent reports by Schunck et al. [114,115] indicate the multiplicity of alkane-induced cytochrome P-450s in *C. maltosa*. The expression of the $P\text{-}450_{Cm\text{-}1}$ and of the $P\text{-}450_{Cm\text{-}2}$ cDNA in *Saccharomyces cerevisiae* resulted in systems that hydroxylated *n*-hexadecane with significantly higher activity than lauric acid. There is also evidence for the existence of two alkane-induced cytochrome P-450 forms in *C. tropicalis* [116,117]; one of these was also expressed in *S. cerevisiae* and had a similar substrate specificity to that of like $P\text{-}450_{Cm\text{-}2}$.

Whereas the cytochrome P-450 alkane hydroxylase system is localized in the endoplasmic reticulum (microsomal fraction) of the eukaryotic cell, the subsequent oxidation of the fatty alcohol proceeds in the peroxisomes producing acyl-CoA, which is either converted to lipid or oxidized by the β-oxidation cycle to yield acetyl-CoA [118]. The formation of peroxisomes is typically increased during growth on *n*-alkanes [119].

In yeasts, originally the involvement of an NAD(P)-dependent alcohol dehydrogenase in alkane-1-ol oxidation was reported [118–122]. However, recent reports now document the presence of alkane-induced, long-chain (fatty) alcohol oxidases in the peroxisomes of different *Candida* species grown on alkanes [123–127]. Similar long-chain fatty alcohol oxidases also occur in filamentous molds [128,129]. Kemp et al. [130,131] have described a fatty alcohol oxidase in the microsomal fraction of *C. tropicalis* that was light-sensitive, and this may be why this enzyme has been missed in the past. In microsomes of the biosurfactant-producing yeasts *Candida bombicola* and *Candida apicola*, two constitutive alcohol oxidases and a further inducible one were described that were able to oxidize alkane-1-ols, alkane-α,ω-diols and, with the enzyme from *C. apicola*, also ω-hydroxyfatty acids [132,133]. These enzymes strongly differ in their substrate specificity from methanol (alcohol) oxidases. The respective membrane-bound aldehyde dehydrogenase are NAD(P) dependent [123,126,129,130,131].

Fatty alcohol- and aldehyde-oxidizing enzymes are regulated in the same manner as reported for cytochrome P-450 alkane hydroxylases, indicating similar regulation as occurs with the *OCT* plasmid of *Pseudomonas putida* PpG6 (ATCC 17633) where the *n*-alkane-induced *alk S* gene product controls the transcription of the *alk BAC* operon encoding both alkane hydroxylases, alcohol dehydrogenase, and aldehyde dehydrogenase [87,120–122]. These enzymes are localized in the cytoplasmic membrane. It is postulated that the alkane passes through the outer membrane by diffusion and is then oxidized to the corresponding fatty acid, which enters the cytoplasm [123]. This alkane-oxidizing complex would be able to regulate the alkane consumption of the cell. Alkane-induced bacterial alcohol and aldehyde dehydrogenases are typically NAD(P) dependent and are either soluble or membrane bound. Recently, the involvement of NAD(P)-independent and PQQ-dependent enzymes in strains of *P. putida, A. calcoaceticus, Acetobacter rances*, for example, has been reported [134–136]. A further degradative path of alkanes via alkenes by action of NAD(P)-dependent dehydrogenase may be an alternative route under anaerobic conditions [94].

The metabolism of alkenes has been recently reviewed by Hartmans et al. [137]. The major patterns of initial attack are given in Fig. 8. These may include a conventional oxygenase attack upon the terminal methyl group providing the corresponding alcohol and acid and also oxidation either across the double bond to give the corresponding epoxide or across the double bond to the corresponding diol [92].

C. Formation of Hydroxy Fatty Acids and Dicarboxylic Fatty Acids

Monoterminal oxidation of alkanes may be followed by hydroxylation of the remaining methyl terminus to give either or both α,ω-dioic acids and ω-hydroxy fatty acids, which have been detected mainly in yeasts, molds but also to a minor extent in bacteria [93,99]. Yi and Rehm [138–140] demonstrated in a mutant strain of *C. tropicalis* the formation of dodecanedioic acid by both terminal and diterminal oxidation (via the α,ω-diol derivative see Fig. 7). Similar observations

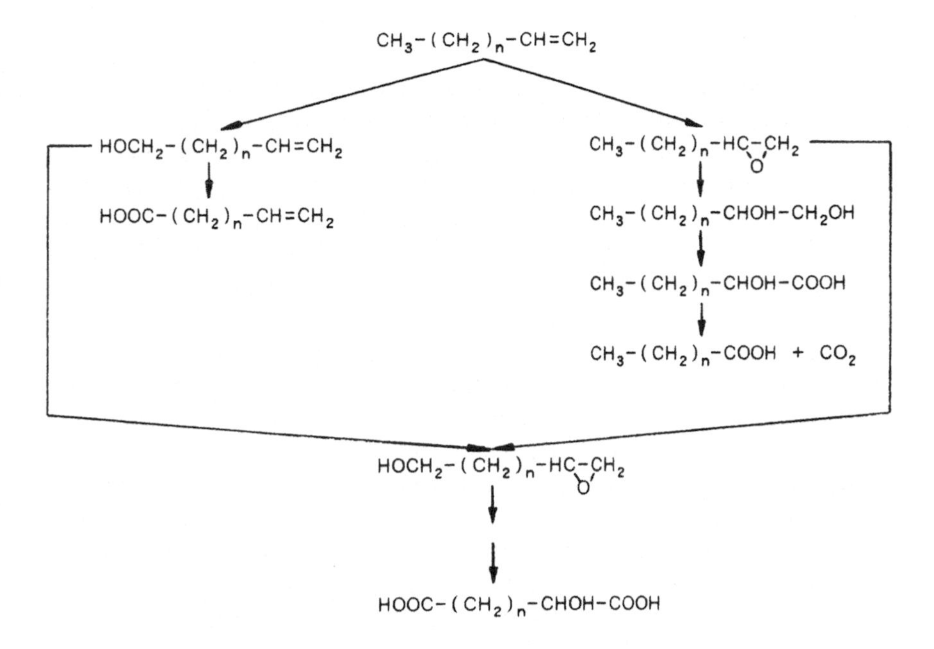

FIG. 8 Basic metabolic pathways for degradation of alkenes. (From Ref. 94.)

have been made in *Mortierella isabellina* [141], *Candida maltosa* [124,142] and the noncytochrome P-450-dependent in *Rhodococcus rhodochrous* [143]. The conversion of oleic acid to 1,18-octadec-9-enedioic acid by *C. tropicalis* has also been reported [144]. Similar properties of the fatty acid ω-hydroxylase may also be assumed for the sophorolipid-producing strains of *C. bombicola* and of *C. apicola*, which synthesize the corresponding ω- and [ω-1]-hydroxy fatty acid constituents from alkanes of different chain lengths starting with C_{14} [145] (cf. Section IV.C).

The existence of different cytochrome P-450 alkane hydroxylases in *C. maltosa* and *C. tropicalis* with different specificities toward alkane and lauric acid hydroxylation, respectively [114–117], support the assumption that the diterminal hydroxylation of the corresponding fatty acid is carried out by another cytochrome P-450 hydroxylase other than that involved in the primary attack on alkane. Both cytochrome P-450 hydroxylases were induced by *n*-alkanes.

The capability of fatty alcohol oxidases or alcohol dehydrogenases to oxidize other hydroxy compounds, for example, 2-alkanols, 1,2-alkanediols, α,ω-alkanediols, and ω-hydroxy fatty acids, is essential for the diterminal (and the subterminal) pathways. The activities of alcohol oxidases from *C. bombicola*, for example, are one order of magnitude lower toward the diols than with straight-chain alcohols [132]. There are no reports on substrate specificity of the subsequent aldehyde dehydrogenases toward the expected reaction products. (Aldehydes, being more unstable than alcohols, tend to autooxidize during isolation procedures or be converted enzymatically *in situ* to fatty acids.) Both the long-chain monocarboxylic and dicarboxylic acids undergo subsequent degradation by the β-oxidation cycle; for example Hill et al. [146] using *C. tropicalis* S_{76} reported the additional appearance of 3-hydroxy fatty acids and of dioic acids with shortened carbon chains due to degradation by β-oxidation and the formation of unsaturated fatty acids. Similar findings have been made with *Corynebacterium* HO1N [147].

The appearance of L-[ω-1]-hydroxy fatty acids, derived from the alkane carbon source, in the sophorolipids produced by *C. bombicola* [145] may suggest an additional specific hydroxylation reaction. This assumption is supported by the ability of the growth conditions to alter the ratio of the [ω-1] and the ω-hydroxylated fatty acid in the sophorolipids of *C. apicola* [148]. In *Bacillus megaterium*, which does not degrade alkanes, the cytochrome $P\text{-}450_{\omega\text{-}2}$ enzyme was partially purified and characterized as a soluble fatty acid hydroxylase that could catalyze [ω-1]-, [ω-2]-, and [ω-3]-hydroxylation but not terminal hydroxylation of saturated long-chain fatty acids, alcohols, and amides. This activity was reviewed by Slingar and Murray [149].

The degradation of fatty acids, and also hydroxy fatty acids and dioic fatty acids, proceeds via β-oxidation. In *Yarrowia (Candida) lipolytica* one of the two acyl-CoA-synthetases, synthetase II, is induced by growth of the yeasts on oleic acid [150] and is able to activate not only fatty acids but also ω-hydroxy fatty acids and dioic fatty acids as well. This enzyme is localized in the peroxisomes and serves to provide acyl-CoA for β-oxidation. The other synthetase, synthetase I, is responsible for producing acyl-CoA to be used for lipid synthesis. This is located in both mitochondria and microsomes. For the production of dicarboxylic and ω-hydroxy fatty acids by yeasts, mutant strains can be produced that are partially blocked in alkane-induced peroxisomal β-oxidation [46,99]. The massive production of ω- and [ω-1]- hydroxy fatty acids in the sophorolipid-producing *C. bombicola* and *C. apicola* may indicate a low activity of the β-oxidation cycle enzymes or of the acyl-CoA synthetases that would initiate β-oxidation. Alternatively, glycolipid synthesis in these yeasts may be rigidly compartmentalized with little ω-hydroxy fatty acid becoming available for β-oxidation in the peroxisomes.

IV. BIOSYNTHESIS OF ACYLATED COMPOUNDS

Molecular biosurfactants owe their particular properties to the amphipathic nature of their molecular structure, that is, both polar and apolar moieties are combined in a single compound. In the case of microbial surfactants, the apolar portion of

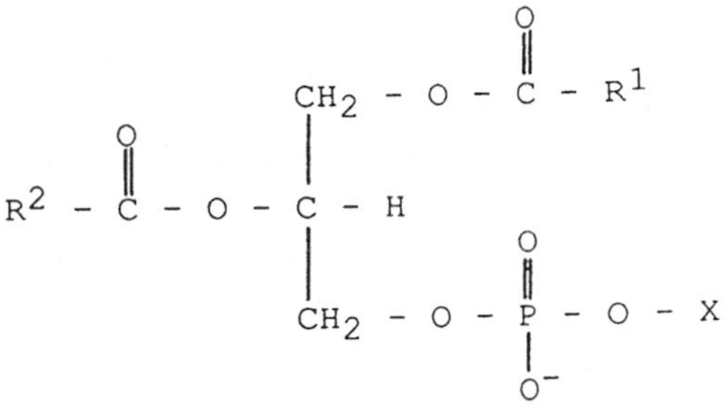

```
                              O
                              ‖
                 CH2 - O - C - R1
        O         |
        ‖         |
R2 - C - O - C - H
                  |           O
                  |           ‖
                 CH2 - O - P - O - X
                              |
                              O-
```

FIG. 9 Generalized phospholipid structures. R^1,R^2 alkyl substituents are usually C_{15} or C_{17}. X could be H, phosphatidic acid; $-CH_2{\cdot}CH_2{\cdot}NH_2$, phosphatidyl ethanolamine; $-CH_2{\cdot}CH(NH_2)COOH$, phosphatidyl serine; $-CH_2{\cdot}CH_2{\cdot}N(CH_3)_3$, phosphatidyl choline (lecithin); $-CHOH{\cdot}CHOH{\cdot}CH_2OH$, phosphatidyl glycerol; –inositol, phosphatidyl inositol; –phosphatidylglycerol, diphosphatidylglycerol (cardiolipin). (For further details see Refs. 2 and 4.)

the molecule often consists of one or more long-chain acyl groups. The polar moiety may be supplied from a variety of sources. These would include sugars and polar portions of compounds such as phospholipids (Fig. 9) and various esters.

Previous sections in this review have described how fatty acids are synthesized *de novo* or produced as a result of hydrocarbon oxidation. In this section, a brief survey is presented of the various classes of microbial lipids into which these fatty acids may be incorporated.

It is important to appreciate that fatty acids themselves are too toxic, because of their action against cell membranes, to be tolerated beyond a very low concentration within a cell, although they may, and do, occur as free fatty acids extracellularly. Within the cell, including the outer envelope, the fatty acids must be esterified in some manner. This may be as the thioester of CoA, which would precede degradation via the usual β-oxidation cycle or the less common α-oxidation cycle [4,10]. Alternatively, the fatty acid may be more permanently esterified to glycerol to give the mono-, di-, and tri-acylglycerols of which the latter is the commonest storage form of lipid in eukaryotic cells. A wide variety of esterified forms of fatty acids can occur in microorganisms. Some of these are used below as examples where they are relevant as potential surfactants.

A. Neutral Lipids

1. Triacylglycerols

Triacylglycerols (or triglycerides) are the common form of storage lipid found in all eukaryotic cells: yeasts, molds, lower and higher plants, and animals. They are the oils and fats of everyday food materials. They are formed and stored intracellularly often in the form of discrete droplets [2,151] and fulfill a storage function for carbon, energy, and even water (hence the hump on the camel!). They are formed when the supply of carbon is plentiful and broken down when the carbon becomes exhausted [151,152]. Triacylglycerols do not have any biosurfactant properties. In oleaginous yeasts and molds, triacylglycerols can accumulate up to 80% of the cell biomass [6,7,153]. The pathway of biosynthesis of triacylglycerols (see Fig. 10) is inextricably linked to the biosynthesis of phospholipids (as discussed in Sec. IV.B).

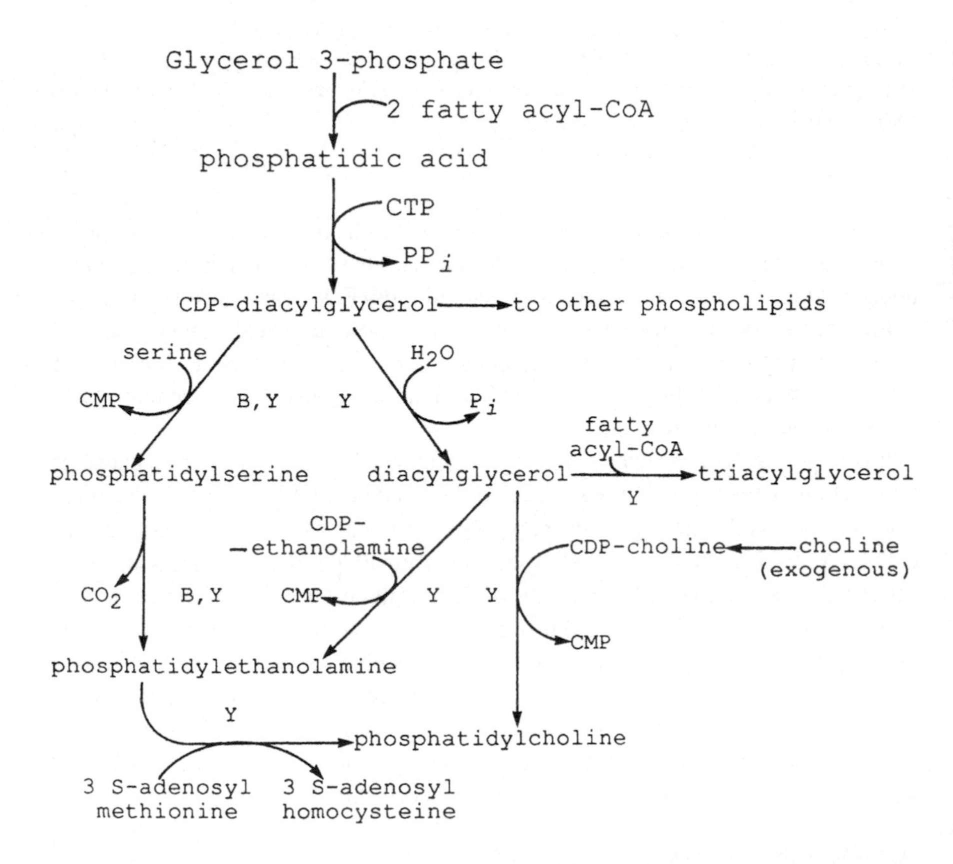

FIG. 10 Biosynthesis of triacylglycerol and phospholipids. B, bacterial route; Y, yeast (and probably all eukaryotic cells) route. For structures see Fig. 9. CMP, CDP, and CTP, cytosine mono-, di-, and tri-phosphate, respectively; P$_i$, inorganic phosphate; PP$_i$, pyrophosphate.

2. Esters

Fatty acids, beside forming triesters of glycerol (i.e., triacylglycerols as above), may form simple monoesters [4] with a variety of alcohols. Such esters are formed in appreciable amounts in *Acinetobacter* spp. and related bacteria such as *Moraxella* (*Branhamella*) [154,155] and occur in some eukaryotic algae. They are formed constitutively up to 5% of the biomass when the cells are grown on acetate or succinate (these bacteria do not usually grow on glucose) but are increased up to 10% of the biomass when the cells are grown on alkanes [156–158].

$$CH_3\cdot(CH_2)_n CO\cdot O(CH_2)_m\cdot CH_3$$

n=14,16 or 18 m=14,16 or 18

[4]

Attempts have been made to produce diunsaturated wax esters where there is a single double bond in both the acyl and alcohol components of the ester [159,160]. This would, in effect, create a jojoba oil-like lipid. (For further details see either Refs. 7 or 160.) Such esters probably have a minimal role in alkane solubilization or in any surfactant function.

The biosynthesis of these esters has been investigated by Lloyd and Russell [161] using *Micrococcus cryophilus*. Two pathways exist depending on whether the organism has to synthesize fatty acids *de novo* (from acetate) or is utilizing exogenous alkanes or fatty acids. The fatty alcohol component is probably formed by reduction of the fatty acyl-CoA ester, which has been shown to occur in an anaerobic bacterium [162].

Diesters [5] have been reported as extracellular lipids in a *Corynebacterium* sp. grown on decane [163] but no further developments in this direction appear to have occurred.

$$CH_3(CH_2)_8CH_2O\cdot OC(CH_2)_8CO\cdot OCH_2(CH_2)_8CH_3$$

[5]

The bacterial polyester, poly-β-hydroxybutyrate [6], is produced in numerous bacteria [71,164] but is the equivalent of the triacylglycerol storage lipid of higher organisms. As such it has no surfactant properties but is of commercial interest (being produced by ICI plc of the United Kingdom under the trade name of Biopol) as a useful biodegradable polymer.

Other esters of fatty acids are discussed in Secs. IV. through IV.E.

$$H\left[O-\underset{|}{\overset{CH_3}{C}}H-CH_2-\overset{O}{\overset{\|}{C}}\right]_n OH$$

n = 600 to 2500

[6]

B. Phospholipids

Phospholipids are ubiquitous components for all living microorganisms, as they form the major constitutent of all cell membranes—both of the outer cell envelope and of many intracellular structures. The structures of the principle acylglycerophospholipids are given in Fig. 9. Variations on these structures, including ether-linked glycerophospholipids and phosphonolipids, occur and are detailed in other reviews [4,165].

The concentration of phospholipids in microbial cells appears to be approximately constant at between 2% and 4% of the cell dry weight irrespective of the growth conditions [166] and is only slightly increased to 5% or 6% following growth of microorganisms on alkanes. The highest contents, up to 16% have been recorded in bacteria grown anaerobically on methane where extensive intracellular membranes are developed [167].

Extracellular phospholipids may be produced when hydrocarbon substrates are used and such molecules do have slight surfactant properties. For example, Siporin and Cooney [168] identified phosphatidylcholine, phosphatidylserine, phosphatidylethanolamine, and a cardiolipinlike compound in the growth medium of *Cladosporium resinae* grown on *n*-alkanes. In each case, dodecanoic acid was the most common alkyl substituent. Similar observations have been made by Finnerty when analyzing the extracellular lipid constituents of *Acinetobacter* sp. [164].

Lysophospholipids, that is phospholipids without an acyl group at either the 1- or 2-position (see Fig. 9), are much better surfactive materials than phospholipids themselves. Indeed, lysolecithin (i.e., monoacylphosphatidylcholine) is used commercially as an emulsifying agent. However, it is by no means certain that such compounds are produced in any significant amounts even when microorganisms are grown on water-insoluble substrates that would require emulsification. Finnerty [see 170 for review] found that 2%–12% of the total cellular phospholipids (approximately 4% of the cell biomass) in *Acinetobacter* were lysophospholipids: monodeacyl(lypo)cardiolipin and dideacyl(lyso)cardiolipin (see Fig. 9). Such lysophospholipids would be produced by the action of a phospholipase A_1 or A_2, depending on whether the 1-acyl or 2-acyl group, respectively, was removed. However, stringent precautions have to be taken to deactivate such enzymes in harvested microorganisms prior to their solvent extraction, as these enzymes function admirably well in organic solvents. Whether lysophospholipids are of functional significance in any microorganism must remain an intriguing, yet open, question in spite of their known emulsifying abilities.

Phospholipids are synthesized by two principle routes depending on whether they are in bacteria or yeasts (Fig. 10). The bacterial route synthesizes primarily phosphatidylserine and phosphatidylethanolamine; phosphatidylcholine is not normally found in bacterial membranes, although there are some exceptions [2,166]. In yeasts, and presumably in other eukaryotic microorganisms as well, diacylglycerol is a key intermediate. This can be used directly for the biosynthesis of triacylglycerols or be directly converted to phosphatidylethanolamine using ethanolamine derived by decarboxylation of serine. The diacylglycerol may also be converted directly to phosphatidylcholine (via CDP-choline) when choline is provided exogenously. If choline is not so available, the phosphatidylethanolamine is methylated, with *S*-adenosylmethionine, three times in succession, to give phosphatidylcholine. This route may also occur in a limited number of bacteria; the intermediates, mono- and di-methylphosphatidylethanolamines, can also be found as phospholipids in their own right in a few bacteria [166]. Phosphophatidylcholine is thus readily formed by a variety of routes in eukaryotic cells, where consequently it is often the principal phospholipid component of the membrane structures.

The collective pathways have been extensively reviewed by Pieringer [171]; the regulation of phospholipid biosynthesis in *E. coli* has been reviewed by Boom and Cronan [14] and the subject of phospholipid biosynthesis in yeast, with emphasis on the molecular genetics, by Carman and Henry [172]. Microbial sources of phospholipids were summarized by Ratledge [166].

C. Glycolipids

Glycolipid biosurfactants are commonly considered as sugar-containing lipids in which both moieties may be linked either glycosidically (i.e., as an ether or more correctly as a hemiacetal link) as in the sophorose, rhamnose, and cellobiose lipids, or via acylation (i.e., as an ester link) as in the acylpolyols, trehalose lipids, and sugar mycolates (discussed separately in Sec. IV.D). The biosynthesis pathways to these lipids must include the generation of both the lipophilic (fatty acid) and the hydrophilic (carbohydrate) moiety. Biosynthesis of biosurfactants, in general, and especially for glycolipid biosynthesis, may fall into one of the following pathways according to Haferburg et al. [48] and Syldatk and Wagner [173]:

Both the carbohydrate and the lipid moiety are synthesized independently of the growth substrate.
The synthesis of the lipid moiety depends on the hydrophobic carbon source, that is, it is a derivative of the hydrophobic carbon source and the carbohydrate is synthesized *de novo.*
The carbohydrate moiety reflects the carbon source being used for growth or maintenance and the lipid moiety is synthesized *de novo.*
The synthesis of both residues depends on the carbon substrates being used.

Examples of the latter pathway are not known so the following discussion will be confined to the other three possibilities.

Sophorose lipids are only synthesized by yeasts of the genus *Candida: Torulopsis magnoliae* [174] and *Torulopsis apicola* [175]. *Torulopsis gropengiesseri* [176] and *Torulopsis bombicola* [177] have now been reclassified as species of *Candida* thus joining *Candida bogoriensis* [178,179], which remains unchanged.

The constitutive carbohydrate backbone of these extracellularly glycolipids is sophorose (2′-*O*-β-D-glucopyranosyl-β-D-glucopyranose), which may be acetylated in positions 6 and 6′ and is linked in position 1 with a hydroxy fatty acid (Fig. 11). Although the fatty acid moiety, 13-hydroxydocosanic acid, of the *C. bogoriensis* lipid (Fig. 12) is produced only when hydrophilic carbon sources are used, the nature of ω- and [ω-1]-hydroxy fatty acids in the other sophorolipids depends on the composition of the hydrophobic carbon source used for growth, which may be a saturated or unsaturated compound [145,180]. Additionally, the glycolipids (with exception of that of *C. bogoriensis*) appear as mixtures of several (at least six) sophorolipids that differ in degree of saturation and unsaturation, in different types of lactonization (1,6; 1,6″; and 1,4″), and in different degrees of acetylation [181–184]. The number of individual sophorolipids was decreased when organic acids, and in particular citrate, or amino acids were added to the medium [185] giving 80% of the total glycolipids in the lactonized form (Fig. 11b).

Growth experiments with alkanes and derivatives of different chain length have revealed direct incorporation of these compounds into the glycolipids of *Torulopis* strains [180]. Similar results were obtained with methyl esters of saturated and unsaturated fatty acids with chain lengths in the range between C_{13} and C_{20} and with halogenated derivatives of fatty acids and alkanes [182]. Monooxygenase systems (see Secs. III.B and C) are evidently responsible for both the hydroxylation of the alkane and also for the hydroxylation of the opposite terminus in either the ω or [ω-1]-position. With whole cells of *Torulopsis* sp. strain 319-67, the stereospecific hydroxylation of octadecanoate to 17-L-hydroxyoctadecanoic acid was shown by incorporation of $^{18}O_2$, but not of $H_2 18O$, into the hydroxyl group [186,187]. Heinz et al. [187] reported the conversion of oleic acid to

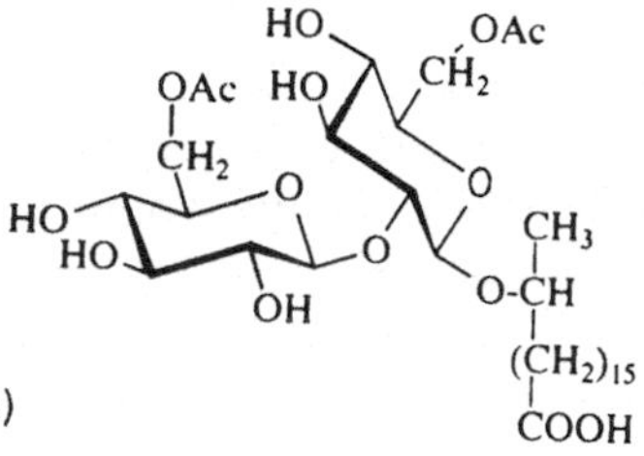

(a)

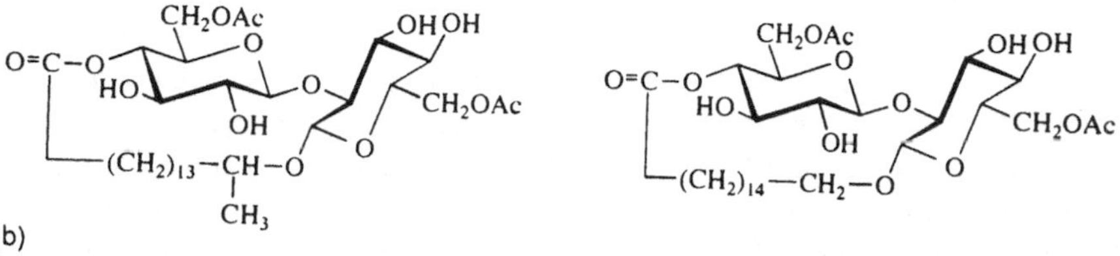

(b)

FIG. 11 (a) Acidic sophorose lipid of *Torulopsis (Candida) bombicola* and (b) lactonic sophorosides of *Torulopsis (Candida) apicola* IMET 43747. [(a) From Ref. 183; (b) from Ref. 184.]

17-hydroxyoleic acid by a mixed function oxidase in a cell-free system of *T. bombicola* in the presence of NADPH. This reaction was inhibited by CO.

Kleber et al. [188] detected cytochrome P-450 hydroxylase in *T. apicola* IMET 43747 grown on glucose, on hexadecane, or on a mixture of both, coinciding with the beginning of glycolipid production. No repression of cytochrome P-450 hydroxylase occurred in the presence of glucose, which is a key difference from alkane monooxygenase systems of other yeasts (cf. Sec. III.B). More recent

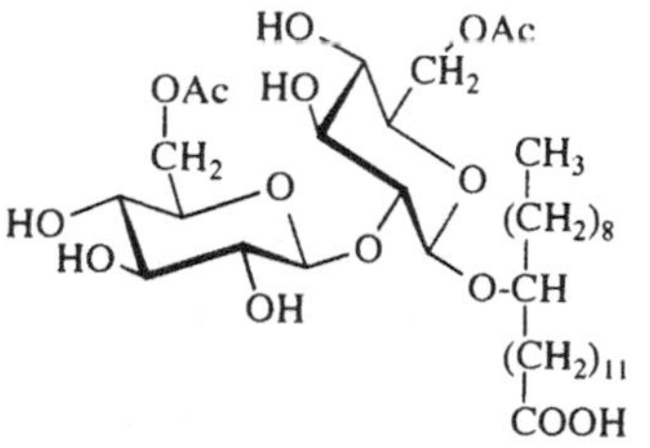

FIG. 12 Sophorose lipid of *Candida bogoriensis*. (From Ref. 174.)

results [189] with the same strain confirm the lack of a repressible alkane hydroxylase even after alkanes have been totally consumed and glucose is the sole remaining carbon source. By adding [1-^{13}C]palmitic acid to either growing (logarithmic) or glycolipid-producing (stationary) cells of *T. apicola*, Weber et al. [190] could show by NMR spectroscopy that the recovery of the ^{13}C in the sophorolipid was qualitatively the same as that when the compounds were added to stationary phase cells (Table 2). However, the stationary phase cells incorporated [1-^{13}C]hexadecan-1-ol and [1-^{13}C]hexadecanoic acid faster than they oxidized the unlabeled hexadecane as revealed by the enhancements of 10.5 and 13.0 for the alcohol and the acid, respectively (Table 2). In both experiments the ^{13}C of the alcohol was found in position 1, thus excluding the diol as a possible intermediate (see Fig. 7).

These results suggest that two different monooxygenase systems are involved in the formation of hydroxyhexadecanoic acids. Furthermore, selective ω-hydroxylation is principally expressed in logarithmic phase cells whereas in stationary phase cells an ω- or [ω-1]-hydroxylating monooxygenase system must be induced. The ratio of ω- to [ω-1]-hydroxylated lactonized products (Fig. 11) may be shifted from 1:1 to 1:2.3 [148,184]. On the other hand, Jones [191] described three different metabolic routes in *T. gropengisseri* for the conversion of long-chain methyl branched alkanes to ω- and/or [ω-1]-hydroxyalkan-1-ol, and also to dicarboxylic fatty acids, which were not detected in *T. apicola* [189]. Furthermore, Jones [191] assumed all the alkanoic acid oxidations in *T. gropengisseri* were catalyzed by the same enzyme.

Hommel and Ratledge [132] reported two constitutive microsomal fatty alcohol oxidases in *T. bombicola* ATCC 22214 that were not repressed by glucose and were able to oxidize long-chain alcohols and diols but not ω-hydroxy fatty acids. More recently a microsomal fatty alcohol oxidase has been detected in *T. apicola*

TABLE 2 ^{13}C-Signal Appearance in the Lactonic Sophorolipid of *Torulopsis apicola* IMET 43747 after Addition of Different ^{13}C-Labeled Hydrophobic Carbon Sources to Cultures Growing on Mixed Substrates (glucose and hexadecane)

	Exponential cell		Stationary cell	
Labeled compound	Position of ^{13}C signal	Enrichment ratio	Position of ^{13}C signal	Enrichment ratio
[1-^{13}C]-hexadecane	1 and 16	2 * 4.6	1 and 16	2 * 4.6
[1-^{13}C]-hexadecanol	1	8.41	1	10.5
[1-^{13}C]-palmitic acid	1	10.6	1	13.0

Source: Ref. 190.

[133]. However, this enzyme is induced in the transition from the logarithmic growth phase into the stationary phase. Long-chain alcohols and diols were oxidized as were hydroxy fatty acids but at a lower rate. The induction occurred whether glucose or hydrocarbons, or a mixture of both, were used as carbon sources. In the same manner, a microsomal fatty aldehyde dehydrogenase has been found in *T. apicola* [133].

The proposed scheme of the oxidation of the alkane in *T. apicola* IMET 43747 is given in Fig. 13. From the experiments with ^{13}C-labeled compounds and also enzymatic studies, the induction of the enzyme(s) involved in the hydroxylation of the fatty acid will be the overall rate-limiting step in *T. apicola*. The yields of glycolipid synthesized under mixed substrate cultivations are determined by the oxidation status of the hydrophobic substrate; that is, with triacylglycerols, significantly higher yields of sophorolipid production occur [185] than with *n*-alkanes, which could indicate that one of the prior reactions (see Fig. 13) is

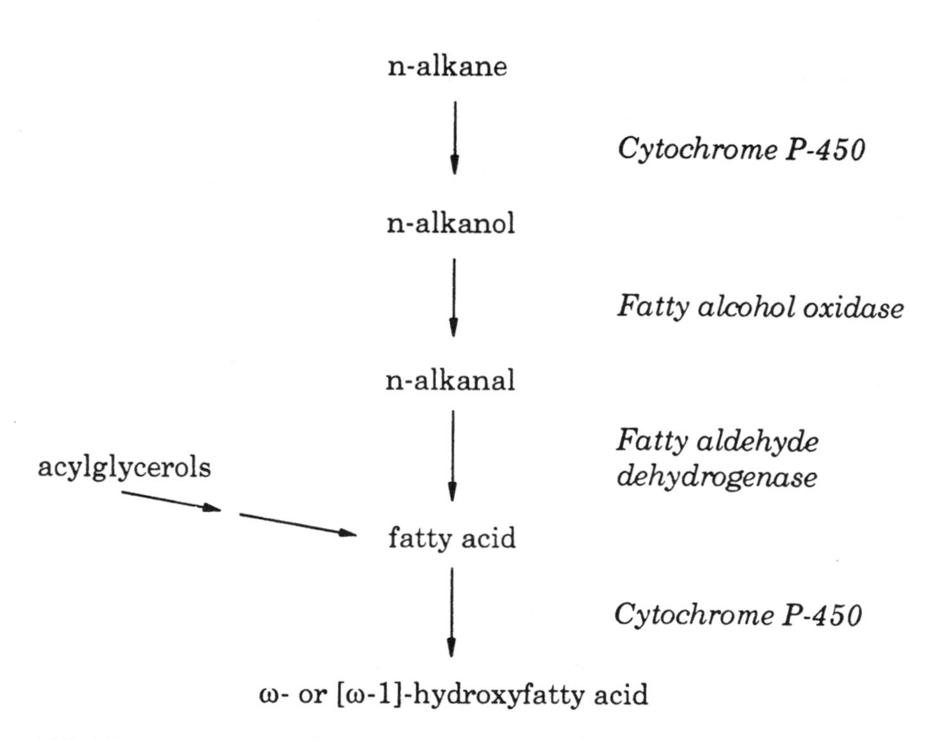

FIG. 13 Proposed scheme of the microsomal oxidation reactions of long-chain alkanes like *n*-hexadecane toward hydroxypalmitic acid in *Torulopsis (Candida) apicola* IMET 43747. (According to Refs. 131, 188, and 190.)

now rate-limiting. Similar kinetics of production by sophorose lipid strains of *T. bombicola* [181,192,193] suggest comparable pathways and regulation as in *T. apicola*.

Attempts to alter the basic structure of the carbohydrate moiety in sophorolipids by substituting other sugars for glucose failed with resting cells of *T. bombicola* [194] indicating highly specific enzymatic sequences of the *de novo* synthesis of sophorose and its subsequent linkage to the hydroxy fatty acid. From this, one can conclude that the appropriate strains synthesize both the lipid and the sugar moiety *de novo* when using carbohydrates and also in the logarthmic phase of mixed substrate cultivation. In the presence of lipophilic carbon sources only, the sophorose is formed *de novo*.

In *C. bogoriensis*, which carries out the complete *de novo* synthesis of its sophorose liquid, two indistinguishable glycosyltransferases (I and II) responsible for establishment of glucosidic linkages in the sophorose lipid have been characterized [179,195–198]. Both were at maximal activity in cells about to enter the stationary phase of growth. Both cytosolic enzymes were depressed by growth of the yeast on low glucose concentrations. Transferases I and II could not be separated by affinity chromatography nor by isoelectric focusing; their molecular sizes were both 52,000 Daltons and had identical Arrhenius plots and inhibition phenomena indicating that both enzyme activities are probably part of the same polypeptide chain [198]. The synthesis of the sophorolipid involves stepwise transfer of glucose residues from UDP-glucose, first to 13-hydroxydocosanoic acid (HDA) to form 13-(β-D-glucopyranosyl)hydroxydocosanoic acid (methyl GlcHDA) and finally to form 13-[C2′-*O*-β-D-glycopyranosyl-β-D-glucopyranosyl)oxy]docosanoic acid (Glu2HDA) (cf. Fig. 14).

$$\text{UDPglucose} + \text{HDA} \xrightarrow{\text{glycosyltransferase I}} \text{GlcHDA}$$

$$\text{GlcHDA} + \text{UDPglucose} \xrightarrow{\text{glycosyltransferase II}} \text{Glc}_2\ \text{HDA}$$

$$\text{Glc}_2\ \text{HDA} + \text{acetyl CoA} \xrightarrow{\text{acetyltransferase}} \text{Ac}_2\ \text{Glc}_2\ \text{HDA} + \text{AcGlc}_2\ \text{HDA}$$

FIG. 14 Sequence of glycosyltransfer in *Candida bogoriensis*. (From Refs. 196 and 197.)

In the same yeast, an acetyl-CoA-dependent acetyltransferase with a relative molecular weight of 500,000 has been purified [199]. The expression of this enzyme was not affected by low glucose concentrations. The enzyme catalyzed the acetylation reaction in the 6′ and 6″ positions of Glc_2HDA (cf. Fig. 12) producing the 6′,6″-diacetate and the monoacetate in a product ratio of 5:1. In late stationary cultures, Ac_2Glc_2HDA was slowly converted to $AcGlc_2$ and Glc_2HDA by action of a nonspecific esterase in cells [179,200].

Based on the structural elucidation of the mixture of sophorolipids produced by *T. bombicola* ATCC 22214, Asmer et al. [181] proposed a scheme (Fig. 15) of the last steps in biosynthesis of the lactonic diacetylated lipid in which, in principle, both the acetylation and lactonization steps should be involved.

How the generated products are excreted from the cell and what is the localization of the biosynthetic enzymes remains unclear. Using protoplasts of *T. apicola*,

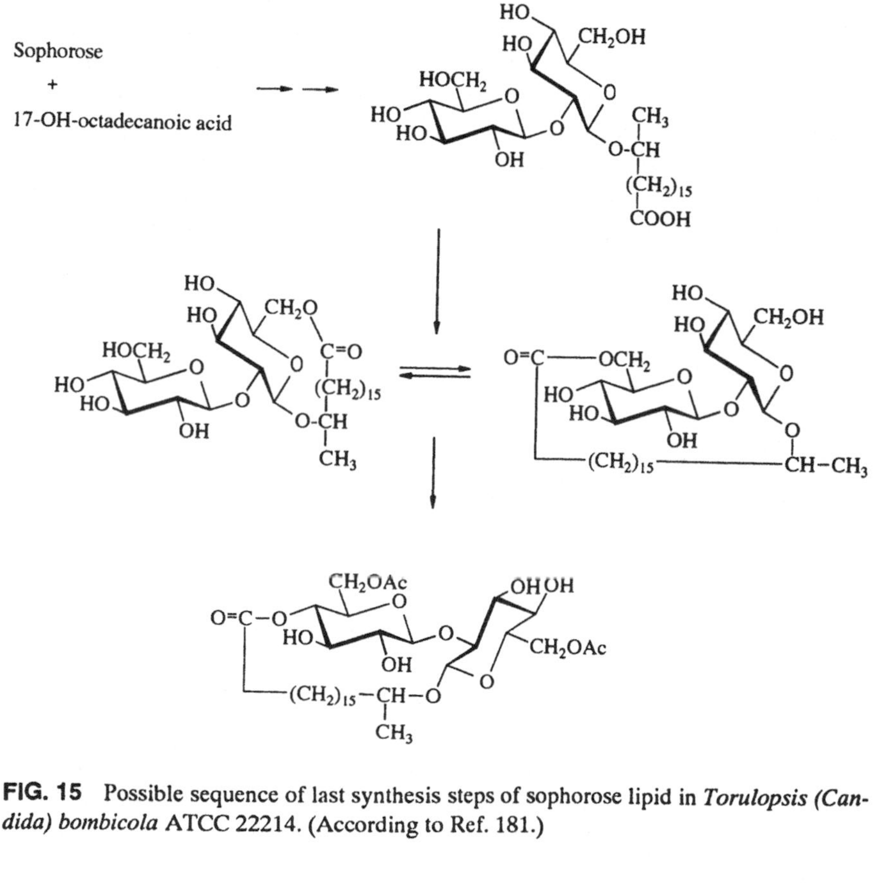

FIG. 15 Possible sequence of last synthesis steps of sophorose lipid in *Torulopsis (Candida) bombicola* ATCC 22214. (According to Ref. 181.)

Rilke et al. [201] have shown the probable existence of periplasmic vesicles derived from the cytoplasmic membrane that became completely separated from the cytoplasmic membrane after protoplasting. In *Serrartia marcescens*, biosurfactants have been reported to be the main lipid components of extracellular vesicles [202] and thus by inference these vesicles in *T. apicola* could be the site of sophorolipid formation and accumulation. The potentially deleterious effect of the sophorolipids within the cell would then be obviated, as such vesicles could secrete the surfactant by a process akin to exocytosis.

The only known fungal glycolipids are produced by *Ustilage maydis* PRL-617 and are the anionic cellobiose lipids, so-called ustilagic acid (Fig. 16) [203–206]. They were produced after growth on glucose and vegetable oil. The analysis of the constituents indicated that the sugar and both fatty acids are synthesized *de novo* from C_3 and from C_2 fragments, respectively [207], even though glucose is being used as the carbon source. Recently, Frautz et al. [208] reported the synthesis of similar lipids in *U. maydis* ATCC 14826. Although the ester-linked fatty acid (Fig. 16) could be changed by using other carbon sources, such as coconut oil, the ether-linked fatty acid remained always as 15,16-dihydroxyhexadecanoic acid.

Biotechnological processes aiming to produce rhamnolipids by growing and resting cells of different strains of *Pseudomonas aeruginosa* have been reviewed by Haferburg et al. [48], Lang and Wagner [50], and Syldatk and Wagner [73]. The sugar moiety is represented by one or two rhamnose units, glycosidically linked, and one or two units of fatty acid (Fig. 17) [209,210]. This structure is not subject to significant alterations indicating *de novo* synthesis

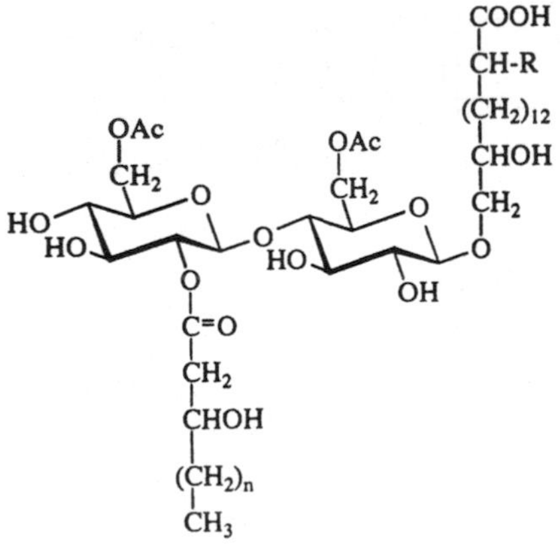

COOH
CH-R
$(CH_2)_{12}$
OAc OAc CHOH
CH_2 CH_2 CH_2
HO HO O HO O OH O
O
C=O
CH_2
CHOH
$(CH_2)_n$
CH_3

FIG. 16 Cellobiose lipid of the corn smut fungus *Ustilago maydis* PRL 119 with R = H or OH and n = 2 or 4. (From Ref. 208.)

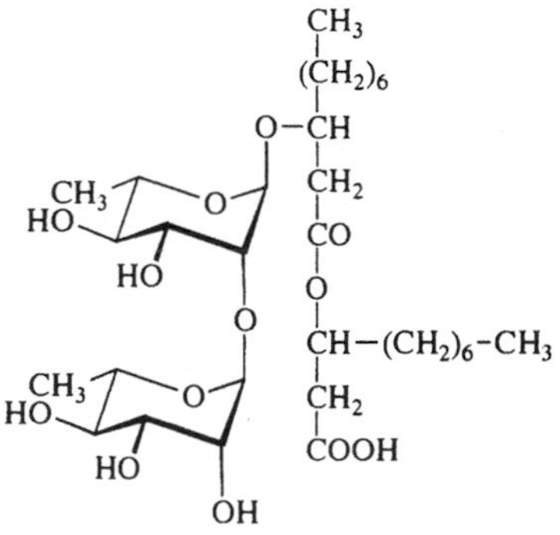

FIG. 17 Rhamnose lipid RH1 of *Pseudomonas aeruginosa* DSM 2874. (From Ref. 210.)

of the entire lipid irrespective of carbon source being used. The only exception of the general structure was reported by Yamaguchi et al. [211], where additional decenoyl moieties were esterified in positions 2′ and 2″ of the mono- and di-rhamnosyl unit.

Early work [212-214] on *de novo* biosynthesis of this lipid, showed that rhamnose was derived by condensation of two C_3 and C_2 units but that the fatty acid was synthesized, as may have been expected, only from acetate. Studies of the biosynthetic pathway of TDP-L-rhamnose, which is derived from TDP-D-glucose, suggest the former nucleotide is the precursor of the rhamnolipids, as *P. aeruginosa* lacks the corresponding UDP-compound [215]. This route was confirmed by Burger et al. [216] using ^{14}C-labeled TDP-rhamnose, β-hydroxydecanoic acid, and derivatives. In stationary cells of *P. aeruginosa* ATCC 7700, two different specific rhamnosyl transferases, TDP-L-rhamnose:β-hydroxydecanoyl-β-hydroxydecanoate rhamnosyl transferase and TDP-L-rhamnose: L-rhamnose-β-hydroxydecanoyl-β-hydroxydecanoate rhamnosyl transferase, catalyze the stepwise rhamnosyl transfer reaction (Fig. 18). Both enzymes accept only TDP-L-rhamnose. The transferase activities are localized in both the membrane fraction and the cytosolic fraction. Each has been separated and partially purified [216].

The synthesis of β-D-hydroxydecanoic acids proceeds independently of the carbon source via the usual fatty acid synthetase. The exact pathway of synthesis of the β-hydroxydecanoyl-β-hydroxydecanoyl unit (Fig. 18, step 1) could not be confirmed. Enzymes catalyzing this CoA-dependent reaction appear to be labile. Whether only one or two activated fatty acid units are involved in the rhamnolipid synthesis could not be determined [216].

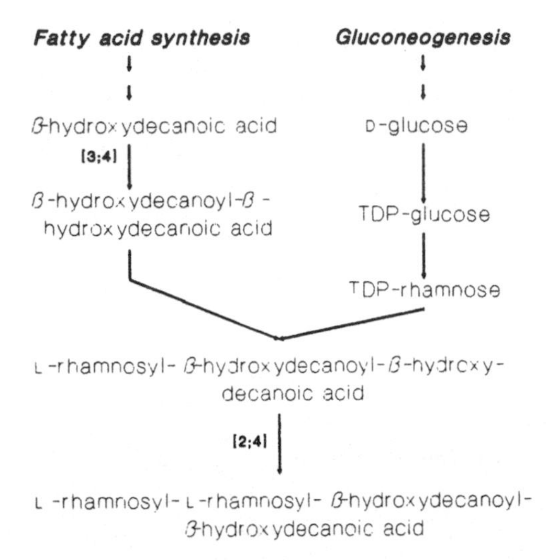

FIG. 18 Scheme of biosynthesis of rhamnolipid by *Pseudomonas aeruginosa* [212–214] which is identical with RH1 [210]. Numbers indicate alterations of metabolic path providing the corresponding derivatives of RH1 [210]. [2] L-rhamnosyl-β-hydroxydecanoyl-β-hydroxydecanoic acid (RH2); [3] L-rhamnosyl-L-rhamnosyl-β-hydroxydecanoic acid (RH3); [4] L-rhamnosyl-β-hydroxydecanoic acid (RH4).

A direct but unexplained correlation of rhamnolipid production and specific activity of glutamine synthetase has been reported [217]. This latter enzyme showed maximum activity at the end of exponential phase and remained so throughout rhamnolipid production. Mulligan et al. [218] reported increased production with a chloramphenicol-tolerant mutant of *P. aeruginosa*. These authors demonstrated that *P. aeruginosa* undergoes two distinct types of metabolism on complex media: exponential growth is linked with amino acid catabolism and stationary growth is linked with glucose metabolism. This metabolic shift is expressed by induction of transhydrogenase and by glucose-6-phosphate dehydrogenase. At the transition point of this reverse diauxie, biosurfactant production is initiated.

D. Acylpolyols

In contrast to glycolipids in which the carbohydrate and the hydroxy fatty acid moieties are linked glycosidically, in acyl-polyols the sugars or sugar alcohols representing the hydrophilic part of the molecule and the hydrophobic long-chain fatty acids are linked via ester bonding. Whereas the former group is found in bacteria, yeasts, and fungi, ester linkages between sugars and fatty acids are predominantly found only in the acylated sugars of strains of the actinomycetes: *Mycobacteriuim, Corynebacterium, Brevibacterium, Norcadia, Rhodococcus,* and so on [for reviews see Refs. 36 and 48]. The appearance of such extracellular lipids has been studied in connection with growth on water-insoluble carbon sources [219–223]. The presence of acylpolyols in the cell wall of the mycobacteria and related bacteria is well established [35,43,223,224]. Examples of the mycolic acid (see Fig. 5) esters of trehalose are given in Fig. 19. A survey of the different trehalose lipids found in actinomycetes is given by Ioneda [225]. In principle, all hydroxy groups of trehalose may be linked with either short-chain fatty acids or corynomycolic or mycolic acids (see Sec. II.D). Recently, the occurrence of trehalose 2,3,6′-trimycolate, which contained only C_{62-74} polyunsaturated acids, has been reported [226]. Further unusual trehalose esters, such as α,α-trehalose-2,3,4,2′-tetraester (Fig. 19b) with two decanoyl, one octanoyl, and one succinoyl residues, was reported by Ristau and Wagner in *Rhodococcus erythropolis* grown under nitrogen-limited conditions with *n*-alkanes [227]. The same residues, however, occurred at the 2, 3, 6, and 2′ positions of the glycolipid in *Mycobacterium paraffinicum* grown on *n*-alkanes [228]. A succinoyl trehalose lipid produced by *R. erythropolis* has also been found [229].

The synthesis of the sugar residue, trehalose-6-phosphate, is catalyzed by a trehalose-6-phosphate synthetase (22), which links two α-D-glycopyanosyl units

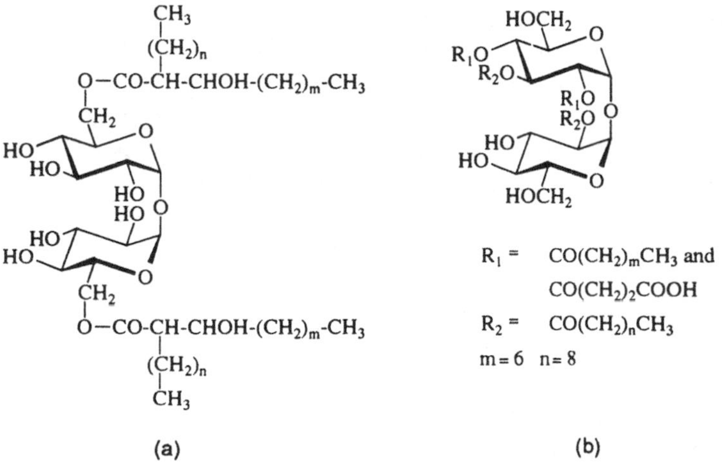
CH3
(CH2)n
O–CO-CH-CHOH-(CH2)m-CH3
CH2
O
HO
HO
HO
HO O
HO
HO
O
CH2
O–CO-CH-CHOH-(CH2)m-CH3
(CH2)n
CH3
HOCH2
O
R1O
R2O
R1O
R2O O
HO
HO
O
HOCH2
R1 = CO(CH2)mCH3 and
CO(CH2)2COOH
R2 = CO(CH2)nCH3
m=6 n=8
(a)
(b)

FIG. 19 Trehalose esters of *Rhodococcus erythropolis*: (a) Trehalose 6,6′-dicorynomycolate of growing cells ($m+n$ = 27-31) [58] and (b) trehalose tetraester of resting cells [227].

at C1 and C1′ [230]. UPD-glucose and glucose 6-phosphate probably act as the immediate precursors [231,232]:

$$\text{UDP-glucose + glucose 6-phosphate} \rightarrow \text{trehalose 6-phosphate + UDP} \quad (21)$$

The reaction product is subsequently hydrolyzed by trehalose-6-phosphate phosphatase at the same rate as it is produced [230].

Synthesis of trehalose mycolates proceeds from the trehalose 6-phosphate where various esters have been found in *Corynebacterium diphtheriae* and other bacteria after pulse labeling with [^{14}C]palmitate [233–235] or [^{14}C]acetate [236]. Recently, Shimakata et al. [237] demonstrated that trehalose plays an important role in mycolate ester synthesis in *Bacteroinema matruchotii*, as newly synthesized mycolic acid was mainly in form of trehalose monomycolate rather than of free mycolate. Synthesis of trehalose dimycolate was slower than that of trehalose monomycolate. The appearance of glycosylated intermediates in other bacteria, such as glucose monomycolates [238,239] or mycolyl acetyl trehalose [236,240], suppose a widespread trehalose-linked, trehalose mycolate synthesis.

Kretschmer and Wagner [231] elucidated the *n*-alkane-induced biosynthetic pathway of the trehalose mycolates in *R. erythropolis*. This pathway (Fig. 20) differs from that discussed above by appearance of free corynomycolic acid intermediates: the corynomycolic acid moiety and the trehalose moiety therefore are synthesized independently and are subsequently esterified. Trehalose 6-phosphate, which accumulates in these cells, and not trehalose, is condensed to the appropriate sugar corynomycolate. By comparison of the individual *in vitro* enzyme activities, the rate of this reaction probably limits the overall biosynthetic flux. Thereafter, the addition of precursors, such as glucose or free corynomycolates, improves the biosynthetic productivity [241].

Beside the appearance of trehalose monomycolates and trehalose dimycolates, arabinose residues esterified to mycolic acids have been reported as cell wall components in different *Nocardia* and related microorganisms [242]. Additionally, *Nocardia rhodochrous*, grown in a glucose-supplemented medium, produced a high yield of a 6-(C_{40}-C_{46})nocardomycoloylglucose cell wall constituent [243].

In other corynebacteria and mycobacteria, acylglucoses have been detected as major components of the soluble lipid [239]. Suzuki et al. [244] and Itoh and Suzuki [245] reported the synthesis of sucrose and fructose mono- and dimycolates by *Arthrobacter paraffineus* KY 4303 and several related bacteria after growth on the respective sugar. When grown on *n*-alkanes, this strain produced trehalose mono- and dimycolates [246]. *Arthrobacter* sp. DSM 2567 synthesized a series of mycolic acid esters when incubated with various mono-, di-, or tri-saccharides [247]. These transesterification reactions are

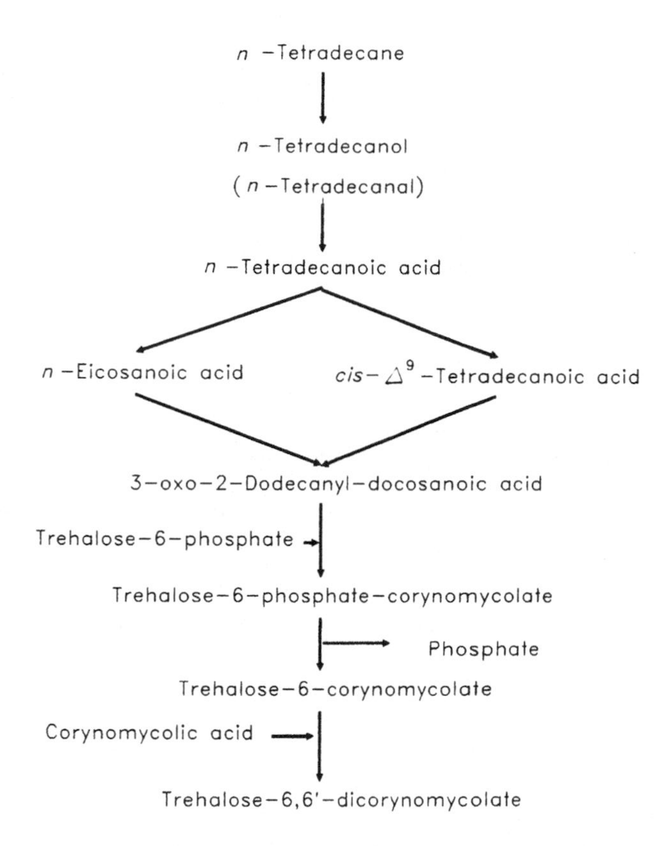

FIG. 20 Metabolic path of synthesis of mono- and dicorynomycolates by *Rhodococcus erythropolis* growing on *n*-tetradecane. (From Ref. 231.)

therefore nonspecific with regard to the carbohydrate but are regioselective with regard to which position of the sugars are acylated [248]. Sugars that have a hydroxyl group or other substituent on the 1-β- or 4-β- position prevent esterification.

Different genera of the actinomycetes may use different sequences of reactions leading to different sugar mycolates. These may differ with respect to sugar components that may be used as well as to the sequence of esterification reactions.

Detailed information is not always available about the biosynthetic pathways of other acylpolyols. For the diacyl mannosylerythritol, found in both *Shizonella melanogamma* [249] and in different *Candida* spp. [250–252] (Fig. 21), 4-*O*-β-D-mannosyl-D-erythritol was suggested as the direct precursor [253]. This becomes acylated when the organism grows on alkanes or triacylglycerols.

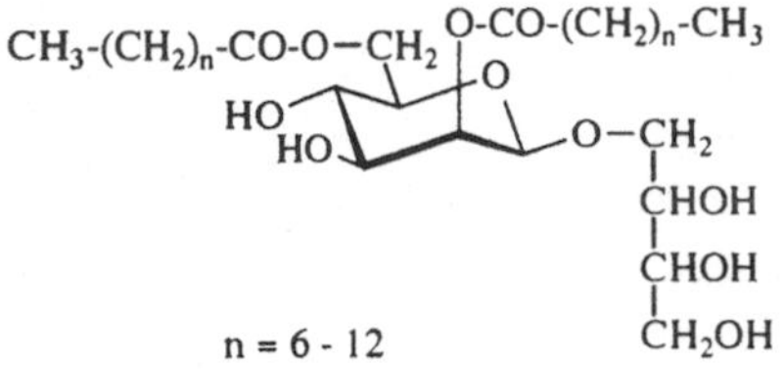

CH_3-$(CH_2)_n$-CO-O-CH_2 O-CO-$(CH_2)_n$-CH_3
O
HO
HO O-CH_2
CHOH
CHOH
n = 6 - 12 CH_2OH

FIG. 21 Diacyl mannosylerythritol lipid from *Candida* sp. (From Ref. 250.)

E. Acylpeptides

Growth of *Bacillus subtilis* on water-soluble carbon sources is not affected by addition of hydrocarbon, but the production of surfactin is suppressed [254]. Surfactin is one of several proteolipid biosurfactants produced by bacteria. Its structure (Fig. 22) was elucidated by Arima et al. [255]. The lipid moiety is a β-hydroxy-*iso*-fatty acid (C_{13-15}). Other examples of proteolipids, which belong to a family of cyclic peptides usually with 7- or 8-membered rings and containing one single ester bond formed by lactonization of β-hydroxy fatty acids or β-amino fatty acid [256], are iturin A [257], which is coproduced with surfactin [258], bacillomycin L [259,260], and mycosubtilisin [257]. In these, the hydroxy acid of surfactin is substituted by β-amino fatty acids ranging from C_{14} to C_{17}. Usually the lipopeptides appear as mixtures of closely related compounds with a slight variation in the lipid moiety. For example, the lipid moiety of bacillomycin is composed of the following β-amino fatty acids such as *n*-C_{14} (39%), *iso*-C_{15} (25%), *anteiso*-C_{15} (15%), *iso*-C_{16} (10%), and *n*-C_{16} (6%) [261]. Recently the formation of viscosin by *Pseudomonas fluorescens* [262] and of serratamolide by *Serratia marcescens* have been reported [263] but information concerning their biosynthesis is scant. Only the biosynthesis of surfactin has been studied in any detail. Kluge et al. [264], for example, demonstrated the incorporation of ^{14}C-acetate into surfactin being partially converted into the leucine residues. The same authors reported the incorporation of various ^{14}C-amino acids directly into the product. The *de novo* synthesis of the peptide moiety will be discussed further in Sec. V.A. The ester linkage of the lactone is assumed to be formed by a specific enzymatic

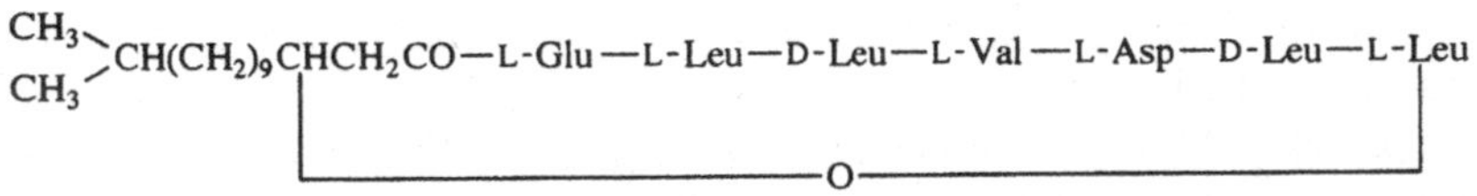

$(CH_3)_2$>$CH(CH_2)_9CHCH_2CO$—L-Glu—L-Leu—D-Leu—L-Val—L-Asp—D-Leu—L-Leu
O

FIG. 22 Surfactin from *Bacillus subtilis*. (From Ref. 255.)

reaction, which should be similar to peptide bond formation as reported for enniatin biosynthesis [265].

V. BIOSYNTHESIS OF NONLIPID COMPONENTS OF SURFACTANTS

The synthesis of nonlipid moieties of biosurfactants is normally connected with common metabolic pathways and can thus be found in most textbooks of biochemistry. Sometimes special reaction sequences are involved in the synthesis of the carbohydrate or the peptide residues.

A. Peptides

The synthesis of the peptide moiety of lipopeptides has not been extensively investigated. By comparison with other bioactive (antibiotic) peptides of related structures [256,266,267] peptide synthesis probably occurs without the involvement of messenger and ribosomal RNA [268].

Two principal enzyme systems catalyzing the nucleic acid-independent synthesis of peptides have so far been elucidated. Gramicidin S and tyrocidines produced by *Bacillus brevis* are characteristic examples of the peptide synthesis occurring completely on multifunctional polypeptide chains that are stable complexes of several bifunctional polypeptides, some of which may be bifunctional [269,270]. Activation of the individual amino acid substrates proceeds by a two-step reaction involving aminoacyl adenylates and thioesters. A 4′-phosphopantotheine arm acts as an internal vector transport system within the enzyme structure. The covalently bound intermediates cannot be exchanged with the free amino acids. A series of transpeptidation and transthiolation reactions then form the basis of chain elongation to yield the final peptide. The sequence of the peptide is determined by the structural organization of activation domains on the multifunctional polypeptide chain [268].

The synthesis of glutathione or mycobacillin [271,272] represents the second type of nonribosomal peptide synthesis. Here, carboxyl transfer proceeds from aminoacyl phosphate intermediates by a one-step mechanism. Intermediate peptides may be released into the reaction medium. Glutathione, which is a tripeptide, is formed by the cooperation of two freely interacting or weakly associated enzymes.

Recently, Kluge et al. [264] reported continuous surfactin biosynthesis in *B. subtilis* ATCC 21332 when protein biosynthesis was completely inhibited by chloramphenicol. They identified and partially purified an enzyme catalyzing ATP-P_i-exchange reactions that were mediated specifically by the amino acid constituents of surfactin; that is, the substrate amino acids were simultaneously activated as aminoacyl phosphates. This pattern was consistent with a peptide-synthesizing system, which activates its substrate simultaneously as aminoacyl phosphates.

For further and fuller details on the proposed mechanisms for assembly of amino acids into short-chain, straight as well as cyclic peptides, readers should consult the review by Kleinkauf and von Döhren [266].

B. Carbohydrates and Gluconeogenesis

1. Carbohydrates

Glucose is the most ubiquitous hexose in nature, forming the extensive biopolymers of cellulose and starch and also of glycogen (often referred to as animal starch) and various glucans, which occur as wall structures in yeasts. Photosynthesis in plants and algae results in the formation of both glucose and the related keto sugar, fructose, an end-product carbohydrate. These two sugars may then be linked together to form sucrose, which is then frequently found as the storage sugar in many plant systems. Thus adequate mechanisms exist in nature for the formation of carbohydrates, principally glucose and fructose. Microorganisms, perhaps not surprisingly, are usually well adapted to use these two sugars and other hexoses such as mannose and galactose. Readers wishing to know more about pathways of glucose breakdown in microorganisms should consult any standard textbook on biochemistry, which will also delineate the biochemical connections between the principal hexoses. A brief outline is given later in Fig. 25.

Glucose may therefore be regarded as the starting point for most microbial fermentations, not only because of it being the most universally used carbon source but also because its natural abundance means that it is often the cheapest source of carbon. Formation of the disaccharides found in many surfactants (trehalose, sophorose and cellobiose, which are all diglucoses) follows the pattern established [273] for sucrose biosynthesis:

$$\text{Glucose} + \text{UTP} \rightarrow \text{UDP-glucose} + P_i \tag{22}$$

$$\text{UDP-glucose} + \text{fructose 6-phosphate} \rightarrow \text{sucrose 6-phosphate} + \text{UDP} \tag{23}$$

$$\text{Sucrose 6-phosphate} + H_2O \rightarrow \text{sucrose} + P_i \tag{24}$$

Thus Liu et al. [274] established that trehalose is synthesized in *Mycobacterium smegmatis* and *Nocardia* sp. by the sequence given above.

In *Streptomyces* spp., however, GDP-glucose is used as the starting substrate:

$$\text{GDP-glucose} + \text{glucose 6-phosphate} \rightarrow \text{trehalose 6-phosphate} + \text{GDP} \tag{25}$$

In the synthesis of the disaccharides, a specific synthetase (reactions 23 and 25) acts to ensure that the correct regiospecific condensation occurs. The phosphorylated disaccharide will then be used as the activated sugar for the formation of the glycolipid surfactant (see also Sec. IV.D). Small, or sometimes large, amounts of the free disaccharide will be produced by the action of a phosphatase (reaction 24).

Rhamnose is an unusual sugar but occurs in the rhamnolipid surfactants of the pseudomonads (see Fig. 17). It is 6-deoxymannose and is synthesized in several bacteria [275] by an NADPH-linked dTDP-D-glucose oxidoreductase (Fig. 23). This is a multienzyme reaction and involves not only the reduction at C6 but also inversion of the configuration at C3, C4 and C5 of glucose. A related CDP-glucose oxidoreductase occurs in *Salmonella* sp. but produces the unusual sugars 3,6-dideoxy-D-arabinohexose (tyvelose) and 3,6-dideoxy-D-xylohexose (abequose) [275]. Of these, only rhamnose is involved in glycolipid formation.

The dTDP-rhamnose (Fig. 23) is used directly as the substrate for condensation with the lipid moiety (see Fig. 18). The second rhamnose is added to the rhamnolipid again presumably being activated as the dTDP derivative.

Oligosaccharide lipids are apparently rare in nature [276] but presumably could be produced by successive condensation of a UDP-sugar on to the first formed disaccharide. The pentasaccharide lipid formed by *Nocardia corynebacterioides* [276], however, consists of two trehalose units attached to positions 1 and 6 of a central glucose and suggests that trehalose, or the trehalose lipid, is used as the biosynthetic building unit for the oligosaccharide lipid. This would seem to be a more simplistic view than what was stated above; activation of one of the carbohydrates must occur to achieve condensation. Simple linkage of a carbohydrate to a lipid would not provide sufficient activation energy to allow such a condensation to occur.

2. Gluconeogenesis

When noncarbohydrate materials, such as alkanes, fatty acids, or even acetate or ethanol, are used as substrates for microbial growth, sugars must be synthesized

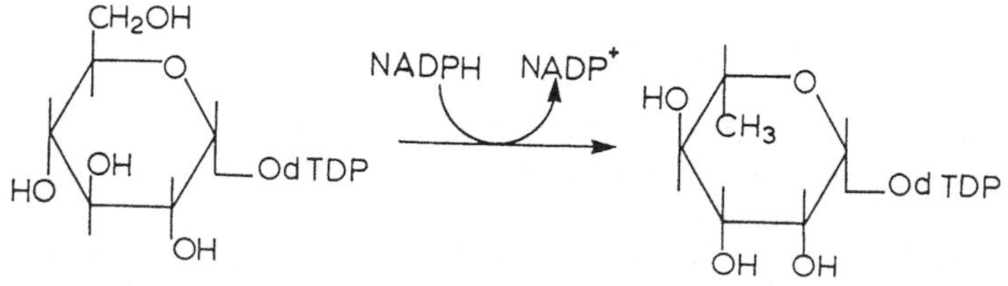

FIG. 23 Formation of rhamnose (dTDP = deoxythymidine phosphate). (From Ref. 275.)

from the pool of intermediates in the central pathways of metabolism. Such sugars (hexoses, pentoses, tetroses, and trioses) are all needed either for synthesis of structural entities of the cell or for the biosynthesis of amino acids, proteins, purines, pyrimidines, and nucleic acids. Such pathways are given in detail in all biochemistry textbooks and in synoptic form in a recent review [277].

The synthesis of sugars from acetyl-CoA is essentially the reverse of the glycolytic sequence of sugar breakdown (see Fig. 25) and is termed gluconeogenesis. Two new enzymes, however, are required to achieve this as the reactions catalyzed by the glycolytic enzymes, pyruvate kinase and phosphofructokinase, are not reversible. The new enzymes are phosphoenolypyruvate carboxykinase and fructose biphosphatase:

$$\text{Oxaloacetate} + \text{ATP} \rightarrow \text{phosphoenolypyruvate} \tag{26}$$

$$\text{Fructose 1,6-biphosphate} + H_2O \rightarrow \text{fructose 6-phosphate} + P_i \tag{27}$$

These reactions are shown in the context of the flux of carbon from alkanes to cell intermediates in Fig. 24.

The key enzymes for cells to be able to handle acetate, whether used per se as a carbon source, derived from ethanol (by direct oxidation), or derived from degradation of long-chain fatty acids, are isocitrate lyase and malate synthase:

$$\text{Isocitrate} \rightarrow \text{glyoxylate} + \text{succinate} \tag{28}$$

$$\text{Glyoxylate} + \text{acetyl-CoA} \rightarrow \text{malate} + \text{CoA} \tag{29}$$

These two enzymes serve to generate a C4 unit (malate) from the C2 acetate and are induced only when the cell is growing on acetate or alkanes. The key functional role of these two enzymes in the overall context of the flux of carbon from alkanes to the various metabolites is again shown in Fig. 24. Further details concerning the compartmentalization that must occur to separate the activities of the mitochondrion (principally those of the citric acid cycle and oxidative phosphorylation for ATP production) and of the peroxisome where fatty acid degradation occurs have been presented by Boulton and Ratledge [54,95].

VI. REGULATORY MECHANISMS RELATED TO SURFACTANT PRODUCTION

The process of microbial lipogenesis has been described in some detail elsewhere [95] and a brief account of how this process is now thought to take place was given in Sec. II. An overall view of the flow of carbon from glucose and other sugar

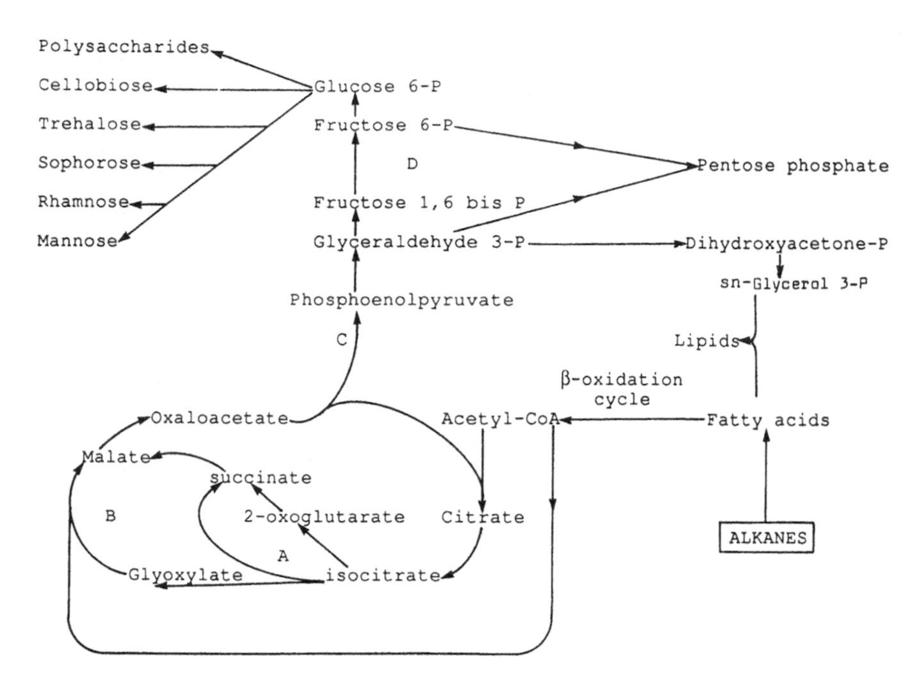

FIG. 24 Outline of intermediary metabolism relating to biosurfactant precursor synthesis from hydrocarbon substrates. Key enzymes are A, isocitrate lyase; B, malate synthase; C, phosphoenolpyruvate carboxykinase; D, fructose-1,6-bisphosphatase.

sources into lipids is given in Fig. 25. [For details concerning the flow of carbon into all the intermediates of the cell, the reader should consult Ref. 277.]

The process of lipogenesis, that is of lipid formation, can be readily stimulated in a number of microorganisms by depriving them of an essential nutrient other than carbon (usually this is nitrogen) and ensuring a continuous supply of carbon. Under such conditions and with an oleaginous microorganism, extensive lipid formation may occur. Increased production of extracellular lipids and of biosurfactants can occur under such conditions [278,279]. However, it is clear that the most commonly used procedure for ensuring high productivity of surfactants is to grow the appropriate microorganisms on a water-insoluble substrate, such as an alkane or fatty acid. Not only are these molecules then the direct precursors of the surfactants, but the surfactants themselves then apparently serve to increase the availability of the insoluble substrates to the cell (see Sec. III.A). Thus, the process accelerates to the point where some internal enzyme activity becomes the overall rate-limiting step. In the absence of the alkane, some surfactant-producing

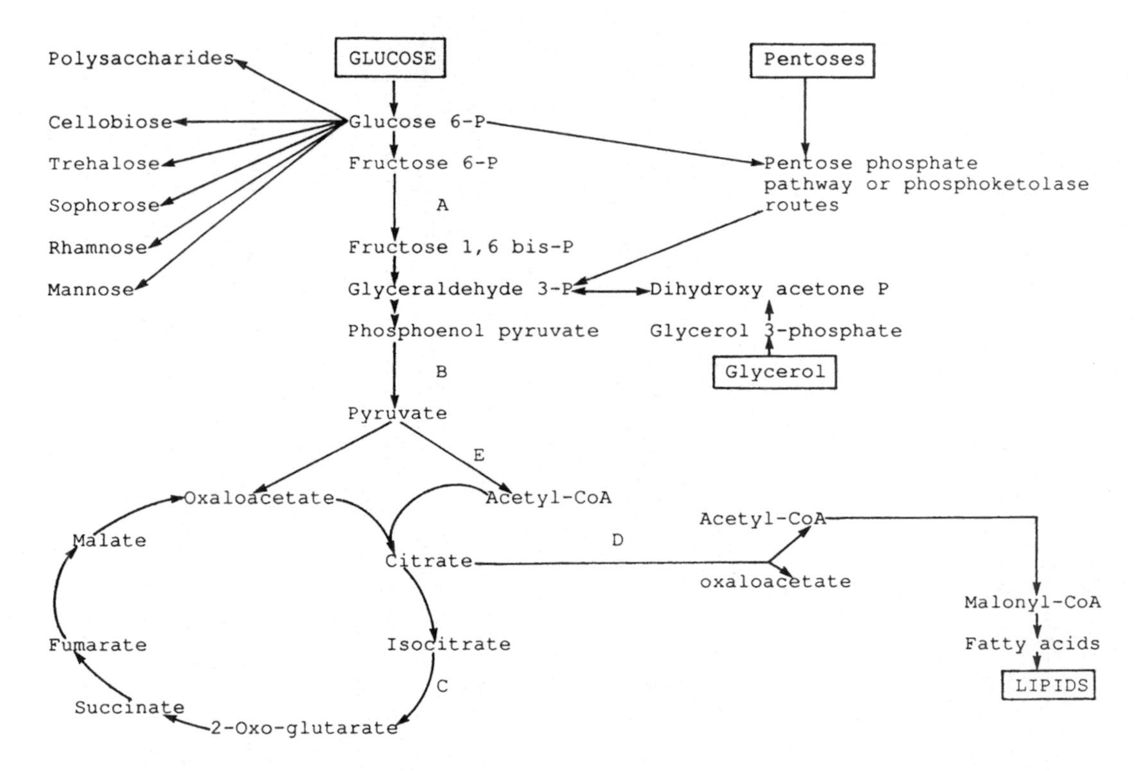

FIG. 25 Intermediary metabolism relating to biosurfactant precursor synthesis from carbohydrate substrates. Key enzymes for controlling flux of carbon: A, phosphofructokinase; B, pyruvate kinase; C, isocitrate dehydrogenase; D, ATP:citrate lyase (oleaginous yeasts and molds only). Reaction E, pyruvate dehydrogenase occurs intramitochondrially in eukaryotes thus requiring an exit mechanism for acetyl units back into the cytoplasm (see text); this is not necessary in bacteria.

organisms when grown just on glucose now produce only the carbohydrate moiety of the surfactant: in this way a strain of *Candida* forming mannosylerythirol lipid (Fig. 21) when grown on glucose in place of alkanes or plant oils, now produced just mannosylerythritol [280]. Other strains of both yeast and bacteria may however synthesize surfactant molecules regardless of what substrate—water-soluble or water-insoluble—they are grown on.

It can be readily appreciated from a comparison of Figs. 24 and 25 that the metabolism of alkanes and fatty acids is completely different from the metabolic sequence that occur during cell growth on glucose or other sugars. The two pathways, in some respects, are mutually opposed. In glucose-grown cells, glucose is degraded and fatty acids are synthesized, whereas in alkane-grown cells, fatty acids are degraded and glucose must be synthesized. This, however, is a simplistic view as all the other reactions of the cell for synthesis of amino acids, proteins, nucleic acids, and so on remain more or less unchanged. Nevertheless, as with all cell activities, there must be close coordination of catabolic (breaking down) reactions and anabolic (building up) reactions. Such coordination is achieved by an extensive system of metabolically interlocking regulatory events. Activities of particular enzymes may be inhibited or activated by metabolic intermediates that may begin to accumulate or whose concentration falls below a critical point. Such regulatory molecules not only include the metabolites of glucose or alkanes but also compounds such as ATP, ADP, $NAD(P)^+$, and NAD(P)H.

The synthesis of enzymes themselves may, and is, a closely controlled event: enzymes may be constitutively expressed; that is, they are always present in the cell no matter what growth substrate or conditions are being used. Alternatively they may be only formed when a particular growth substrate is used: for example, isocitrate lyase (see Fig. 24) is induced only when cells are grown on alkanes, fatty acids, or acetate. This enzyme is therefore essential for the cell to be able to deal with acetyl-CoA and convert it into a C_4 unit. When cells are grown on glucose the enzyme is superfluous to requirements and so its synthesis is now repressed. [Further details of the various means that exist for controlling cellular activity, in a general manner, may be gained from Ref. 277.]

When cells are grown on a mixture of alkanes and glucose, a complex array of metabolic control mechanisms must occur. These have been very poorly investigated at the level of the control pathways of metabolism. In general, however, the presence of a fatty acid, if not alkanes, in the culture medium effectively stops further *de novo* synthesis of fatty acids. This is achieved by the incoming fatty acids creating a pool of fatty acyl-CoA esters, which then first inhibit the key enzymes of fatty acid biosynthesis [6,8] and then, at a slower rate, cause the repression of these enzymes. Inhibition is therefore a rapid control mechanism; repression is a slower process. Both are essential for the overall economy of the cell. The cell when faced with dual substrates attempts to take the maximum advantage of what it is offered and uses any gratuitously supplied, preformed carbon skeleton for immediate biosynthetic reactions. The presence of exogenous fatty acids to a glucose-growing culture will not induce the isocitrate lyase enzyme of the glyoxylate cycle (Fig. 24) or the key enzymes of glyconeogenesis (see reactions 26 and 27) as long as glucose is above a critical concentration. Although there is a considerable amount of information about certain aspects of regulatory control mechanisms, our status of knowledge of the regulatory mechanisms involved in the biosynthesis of biosurfactants and their secretion is rather limited. The interplay between two metabolic routes (Figs. 24 and 25), the supply of polar moieties of the molecule, like carbohydrates or amino acids, and the acyl moieties raise the question of general regulatory metabolic principles that still cannot be adequately answered. Enzymes participating in the immediate biosynthesis of the surfactants are not well-characterized in their kinetic properties and in how they may be induced or repressed. Investigations at the genetic level of biosurfactant synthesis have not been published, although this has been a very powerful method for unravelling certain aspects of metabolism. Strains mutated in the respective biosynthetic pathways seem difficult to obtain [281,282] making it far from easy to unravel the necessary regulatory controls.

In general, growth and product formation proceed as separate events: in the exponential phase of growth of cells, there is often only a very low rate of surfactant production [66]; overproduction of the surfactant then occurs as the cells cease to grow. Although this might suggest that the surfactant is not involved in alkane assimilation, it could be argued that the amount of surfactant needed to stimulate alkane solubilization and uptake (as in Fig. 6) will be very small as the surfactant is not, as far as we know, consumed by the cells in the uptake process. Like the production of other metabolites, once the signal to begin surfactant production has been given by the presence of the alkane, then surfactant production will continue in an unregulated manner until the signal to stop is received. As that signal—the disappearance of the last alkane droplet—never arrives, the cells continue surfactant production even after they have ceased to grow. The emulsification of the residual alkanes by the surfactant must continue beyond what the cell requires but the cell has no means of halting this needless production of the surfactant. Regulation must perforce be occurring to control bulk uptake of the alkane into the cell, perhaps by feedback inhibition by the intracellular pool of fatty acyl-CoA; but nevertheless, at the end of growth, cells are found with excess (i.e., unused) alkanes now contained within them [see Ref. 54 for typical electron micrographs]. However this is not the only interpretation that can be made on the basis of the limited data that is available, and, indeed, in the final paragraphs of this section, we argue that surfactant production may be nothing more than a method of removing unwanted and surplus fatty acids.

Biosurfactant yields may be increased and the composition of the product mixture may be altered by addition of carboxylic acids, such as by citrate, which do not serve as carbon sources [185]. Similar observations were reported for other yeast strains [283] and *B. subtilis* [284]. Additionally, the concentration of several trace elements can be important for optimum surfactant production. The different types of limitation and cultivation conditions that can affect the biosynthesis of surfactants have been reviewed by Syldatk and Wagner [173]. Although there are exceptions as pointed out above, enzymes involved in these metabolic pathways are normally repressed in cells actively growing on glucose. Results obtained for different enzymes assumed to be involved in sophoroplipid biosynthesis in *C. apicola* and *C. bogoriensis* were discussed in Sec. IV.A. Limitation of nitrogen and phosphate may lead to the derepression (i.e., induction) of the enzymes involved in secondary metabolite formation [285], of which surfactants may be considered to be an example and for which as an example the enzymes for sophorolipid formation are not fully active until growth has ceased [286,287].

Principles of regulation of the individual different pathways of secondary metabolism have been reviewed by Malik [287]. The appropriate microorganisms possess, in addition to the well-regulated mechanisms of primary metabolism, additional pathways to channel the flux of accumulated intermediates and primary metabolites into secondary metabolite formation. (Most work in this area has of course mainly considered aspects of antibiotic synthesis.) The directed biosynthesis of appropriate secondary metabolites uses the capability of the microorganism to include precursors of the product and also structurally homologous substrates into the metabolic flux resulting in enhanced product yields and new types of secondary metabolites [285]. The effect of precursor addition to biosurfactants synthesis is well-documented (see Sec. IV and Ref. 173). Additionally, high concentrations of carbon source can also increase the yield of product, but this presumably may be attributed to a mass action effect and to the signal for the cells to cease surfactant production never being received (see above).

It may therefore be assumed that high yields of biosurfactants (at least for most surfactants containing sugar components) can be attributed to biosynthesis being directed or controlled by the concentration of one or more of the precursors. Following from this assumption, the overproduction of surface-active compounds could be argued to be not as a prerequisite for growth on alkanes but as the result of alkane assimilation and may be viewed as one way of neutralizing the fatty acids derived from the alkane which the cell cannot deal with in any other way. We would stress, however, that this hypothesis still requires substantiation and it is by no means certain whether this view will be found to be correct or whether the more orthodox view that surfactant molecules are an intrinsic part of alkane uptake (Fig. 6) will be upheld.

REFERENCES

1. P. E. Milson and J. L. Meers, in *Comprehensive Biotechnology*, (C. L. Cooney and A. E. Humphrey, eds.), Pergamon Press, New York, 1984, pp. 655–680.
2. C. Ratledge and S. G. Wilkinson (eds.), *Microbial Lipids*, Vols. 1 and 2, Academic Press, London, 1988, 1989.
3. J. D. Weete, *Lipid Biochemistry of Fungi and Other Organisms*, Plenum Press, New York, 1980.
4. J. L. Harwood and N. L. Russell, *Lipids in Plants and Microbes*, George Allen and Unwin, London, 1984.
5. M. I. Gurr and J. L. Harwood, *Lipid Biochemistry*, Chapman and Hall, London, 1991.
6. C. Ratledge and C. T. Evans, in *The Yeasts* (D. H. Rose and J. Harrison, eds.), Vol. 3, Academic Press, London, 1989, pp. 367–455.
7. C. Ratledge, in *Microbial Lipids* (C. Ratledge and S. G. Wilkinson, eds.), Vol. 2, Academic Press, London, 1989, pp. 567–668.
8. C. Ratledge, in *World Conference on Biotechnology for the Fats and Oils Industry* (T. H. Applewhite, ed.), American Oil Chemists' Society, Champaign, IL, 1988, pp. 7–16.
9. R. Sheridan, C. Ratledge, and P. A. Chalk, *FEMS Microbiol. Lett., 69*: 165–170 (1990).
10. E. Schweizer, in *Microbial Lipids* (C. Ratledge and S. G. Wilkinson, eds.), Vol. 2, Academic Press, London, 1989, pp. 3–50.
11. M. Sumper and C. Riepertinger, *Eur. J. Biochem. 29*: 237–248 (1972).
12. M. Mishina, T. Kamiryo, A. Tanaka, S. Fukui, and S. Numa, *Eur. J. Biochem. 71*: 295–300 (1976).
13. M. L. Ernst-Fonberg and J. S. Wolpert, *Meth. Enzymol. 71*: 60–73 (1981).
14. T. V. Boom and J. E. Cronan, *Annu. Rev. Microbiol. 43*:317–343 (1989).
15. S. Jackowski, C. M. Murphy, J. E. Cronan, and C. O. Rock, *J. Biol. Chem. 264*: 7624–7629 (1989).
16. E. Schweizer, E. Muller, L. M. Roberts, M. Schweizer, J. Rosch, P. Weisner, J. Beck, D. Statmann, and I. Zaumer, *Fat Sci. Technol. 89*: 570–577 (1987).
17. E. Schweizer, H. Kottig, R. Regler, and G. Rottner, *J. Basic Microbiol. 28*: 283–292 (1988).
18. J. Elovson, *J. Bacteriol. 124*: 524–533 (1975).
19. P. Weisner, J. Beck, K. F. Beck, S. Ripka, G. Muller, S. Lucke, and E. Schweizer, *Eur. J. Biochem. 177*: 69–79 (1988).
20. H. Okuyama, M. Saito, V. C. Joshi, S. Gunsberg, and S. J. Wakil, *J. Biol. Chem. 254*: 12281–12284 (1979).
21. D. Tyrrell and J. Weatherstone, *Can. J. Microbiol. 22*: 1058–1060 (1976).
22. A. Kendrick and C. Ratledge, *Lipids 27*: 15–20 (1992).
23. C. Ratledge, *The Biology of the Mycobacteria* (C. Ratledge and J. Stanford, eds.), Vol. 1, Academic Press, London, 1982, pp. 53–92.
24. P. K. Stumpf, *World Conference on Biotechnology for the Fats and Oils Industry* (T. H. Applewhite, ed.), American Oil Chemists' Society, Champaign, IL, 1988, pp. 1–6.
25. G. Ferrante and M. Kates, *Can. J. Biochem. Cell Biol. 61*: 1191–1196 (1983).
26. A. Kendrick and C. Ratledge, *Eur. J. Biochem.*, in press.
27. S. L. Neidleman, *Biotech. Gen. Eng. Rev. 5*: 245–268 (1987).
28. Y. Ito, Y. Oh-hashi, and H. Okuyama, *J. Biochem. 99*: 1713–1718 (1986).

29. A. H. Rose, in *Microbial Lipids* (C. Ratledge and S. G. Wilkinson, eds.), Vol. 2, Academic Press, London, 1989, pp. 255–278.
30. C. P. Kurtzmann, R. F. Vesonder, and M. J. Smiley, *Mycologia 66*: 586–587 (1974).
31. S. G. Wilkinson, *Microbial Lipids* (C. Ratledge and S. G. Wilkinson, eds.), Vol. 1, Academic Press, London, 1988, pp. 299–488.
32. T. Fuji and K. Tonomura, *Agric. Biol. Chem. 35*: 1188–1193 (1971).
33. K. Soda, *World Conference on Biotechnology for the Oils and Fats Industry* (T. H. Applewhite, ed.), American Oil Chemists' Society, Champaign, IL, 1988, pp. 178–179.
34. T. Kaneda, *J. Biol. Chem. 238*: 1229–1235 (1963).
35. Y. Akamatsu and J. H. Law, *J. Biol. Chem. 245*: 709–713 (1970).
36. P. J. Brennan, *Microbial Lipids* (C. Ratledge and S. G. Wilkinson, eds.), Vol. 1, Academic Press, London, 1988, pp. 203–298.
37. T. Shimakata, M. Iwaki, and T. Kusaka, *Arch. Biochem. Biophys. 229*: 329–339 (1984).
38. D. E. Minnikin, *The Biology of the Mycobacteria* (C. Ratledge and J. L. Stanford, eds.), Vol. 1, Academic Press, London, 1982, pp. 95–184.
39. G. Laneelle, *Acta Leprologica, 7*(suppl. 1): 65–73 (1989).
40. M. P. Lechevalier, *C.R.C. Crit. Rev. Microbiol. 5*: 109–210 (1977).
41. M. B. Goren and P. J. Brennan, *Tuberculosis* (G. P. Youmans, ed.), W. B. Saunders, Philadelphia, 1979, pp. 69–193.
42. K. Takayama and N. Qureshi, in *The Mycobacteria—A Sourcebook*, Part A (G. P. Kubica and L. G. Wayne, eds.), Marcel Dekker, New York, 1984, pp. 315–344.
43. D. E. Minnikin and A. G. O'Donnell, in *The Biology of the Actinomycetes* (M. Goodfellow, M. Mordarski, and S. T. Williamson, eds.), Academic Press, London, 1984, pp. 337–388.
44. C. Lacave, M. A. Laneelle, and G. Laneelle, *Biochim. Biophys. Acta 1042*: 315–323 (1990).
45. A. Einsele, in *Biotechnology* (H.-J. Rehm, and G. Reed, eds.), Vol. 3, Verlag Chemie, Weinheim, 1983, pp. 43–81.
46. J. Schindler, F. Meussdoerffer, and H. Giesel-Buhler, *Forum Mikrobiol. 5*: 274–281 (1990).
47. C. Ratledge, in *Biodeterioration* (D. R. Houghton, R. N. Smith, and H. O. W. Eggins, eds.), Vol. 7, Elsevier, London, 1988, pp. 219–236.
48. D. Haferburg, R. Hommel, R. Claus, and H.-P. Kleber, *Adv. Biochem. Eng./Biotechnol. 33*: 53–93 (1986).
49. J. E. Zajic and C. L. Panchal, *Crit. Rev. Microbiol. 5*: 39–66 (1976).
50. S. Lang and F. Wagner, in *Biosurfactants and Biotechnology* (N. Kosaric, W. L. Cairns, and N. C. C. Gray, eds.), Marcel Dekker, New York, 1987, pp. 21–45.
51. D. L. Gutnick and W. Minas, *Biochem. Soc. Trans. 15*: 22S–35S (1987).
52. D. L. Gutnick and Y. Shabtai, in *Biosurfactants and Biotechnology* (N. Kosaric, W. L. Cairns, and N. C. C. Gray, eds.), Marcel Dekker, New York, 1987, pp. 211–247.
53. D. G. Cooper and J. E. Zajic, *Adv. Appl. Microbiol. 26*: 229–253 (1980).
54. C. A. Boulton and C. Ratledge, in *Topics Enz. Ferment. Biotechnol.* (A. Wiseman, ed.), Vol. 9, John Wiley, New York, 1984, pp. 11–77.
55. C. Syldatk, U. Matulovic, and F. Wagner, *Biotech. Forum 1*: 58–66 (1984).
56. G. Goma, A. Pareilleux, and G. Durand, *J. Ferment. Technol. 51*: 616–621 (1976).

57. L. Eastcott, W. Y. Shiu, and D. Mackay, *Oil Chem. Pollut. 4*: 191–216 (1988).
58. P. Rapp, H. Bock, V. Wray, and F. Wagner, *J. Gen. Microbiol. 115*: 491–503 (1979).
59. B. Ramsay, J. McCarthy, L. Guerra-Santos, O. Käppeli, and A. Fiechter, *Can. J. Microbiol. 34*: 1209–1212 (1988).
60. M. Rosenberg, E. Rosenberg, and D. Gutnick, in *Microbial Adhesion to Surfaces* (R. C. W. Berkley, J. M. Lynch, J. Helling, P. R. Rutter, and E. Vincent, eds.), Society of Chemical Industry, London, 1980, pp. 541–542.
61. K. Hisatsuka, T. Nakahara, N. Sano, and K. Yamada, *Agric. Biol. Chem. 35*: 686–692 (1971).
62. S. Itoh, H. Honda, F. Tomita, and T. Suzuki, *J. Antibiot. 24*: 885–859 (1971).
63. S. Itoh and T. Suzuki, *Agric. Biol. Chem. 36*: 2233–2235 (1972).
64. S. Ito and S. Inoue, *Appl. Environ. Microbiol. 43*: 1278–1263 (1982).
65. S. Ito, M. Kinta, and S. Inoue, *Agric. Biol. Chem. 44*: 2221–2223 (1982).
66. R. Hommel, O. Stüwer, W. Stuber, D. Haferburg, and H.-P. Kleber, *Appl. Microbiol. Biotechnol. 26*: 199–205 (1987).
67. R. K. Hommel, *Biodegradation 1*: 107–119 (1990).
68. A. Oberbremer and R. Müller-Hurtig, *Appl. Microbiol. Biotechnol. 31*: 582–586 (1989).
69. A. Oberbremer, R. Müller-Hurtig, and F. Wagner, *Appl. Microbiol. Biotechnol. 32*: 485–469 (1990).
70. D. G. Cooper and B. G. Goldenberg, *Appl. Environ. Microbiol. 47*: 173–176 (1987).
71. Z. Duvnjak, D. G. Cooper, and N. Kosaric, *Biotechnol. Bioeng. 24*: 165–175 (1982).
72. Z. Duvnjak and N. Kosaric, *Biotechnol. Lett. 7*: 793–796 (1985).
73. W. R. Finnerty and M. E. Singer, in *Organisation of Procaryotic Cell Membranes* (B. K. Ghosh, ed.), CRC Press, Boca Raton (1985), pp. 1–44.
74. O. Käppeli and A. Fiechter, *J. Bacteriol. 131*: 917–921 (1977).
75. C. C. L. Scott and W. R. Finnerty, *J. Gen. Microbiol. 94*: 342–350 (1976).
76. M. N. Meisel, G. A. Medvedova, and T. M. Kozlova, *Mikrobiologiya 45*: 844–851 (1976).
77. M. Rosenberg, E. A. Bayer, J. Delarea, and E. Rosenberg, *Appl. Environ. Microbiol. 44*: 929–937 (1982).
78. O. Käppeli and W. R. Finnerty, *J. Bacteriol. 140*: 707–711 (1979).
79. P. Borneleit, T. Hermsdorf, R. Claus, P. Walther, and H.-P. Kleber, *J. Gen. Microbiol. 134*: 1983–1992 (1988).
80. E. Rosenberg, *CRC Crit. Rev. Biotechnol. 3*: 109–132 (1986).
81. N. D. Lindley and M. T. Heydeman, *J. Gen. Microbiol. 129*: 2301–2305 (1983).
82. N. D. Lindley and M. T. Heydeman, *J. Gen. Microbiol. 132*: 751–756 (1986).
83. J. B. Bassel and R. K. Mortimer, *Curr. Genetics 9*: 579–586 (1985).
84. P. W. Kirk and A. S. Gordon, *Mycologia 80*: 776–782 (1988).
85. W. Fischer, B. Büchner, and H. W. Meyer, *Z. Allg. Mikrobiol. 22*: 227-236 (1982).
86. C. C. L. Scott and W. R. Finnerty, *J. Bacteriol. 127*: 481–489 (1976).
87. M. E. Singer and W. R. Finnerty, in *Petroleum Microbiology* (R. M. Atlas, ed.), Macmillan, New York, 1984, pp. 1–59.
88. H. Müller, A. Naumann, R. Claus, and H.-P. Kleber, *Z. Allg. Mikrobiol. 23*: 645–651 (1983).
89. H. Sorger, H. Aurich, B. Fricke, and J. Vorisek, *J. Basic Microbiol. 26*: 541–546 (1986).

90. R. J. Watkinson, in *Hydrocarbons in Biotechnology* (D. E. F. Harrison, I. J. Higgins, and R. J. Watkinson, eds.), Heyden, London, 1980, pp. 11–24.
91. J. Casey, R. Dobb, and G. Mycock, *J. Gen. Microbiol. 136*: 1197–1202 (1990).
92. L. N. Britton, in *Microbial Degradation of Organic Compounds* (D. T. Gilson, ed.), Marcel Dekker, New York, 1984, pp. 89–129.
93. H. J. Rehm and I. Reiff, *Adv. Biochem. Engin. 19*: 175–215.
94. R. J. Watkinson and P. Morgan, *Biodegradation 1*: 79–92 (1990).
95. C. A. Boulton and C. Ratledge, in *Surfactant Science Series. Biosurfactants and Biotechnology* (N. Kosaric, W. L. Cairns, and W. L. Gray, eds.), Vol. 25, Marcel Dekker, New York, 1987, pp. 47–87.
96. J. A. Peterson, D. Basu, and M. J. Coon, *J. Biol. Chem. 241*: 5162–5164 (1966).
97. H. Aurich, D. Sorger, and O. Asperger, *Acta Biol. Med. Germ. 35*: 443–451 (1976).
98. R. Claus, O. Asperger, and H.-P. Kleber, *Z. Allg. Mikrobiol. 24*: 695–704 (1979).
99. M. Bühler and J. Schindler, in *Biotechnology* (H.-J. Rehm and G. Reed, eds.), Vol. 6a, Verlag Chemie, Weinheim, 1984, pp. 329–385.
100. H.-G. Müller, W.-H. Schunck, and E. Kärgel, in *Frontiers in Biotechnology* (K. Ruckpaul and H. Rein, eds.), Vol. 4, Akademie Verlag, Berlin, 1991, pp. 87–126.
101. O. Asperger, A. Naumann, and H.-P. Kleber, *FEMS Microbiol. Lett. 11*: 309–312 (1981).
102. R. Müller, O. Asperger, and H.-P. Kleber, *Biomed. Biochim. Acta 48*: 243–254 (1989).
103. O. Käppeli, *Microbiol. Rev. 50*: 244–258 (1986).
104. M. Gilewicz, M. Zacek, J.-C. Bertrant, and E. Azoulay, *Can. J. Microbiol. 25*: 201–206 (1979).
105. S. Mauersberger, W.-H. Schunck, and H.-G. Müller, *Z. Allg. Microbiol. 21*: 313–321 (1981).
106. M. Takagi, K. Moriya, and K. Yano, *Cell. Mol. Biol. 25*: 363–369, 371–375 (1980).
107. M. Wiedmann, B. Wiedmann, E. Kärgel, W.-H. Schunk, and H.-G. Müller, *Biochem. Biophys. Res. Commun. 136*: 1148–1154 (1986).
108. A. P. Ilchenko, S. Mauersberger, R. N. Matyashova, and A. B. Lozinov, *Mikrobilogiya 49*: 452–458 (1980).
109. S. Mauersberger and R. N. Matyashova, *Mikrobiologiya 49*: 571–577 (1980).
110. S. Mauersberger, W.-H. Schunck, and H.-G. Müller, *Appl. Microbiol. Biotechnol. 27*: 565–582 (1984).
111. W.-H. Schunck, S. Mauersberger, E. Kärgel, J. Huth, and H.-G. Müller, *Arch. Microbiol. 147*: 245–248 (1987).
112. F. K. Gmünder, O. Käppeli, and A. Fiechter, *Eur. J. Appl. Microbiol. Biotechnol. 12*: 135–142 (1981).
113. D. Sanglard, O. Käppeli, and A. Fiechter, *J. Bacteriol. 157*: 297–302 (1984).
114. W.-H. Schunck, E. Kärgel, B. Gross, B. Wiedmann, S. Mauersberger, K. Kopke, U. Kiessling, M. Strauss, M. Gasestel, and H.-G. Müller, *Biochem. Biophys. Res. Commun. 161*: 843–850 (1989).
115. W.-H. Schunk, F. Vogel, B. Gross, E. Kärgel, S. Mauersberger, K. Köpke, C. Gengnagel, and H.-G. Müller, *Eur. J. Cell Biol.* in press.
116. D. Sanglard and J. C. Loper, *Gene 76*: 121–136 (1989).

117. D. Sanglard and A. Fiechter, *FEBS Lett. 256*: 128–134 (1989).
118. T. Yamada, H. Nouva, S. Kawamoto, A. Tanaka, and S. Fukui, *Arch. Microbiol. 128*: 145–151 (1980).
119. M. Osumi, F. Fujuzumi, N. Yamada, T. Nagatami, T. Teranishi, A. Tanaka, and S. Fukui, *J. Ferment. Technol. 53*: 244–248 (1975).
120. J. M. Lebeault, B. Roche, Z. Duvnjak, and E. Azoulay, *Biochim. Biophys. Acta 220*: 386–395 (1970).
121. M. Gallo, B. Roche-Penverne, and E. Azoulay, *FEBS Lett. 46*: 78–82 (1974).
122. G. P. Saposhnikova and A. B. Lozinov, *Mikrobioloiya 47*: 682–688 (1978).
123. S. Mauersberger, E. Kärgel, R. N. Matyashova, and H.-G. Müller, *J. Basic Microbiol. 27*: 565–582 (1987).
124. R. Blasig, S. Mauersberger, P. Riege, W.-H. Schunck, W. Jockisch, P. Franke, and H.-G. Müller, *Appl. Microbiol. Biotechnol. 28*: 589–598 (1988).
125. A. P. Ilchenko, *Mikrobiologiya 53*: 903–907 (1984).
126. V. I. Krauzova, A. P. Ilchenko, A. A. Sharyshew, and A. B. Lozinov, *Biokjimiya 50*: 726–732 (1985).
127. V. I. Krauzova, T. N. Kuvichkina, A. A. Sharyshew, A. A. Romanova, and A. B. Lozinov, *Biokjimiya 51*: 23–27 (1986).
128. J. Savitha and C. Ratledge, *FEMS Microbiol. Lett. 80*: 221–224 (1991).
129. G. D. Kemp, Ph.D. Thesis, University of Hull, UK, 1990.
130. G. D. Kemp, F. M. Dickinson, and C. Ratledge, *Appl. Microbiol. Biotechnol. 29*: 370–374 (1988).
131. G. D. Kemp, F. M. Dickinson, and C. Ratledge, *Appl. Microbiol. Biotech. 32*: 461–464 (1990).
132. R. Hommel and C. Ratledge, *FEMS Microbiol. Lett. 70*: 183–186 (1990).
133. R. Hommel, D. Lassner, J. Weiss, and H.-P. Kleber, in press.
134. H. Tauchert, M. Roy, W. Schöpp, and H. Aurich, *Z. Allg. Mikrobiol. 15*: 675–680 (1975).
135. H. Tauchert, M. Grunow, and H. Aurich, *Z. Allg. Mikrobiol. 18*: 675–680 (1978).
136. R. Hommel and H.-P. Kleber, *J. Gen. Microbiol. 136*: 1705–1711 (1990).
137. S. Hartmans, J. A. M. de Bont, and W. Harder, *FEMS Microbiol. Rev. 63*: 235–264 (1989).
138. Z.-U. Yi and H.-J. Rehm, *Appl. Microbiol. Biotechnol. 14*: 254–258 (1982).
139. Z.-U. Yi and H.-J. Rehm, *Appl. Microbiol. Biotechnol. 15*: 144–146 (1982).
140. Z.-U. Yi and H.-J. Rehm, *Appl. Microbiol. Biotechnol. 15*: 175–179 (1982).
141. H.-J. Rehm, L. Hormann, and I. Reiff, *Acta Biotechnol. 3*: 279–288 (1983).
142. R. Blasig, J. Huth, P. Franke, P. Borneleit, W.-H. Schunck, and H.-G. Müller, *Appl. Microbiol. Biochem. 31*: 571–576 (1989).
143. N. R. Woods and J. C. Murrell, *J. Gen. Microbiol. 135*: 2335–2344 (1989).
144. Z.-U. Yi and H.-J. Rehm, *Appl. Microbiol. Biotechnol. 28*: 520–526 (1988).
145. A. P. Tulloch, in *Glycolipid Methodology* (L. A. Wittling, ed.), American Oil Chemists' Society, Champaign, IL, 1976, pp. 329–344.
146. F. F. Hill, I. Venn, and K. L. Lukas, *Appl. Microbiol. Biotechnol. 24*: 168–174 (1986).
147. N. Broadway, Ph.D. Thesis, University of Hull, UK, 1991.
148. R. Hommel, S. Kirste, L. Weber, and H.-P. Kleber. German Patent DD298273 (1992).

149. S. G. Slingar and R. I. Murray, in *Cytochrome P-450* (P. R. Ortiz de Montellano, ed.), Plenum Press, New York, 1986, pp. 429–503.
150. K. Hosaka, M. Mishina, T. Kamiryo, and S. Numa, *Meth. Enzymol. 72*: 325–333 (1981).
151. J. E. Holdsworth, M. Veenhuis, and C. Ratledge, *J. Gen. Microbiol. 134*: 2907–2915 (1988).
152. J. E. Holdsworth and C. Ratledge, *J. Gen. Microbiol. 134*: 339–346 (1988).
153. C. Ratledge, in *World Conference on Emerging Technologies in the Fats and Oils Industry* (A. R. Baldwin, ed.), American Oil Chemists' Society, Champaign, IL, 1986, pp. 318–330.
154. N. J. Russell, *J. Gen. Microbiol. 80*: 217–225 (1974).
155. L. M. Fixter, M. N. Nagi, J. G. McCormack, and C. A. Fewson, *J. Gen. Microbiol. 132*: 3147–3157 (1986).
156. R. A. Makula, P. A. Lockwood, and W. R. Finnerty, *J. Bacteriol. 121*: 250–258 (1975).
157. W. R. Finnerty, in *Biotechnology for the Oils and Fats Industry* (C. Ratledge, P. Dawson, J. Rattray, eds.), American Oil Chemists' Society, Monograph no. 11. Champaign, IL, 1984, pp. 199–215.
158. H. Miller and B. Voigt, *Acta Biotechnol. 2*: 155–160 (1982).
159. S. L. Neidleman and J. L. Erwin, US Patent 4567144, 1986.
160. J. L. Erwin, J. Geigert, S. L. Neidleman, and J. Wadsworth, in *Biotechnology for the Oils and Fats Industry* (C. Ratledge, P. Dawson, J. Rattray, eds.) American Oil Chemists' Society, Champaign, IL, 1984, pp. 217–222.
161. G. M. Lloyd and N. J. Russell, *J. Gen. Microbiol. 129*: 2641–2647 (1983).
162. L. Wall, A. Rodriguez, and E. Meighen, *J. Biol. Chem. 261*: 15981–15988.
163. P. Bacchin, A. Robertiello, and A. Viglia, *Appl. Microbiol. 28*: 737–741 (1974).
164. A. J. Anderson and E. A. Dawes, *Microbiol. Rev. 54*: 450–472 (1990).
165. C. Ratledge and S. G. Wilkinson, in *Microbial Lipids* (C. Ratledge and S. G. Wilkinson, eds.), Vol. 1, Academic Press, London, 1988, pp. 23–53.
166. C. Ratledge, in *Lecithin: Sources, Manufacture and Uses*, (B. F. Szuhaj, ed.), American Oil Chemists' Society, Champaign, IL, 1989, pp. 72–96.
167. U. Smith and D. W. Ribbons, *Arch. Mikrobiol. 74*: 116–122 (1970).
168. C. Siporin and J. J. Cooney, *Appl. Microbiol. 29*: 604–609 (1975).
169. W. R. Finnerty, *Trends Biochem. Sci. 2*: 73–76 (1977).
170. W. R. Finnerty, in *Microbial Lipids* (C. Ratledge and S. G. Wilkinson, eds.), Vol. 2, Academic Press, London, 1989, pp. 525–566.
171. R. A. Pieringer, in *Microbial Lipids* (C. Ratledge and S. G. Wilkinson, eds.), Vol. 2, Academic Press, London, 1984, pp. 51–114.
172. G. M. Carman and S. A. Henry, *Annu. Rev. Biochem. 58*: 635–639 (1989).
173. C. Syldatk and F. Wagner, in *Biosurfactants and Biotechnology* (N. Kosaric, W. L. Carins, and N. C. C. Gray, eds.), Marcel Dekker, New York, 1987, pp. 89–120.
174. P. A. J. Gorin, J. F. T. Spencer, and A. P. Tulloch, *Can. J. Chem. 39*: 846–855 (1961).
175. A. P. Tulloch and J. F. T. Spencer, *Can. J. Chem. 46*: 1523–1528 (1968).
176. D. J. Jones, *J. Chem. Soc. (C)* 479–484 (1967).
177. J. F. T. Spencer, P. A. J. Gorin, and A. P. Tulloch, *Antonie van Leeuwenhoek 36*: 129–133 (1970).

178. A. P. Tulloch, J. F. T. Spencer, and M. H. Deinema, M. H., *Can. J. Chem. 46*: 345–348 (1968).
179. T. W. Esders and R. J. Light, *J. Biol. Chem. 247*: 1375–1386 (1972).
180. J. F. T. Spencer, D. M. Spencer, and A. P. Tulloch, in *Secondary Products of Metabolism, Economic Microbiology* (A. H. Rose, ed.), Vol. 3, Academic Press, London, 1979, pp. 523–540.
181. H.-J. Asmer, S. Lang, F. Wagner, and V. Wray, *J. Am. Oil Chem. Soc. 65*: 1460–1466 (1988).
182. D. F. Jones and R. Howe, *J. Chem. Soc. (C)* 2816–2821 (1968).
183. A. P. Tulloch, A. Hill, and J. F. T. Spencer, *Can. J. Chem. 46*: 3337–3351 (1968).
184. L. Weber, J. Stach, G. Haufe, R. Hommel, and H.-P. Kleber, *Carbohydrate Res. 206*: 13–19 (1990).
185. O. Stüwer, R. Hommel, D. Haferburg, and H.-P. Kleber, *J. Biotechnol. 6*: 259–269 (1987).
186. E. Heinz, A. P. Tulloch, and J. F. T. Spencer, *J. Biol. Chem. 244*: 882–888 (1969).
187. E. Heinz, A. P. Tulloch, and J. F. T. Spencer, *Biochim. Biophys. Acta 202*: 49–55 (1970).
188. H.-P. Kleber, O. Asperger, O. Stüwer, B. Stüwer, and R. Hommel, in *Cytochrome P-450: Biochemistry and Biophysics* (I. Schuster, ed.), Taylor and Francis, London, 1989, pp. 169–172.
189. R. Hommel, S. Stegner, C. Ziebolz, L. Weber, and H.-P. Kleber, *Microbiol. Lett. 45*: 41–47 (1990).
190. L. Weber, C. Döge, G. Haufe, R. Hommel, and H.-P. Kleber, *Biocatalysis 5*: 267–272 (1992).
191. D. F. Jones, *J. Chem. Soc. (C)* 2809–2815 (1968).
192. D. G. Cooper and D. A. Paddock, *Appl. Environ. Microbiol. 47*: 173–176 (1984).
193. S. Ito and S. Inoue, *Appl. Environ. Microbiol. 43*: 1278–1283 (1982).
194. U. Göbbert, S. Lang, and F. Wagner, *Biotechnol. Lett. 6*: 225–230 (1984).
195. T. W. Esders and R. J. Light, *J. Biol. Chem. 247*: 7494–7497 (1972).
196. T. W. Esders and R. J. Light, *J. Lipid Res. 13*: 663–671 (1972).
197. A. J. Culter and R. J. Light, *J. Biol. Chem. 254*: 1944–1950 (1979).
198. T. B. Breithaupt and R. J. Light, *J. Biol. Chem. 257*: 9622–9628 (1982).
199. M. L. Bucholtz and R. J. Light, *J. Biol. Chem. 252*: 424–430 (1976).
200. M. L. Bucholtz and R. J. Light, *J. Biol. Chem. 252*: 431–437 (1976).
201. O. Rilke, A. Baum, R. Hommel, J. Weiss, and H.-P. Kleber, *World J. Microbiol. Biotechnol. 8*: 14–20 (1992).
202. T. Matsuyama, T. Murakami, M. Fujita, S. Fujita, and I. Yano, *J. Gen. Microbiol. 132*: 865–875 (1986).
203. R. U. Lemieux, *Can. J. Chem. 29*: 415–425 (1951).
204. R. U. Lemieux, *Can. J. Chem. 31*: 396–417 (1953).
205. R. U. Lemieux, J. A. Thorn, and H. F. Bauer, *Can. J. Chem. 31*: 1054–1059 (1953).
206. S. S. Bhattacharjee, R. H. Haskins, and P. A. J. Gorin, *Carbohydrate Res. 13*: 235–246 (1970).
207. B. Boothroyd, J. A. Thorn, and R. H. Haskins, *Can. J. Biochem. Physiol. 33*: 289–296 (1955).
208. B. Frautz, S. Lang, and F. Wagner, *Biotechnol. Lett. 11*: 757–762 (1986).

209. C. Syldatk, S. Lang, U. Matulovic, and F. Wagner, *Z. Naturforschung 40c*: 61–67 (1985).
210. C. Syldatk, S. Lang, F. Wagner, V. Wray, and L. Witte, *Z. Naturforschung 40c*: 51–60 (1985).
211. M. Yamaguchi, M. Sato, and K. Yamada, *Chem. Ind. 17*: 741–742 (1976).
212. G. Hauser and M. L. Karnovsky, *J. Bacteriol. 68*: 645–654 (1954).
213. G. Hauser and M. L. Karnovsky, *J. Biol. Chem. 224*: 91–105 (1957).
214. G. Hauser and M. L. Karnovsky, *J. Biol. Chem. 233*: 287–291 (1958).
215. L. Glaser and S. Kornfeld, *J. Biol. Chem. 236*: 1795–1799 (1961).
216. M. M. Burger, L. Glaser, and R. M. Burton, *J. Biol. Chem. 238*: 2595–2602 (1963).
217. C. N. Mulligan and B. F. Gibbs, *Appl. Environ. Microbiol. 55*: 3016–3019 (1989).
218. C. N. Mulligan, G. Mahmourides and B. F. Gibbs, *J. Biotechnol. 12*: 37–44 (1989).
219. J. Atkit, D. G. Cooper, K. I. Manninen, and J. E. Zajic, *Curr. Microbiol. 6*: 145–150 (1981).
220. D. G. Cooper, J. E. Zajic, and D. F. Gerson, *Appl. Environ. Microbiol. 37*: 4–10 (1979).
221. D. F. Gerson and J. E. Zajic, *Antonie van Leeuwenhoek 45*: 81–94 (1979).
222. C. R. Macdonald, D. G. Cooper, and J. E. Zajic, *Appl. Environ. Microbiol. 41*: 117–123 (1981).
223. C. Asselineau and J. Asselineau, *Prog. Chem. Fats Other Lipids 16*: 59–99 (1978).
224. J. Asselineau, *The Bacterial Lipids*, Hermann, Paris and Holden-Day, San Francisco, 1966.
225. T. Ioneda, in *Biological and Biomedical Aspects of Actinomycetes* (L. Ortiz-Ortiz, L. F. Bojalil, and V. Yakaleff, eds.), Academic Press, Orlando, (1984), pp. 239–249.
226. I. Tomiyasu, J. Yoshinaga, F. Kurano, Y. Kato, K. Kaneda, S. Imaizumi, and I. Yano, *FEBS Lett. 203*: 239–242 (1986).
227. E. Ristau and F. Wagner, *Biotechnol. Lett. 5*: 95–100 (1983).
228. S. G. Batrakov, B. V. Rozynov, T. V. Koronelli, and L. D. Bergelson, *Chem. Phys. Lipids 29*: 241–248 (1981).
229. Y. Uchida, R. Tsuchiya, M. Chino, J. Hirano, and T. Tabuchi, *Agric. Biol. Chem. 53*: 757–761 (1989).
230. S. H. Loomis, K. A. C. Madin, and J. H. Cowe, *J. Exp. Zool. 211*: 311–320 (1980).
231. A. Kretschmer and F. Wagner, *Biochim. Biophys. Acta 753*: 306–313 (1983).
232. A. Kretschmer, H. Bock, and F. Wagner, *Appl. Environ. Microbiol. 44*: 864–870 (1982).
233. J.-C. Prombe, R. W. Walker, and C. S. Lacave, *C. R. Acad. Sci. (C) 278*: 1065–1068 (1974).
234. A. Ahibo-Coffy, H. Aureleel, C. Lacave, J.-C. Prombe, G. Puzo, and A. Savagnac, *Chem. Phys. Lipids 22*: 185–195 (1978).
235. G. Puzo, G. Tissie, H. Aurelle, C. Lacave, and J.-C. Pombe, *Eur. J. Biochem. 98*: 99–105 (1979).
236. K. Takayama and E. L. Armstrong, *Biochemistry 15*: 441–447 (1975).
237. T. Shimakata, T. Kimiko, T. Kusaka, and H. I. Shizukuishi, *Arch. Biochem. Biophys. 238*: 497–508 (1985).
238. H. Okazaki, H. Sugino, T. Kanzaki, and H. Fukuda, *Agric. Biol. Chem. 43*: 767–770 (1969).

239. P. J. Brennan, D. P. Lahane, and D. W. Thomas, *Eur. J. Biochem. 13*: 117–123 (1970).
240. K. Takayama and E. L. Armstrong, *J. Bacteriol. 130*: 569–570 (1977).
241. A. Kretschmer, Ph.D. Thesis, Technical University of Braunschweig, FRG, 1981.
242. T. Ioneda, C. L. Silva, and J.-L. Gesztesi, *Zbl. Bakt. Suppl. 11*: 401–406 (1981).
243. M. C. Z. Teixeira, T. Ioneda, and J. Asslineau, *Chem. Phys. Lipids 37*: 155–164 (1985).
244. T. Suzuki, H. Tanaka, and S. Itoh, *Agric. Biol. Chem. 38*: 557–563 (1974).
245. S. Itoh and T. Suzuki, *Agric. Biol. Chem. 38*: 1443–1449 (1974).
246. T. Suzuki, H. Tanaka, I. Matsubara, and S. Kinoshita, *Agric. Biol. Chem. 33*: 1619–1627 (1969).
247. Z.-Y. Li, S. Lang, F. Wagner, L. Witte, and V. Wray, *Appl. Environ. Microbiol. 48*: 610–617 (1984).
248. U. Göbbert, A. Schmeichel, S. Lang, and F. Wagner, *J. Am. Oil Chem. Soc. 65*: 1519–1525 (1988).
249. G. Deml, T. Anke, F. Oberwirker, B. M. Giannetti, and W. Steglich, *Phytochemistry 19*: 83–87 (1980).
250. H. Kawashima, T. Nakahara, M. Ozaki, and T. Tabuchi, *J. Ferment. Technol. 61*: 143–149 (1983).
251. D. Kitamoto, K. Haneishi, T. Nakahara, and T. Tabuchi, *Agric. Biol. Chem. 54*: 37–40 (1990).
252. D. Kitamoto, S. Akiba, C. Hioko, and T. Tabuchi, *Agric. Biol. Chem. 54*: 31–36 (1990).
253. T. Kobayashi, S. Ito, and K. Okamoto, *Agric. Biol. Chem. 51*: 1715–1716 (1987).
254. D. C. Cooper, *AOCS Monogr. 11*: 281–287 (1984).
255. K. Arima, A. Kakinuma, and G. Tamura, *Biochem. Biophys. Res. Commun. 31*: 488–494 (1968).
256. H. Kleinkauf and H. von Döhren, *Curr. Topics Microbiol. Immunol. 91*: 129–177 (1981).
257. F. Peypoux, G. Michel, and L. Delcambe, *Eur. J. Biochem. 63*: 391–398 (1976).
258. C. Sandrin, F. Peypoux, and G. Michel, *Biotechnol. Appl. Biochem. 12*: 370–375 (1990).
259. F. Besson, F. Peypoux, G. Michel, and L. Delcambe, *Eur. J. Biochem. 77*: 61–67 (1977).
260. F. Peypoux, M.-T. Pommier, B. C. Das, F. Besson, L. Delcambe, and G. Michel, *J. Antibiot. 37*: 1600–1604.
261. M. Javaheri, G. E. Jenneman, M. McInerney, and R. M. Knapp, *Appl. Environ. Microbiol. 50*: 698–700 (1985).
262. T. R. Neu, T. Härtner, and K. Poralla, *Appl. Microbiol. Biotechnol. 32*: 518–520 (1990).
263. T. Matsuyama, M. Fujita, and I. Yano, *FEMS Microbiol. Lett. 28*: 125–129 (1985).
264. B. Kluge, J. Vater, J. Salnikow, and K. Eckart, *FEBS Lett. 231*: 107–110 (1988).
265. R. Zocher and H. Kleinkauf, *Biochem. Biophys. Res. Commun. 81*: 1162–1167 (1978).

266. H. Kleinkauf, and H. von Döhren, in *Biotechnology* (H. Pape and H. J. Rehm, eds.), Vol. 4, VCH, Weinheim, 1986, pp. 283–307.
267. E. Katz and A. L. Demain, *Bacteriol. Rev. 441*: 449–474 (1977).
268. J. Vater, *Progr. Colloid Polymer Sci. 72*: 12–18 (1986).
269. H. Kleinkauf and H. Koischwitz, in *Multifunctional Proteins* (H. Bisswanger and E. Schmincke-Ott, eds.), Wiley, New York, 1980, pp. 217–233.
270. O. Froyshov, T. L. Zimmer, and S. G. Laland, *Int. Rev. Biochem. Amino Acids Protein Biosynth. II 18*: 49–78 (1978).
271. A. Meister and S. S. Tate, *Annu. Rev. Biochem. 45*: 559–604 (1976).
272. S. K. Ghosh, S. Majumder, N. K. Mukhopadhyay, and S. K. Bose, *Biochem. J. 230*: 785–789 (1985).
273. H. Nikaido and W. Z. Hassid, *Adv. Carbohydr. Chem. Biochem. 26*: 351–483 (1971).
274. C. Liu, B. W. Patterson, D. Lapp, and A. D. Elbein, *J. Biol. Chem. 244*: 3728–3731 (1969).
275. L. Glazer and H. Zarkowsky, in *The Enzymes* (P. D. Boyer, ed.), Vol. 5, Academic Press, New York, 1971, pp. 465–480.
276. M. Powalla, S. Lang, and V. Wray, *Appl. Microbiol. Biotechnol. 31*: 473–479 (1989).
277. C. Ratledge, in *Basic Biotechnology* (J. Bu'Lock and B. Kristiansen, eds.), Academic Press, London, 1987, pp. 11–55.
278. A. Persson and G. Molin, *Appl. Microbiol. Biotechnol. 26*: 439–442 (1987).
279. A. Persson, E. Österberg, and M. Dostalek, *Appl. Microbiol. Biotechnol. 29*: 1–4 (1988).
280. T. Kobayashi, S. Ito, and K. Okamoto, *Agric. Biol. Chem. 51*: 1715–1716 (1987).
281. M. A. Laurila, Ph.D. thesis, ETH, Zürich, Switzerland, 1985.
282. A. Baum, Ph.D. thesis, University of Leipzig, FRG, 1991.
283. R. V. Kutcher, O. Y. Lesik, E. V. Karpenko, and S. A. Eliseev, *Doklady AN Ukr. SSR 3*: 72–74 (1990) (in Russian).
284. S. A. Eliseey, A. N. Shulga, and E. V. Karpenko, *Mikrobiol. J. 52*: 41–44 (1990) (in Russian).
285. H. Pape and H.-J. Rehm (eds.), *Biotechnology*, Vol. 4, VHC, Weinheim, 1986.
286. V. Behal, *Trends Biochem. Sci. 11*: 88–91 (1986).
287. V. S. Malik, *Adv. Appl. Microbiol. 28*: 27–115 (1982).

2

Production of Biosurfactants

JITENDRA D. DESAI Research Centre, Indian Petrochemicals Corporation Ltd., Baroda, India

ANJANA J. DESAI Department of Microbiology and Biotechnology Centre, The M. S. University of Baroda, Baroda, India

I. INTRODUCTION

Surfactants possess both hydrophilic and hydrophobic (generally hydrocarbon) structural moieties, which in turn impart many unusual properties, including an ability to lower the surface tension of water. The ability of the surfactant to reduce the surface tension of water depends on its molecular structure. Synthetic surfactants are commonly produced using a variety of organic chemistry methods, depending on the type and structure of the molecule desired. The commercial importance of surfactants is evident from the increasing trend in their production and a variety of industrial applications. The surfactant industry has grown about 300% within the U.S. chemical industry during the last decade. The U.S. surfactant industry shipments in 1989 were approximately $3.65 billion, 14% higher than the previous year [1]. United States surfactant production in 1989 was 7.6 billion lb. and worldwide surfactant production was estimated to be 15.5 billion lb. [1,2].

Interest in biosurfactants has increased considerably in recent years, as they are potential candidates for many commercial applications in the petroleum, pharmaceuticals, and food processing industries [3–7]. The term *biosurfactant* has been used very loosely and refers to any usable and isolatable compound obtained from microorganisms that has some influence on interfaces. Thus, it is also used for emulsifying and dispersing agents that do not significantly lower the surface tension of water or exhibit other properties of a classical surfactant.

Many biosurfactants lower the interfacial tension between oil and brine to less than 0.01 mN/m and are potential candidates for enhancing the oil recovery process [3,8,9]. Some of them also act as potential deemulsifiers [10]. Biosurfactants have special advantages over their chemically manufactured counterparts because of their lower toxicity, biodegradable nature [11], effectiveness at extreme temperatures, pH, and salinity [12], and ease of synthesis. In addition, they possess surface-active properties differing in many cases from synthetic surfactants. Several reviews and monographs have appeared in literature on the properties [6,13,14], chemistry [3,4,7,15,16], biosynthesis [16,17], and applications of biosurfactants [5,18,19]. In this chapter, efforts are made to give a brief account of the sources and properties of various classes of biosurfactants and to review the present status of biosurfactant production.

II. SOURCE, CHARACTERISTICS, AND PROPERTIES OF BIOSURFACTANTS

Owing to the large surface-to-volume ratio and diverse biosynthetic capabilities, microbes are promising candidates in the search for enlarging our present range of surfactants. Many microbes appear to produce a complex mixture of biosurfactants, particularly during their growth on water-immiscible substrates. Among microbes, a majority of biosurfactants are found to be produced by bacteria. Generally, biosurfactants are microbial metabolites with the typical amphiphilic structure of a surfactant, where the hydrophobic moiety is either a long-chain fatty acid, hydroxy fatty acid, or α-alkyl-β-hydroxy fatty acid and the hydrophilic moiety can be a carbohydrate, an amino acid, a cyclic peptide, a phosphate, a carboxylic acid, alcohol, etc. Physical and chemical properties, surface tension reduction, and stability of the emulsion formed are very important in the search for a potential biosurfactant. These properties are used in evaluating biosurfactants and in screening potential microorganisms for biosurfactant production.

Synthetic surfactants are usually classified according to the nature of their polar group. However, microbial surfactants are commonly differentiated on the basis of their biochemical nature and the microbial species producing them. The important surfactant types and the producing microbial species are listed in Table 1. Major classes of biosurfactants include (1) glycolipids, (2) phospholipids and fatty acids, (3) lipopeptide/lipoproteins, (4) polymeric surfactants, and (5) particulate surfactants.

A. Glycolipids

Glycolipids, the most commonly isolated and studied biosurfactants, are carbohydrates in combination with long-chain aliphatic acids or hydroxy aliphatic acids. Glycosyl diglycerides present in the cell membrane of a wide variety of bacteria are the most common glycolipids. The best examples of glycolipids studied from the point of view of surfactant characterization and properties are (1) trehalose lipids, (2) rhamnolipids, and (3) sophorolipids.

1. Trehalose Lipids

Several structural types of trehalose lipids are found to be widely distributed. Disaccharide trehalose linked at C6 and C6′ to mycolic acids are associated with the cell wall structure of most species of the genera *Mycobacterium, Nocardia*, and *Corynebacterium*. Mycolic acids are long-chain α-branched β-hydroxy fatty acids and the chain length synthesized is a characteristic of producing genera. Trehalose diester produced by *Rhodococcus erythropolis* and *Arthrobacter paraffineus* have been extensively studied by Rapp et al. [20] and Suzuki et al. [21]. *Mycobacterium phlei* [22] produces trehalose 6-*O* monomycolates, whereas *Mycobacterium fortuitum* and *Micromonospora* sp. F3 produce trehalose esters of straight-chain nonhydroxylated C_{16} and C_{18} fatty acids [23,24]. On the other hand, *Mycobacterium smegmatis* produces trehalose esters fully acetylated by polyunsaturated acids [25]. Production of novel nonionic trehaloselipid from *Mycobacterium paraffinicum* [26] and anionic trehaloselipid from *R. erythropolis* [27] have been isolated and characterized. Production of mono-, di-, and trisaccharides have

TABLE 1 Major Types of Biosurfactants Produced by Microorganisms

Biosurfactant type	Producing microbial species	References
A. Glycolipids		
Trehalose mycolates	*Rhodococcus erythropolis*	20
	Arthrobacter paraffineus	21
	Mycobacterium phlei	22
Trehalose esters	*Mycobacterium fortitum*	23
	Micromonospora spp.	24
	Mycobacteriun smegmatis	25
	Mycobacterium paraffinicum	26
	Rhodococcus erythropolis	27
Mycolates of mono-, di-,	*Corynebacterium diphtheriae*	
and trisaccharide	*Mycobacterium smegmatis*	28
	Arthrobacter spp.	29–31
Rhamnolipids	*Pseudomonas* spp.	15, 32–34
Sophorolipids	*Torulopsis bombicola*	35, 37
	Torulopsis petrophilum	
	Torulopsis apicola	38–40
	Candida spp.	41
B. Phospholipids and Fatty Acids		
Phospholipids and	*Candida* spp.	
fatty acids	*Corynebacterium* spp.	
	Micrococcus spp.	
	Acinetobacter spp.	7, 42–47
Phospholipids	*Thiobacillus thiooxidans*	45
	Aspergillus spp.	46
C. Lipopeptides and Lipoproteins		
Gramicidens	*Bacillus brevis*	48
Polymyxins	*Bacillus polymyxa*	49
Ornithine–lipid	*Pseudomonas rubescens*	50
	Thiobacillus thiooxidans	51
Cerilipin	*Gluconobacter cerinus*	52
Lysin–lipid	*Agrobacterium tumefaciens*	53
	Streptomyces sioyaensis	54
Surfactin. subtilysin	*Bacillus subtilis*	55, 57
Peptide–lipid	*Bacillus licheniformis*	58
D. Polymeric Surfactants		
Lipoheteropolysaccharide	*Arthrobacter calcoaceticus RAG-1*	59–62
Heteropolysaccharide	*A. calcoaceticus* A2	67, 68
Polysaccharide–protein	*A. calcoaceticus* strains	69, 70
	Candida lipolytica	72–74
Manno–protein	*S. cerevisiae*	75
Carbohydrate–protein	*Candida petrophilum*	76
	Endomycopsis lipolytica	77
Mannan–lipid complex	*Candida tropicalis*	78, 79
Mannose/erythrose–lipid	*Shizonella melanogramma*	80
	Ustilago maydis	81
Carbohydrate–protein-	*Pseudomonas* spp.	82–86
lipid complex	*Pseudomonas fluorescens*	87
	Debaryomyces polymorphus	88
E. Particulate Biosurfactants		
Membrane vesicles	*Acinetobacter* sp. H01-N	91
Fimbriae	*A. calcoaceticus*	7, 63
Whole cells	Variety of microbes	7, 89, 92

been reported using various species of *Corynebacteria, Mycobacteria*, and *Arthrobacter* [28–30].

Wagner and coworkers have extensively studied the surface and interfacial activities of trehalose lipids. Trehalose lipid from *Rhodococcus erythropolis* has been shown to reduce surface tension to 25–30 mN/m and interfacial tension to 1 mN/m [15]. Corynemycolates of various mono- and disaccharides from *Arthrobacter* spp. have also been reported to lower the surface tension to 33–40 mN/m and interfacial tension to 1–5 mN/m [15,29].

2. Rhamnolipids

Certain species of *Pseudomonas* are known to produce large amounts of glycolipids containing one or two molecules of rhamnose linked to one or two molecules of β-hydroxydecanoic acid. Edwards and Hyashi [31] and Hisatsuka et al. [32] have reported formation of glycolipid, type R-1, containing two rhamnose and two β-hydroxydecanoic acid units by *Pseudomonas aeruginosa*. Subsequently, Itoh and Suzuki [33] presented evidence for the production of a second kind of rhamnolipid (R-2) containing only one rhamnose unit. The results reported by Lang and Wagner [15] and Parra et al. [34] have demonstrated the production of novel rhamnolipids under various cultivation conditions by *Pseudomonas* spp.

There are now several lines of evidences available to show that depending on the pH and salt concentration, pure rhamnolipids from *Pseudomonas* spp. can lower the interfacial tension against *n*-hexadecane to around 1 mN/m and surface tension to 25–30 mN/m [15,33,34].

3. Sophorolipids

Production of biosurfactants heavier than water and consisting of the dimeric carbohydrate sophorose linked to long-chain hydroxycarboxylic acids have been reported using *Torulopsis bombicola* [35–37], *Torulopsis petrophilum* [38], and *Torulopsis apicola* [39]. These biosurfactants are a mixture of at least six to nine different hydrophobic sophorosides. Recently, Hommel et al. [40] have investigated the production of a mixture of water-soluble sophorolipids from yeasts. Culter and Light [41] showed that *Candida bogoriensis* produces glycolipids in which sophorose is linked to docosanoic acid diacetate. More details on extracellular glycolipids of yeasts are documented [7, 15, 42].

Although, sophorolipids lower surface and interfacial tension, they are not effective emulsifying agents [35]. Both lactonic and acidic sophorolipids lowered the interfacial tension between *n*-hexadecane and water from 40 to 5 mN/m and showed a remarkable stability toward pH and temperature changes [15, 38].

B. Phospholipids and Fatty Acids

Certain hydrocarbon-degrading bacteria and yeasts produce appreciable amounts of phospholipids and fatty acids when grown on *n*-alkanes [25, 44, 74]. These surfactants are able to produce optically clear microemulsions of alkanes in water. *Thiobacillus thiooxidans* [45] produces a quantitative amount of phospholipids, which has a role in the wetting of elemental sulfur. Miyazima et al. [46] reported the production of phospholipids by *Aspergillus* sp. grown on hydrocarbons.

Extracellular free fatty acids produced by microorganisms grown on alkanes also show surfactant activity. The important candidates are saturated fatty acids in the range of C_{12} to C_{14} and the complex fatty acids containing hydroxyl groups and alkyl branches [12, 47]. *Arthrobacter* AK-19 [43] and *P. aeruginosa* 44T1 [44] have been shown to accumulate up to 40%–80% w/w lipid when cultivated on hexadecane and olive oil, respectively.

C. Peptides and Amino Acid Containing Lipids

Decapeptide antibiotics (Gramicidins) and lipopetide antibiotics (Polymyxins) produced by *Bacillus brevis* [48] and *B. polymyxa* [49], respectively, possess remarkable surface-active properties. Similarly, peptide-containing lipids exhibit biosurfactant activity. They include ornithine-containing lipids from *P. rubescens* [50] and *T. thiooxydans* [51], cerilipin, and ornithine- and taurine-containing lipid from *Gluconobacter cerinus* IFO 3267 [52], and lysine-containing lipids from *Agrobacterium tumefaciens* IFO 3058 [53] and *Streptomyces sioyaensis* [54].

Surfactin, a cyclic lipopetide, reported first by Arima et al. [55] in *B. subtilis* ATCC-21332, is one of the most effective biosurfactants known so far. It is capable of lowering the surface tension from 72 to 27.9 mN/m at a concentration as low as 0.005% [55]. The ability of surfactin to lyse red blood cells is of limited use, but this discovery has led to the development of a quick method for the screening of biosurfactant-producing microbes [56]. Production of surfactant by *B. subtilis* QMB [57] and lipopeptide surfactant, lichenysin by *B. licheniformis* JF2 [58, 179] with similar structural and physicochemical properties to surfactin have been reported. *B. licheniformis* also produce several other surface active agents which act synergistically and exhibit excellent temperature, pH and salt stability [179]. A surfactant, BL-86 produced by *B. licheniformis* 86, lowered surface tension of water to 27 dynes/cm and interfacial tension between water and *n*-hexadecane to 0.36 dynes/cm [180]. The surfactant is stable to a wide range of pH, temperatures, and NaCl concentrations [180] and promoted dispersion of colloidal 3-silicon carbide and aluminum nitride slurries far better than commercial agents [181]. Recently, Horowitz and Griffin [182] carried out a detail structural analysis of surfactant BL-86 and found that it is a mixture of lipopeptides with the major components ranging in size from 979 to 1091 daltons with varying in increments of 14 daltons. There are seven amino acids per molecule while, lipid portion is composed of 8 to 9 methylene groups and a mixture of linear and branched tails.

D. Polymeric Biosurfactants

High molecular weight biopolymers generally exhibit useful properties, such as high viscosity, tensile strength, and resistance to shear. It is, therefore, not surprising that polymeric biosurfactants have found a variety of industrial uses. The best studied of these biosurfactants are emulsan, liposan, mannoprotein, and other polysaccharide-protein complexes.

1. Emulsan

Acinetobacter calcoaceticus RAG-1 has been shown to produce a potent extracellular polymeric bioemulsifier called emulsan [59]. Emulsan has been characterized as a polyanionic amphipathic heteropolysaccharide [59]. The heteropolysaccharide backbone contains repeating trisaccharide of *N*-acetyl-D-galactosamine, *N*-acetylgalactosamine uronic acid, and an unidentified *N*-acetyl amino sugar [60]. In addition, fatty acids, which constitute about 10%–15% of the dry weight are shown to be covalently linked to the polysaccharide through *O*-ester linkages [60–62]. Emulsan does not appreciably reduce interfacial tension, but it is a very effective emulsifying agent for hydrocarbons in water even at a concentration as low as 0.001%–0.01%. It is one of the most powerful emulsion stabilizers known today and resists inversion even at a water-to-oil ratio of 1:4 [62, 63].

Emulsan stabilized emulsion, on long standing or centrifugation, gets separated into two layers. The upper cream layer, also known as emulsanosol, contains approximately 70%–75% oil in bulk aqueous phase. Emulsanosols can remain stable for months and can withstand enormous shear without any inversion [64]. Interestingly, addition of small amounts of alkanols or alcohol derivatives has been shown to alter the hydrocarbon substrate specificity of emulsan [65]. The enzyme responsible for the depolymerization of emulsan has been isolated and has been found to act by transelimination [66]. For more details on emulsan refer to the recent reviews [7, 17, 63].

2. Biodispersan

Production of an extracellular, nondialyzable dispersing agent called biodispersan has been reported using *A. calcoaceticus* A2 [67]. The active component of biodispersan is an anionic heteropolysaccharide, with an average molecular weight of 51,400 and four reducing sugars, namely, glucosamine, 6-methyl aminohexose, galactosamine uronic acid, and an unidentified amino sugar [68].

3. Liposan

Extracellular water-soluble emulsifier, designated as liposan, is found to be synthesized by *C. lipolytica* [72, 73]. Cirigliano and Carmen [74] recently purified and elucidated the structure of liposan. It is composed of 83% carbohydrate and 17% protein. The carbohydrate portion is a heteropolysaccharide consisting of glucose, galactose, galactosamine, and galacturonic acid. The partially purified liposan stabilized the emulsion formed between many commercial vegetable oils and water [74].

4. Other Polysaccharide Protein Complexes

The surface-active property of *A. calcoaceticus* BD 4 is due to the production of heteropolysaccharide-containing capsules [69]. These capsules are composed of repeating units of heptasaccharide and are released in the medium during the growth on hydrocarbons. According to Sar and Rosenberg [70], polysaccharides alone showed no emulsification activity, but polysaccharide released with protein during the growth of a parent strain on ethanol or by a mutant strain BD-413 showed potent emulsification activity [70]. They also demonstrated the production of emulsifiers by other eight strains of *A. calcoaceticus.* Although, these emulsifiers have not been purified, they have been found to contain mainly polysaccharide and protein in different proportions. Palejwala and Desai [71] have reported the production of bioemulsifier with carbohydrate as a major component during the growth of a Gram-negative bacterium on ethanol.

Recently, high amounts of mannoprotein emulsifier production by *S. cerevisiae* has been reported by Cameron et al. [75]. The purified emulsifier contains 44% carbohydrate (mannose) and 17% protein. This product emulsifies many oils, alkanes, and organic solvents, and the emulsions are stable at extreme pH, temperatures, and salt concentrations. The emulsifiers from *C. petrophilum* [76] and *E. lipolytica* [77] also contain carbohydrate and protein. Feichter and his group [78, 79] have isolated a mannan-fatty acid complex from alkane-grown *C. tropicalis* that stabilized hexadecane in water emulsions. *Shizonella malanogramma* and *Ustilago maydis* produce a biosurfactant that is characterized as erythritol- and mannose-containing lipid [81]. Hisatsuka et al. [82] isolated an emulsifying protein, called PA, along with rhamnolipid from *P. aeruginosa.* The molecular weight of PA is approximately 14,300, and 51 out of 147 amino acids are serine and threonine [83]. Production of peptidoglycolipid that bears 52 amino acids, 11 fatty acids, and a sugar unit by *P. aeruginosa* P-20 has been described [84]. Recently, Singh and his colleagues [85, 86] have isolated, purified and characterized an emulsifying and solubilizing factor from hexadecane-grown *Pseudomonas* spp. Both protein and carbohydrate are essential for *n*-hexadecane solubilizing activity. Desai et al. [87] reported the production of bioemulsifier by *P. fluorescens* during growth on gasoline. The bioemulsifier is composed of 50%

carbohydrate, 19.6% protein, and 10% lipid. Trehalose and lipid-*o*-dialkyl monoglycerides were the major components of the carbohydrate and lipid, respectively. Similarly, from *C. tropicalis* [88] and *Phormidium* J1 [89], an extracellular bioemulsifier composed of carbohydrate, protein, and lipids has been isolated and characterized. A glycolipopeptide capable of emulsifying water-immiscible organophosphorus pesticides has been isolated from *B. subtilis* FE-2 [90].

E. Particulate Biosurfactants

Accumulation of extracellular membrane vesicles having 20–50 nm dia. and a buoyant density of 1.158 g/cm^3 has been reported in *Acinetobacter* sp. HO1-N cells [91]. The vesicles partition hydrocarbons in the form of microemulsion and play an important role in alkane uptake by the cells. The purified vesicles are composed of protein, phospholipid, and lipopolysaccharide. The vesicles have a phospholipid five times higher and a polysaccharide content 360-fold higher than that observed in the outer membrane of the same organism.

A large variety of microorganisms such as most hydrocarbon degraders, strains of cyanobacteria, and pathogenic bacteria possess surfactant activity [7, 89, 92]. The surface components that contribute to the surfactant activity include M-protein and lipoteichoic acid on *Streptococci* group-A, protein-A of *Staphylococcus aureus*, layer-A of *Aeromonas salmonicids*, prodigiosin of *Sarratia* spp., gramicidins in *B. brevis* spores, and thin fimbriae in *A. calcoaceticus* RAG-1. This has been extensively covered in many reviews [6, 7, 14, 15].

III. BIOSYNTHESIS OF BIOSURFACTANTS

A. General Features of Biosynthesis

Organisms use energy, reducing power and precursors produced by catabolism to synthesize the molecular components they require for growth and reproduction. Microbial metabolism is an integrated process that demands the coordinated activity of several enzymes. This is a conservative process, which under normal circumstances neither wastes energy and/or carbon source for synthesis of substances available in the surrounding medium nor does it overproduce them. Generally, biosurfactants are the microbial metabolites with a typical amphiphilic structure, where the hydrophobic moiety is a long-chain fatty acid, hydroxy fatty acid, or α-alkyl-β-hydroxy fatty acid and the hydrophilic moiety can be a carbohydrate, amino acid, cyclic peptide, phosphate, carboxylic acid, or alcohol. Metabolic pathways involved in the synthesis of these two groups of precursors are diverse and utilize a specific set of enzymes. In many cases, the first enzyme, which is unique to the biosynthetic pathways for the synthesis of these precursors, are regulatory enzymes. Therefore, in spite of the wide spectrum of interfacially active compounds, there are some common features for biosynthesis of biosurfactants and their regulation. It is not within the scope of this chapter to give a detailed description of the regulation of precursor synthesis.

According to Syldatk and Wagner [17], the following different possibilities exist for the synthesis of biosurfactants.

1. *De novo* synthesis of hydrophilic and hydrophobic moieties by two independent pathways followed by their linkage to form a complete biosurfactant molecule.
2. *De novo* synthesis of the hydrophilic moiety and the substrate-dependent synthesis of the hydrophobic moiety and its linkage.
3. *De novo* synthesis of the hydrophobic moiety and the substrate-dependent synthesis of the hydrophilic moiety followed by its linkage.

The synthesis of both hydrophobic and hydrophilic moieties dependent on the substrate used for biosurfactant production and their linkage is also possible. Both lipid and peptide domains have been found to be directly synthesized from carbohydrates in case of herbicollin A. Addition of amino acids or fatty acids in the growth medium affected the yield but not the structure of the surfactant [183]. Hydrophilic moieties in biosurfactants show a greater degree of complexity and involve a number of biosynthetic pathways. Recent studies have shown that in Gramicidin-S, surface active antibiotic, lipopeptide is synthesized non-ribosomally by a multi enzyme complex with the involvement of pantetheine cofactor by a thio-template mechanism [184]. Nakano et al. [185] have presented genetic evidence that two components of surfactin synthesizing enzyme complex of *B. subtilis* are homologous to tyrocidine synthetase I and gramicidin S synthetase. Moreover, the recent biochemical studies demonstrated occurrence of surfactin synthesis via a thiotemplate mechanism [186]. Enzymatic synthesis of surfactin requires ATP, Mg^{++}, precursors and sucrose. The fatty acid component is incorporated only as a acetyl-CoA derivative and L-isomer of amino acids are incorporated in the peptide chain [186].

Hauser and Karnovsky [93, 94] have extensively studied the rhamnolipid synthesis using enzymology and radioactive precursors and proposed a biosynthetic pathway. Wagner and his group [95, 96] have shown that, although the composition of biosurfactant produced by *Pseudomonas* spp. is affected by the carbon substrate in the medium and the cultivation conditions, the hydrocarbon substrate of a different chain length has no effect on the chain length of the fatty acid moiety in glycolipids. A similar observation has been noted in *P. aeruginosa* by Edmonds and Cooney [97]. While investigating the synthesis of glycolipids in a bacterial isolate H-13A grown on alkanes, Finnerty and Singer [4] presented evidence for the qualitative and quantitative variations in fatty acids reflective to the alkane carbon number of the substrate used. Almost identical results were observed during the bioemulsifier production by *Acinetobacter* sp. H01-N using alkane as the substrate [98]. Other examples of *de novo* synthesis of biosurfactant are cellobiose lipid by *Ustilago zeae* [99] and sophorolipid produced by *T. bombicola* [36] from different lipophilic substrates.

During the synthesis of trehalose mono- and dicorynomycolates in *R. erythropolis*, the sugar moiety of the surfactant is *de novo* synthesized and the chain length of the lipid moiety is dependent on hydrocarbon substrate used in the medium [100]. Thus, the biosynthesis of corynomycolates do not proceed by *de novo* synthesis from C_2 units but, by chain elongation. A similar pathway has been found to be operative in *R. erythropolis* for the synthesis of trehalose tetra-esters [101], in *Candida* sp. for the synthesis of mannosylerythritol lipids [102], and in *T. magnoliae* [103] and *N. erythropolis* [47] for extracellular glycolipid synthesis.

Suzuki et al. [21, 30, 105] during the investigation on the influence of substrate on the sugar moiety of the glycolipids synthesized by *A. paraffineus* observed the synthesis of fructolipids when grown on fructose [105] and glucose and sucrose lipids when grown on sucrose [30]. Similar results have been noted in the resting cells of *Arthrobacter* sp. [29].

B. Regulation of Biosurfactant Synthesis

Essentially three major phenomena regulate the overproduction of biosurfactants, namely (1) induction, (2) repression, and (3) nitrogen and multivalent ions mediated effects.

1. Regulation of Biosurfactant Synthesis by Induction

In general, microorganisms produce biosurfactants when they grow at the expense of water immiscible substrates [32, 45, 59, 76, 104, 106]. Chakrabarty [107] reported the production of glycolipid-EM by *P. aeruginosa* SB-30 when grown on alkanes. Mutants incapable of producing biosurfactant fail to utilize alkanes and interestingly the growth was restored by the augmentation of biosurfactant into the growth medium. Neufeld et al. [108] observed the reduction in surface and inter-facial tension of culture broth during the growth of *A. calcoaceticus* on hydro-carbon substrates. *Pseudomonas aeruginosa* [82, 83, 109] and *Candida* spp. [73,102, 110] when grown on *n*-alkanes secrete surface-active substances in the medium. These substances were not detected during the growth on carbohydrates.

Using actidion as an inhibitor, the inducible nature of solubilizing factor produced by *E. lipolytica* has been confirmed by Roy et al. [77]. The yield of sophorolipids increased several fold by addition of long-chain fatty acids, hydrocarbons, or glycerides in the growth medium of *T. magnoliae* [103]. Rapp and his colleagues [20, 100] have observed the induction of trehalolipid synthesis in *R. erythropolis*, when hydrocarbons were used as the sole carbon source. Recently, while studying the aerobic degradation of hydrocarbon mixture by soil microbial population, Oberbremer and Hurtig [111] failed to observe the production of trehalose lipids until most water-soluble naphthalenes were degraded in the first growth phase. The biosurfactant production was observed in the second degradation phase, in which 89% of the hydrocarbons were metabolized. Similar results have been reported in *C. lipolytica* for the production of liposan [73].

2. Regulation by Catabolic Repression

Catabolic repression of biosurfactant synthesis by glucose or primary metabolites is one of the important regulatory mechanisms found to be operative in microorganisms. The function of biosurfactant is related to the hydrocarbon uptake and therefore, they are synthesized predominantly by hydrocarbon degrading microorganisms [7, 13, 37, 112, 113].

The first enzyme in the oxidation of *n*-alkane has been found to be catabolically repressed [114]. *Actinobacter calcoaceticus* [108] and *A. paraffineus* [115] fail to synthesize surface-active compounds when grown on organic acids and glucose, respectively. A factor responsible for *n*-alkane oxidation in *P. aeruginosa* has been found to be synthesized during the growth on hydrocarbons and not on glucose, glycerol, or palmitic acid as the substrates [82, 83]. Hauser and Karnovsky [94, 116] have demonstrated a drastic reduction in rhamnolipid synthesis on addition of glucose, acetate, and tricarboxylic acids during the growth on glycerol. A similar observation was noted for liposan synthesis in *C. lipolytica* [73].

Conversely, surfactin production by *B. subtilis* is observed with glucose as the carbon source and inhibited by the addition of hydrocarbons in the medium [117]. Parra et al. [34] isolated a new strain of *P. aeruginosa* 44T1, which produces rhamnolipid during the growth on glucose and fails to grow or produce surfactants on *n*-alkanes. In their recent studies, Neu and Poralla [118] examined culture supernatants of 126 bacterial strains for their ability to produce bioemulsifiers. Forty-eight strains were found to produce bioemulsifiers during the growth on glucose, but not on hydrocarbons. Recently, Mulligan et al. [119] isolated a mutant strain of *B. subtilis* (Suf-1), which produced a threefold higher biosurfactant in glucose medium. The mutation is found to be located in between *Arg* C4 and *His* A1 on the genetic map.

3. Effects of Nitrogen Sources and Multivalent Cations

A nitrogen source in the medium plays an important role in the cellular metabolism, including the production of surface-active compounds. Among the inorganic salts tested, ammonium salts and urea are preferred nitrogen sources for the biosurfactant production by *A. paraffineus* [120]. Gerson and Zajic [121] demonstrated a nitrogen-mediated shift in the production of biosurfactant. Supplementation of ammonia in the nitrate-containing medium results in the delayed production of biosurfactant in *Corynebacteria*. Mulligan and Gibbs [122] have recently shown a direct relationship between glutamine synthetase and biosurfactant production in *P. aeruginosa* RC-II. Furthermore, ammonia and glutamine at higher concentrations repressed synthesis of the enzyme and biosurfactant.

Several investigators observed rhamnolipid in the fermentation broth of *P. aeruginosa* with the exhaustion of nitrogen and commencement of the stationary growth phase [123–125]. Moreover, addition of a nitrogen source caused inhibition in the rhamnolipid synthesis by resting cells [95, 101]. Formation of new glycolipids under nitrogen limiting conditions by *R. erythropolis* has been observed by Ristau and Wagner [27]. Similarly, amino acid starvation resulted in higher production of emulsan by *A. calcoaceticus* RAG [126]. Guerra-Santos et al. [123] showed that growth of *P. aeruginosa* under noncarbon-limited conditions, use of nitrate, and the omission of yeast extract resulted in high biosurfactant production. The influence of various nitrogen sources on glycolipid production by *N. corynebacteroids* was examined by Powalla and coinvestigators [127] and found that inorganic nitrogen sources gave distinctly higher yields of biosurfactants of which $NaNO_3$ gave the highest production. The production of biosurfactant AP-6 by *P. fluorescens*-378 was found to be independent of the C:N ratio in the culture medium [128].

The limitation of multivalent cations also causes overproduction of biosurfactant in *Pseudomonas* spp. [96, 105, 124]. In *P. aeruginosa* DSM 2659, Guerra-Santos et al. [129] demonstrated that minimized concentrations of respective salts of magnesium, calcium, potassium, sodium, and trace elements yielded high levels of rhamnolipid. Of the trace elements, iron exhibited a major influence; at concentrations of 2 mg/g glucose and above, formation of surfactant did not occur [124]. Iron limitation is known to stimulate the production of biosurfactants in *P. fluorescens* [126, 130]. In contrast, the surfactin production by *B. subtilis* is found to be stimulated by the addition of iron and manganese salts to the medium [117].

Other factors, which are reported to affect the production of biosurfactant include ethambutol, penicillin, chloramphenicol, and EDTA [21, 42, 96, 101, 126]. More details on the regulation of metabolism in relation to biosurfactant production is given in Chapter 1.

C. Genetics of Biosurfactant Synthesis

Although the genetics of biosurfactant synthesis is in its infancy, significant information has recently been reported in the literature. The main problem in understanding the biosurfactant synthesis has been the lack of a suitable method for the screening of surfactant producing microorganisms. Recently, Matsuyama et al. [187] developed a direct thinlayer chromatographic technique for screening of cells producing wetting agents. However, the method has limited use due to its detectable limit. A sensitive drop collapsing test has also been developed recently by Jain et al. [188]. Mulligan et al. [119] developed a simple assay method based on the hemolytic property of biosurfactants. They isolated a mutant strain of *B. subtilis* (*surf* 1) which produced three times higher levels of surfactin than the wild type. The mutation has been found to be located inbetween *Arg* C4 and *His* A1 on the genetic map. Using a similar approach, Nakano and Zuber [184, 189] isolated *nul* mutants and demonstrated that three chromosomal loci are required for surfactin synthesis in *B. subtilis*. The *srf* A locus is a large operon of more than 25 kb of DNA having two open reading frames that may correspond to subunits of surfactin synthetase. Mutations in *srf* A not only affect the lipopeptide synthesis, but also the sporulation, indicating that expression of *srf* A is important in cell differentiation.

Nakano et al. [190] proposed that the *srf* A operon may be involved in the production of pheromone-like peptide factors which are reported to be responsible in the initiation of sporulation in *Bacillus* [191]. The synthesis of these regulatory peptides and surfactin appear to involve the same intermediate. A closely linked *sfp* locus, encodes a 224 amino acid polypeptide has been found to be responsible in converting these intermediates to surfactin. This polypeptide has also been found to lower the transcription of the *srf* A operon indicating that it has a regulatory function in addition to its direct role in surfactin biosynthesis. Mulligan et al. [119] showed that surfactin production is also affected by the regulatory genes *com* A, *com* P and *spo* OK which are located in *srf* B region of the chromosome. The gene product of *com* A and *com* P may form a complex that acts as a positive regulator of *srf* A transcription, in response to the levels of glutamine and glucose in the growth medium [190, 192]. Moreover, it has been found that the *srf* A genes are under the control of an inducible promoter so that the production of surfactin is only dependent on the addition of the inducer isopropyl-3-galactoside (IPTG) in the growth medium.

IV. PRODUCTION OF BIOSURFACTANTS

A. Production of Biosurfactants via the Fermentation Route

Depending upon the nature of the biosurfactant and the producing microorganisms, the following patterns of biosurfactant production by fermentation are possible: (1) growth associated production, (2) production under growth limiting conditions, (3) production by resting/nongrowing cells, and (4) production associated with the precursor augmentation. The pattern of biosurfactant production in

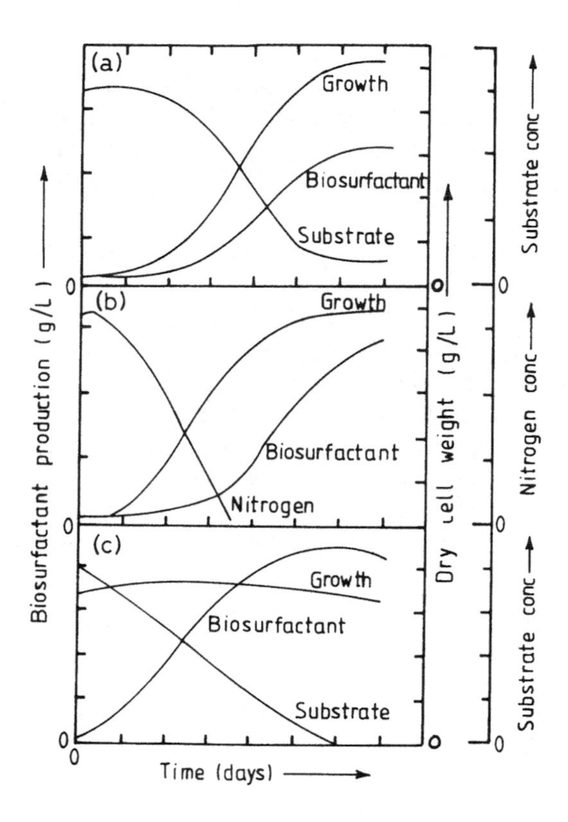

FIG. 1 Pattern of biosurfactant production (a) in a growth associated condition, (b) under growth-limiting conditions, and (c) by resting cells.

the above categories is schematically illustrated in Fig. 1. This section describes the details on biosurfactant production in the above categories and the strategies adopted to improve the yields of biosurfactants.

1. Growth Associated Biosurfactant Production

In this case, there exists a parallel relationship between the substrate utilization, growth, and biosurfactant production (Fig. 1a). The carbon source plays an important role in biosurfactant production [96, 109]. Different carbon sources such as glycerol, glucose, and ethanol could be used for rhamnolipid production in *Pseudomonas* spp., but all were inferior to *n*-alkanes. The chain length of the hydrocarbon used has been found to affect biosurfactant production [96, 109]. Pseudomonads as a group lack the capability to utilize lactose as a carbon source.

Koch et al. [131] constructed *P. aeruginosa* strains by inserting *lac* plasmid from *E. coli.* This strain is found to produce rhamnolipids during the stationary phase of growth in lactose-based minimal medium and whey. Although *P. aeruginosa* 44T1 is able to grow and produce rhamnolipids from glycerol, mannitol, and glucose, olive oil supported the highest amount of growth and biosurfactant production [132].

The normal production of cell-free emulsan by *A. calcoaceticus* RAG-1 is a mixed growth-associated and nongrowth-associated process [61, 133, 134]. Accumulation of emulsanlike polymer on the cell surface during the early exponential phase of growth has been documented [133–135]. Kinetic measurements using ELISA clearly demonstrated that the release of cell-free emulsan is accompanied by a corresponding decrease in cell-associated capsule [126, 133]. The biosurfactant production by *C. hydrocarboclastus* was optimal with linear alkanes of the chain length C_{12} to C_{14} and trehalose lipid by *N. rhodococcus* was dependent on the chain length of the hydrocarbon substrate [100]. Growth-associated production of highly active glycoprotein biosurfactant AP-6 has been reported in *P. fluorescens* strain 378 [136]. Similarly, continuous culture production of emulsan by *A. calcoaceticus* RAG-1 [137] and the fermentative production of surface-active agents from *B. cereus* IAF 346 and *Bacillus* sp. IAF-343 [138] are found to be growth associated.

In the literature, examples can be found in which one carbon source is used for the growth and the other for the production of surface-active compounds. *Arthrobacter paraffenium* ATCC 19558, when grown on glucose added to hexadecane in the medium during the stationary phase of growth, caused significant production of biosurfactant [115]. *Corynebacterium lepus* produced a large amount of biosurfactants when grown on glucose, but it remained cell bound. The surfactant is released from the cells when treated with hexadecane [139]. An addition of lipophilic compounds to the growing culture of *T. magnoliae* increased the biosurfactant production by severalfold [103]. *Torulopsis bombicola* grown on glucose produced substantial amounts of glycolipid by addition of vegetable oils during the growth in the late exponential phase [35, 147]. Stuwer et al. [140] demonstrated that mixed substrate cultivation of *T. apicola* IMET 43747 on glucose and sunflower oil gave glycolipid production up to 90 g/L. Rosenberg et al. [67] have observed that significant amounts of biodispersan began to appear in the culture broth only after *A. calcoaceticus* A2 growth slowed down and continued to accumulate during the stationary growth phase upto 4 g/L. Furthermore, the results led to speculation that biodispersan, which was either synthesized during the exponential growth phase or synthesized *de novo* during the stationary phase, could be cell bound. The extensive studies by Gutnick and his group [126, 133, 135] in *A. calcoaceticus* RAG-1 clearly showed that emulsan or emulsanlike precursors accumulate as capsular material during the exponential growth phase and is released into the medium when the rate of protein synthesis decreases.

The type and addition of the nitrogen source in the growth medium also influence the production of biosurfactants. The biosurfactant production by *A. paraffineus* ATCC 19558 increased by the addition of aspartic acid, asparagine, glycine, or glutamatic acid in the mineral salt medium [120]. Moreover, yeast extract, peptone, and bactotryptone also showed a positive effect on biosurfactant production. A parallel relationship between growth, activity of glutamine synthetase, and biosurfactant production has been observed when *P. aeruginosa* RCII was grown in a nitrate- and peptone-containing medium [122]. Among inorganic nitrogen sources, ammonium salts were preferred to nitrates. In *Corynebacter* sp., nitrate as the sole nitrogen source caused biosurfactant production during the exponential growth phase, whereas addition of ammonium salts caused growth-associated biosurfactant production [121]. Surfactant production by growing cells has been reported to be affected by environmental factors. The pH value of the medium plays an important role in rhamnolipid production by *Pseudomonas* sp. [96] and sophorolipid production by *T. bombicola* [36]. However, no effect on the biosurfactant yield was detected between pH 6.5 and 8 in *P. fluorescence* [127]. The production of biosurfactant by *A. paraffineus* [115] and *Pseudomonas* sp. DSM-2874 [96] is found to be sensitive to the change in temperature. Interestingly, temperature has been found to alter the composition of biosurfactant produced.

Increase in agitation speed results in the reduction of biosurfactant yield due to the effect of shear rate on the growth kinetics of *N. erythropolis* [141]. Recently, Wang and Wang [142] performed extensive studies on the mechanism of biosurfactant accumulation in *A. calcoaceticus* RAG-1. They revealed that the ratio of cell-bound polymer to dry cell is strongly affected by shear force and as the shear stress increases the ratio decreases. In contrast, biosurfactant production in yeasts increased by increasing the agitation speed and aeration rate [42].

Oxygen transfer from gas to liquid is known to be affected by surfactants. Sheppard and Cooper [143] have recently studied the effect of surfactants on surfactin production in cyclone column reactor by *B. subtilis* and concluded that oxygen transfer is one of the key parameters for the process optimization and scale up of surfactin production.

2. Biosurfactant Production by Growing Cells under Growth-Limiting Conditions

The sharp increase in the biosurfactant level as a result of limitation of one or more medium components, as depicted in Fig. 1b, is the unique feature of this category of biosurfactant production. Overproduction of rhamnolipid by *P. aeruginosa* has been demonstrated only when the culture reaches the stationary phase of growth due to limitation of the nitrogen source [28]. Recently, Kosaric et al. [193] reported the production of glycolipid by *Nocardia* SFC-D under nitrogen limitation and observed the correlation between the substrate consumption and surfactant production. Robert et al. [132], while investigating the rhamnolipid production by *Pseudomonas* 44T1, observed that sodium nitrate and olive oil were the best nitrogen and carbon sources, respectively. The production started soon after the culture reached nitrogen limitation at 30 h and continued to increase up to 58 h of fermentation. Ramana and Karanth also noted similar observation when *P. aeruginosa* CFTR-6 was grown on glucose [125]. Singh et al. have demonstrated a significant increase in the bioemulsifier production in *C. tropicalis* IIP-4 by creating nitrogen-limiting conditions [144]. The production of liposan by *C. lipolytica* is also observed in the late stage of hexadecane fermentation [73]. The production of pentasaccharide lipid by *N. corynebacteriods* SM-1 is found to be favored by inorganic nitrogen sources, and sodium nitrate gave maximum specific surfactant production [127]. Moreover, it has been observed that during growth, the initial yield of glycolipid increases greatly after the exhaustion of the nitrogen source and after attaining the stationary growth phase. The production of water-soluble biosurfactant by *T. apicola* has been studied extensively by Hommel et al. [40]. The secretion of exolipid during the growth on *n*-hexadecane started in the middle of the exponential phase and rose sharply in the late exponential growth phase, when nitrogen is nearly exhausted. Furthermore, it was shown that the C:N ratio in the medium plays an important role in biosurfactant production, and a large amount of *n*-hexadecane is found to be incorporated into the surfactant at a higher C:N ratio. This effect is different from those proposed by several other investigators, where nitrogen exhaustion is believed to switch on biosurfactant production [96, 124]. According to Hommel et al. [40], the absolute quantity of nitrogen and not its relative concentration appears to be important in determining an optimum concentration of biomass, whereas the concentration of the hydrophobic carbon source determines the conversion of available carbon into biosurfactants. It has been observed that rhamnolipid production in *P. aeruginosa* by continuous culture using glucose was increased 7- to 10-fold after nitrogen limitations [30, 99]. Nitrogen limitation not only causes the overproduction of biosurfactant but also changes the composition of biosurfactant produced [95]. Growth of *R. erythropolis* under normal condition produces only nonionic trehalose corynomycolates, whereas under nitrogen limitations it produced only anionic trehalose tetraesters. The limitation of multivalent cations like Fe^{++}, Mg^{++}, or Ca^{++} has been found to increase the rhamnolipid production in *Pseudomonas* sp. [101, 105, 124]. A limitation of multivalent cations by the addition of EDTA into the nitrogen-limiting condition and temperature shift further improved biosurfactant production [17, 95]. In their recent observation, Goswamy and Singh [165] showed growth-associated production of glycoprotein and lipoprotein biosurfactants in *Pseudomonas* N-1 and the requirement of calcium ions in pseudosolubilization of hexadecane.

3. Biosurfactant Production by Resting Cells

In this category, the cells used are harvested from the surfactant-producing state and maintained in the same state. Thus, they do not multiply but continue to utilize carbon source for the synthesis of biosurfactants as illustrated in Fig. 1c. Wagner and his co-workers [17, 96, 101] have reported the production of rhamnolipid by resting free and immobilized cells of *Pseudomonas* sp. DSM-2874. Other examples of production of biosurfactants by resting cells include sophorolipid production by *T. bombicola* [36], cellobioselipid production by *Ustilago maydis* [95], and trehalose tetraester production by *R. erythropolis* [17, 95]. Using resting cells of *Arthrobacter* sp. DSM 2567 and various mono-, di-, or trisaccharides as the carbon source, the production of corresponding glycolipids were observed [29]. The highest yield of rhamnolipid from *n*-alkanes was reported by the resting free cells of *Pseudomonas* sp. DSM 2874 at pH 6.6 and 37°C [96]. In contrast to the rhamnolipids synthesized by growing cells, two new rhamnolipids, R3 and R4 were synthesized by the resting cells. The production of these biosurfactants were dependent on the carbon source in the medium and the incubation temperature. By incubating resting cells in phosphate buffer, repeated use of cells for rhamnolipid production is possible. However, biosurfactant production rate was much lower as compared to that with growing cells. Ramana and Karanth reported a twofold increase in rhamnolipid production when *P. aeruginosa* CFTR-6 was transferred after the growth phase into a medium devoid of phosphate [145]. Immobilizing the resting cells of *Pseudomonas* sp. DSM-2874 in hydrophilic polymer, better stability of the cells has been demonstrated [17, 101]. In this case, the product recovery was continuously achieved by XAD-2 adsorption. Using this technique rhamnolipid production in *Pseudomonas* sp. DSM-2874 has been increased by almost 5- to 6-fold. It is proposed that the effect may be due to relieving the product inhibition [101]. The enhancement in production of cell-bound biosurfactant by immobilized *A. calcoaceticus* RAG-1 has been reported [146]. The observed higher production may be due to the near zero shear stress and oxygen-limiting conditions [142]. On the other hand, in *T. bombicola*, no difference was noted between the production of sophorolipid between the growing cells and the resting cells [36]. Production of trehalose tetraester by the resting cells of *R. erythropolis* continued for 100 h and then declined [17, 36]. In this case, the conversion rate of substrate to product was found to be much higher than that observed with growing cells under nitrogen limitation. The production of biosurfactant by resting cells is important from the point of view of product recovery, as in such cases the growth phase and the product formation phases are separated.

4. Biosurfactant Production by Microbial Cells and Addition of Precursors

Significant increase in biosurfactant production by addition of its precursors to the growth medium has been reported by many investigators. Addition of lipophilic compounds to the culture medium of *T. magnoliae* [103] and *T. bombicola* [35, 141] resulted in the higher production of biosurfactants. A large number of observations have provided the conclusion that the carbon source in the medium, particularly the carbohydrate, has great bearing on the type of glycolipid formed. Glucose, fructose, and sucrose lipids are produced by *A. paraffineus* and several species of *Corynebacterium, Nocardia*, and *Brevibacterium* by using the corresponding sugar in the medium [30, 105]. Production of biosurfactants containing different mono-, di-, or trisaccharides are reported when the resting cells of *A. paraffineus* DSM 2567 are incubated in the presence of respective sugar in phosphate buffer at 30°C [29].

B. Production of Biosurfactants via the Biotransformation Route

Over the past few years, considerable attention has been given to the production of biosurfactants via biotransformation for several reasons: (1) there exists a close structural similarity between a group of commercial surfactants, sucrose esters, and glycolipid biosurfactants; (2) it is possible to produce commercial biosurfactants from biomass; (3) it is possible to derive different hydrophobic and hydrophilic moieties through microbial fermentations; and (4) by enzymatic treatment, it is possible to modify the hydrophobic moiety and attach the same to the hydrophilic portion of the biosurfactant.

In the production of commercial surfactants from either agricultural or municipal biomass, the attachment of hydrophobic moieties and the production of hydrophilic carbohydrate components of the surfactants are accomplished by various chemical processes, which are not covered here. Readers may refer to the recent review by Egan [148] for further details.

It is now well established that the lipophilic portion in biosurfactants is almost always the hydrocarbon tail of one or more fatty acids. The microbial production of extracellular lipids having biosurfactant activity has been very well documented [149, 150]. In batch culture, lipid accumulation is usually favored when an excess of principal carbon source over some other limiting nutrient, usually nitrogen, is incorporated in the growth medium (refer to Sec. III.B). When the limiting nutrient is exhausted from the medium, growth declines. However, carbon continues to be transported into the cell, where it may be utilized for lipid biosynthesis. There are examples in the literature to conclude that the fatty acids of the surfactants are synthesized *de novo* in a physiological response to the surfactant properties required for the efficient metabolism of principal substrate. The nature of principal substrate probably exerts the greatest effect on lipid composition. It is now generally accepted that microorganisms cultivated on hydrocarbon substrates produce more lipid than those grown on carbohydrates. Maximum yields of glycolipid production by *T. bombicola* has been observed using a mixture of glucose and vegetable oils [35, 147]. The alkane utilizing *Acinetobacter* sp. HO1-N synthesized a number of unique lipids when grown at the expense of long-chain alkanes, fatty alcohols, fatty acids, and symmetrical long-chain dialkylethers [4, 98]. The literature also supports the formation of different glycolipids as a result of change in the carbon substrate, especially sugars in the growth medium [21, 29, 105]. A variety of medium constituents, such as magnesium, iron, and phosphate, can assume the role of limiting nutrient and, when used at an appropriate concentration, may elicit a response similar to that of nitrogen toward the production of lipid and biosurfactants (see Sec. III.B).

Like carbon substrate, temperature is also one of the important determinants in altering the biosynthesis and distribution of fats and fatty acids. Temperature affects the biosynthesis of lipid beyond the degree of unsaturation. For example, the change in chain length of fatty acids, levels of fatty acid branching and cyclization, and the distribution and relative proportions of different glycolipids and phospholipids due to temperature shifts have been documented [96, 98]. Other factors that influence the nature of microbial lipid and in some cases the quantity of lipid synthesized include pH, oxygen concentration, and salinity. The factors affecting lipid synthesis in relation to the biosurfactant production have been described by Hommel and Ratledge in Chapter 1 and in the recent review [149].

More recently, many studies have indicated that microorganisms and selected enzymes may have a high potential for transformation and modifications of fats and oils. Enzyme systems offer many advantages, including reaction specificity, operation at normal temperature and pressure, and easy product recovery. In particular, consideration has been given to the use of various enzyme systems for specific hydrolysis, esterification, and transesterifications in the preparation of desirable fatty acids and triglycerides. It is noteworthy that the pharmaceutical industries have used the enzymatic activity of certain microbes to achieve the often difficult stereospecific substitutions involved in the chemical synthesis of various steroid drugs. The simplest of the transformation attempts with lipid has been feeding fats or fatty acids to selected yeasts to upgrade the oil quality by desaturating or saturating the component fatty acids [150–152]. The examples of the vast potential of microbes in production of specific fatty acids or tailor-made fats include, the mutants of *C. cloacae* and *C. tropicalis*, which are reported to convert *n*-alkanes to α-ω-dioic acids with a 70% conversion yield and up to 60 g/L content [153]. Production of high amounts of arachidonic and eicosapentaenoic acids by soil isolates *Mortierella elongata* IS-4 and *M. alpina* have been documented [154]. A strain of soil bacterium BMD-120 showed the ability to transform oleic acid to ricinoleic acid [155].

Although many enzymes are involved in the catabolic and anabolic reactions of lipids, lipolytic enzymes such as lipases and phospholipases have been considered most often for industrial application. Lipases, besides being either nonspecific or specific and being able to carry out transesterification reactions, may also act in synthetic mode and have much to offer to the oil and fat industry [156–158]. This enzyme has a wide range of properties depending on its source. Thus, one can find a suitable lipase or lipase-producing organism that fits to a given application. Investigations by Seino [159] and Nagai [160] in collaboration with Dai-ichi Ltd. reported preparation of sucrose, glucose, fructose, and sorbitol esters with oleic, linoleic, and stearic acid using lipase of *Candida cylindracea* at 40°C and at pH 5.4 with a conversion of up to 68%. These studies have opened up the possibility of a biocatalytic approach for the *in vitro* synthesis of biosurfactants from inexpensive sugar and vegetable oils. Another example is the conversion of soybean lecithin, one of the well-accepted emulsifiers, to a new biosurfactant by phospholipase-D from *Str. chromofuscus*. In this case, most of the phosphotidylcholine and phosphotidylethanolamine of lecithin are converted to phosphotidylglyerol, even in the presence of calcium ions [161].

V. PRODUCT RECOVERY

It is well known that most fermentative products are released in a dilute aqueous solution. In many cases, therefore, the downstream process comprises approximately 60% of the total cost of the product. Due to economic considerations, most applications of biosurfactants will have to be either microbial culture broth or crude preparations. Moreover, the interference due to the presence of other materials in the activity of crude biosurfactant preparations is found to be negligible. The choice of method for the recovery of a particular biosurfactant depends on its ionic charge, solubility in water, and whether the product is cell bound or extracellular. The methods used for biosurfactant recovery are listed in Table 2. They include solvent extraction [29, 30, 74, 162], adsorption followed by solvent extraction [35–37, 163], precipitation [38, 57, 59, 117], crystallization [99], centrifugation [42, 94], and foam fractionation [117]. Most biosurfactants are secreted into the medium, and they are isolated from either culture filtrate or supernatant obtained after removal of cells.

Solvent extraction is the most commonly used technique for the recovery of biosurfactants. Solvents used for this purpose include chloroform—methanol and dichloromethane—methanol mixtures, butanol, ethyl acetate, pentane, and hexane. Trehalose lipid produced by a large variety of *Mycobacterium* spp. [22–24] and *A. paraffineus* [29, 30, 105], trehalose corynomycolates and

TABLE 2 Common Methods Employed for the Recovery of Biosurfactants

Methods	References
A. Batch Recovery	
Solvent extraction	22, 27, 29, 30, 74, 87, 162
Crystallization	32, 42, 73, 99, 111, 164
Precipitation	
Ammonium sulfate	59, 67, 71, 90, 133
Acid	55, 58
Ethanol–acetic acid	75
Acetone	77, 85, 86, 88, 165
B. Continuous recovery	
Centrifugation	42, 94
Foam separation and precipitation	117, 118, 119
Diafiltration and precipitation	166
Adsorption	76, 101
Tangential flow filtration	168

tetraesters produced by *R. erythropolis* [12, 20, 27, 47], mono-, di-, and penta-saccharide lipids of *A. paraffineus* [29, 30], and *N. corynebacterioids* [127], cellobiose lipids produced by *Ustilago* spp. [42, 99], sophorolipids from several yeast species [27, 35, 36, 38, 41], liposan from *C. lipolytica* [74] and rhamnolipids of *Pseudomonas* spp. [32, 87, 109] are some of the classic examples of biosurfactant recovery by solvent extraction procedure.

For the extraction of glycolipids produced by *T. bombicola* [35–37], *T. petrophilum* [38], and *T. apicola* [39, 40] charcoal is added to the cell-free supernatant and mixed. Charcoal is removed by filtration and the lipid is desorbed from the dried charcoal by chilled ethyl acetate. Solvent is removed under reduced pressure and glycolipids are dissolved in hot ethanol and crystallized. Stuwer et al. [140] described an easy and cheaper purification process using liquid chromatography on silica gel for the nonionic glycolipid produced by *T. apicola*. A similar technique has been used for the recovery of biosurfactant from *P. fluorescens*, except that acetone was used to extract the biosurfactant [163]. Glycolipid produced by *U. zeae* and mannosylerythritol lipid produced by *Candida* spp. [42, 94] are settled down as heavy oils on centrifugation. These biosurfactants are then extracted in either ethanol or methanol to remove impurities.

It is possible to recover glycolipids as precipitates or crystals from the culture supernatant of *P. aeruginosa* and *U. zeae* [42, 99] by the addition of acid followed by incubation at low temperature. Recently, glycolipids produced by a mixed microbial population [111] and rhamnolipids formed by *P. aeruginosa* [34] and *C. lipolytica* [73, 74] have been isolated by the acidification of culture supernatant and solvent extraction by chloroform–methanol. Marcade et al. [164] used similar technique for the recovery of dihydroxydecenoic acid produced by *Pseudomonas* sp. 41-A2, except the solvent employed was ethyl acetate. Cameron et al. [75] recently described an extraction of a cell-bound bioemulsifier from *S. cerevisiae*, in which cells were suspended in an extraction buffer (pH 7) containing potassium metabisulfite and autoclaved for 3 h at 121°C. Bioemulsifier was precipitated using an ethanol–acetic acid mixture at 4°C and overnight incubation.

Recovery of many surface-active compounds by precipitation has been documented in the literature. Some of the classic uses of ammonium sulfate precipitation include the isolation of emulsan [59, 61, 133] and biodispersan [67] from *Acinetobacter* spp. by Rosenberg and his group, bioemulsifier from a Gram negative bacterium by Palejwala and Desai [71] and emulsifier from *B. subtilis* FE-2 by Patel and Gopinathan [90]. The recovery of surfactin produced by *B. subtilis* [55] and surfactinlike biosurfactant from *B. lichenoformis* [58] have been shown by acid precipitation. Acetone precipitation techniques has been used to recover emulsifying and solubilizing factors produced by *Pseudomonas* spp. [85, 86, 165] and *C. lipolytica* [77] and bioemulsifier from *C. tropicalis* and *Debaryomyces polymorphus* [88]. Recently, Bryant [166] devised an improved method for the isolation of glycolipid from *Rhodococcus* sp. strain H13A by using XM-50 diafiltration and isopropanol precipitation techniques. This procedure gives a purer glycolipid and removes protein impurities.

The methods discussed above for the recovery of biosurfactants are well studied and are classically used in the recovery of many biotechnologically derived products. However, they do not allow continuous removal of biosurfactants during the fermentation cycle. The technique of foam fractionation has gained greater significance as it offers an advantage of continuous *in situ* removal of biosurfactant from the fermentation broth. In the recovery of surfactin produced by *B. subtilis*, foam is collected and the pH of the collapsed foam is adjusted to 2 with concentrated HCl. In this process, proteins and lipids are precipitated and settle down. The supernatant is decanted off and from the residues, surfactin is extracted in dichloromethane [117, 119].

Recently, Neu and Poralla [118] recovered biosurfactant from *Bacillus* sp. in which foam produced was blown out of the fermenter, collected and centrifuged. Biosurfactant is recovered from this supernatant by cold acetone precipitation. Duvnjak and Kosaric [139, 167] observed that *C. lepus* produced large amounts of cell-bound surfactant when grown on glucose. The surfactant is completely released from the cells when they are treated for 2 h with hexadecane. Similarly, the release of biosurfactants from *A. paraffineus* [115] and *R. erythropolis* [27] on hydrocarbon treatment and from *C. tropicalis* on heat treatment [144] have been reported.

Mattei et al. [168, 169] developed a continuous culture device with a tangential flow filtration system by which continuous biosurfactant production up to 3 g/L by a mixed bacterial population has been attained. Continuous product recovery is also demonstrated by adsorbing rhamnolipid produced by *Pseudomonas* sp. [101] and lipopeptide produced by *C. petrophilum* [76]. Continuous removal of biosurfactant during the fermentation increases the cell density in the reactor and relieves the product inhibition. As a result, biosurfactant yield is increased severalfold as in the case of rhamnolipid production by *Pseudomonas* sp. DSM2874 [101] and surfactin production by *Bacillus* spp. [117–119]. In addition, this technique brings down the product recovery and wastewater treatment costs significantly.

VI. FUTURE OUTLOOK AND SUMMARY

Their amphiphilic nature gives surfactant molecules unique properties, and as a result there are many applications in a broad range of industries. It has been observed that there is no industry that does not use surfactants [1, 5]. The total quantity of surfactants produced during 1989 in the United States and over the globe is estimated to be 7.6 billion lb and 15.5 billion lb, respectively [1]. Almost all of these compounds were synthesized chemically.

In spite of there being so many surfactants already available in recent years, interest in surfactants from microbial origin has continued to increase. Microorganisms, because of their large surface-to-volume ratio and diverse synthetic capabilities, are promising candidates for widening the present range of surfactants. Many of the structures and properties of biosurfactants differ from synthetic surfactants, providing new possibilities for industrial applications. On the other hand, the structure of biosurfactants may offer a clue to the chemists in synthesizing the desired surfactant molecules.

Most common biosurfactants are glycolipids in which trehalose, sophorose, or rhamnose is attached to a lipid moiety. Complex biosurfactants such as cyclic lipopeptide (surfactin) produced by *Bacillus subtilis* [55], heteropolysaccharide-protein complex (Emulsan) produced by *A. calcoaceticus* [59–62], and anionic biosurfactant produced by *B. licheniformis* under anaerobic conditions [58, 170] have also been isolated and studied. Although the nature of the biosurfactant produced depends primarily on the producer organism, factors like carbon and nitrogen source, temperature, aeration, and multivalent cations have a profound effect on the production of type and quantity of biosurfactants.

Several industrial applications of biosurfactants have been envisioned. However, at present the greatest potential use is by the oil industry, with low product specifications, so that even whole cell broth could be used. Compared to chemical surfactants in an oil–water emulsification, the requirements of a biosurfactant is much less and they are more selective [7, 135]. Moreover, the different surfactant properties required for a broad range of oil and reservoir conditions can be easily fulfilled by screening suitable organisms and/or altering the growth condition of the producer organism. Hayes et al. [19] have demonstrated the emulsification of Boscan, Venezuelan heavy crude oil, using emulsan in which the viscosity of oil is reduced from 2,000,000 to 100 centipoise. Furthermore, they showed that by this treatment it is feasible to pump this oil to a distance comparable to 26,000 miles in a commercial pipeline. Under the above conditions, the conventional surfactants fail to retain the stability of emulsion. The results of the U.S. Department of Energy funded project to recover oil from underground oil shale showed a 30% increase in total oil yield by trehalose lipids from *N. rhodochrous* [2].

In the personal care market, biosurfactants are attractive because of their low toxicity, excellent moisturizing properties, and skin compatibility. *Torulopsis bombicola* KSM-36 produced sophorolipid at a stable rate of 100–150 g/L using a combined carbon source of palm oil and glucose [147]. Approximately 90-g/L sophorolipid production by *T. apicola* has been reported using glucose and sunflower oil as the substrates [140]. The addition-polymerized product of 1 mole of sophorolipid and 12 moles of propylene glycol have specific compatibility to the skin and found commercial utility as skin moisturizers [161]. Antibiotic effects [163, 171] of biosurfactants and the inhibitory effect toward the growth of AIDS virus in WBC have been reported recently [2]. The pulmonary surfactant, essential for normal respiration, is a phospholipid protein complex; many premature infants suffer respiration failure because of the deficiency of this surfactant. The human gene for production of the protein molecule of this surfactant has been isolated and cloned in bacteria; this has opened up the possibility of large-scale production of this surfactant for medical applications [173].

Biosurfactants have also been used commercially as an additive to dewater fuel grade peat [174, 175]. The clean up of a natural site contaminated with petroleum has been accomplished using biosurfactants or biosurfactant-producing microbes [57, 176, 177]. Applications are also being considered in industries involving pulp and paper [15, 178], coal, textile, and uranium ore processing [179] and ceramic processing [181].

Probably, the most important advantage of biosurfactants over chemical surfactants is their ecological acceptability. Many chemically synthesized surfactants cause ecological problems owing to their resistance to biodegradation, toxicity, and accumulation in natural ecosystems. On the other hand, biosurfactants are biodegradable [11, 66].

The future of biosurfactants will be governed by the net economic gain between its production cost and applicational benefits. In biosurfactant production, the fermentation step is the major operation in determining its cost of production. Generally, biosurfactants are produced by growth on hydrocarbon substrates. However, hydrocarbon substrates are a poor choice from the point of view of feedstock and process costs. Biosurfactant production from water-soluble substrates like glucose, ethanol, and whey as well as industrial effluent have been reported recently [84, 131, 132, 139]. Recent advances in genetic engineering have offered the possibility of altering the substrate requirement and increasing the productivity of the organisms. One such example is the construction of *P. aeruginosa* strain by inserting *lac* plasmid from *E. coli*. Thus, it is possible to produce rhamnolipid from whey, a waste product from dairy industry [131]. The other example is the isolation of a high biosurfactant-yielding *B. subtilis* and *P. aeruginosa* [172]. Although, the genetic analysis of biosurfactant production is currently in its infancy, significant information on the genetics of surfactin production in *B. subtilis* has recently been reported [184, 189–191]. Other significant achievements in the field during the last few years include biosurfactant production by immobilized cells [101, 154, 155] and by continuous culture systems [168] and cheaper, easier, and continuous biosurfactant recovery [101, 166, 168, 169]. The synthesis of tailor-made surfactants by enzymatic catalysis offers a new dimension in biosurfactant production. Of immediate importance is the synthesis of sugar esters by lipase. This class of biosurfactants is receiving renewed interest because of its mildness to skin and eyes. The structure of sucrose monoester is shown in Fig. 2. It can be seen that seven additional hydroxyl groups are available for esterification. Thus, a variety of sugar esters are possible to be synthesized and the enzymatic route of biosurfactant production could offer a new class of economically attractive compounds.

In spite of the above developments, the key factors governing the success of a biosurfactant will be the development of a cheaper process, the use of low-cost raw materials, a high product yield, and superactive, highly specific, and selective biosurfactants for specific applications.

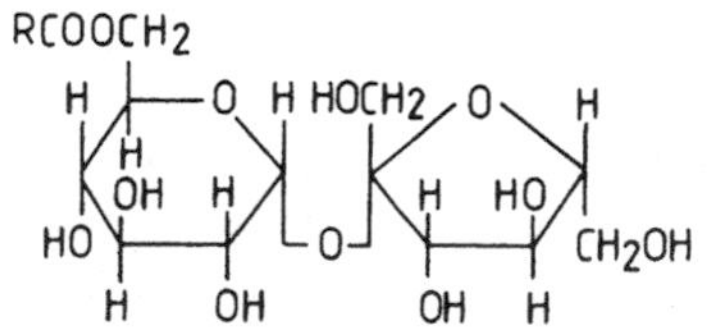

FIG. 2 Structure of sucrose monoester. $RCOOCH_2$, ester linkage with fatty acid.

ACKNOWLEDGMENT

One of us (J. D. D.) wishes to acknowledge the permission granted by the Indian Petrochemicals Corporation Limited for writing this chapter. Without the kind support, encouragement, and constructive criticism from Dr. I. S. Bhardwaj, Director of Research and Development at IPCL, this work would not have been possible.

REFERENCES

1. B. F. Greek, *Chem. Eng. News 68*: 37 (1990).
2. G. Mintz, *Cat. Rev. News Lett. 2*: 8 (1990).
3. W. R. Finnerty and M. E. Singer, *Bio/Technology 1*: 47 (1984).
4. W. R. Finnerty and M. E. Singer, *Dev. Ind. Microbiol. 25*: 31 (1985).
5. N. Kosaric, N. C. C. Gray, and W. L. Cairns, in *Biosurfactants and Biotechnology*, Surfactant Science Series, Vol. 25 (N. Kosaric, W. L. Cairns, and N. C. C. Gray, eds.) Marcel Dekker, New York, 1987, pp. 1–19.
6. J. D. Desai, *J. Sci. Ind. Res. 46*: 440 (1987).
7. E. Rosenberg, *CRC Crit. Rev. Biotechnol. 3*: 109 (1986).
8. M. E. Singer, *Microbes Oil Recovery 1*: 19 (1985).
9. J. L. Shennan and J. D. Levi, in *Biosurfactant and Biotechnology*, Surfactant Sciences Series, Vol. 25, (N. Kosaric, W. L. Cairns, and N. C. C. Gray, eds.) Marcel Dekker, New York, 1987, pp. 163–181.
10. W. L. Cairns, D. G. Cooper, J. E. Zajic, J. M. Wood, and N. Kosaric, *Appl. Environ. Microbiol. 43*: 362 (1982).
11. J. E. Zajic, H. Gignard, and D. F. Gerson, *Biotechnol. Bioeng. 19*: 1303 (1977).
12. A. Kretschmer, H. Bock, and F. Wagner, *Appl. Environ. Microbiol. 44*: 864 (1982).
13. J. E. Zajic and C. J. Panchal, *CRC Crit. Rev. Microbiol. 5*: 39 (1974).
14. J. E. Zajic and W. Seffens, *CRC Crit. Rev. Biotechnol. 1*: 187 (1984).
15. S. Lang and F. Wagner, in *Biosurfactants and Biotechnology*, Surfactant Science Series, Vol. 25, (N. Kosaric, W. L. Cairns, and N. C. C. Gray, eds.), Marcel Dekker, New York, 1987, pp. 21–45.
16. J. E. Zajic and A. Y. Mahomedy, in *Petroleum Microbiology* (R. M. Atlas, ed.) Macmillan, New York, 1984, p. 221.
17. C. Syldatk and F. Wagner, in *Biosurfactants and Biotechnology*, Surfactant Science Series, Vol. 25 (N. Kosaric, W. L. Cairns, and N. C. C. Gray, eds.) Marcel Dekker, New York, 1987, pp. 89–120.
18. D. C. Cooper, *Microbiol. Sci. 3*: 145 (1986).
19. M. E. Hayes, E. Nestaas, and K. R. Hrebenar, *Chemtech. 4*: 239 (1986).
20. P. Rapp, H. Bock, V. Wray, and F. Wagner, *J. Gen. Microbiol. 115*: 491 (1979).
21. T. Suzuki, K. Tanaka, I. Matsuara, and S. Kinosita, *Agr. Biol. Chem. 33*: 1619 (1969).
22. J. C. Prome, C. Lacave, A. Ashibo-Coffy, and A. Savagnac, *Eur. J. Biochem. 63*: 543 (1976).
23. E. Vilkas, A. Adam, and M. Senn, *Chem. Phys. Lipids 2*: 55 (1968).
24. H. Tabaud, H. Tisnovska, and E. Vilkas, *Biochemie 53*: 55 (1971).

25. C. Asselineau, H. Montrozier, J. C. Prome, A. Savagnac, and M. Wiby, *Eur. J. Biochem. 28*: 102 (1972).
26. S. G. Batrakov, B. V. Rozynov, T. V. Koronelli, and L. D. Bergelson, *Chem. Phys. Lipids. 29*: 241 (1981).
27. E. Ristau and F. Wagner, *Biotechnol. Lett. 5*: 95 (1983).
28. P. J. Brennan, D. P. Lehane, and D. W. Thomas, *Eur. J. Biochem. 13*: 117 (1970).
29. Z. Y. Li, S. Lang, F. Wagner, L. Witte, and V. Wray, *Appl. Environ. Microbiol. 48*: 610 (1984).
30. T. Suzuki, H. Tanaka, and S. Itoh, *Agr. Biol. Chem. 38*: 557 (1974).
31. J. R. Edward and J. A. Hayashi, *Arch. Biochem. Biophys. 111*: 415 (1965).
32. K. Hisatsuka, T. Nakahara, N. Sano, and K. Yamada, *Agr. Biol. Chem. 35*: 686 (1981).
33. S. Itoh and T. Suzuki, *Agr. Biol. Chem. 36*: 2233 (1972).
34. J. L. Parra, J. Guinea, M. A. Manresa, M. Robert, M. E. Mercade, F. Comelles, and M. P. Bosch, *J. Am. Oil. Chem. Soc. 66*: 141 (1989).
35. D. G. Cooper and D. A. Paddock, *Appl. Environ. Microbiol. 47*: 173 (1984).
36. U. Gobbert, S. Lang, and F. Wagner, *Biotech. Lett. 6*: 225 (1984).
37. S. Inoue and S. Itoh, *Biotech. Lett. 4*: 3 (1982).
38. D. G. Cooper and D. A. Paddock, *Appl. Environ. Microbiol. 46*: 1426 (1983).
39. P. Tulloch, A. Hill, and J. F. T. Spencer, *Chem. Commun.* 584 (1967).
40. R. Hommel, O. Stuwer, W. Stuber, D. Haferburg, and H. P. Kleber, *Appl. Microbiol. Biotechnol. 26*: 199 (1987).
41. A. J. Cutler and R. J. Light, *J. Biol. Chem. 254*: 1944 (1979).
42. J. F. T. Spencer, D. M. Spencer, and A. P. Tulloch, in *Economic Microbiology*, Vol. 3 (A. H. Rose, ed.) Academic Press, New York, 1979, pp. 523–540.
43. M. Wayman, A. D. Jenkins, and A. G. Kormady, in *Biotechnology for the Oil and Fat Industry* (C. Ratledge, P. S. S. Dawson and J. B. M. Rattaray, eds.), American Oil Chemistry Society, Champaign, IL, 1984, p. 129.
44. C. Andre, M. J. Espuny, M. Robert, M. E. Mercade, A. Manresa, and J. Guinea, *Antonie van Leeuwenhoek J. Microbiol.*, in press.
45. J. L. Beeba and W. W. Umbreit, *J. Bacteriol. 108*: 612 (1971).
46. M. Miyazima, M. Iida, and H. Iizuka, *J. Ferment. Technol. 63*: 219 (1985).
47. C. R. MacDonald, D. G. Cooper, and J. E. Zajic, *Appl. Environ. Microbiol. 41*: 117 (1981).
48. M. Marahiel, W. Denders, M. Krause, and H. Kleinkauf, *Eur. J. Biochem. 99*: 49 (1979).
49. T. Suzuki, K. Hayashi, K. Fujikawa, and K. Tsukamoto, *J. Biochem. (Tokyo) 57*: 226 (1965).
50. S. G. Wilkinson, *Biochim. Biophys. Acta. 270*: 1 (1972).
51. H. W. Knoche and J. M. Shively, *J. Biol. Chem. 247*: 170 (1972).
52. Y. Tahara, M. Kameda, Y. Yamada, and K. Kondo, *Agr. Biol. Chem. 40*: 243 (1976).
53. Y. Tahara, Y. Yamada, and K. Kondo, *Agr. Biol. Chem. 40*: 1449 (1976).
54. J. Kawanami, A. Kimura, and H. Otsuka, *Biochim. Biophys. Acta. 152*: 808 (1968).
55. K. Arima, A. Kakinuma, and G. Tamura, *Biochem. Biophys. Res. Commun. 31*: 488 (1968).
56. C. N. Mulligan, D. G. Cooper, and R. J. Neufeld, *J. Ferment. Technol. 62*: 311 (1984).
57. W. Berheimer and L. S. Avigad, *J. Gen. Microbiol. 61*: 361 (1970).

58. M. Javaheri, G. E. Jenneman, M. J. McInnerney, and R. M. Knapp, *Appl. Environ. Microbiol. 50*: 698 (1985).
59. E. Rosenberg, A. Zuckerberg, C. Rubinovitz, and D. L. Gutnick, *Appl. Environ. Microbiol. 37*: 402 (1979).
60. A. Zukerberg, A. Diver, Z. Peeri, D. L. Gutnick, and E. Rosenberg, *Appl. Environ. Microbiol. 37*: 414 (1979).
61. Y. Shabtai and D. L. Gutnick, *Appl. Environ. Microbiol. 49*: 192 (1985).
62. I. Belsky, D. L. Gutnick, and E. Rosenberg, *FEBS Lett. 101*: 175 (1979).
63. D. L. Gutnick and Y. Shabtai, in *Biosurfactants and Biotechnology*, Surfactant Science Series, Vol. 25 (N. Kosaric, W. L. Cairns, and N. C. C. Gray, eds.). Marcel Dekker, New York, 1987, pp. 211–246.
64. Z. Zosim, D. L. Gutnick, and E. Rosenberg, *Biotech. Bioeng. 24*: 281 (1982).
65. Z. Zosim, E. Rosenberg, and D. L. Gutnick, *Coll. Polymer Scci. 264*: 218 (1986).
66. Y. Shoham, M. Rosenberg, and E. Rosenberg, *Appl. Environ. Microbiol. 46*: 573 (1983).
67. E. Rosenberg, C. Rubinovitz, A. Gottlieb, S. Rosenhak, and E. Z. Ron, *Appl. Environ. Microbiol. 54*: 317 (1988).
68. E. Rosenberg, C. Rubinovitz, R. Legmann, and E. Z. Rost, *Appl. Environ. Microbiol. 54*: 323 (1988).
69. N. Kaplan and E. Rosenberg, *Appl. Environ. Microbiol. 44*: 1335 (1982).
70. N. Sar and E. Rosenberg, *Curr. Microbiol. 9*: 309 (1983).
71. S. Palejwala and J. D. Desai, *Biotechnol. Lett. 11*: 115 (1989).
72. O. Kappeli and A. Fiechter, *J. Bacteriol. 131*: 917 (1977).
73. M. C. Cirigliano and G. M. Carman, *Appl. Environ. Microbiol. 48*: 747 (1984).
74. M. C. Cirigliano and G. M. Carman, *Appl. Environ. Microbiol. 50*: 846 (1985).
75. D. R. Cameron, D. G. Cooper, and R. J. Neufeld, *Appl. Environ. Microbiol. 54*: 1420 (1985).
76. J. Iguchi, T. Takeda, and H. Okasava, *Agr. Biol. Chem. 33*: 1657 (1969).
77. P. K. Roy, H. D. Singh, S. D. Bhagat, and J. N. Baruah, *Biotechnol. Bioeng. 21*: 955 (1979).
78. O. Kappeli, P. Walther, M. Mueller, and A. Fiechter, *Arch. Microbiol. 138*: 279 (1984).
79. O. Kappeli, M. Muller, and A. Fiechter, *J. Bacteriol. 133*: 952 (1984).
80. G. Demi, T. Anke, F. Oberwinker, B. M. Giannetti, and W. Steglich, *Phytochem. 19*: 83 (1980).
81. A. L. Fluharty and J. S. O'Brien, *Biochemistry 8*: 2627 (1969).
82. K. Hisatsuka, T. Nakahara, and K. Tamada, *Agr. Biol. Chem. 36*: 1361 (1972).
83. K. Hisatsuka, T. Nakahara, Y. Minoda, and K. Yamada, *Agr. Biol. Chem. 41*: 445 (1977).
84. T. V. Koronelli, T. I. Komarova, and Y. V. Denisov, *Mikrobiologya 52*: 767 (1983).
85. S. S. Chameotra and H. D. Singh, *J. Ferment. Bioeng. 69*: 341 (1990).
86. P. G. Raddy, H. D. Singh, M. G. Pathak, S. D. Bhagat, and J. N. Baruah, *Biotech. Bioeng. 25*: 387 (1983).
87. A. J. Desai, K. M. Patel, and J. D. Desai, *Curr. Sci. 57*: 500 (1988).
88. M. Singh and J. D. Desai, *Ind. J. Exp. Biol. 27*: 224 (1989).
89. A. Fattom and M. Shilo, *FEMS Microbiol. Ecol. 1*: 1 (1985).
90. M. N. Patel and K. P. Gopinathan, *Appl. Environ. Microbiol. 52*: 1224 (1986).

91. O. Kappeli and W. R. Finnerty, *J. Bacteriol. 140*: 707 (1979).
92. H. Miorner, G. Johansson, and G. Kronvall, *Infect. Immunol. 39*: 336 (1983).
93. G. Hauser and M. L. Karnovsky, *J. Biol. Chem. 229*: 91 (1957).
94. G. Hauser and M. L. Karnovsky, *J. Biol. Chem. 233*: 287 (1958).
95. C. Syldatk, S. Lang, F. Wagner, V. Wray, and L. Witte, *Z. Naturforsch. 40c*: 1 (1985).
96. C. Syldatk, S. Lang, U. Matulovic, and F. Wagner, *Z. Naturforsch. 40c*: 61 (1985).
97. P. Edmonds and J. J. Coonay, *J. Bacteriol. 98*: 16 (1969).
98. S. L. Neidleman and J. Geigert, *J. Am. Oil. Chem. Soc. 61*: 290 (1984).
99. B. Boothroyd, J. A. Thorn, and R. H. Haskins, *Can. J. Biochem. Physiol. 33*: 289 (1955).
100. P. Rapp, H. Bock, E. Urban, and F. Wagner, in *Deschema Monographien*, Vol. 81 (H. J. Rehm, ed.), Verlag Chemie, Weinheim (1977).
101. C. Syldatk, H. Matulovic, and F. Wagner, *Biotech Forum 1*: 53 (1984).
102. H. Kaweshima, T. Nakahara, M. Oogaki, and T. Tabuchi, *J. Ferment. Technol. 61*: 143 (1983).
103. A. P. Tulloch, J. F. T. Spencer, and A. J. Gorin, *Can. J. Chem. 40*: 1326 (1962).
104. J. M. Shively and A. A. Benson, *J. Bacteriol. 94*: 1679 (1967).
105. S. Itoh and T. Suzuki, *Agr. Biol. Chem. 38*: 1443 (1974).
106. O. Kappeli and W. R. Finnerty, *Biotechnol. Bioeng. 22*: 495 (1980).
107. A. M. Chakrabarty, *Trends Biotechnol. 3*: 32 (1985).
108. R. J. Neufeld, J. E. Zajic, and D. F. Gerson, *J. Ferment. Technol. 61*: 315 (1983).
109. M. Yamaguchi, A. Sato, and A. Yukuyama, *Chem. Ind. 17*: 741 (1976).
110. S. Fukui and A. Tanaka, *Adv. Biochem. Eng. 19*: 217 (1981).
111. A. Oberbremer and R. M. Hurtig, *Appl. Microbiol. Biotechnol. 31*: 582 (1989).
112. S. Ito and S. Inoue, *Appl. Environ. Microbiol. 43*: 1278 (1982).
113. M. Singh and J. D. Desai, *J. Sci. Ind. Res. 45*: 413 (1986).
114. A. Dalhoff and H. J. Rehm, *Eur. J. Appl. Microbiol. 3*: 203 (1979).
115. Z. Duvnjak, D. G. Cooper, and N. Kosaric, *Biotechnol. Bioeng. 24*: 165 (1982).
116. G. Hauser and M. L. Karnovsky, *J. Bacteriol. 68*: 645 (1954).
117. D. G. Cooper, C. R. MacDonald, S. J. B. Duff, and N. Kosaric, *Appl. Environ. Microbiol. 42*: 408 (1981).
118. T. R. Neu and K. Poralla, *Appl. Microbiol. Biotechnol. 32*: 521 (1990).
119. C. N. Mulligan, T. Y. K. Chow, and B. F. Gibbs, *Appl. Microbiol. Biotechnol. 31*: 486 (1989).
120. Z. Duvnjak, D. G. Cooper, and N. Kosaric, in *Microbial Enhanced Oil Recovery* (J. E. Zajic, D. G. Cooper, T. R. Jack, and N. Kosaric, eds.), Pennwell Books, Tulsa, OK (1983).
121. D. F. Gerson and J. E. Zajic, *Antonie van Leeuwenhoeck 45*: 81 (1979).
122. C. N. Malligan and B. F. Gibbs, *Appl. Environ. Microbiol. 55*: 3016 (1989).
123. L. Guerra-Santos, O. Kappeli, and A. Fiechter, in *Abstract Papers of the 3rd European Congress of Biotechnology*, Vol. 1, Verlag Chemie, Weinheim, 1984, p. 507.
124. L. Guerra-Santos, O. Kappeli, and A. Fiechter, *Appl. Environ. Microbiol. 48*: 301 (1984).
125. K. V. Ramana and N. G. Karanth, *J. Chem. Technol. Biotechnol. 45*: 249 (1989).

126. C. Rubinovitz, D. L. Gutnick, and E. Rosenberg, *J. Bacteriol. 152*: 126 (1982).
127. M. Powalla, S. Lang, and V. Wray, *Appl. Microbiol. Biotechnol. 31*: 473 (1989).
128. A. Persson, G. Molin, N. Andersson, and J. Sjoholm, *Biotechnol. Bioeng. 36*: 252 (1990).
129. L. Guerra-Santos, O. Kappeli, and A. Fiechter, *Appl. Microbiol. Biotechnol. 24*: 443 (1986).
130. A. Persson, G. Molin, and C. Weibull, *Appl. Environ. Microbiol. 56*: 686 (1990).
131. A. K. Koch, J. Reiser, O. Kappeli, and A. Fiechter, *Biotechnology 6*: 1335 (1988).
132. M. Robert, M. E. Mercade, M. P. Bosch, J. L. Parra, M. J. Espuny, M. A. Manresa, and J. Guinea, *Biotech. Lett. 11*: 871 (1989).
133. S. Goldman, Y. Shabtai, C. Rubinovitz, E. Rosenberg, and D. L. Gutnick, *Appl. Environ. Microbiol. 44*: 165 (1982).
134. D. L. Gutnick, E. A. Bayer, C. Rubinovitz, O. Pines, Y. Shabtai, S. Goldman, and E. Rosenberg, *Adv. Biotechnol. 11*: 455 (1980).
135. O. Pines, E. A. Bayer, and D. L. Gutnick, *J. Bacteriol. 154*: 893 (1982).
136. A. Persson and W. S. Hu, *Appl. Microbiol. Biotechnol. 29*: 1 (1988).
137. T. K. Ng and W. S. Hu, *Appl. Microbiol. Biotechnol. 31*: 480 (1989).
138. D. G. Cooper and B. G. Goldenberg, *Appl. Environ. Microbiol. 53*: 224 (1987).
139. Z. Duvnjak and N. Kosaric, *Biotechnol. Lett. 7*: 793 (1985).
140. O. Stuwer, R. Hommel, D. Haferburg, and H. P. Kleber, *J. Bacteriol. 6*: 259 (1987).
141. A. Margaritis, K. Kennedy, J. E. Jazic, and D. F. Gerson, *Dev. Ind. Microbiol. 20*: 623 (1979).
142. S. D. Wang and D. I. C. Wang, *Biotechnol. Bioeng. 36*: 402 (1990).
143. J. D. Sheppard and D. G. Cooper, *J. Chem. Tech. Biotechnol. 48*: 325 (1990).
144. M. Singh, V. Saini, D. K. Adhikari, J. D. Desai, and V. R. Sista, *Biotechnol. Lett. 12*: 743 (1990).
145. K. V. Ramana and N. G. Karanth, *Biotechnol. Lett. 11*: 437 (1989).
146. S. D. Wang and D. I. C. Wang, *Biotechnol. Bioeng. 34*: 1261 (1989).
147. S. Itoh, *Fat. Sci. Technol. 89*: 470 (1987).
148. P. A. Egan, *Chem. Technol. 12*: 758 (1989).
149. C. Ratledge and S. G. Wilkinson, in *Microbial Lipids*, Vol. 1 Academic Press, New York, 1988.
150. D. Haferburg, R. Hommel, R. Claus, and H. P. Kleber, *Adv. Biochem. Eng. Biotech. 33*: 53 (1987).
151. D. Montel, R. Ratomahenina, P. Glazy, M. Pina, and J. Graille, *Biotechnol. Lett. 7*: 733 (1985).
152. C. Ratledge, *J. Am. Oil Chem. Soc. 64*: 1647 (1987).
153. F. F. Hill, I. Venn, and K. L. Lukas, *Appl. Microbiol. Biotechnol. 24*: 168 (1986).
154. S. Shimizu, K. Akimoto, H. Kawashima, Y. Shinmen, and H. Yamada, *J. Am. Oil. Chem. Soc. 66*: 237 (1989).
155. H. Yamada, *J. Am. Oil Chem. Soc. 64*: 1612 (1987).
156. G. Lazar, A. Weiss, and R. D. Schmid, in *Proceedings of the World Conf. in Emerging the Technology of the Fats and Oil Industry* (A. R. Baldwin, ed.), American Oil Chemistry Society, Champaign, IL, 1986, p. 346.
157. J. Kurashige, N. Matsuzaki, and K. Makabe, *J. Disp. Sci. Technol. 10*: 531 (1989).

158. A. R. Macrac and R. C. Hammond, *Biotechnol. Gen. Eng. Rev. 3*: 193 (1986).
159. H. Seino, T. Uchibori, T. Nishitani, and S. Inamasu, *J. Am. Oil Chem. Soc. 61*: 1761 (1984).
160. H. Nakata, K. Jinno, Y. Chikazawa, I. Morita, N. Nishio, M. Hayashi, and S. Nagai, *Proc. Soc. Ferment. Technol. Japan 28*: 78 (1986).
161. T. Yamane, *J. Am. Oil. Chem. Soc. 64*: 1657 (1987).
162. D. G. Cooper and J. E. Zajic, *Adv. Appl. Microbiol. 26*: 229 (1980).
163. T. R. Neu, T. Hartner, and K. Poralla, *Appl. Microbiol. Biotechnol. 32*: 518 (1990).
164. M. E. Mercade, M. Robert, M. J. Espuny, M. P. Bosch, M. A. Manresa, J. L. Parra, and J. Guinea, *J. Am. Oil. Chem. Soc. 65*: 1915 (1988).
165. P. Goswami and H. D. Singh, *Biotechnol. Bioeng. 37*: 1 (1991).
166. F. O. Bryant, *Appl. Environ. Microbiol. 56*: 1494 (1990).
167. Z. Duvnjak and N. Kosaric, *Biotechnol. Lett. 3*: 583 (1981).
168. G. Mattei and J. C. Bertrand, *Biotechnol. Lett. 7*: 217 (1985).
169. G. Mattei, E. Rambeloarisoa, G. Giusti, J. F. Rontani, and J. C. Bertrand, *Appl. Microbiol. Biotechnol. 23*: 302 (1986).
170. J. A. Ramsay, D. G. Cooper, and R. J. Neufeld, *Geomicrobiol. J. 7*: 155 (1989).
171. S. Lang, E. Katsiwela, and F. Wagner, *Fat. Sci. Technol. 91*: 363 (1989).
172. A. K. Koch, O. Kappeli, A. Fiechter, and K. Reiser, *J. Bacteriol. 173*: 4212 (1991).
173. R. T. While, D. Damm, J. Miller, K. Spratt, J. Schilling, S. Howgood, B. Benson, and B. Cordell, *Nature 317*: 361 (1985).
174. D. G. Cooper, D. Pillon, C. N. Mulligan, and J. D. Sheppard, *Fuel 65*: 255 (1986).
175. D. G. Cooper, E. R. A. Eccles, and J. D. Sheppard, *Can. J. Chem. Eng. 66*: 393 (1988).
176. S. Harvey, I. Elashvili, J. J. Valdes, D. Kamely, and A. M. Chakrabarty, *Biotechnology 8*: 228 (1990).
177. R. Bartha, *Microbiol. Ecol. 12*: 155 (1986).
178. E. Rosenberg, Z. Schwartz, and A. Tenenbaum, *J. Disp. Sci. Technol. 10*: 241 (1989).
179. M. J. McInerney, M. Javaheri, and D. P. Nagle, *J. Ind. Microbiol. 5*: 95 (1990).
180. S. Horowitz, J. N. Gilbert, and W. M. Griffin, *J. Ind. Microbiol. 6*: 343 (1990).
181. S. Horowitz and J. K. Currie, *J. Disp. Sci. Technol. 11*: 637 (1990).
182. S. Horowitz and W. M. Griffin, *J. Ind. Microbiol. 7*: 45 (1991).
183. M. Greiner and G. Winkelman, *Appl. Microbiol. Biotechnol. 34*: 565 (1991).
184. M. M. Nakano and P. Zuber, *CRC Crit. Rev. Biotechnol. 10*: 223 (1990).
185. M. M. Nakano, N. Corbell, and P. Zuber, *Mol. Gen. Genet.* (in press) (1991).
186. C. Ullrich, B. Kluge, Z. Palacz, and J. Vater, *Biochem. 30*: 6503 (1991).
187. T. Matsuyama, M. Sogawa, and I. Yano, *Appl. Environ. Microbiol. 53*: 1186 (1991).
188. D. K. Jain, D. L. C. Thompson, H. Lee, and J. T. Trevors, *J. Microbiol. Methods 13*: 271 (1991).
189. M. M. Nakano and P. Zuber, *J. Bacteriol. 170*: 5662 (1988).
190. M. M. Nakano, R. Magnuson, A. Myers, J. Curry, A. D. Grossman, and P. Zuber, *J. Bacteriol. 173*: 1770 (1991).
191. A. Grossman, *Cell 65*: 5 (1991).
192. M. M. Nakano and P. Zuber, *J. Bacteriol. 173*: 5487 (1991).
193. N. Kosaric, H. Y. Choi, and R. Bhaszczyk, *Tenside. Surf. Det. 27*: 294 (1990).

3

Prospects and Limits for the Production of Biosurfactants Using Immobilized Biocatalysts

MARTIN SIEMANN and FRITZ WAGNER Institute of Biochemistry and Biotechnology, Technical University of Braunschweig, Braunschweig, Germany

I. INTRODUCTION

Industrial processes with microorganisms are generally based on the exploitation of free, growing, or resting cells. However, in the last years, many publications seemed to indicate that immobilized cell systems are gaining importance in the development of biotechnological processes [1].

Immobilization of cells involves the fixation on or in an inert support by one of three principal methods [2–6].

Binding of cells to a solid support by surface adsorption or covalent binding.
Physical retainment of cells by using membranes or fiber systems.
Physical entrapment of cells into a porous matrix.

Common advantages of immobilized cells over free cells could be termed as following [1]:

Products can be recovered more easily as they are not contamined by biomass.
If necessary high cell densities can be employed.
Their use generates less biomass and, hence, processes involving immobilized cells are more efficient with respect to substrate and biomass utilization.
They can be used in continuous processes in which the cells, also at high dilution rates, are retained in the bioreactor.

Along with all the above advantages one should not forget disadvantages, such as the cost of immobilization or mass-transfer limitations created by the immobilization support and the use of high cell densities. These problems have to be solved before commercializing this system.

For the past 15 years the Department of Biotechnology at the Technical University of Braunschweig has been involved in the production of biosurfactants [7–9], including the production of anionic biosurfactants, such as the rhamnolipids [10–13]. Naturally, at the beginning of these investigations, the production of rhamnolipids as it was done by other institutions was analyzed and later optimized in free cell systems [14]. In 1983, we started research on the production of rhamnolipids with immobilized cell systems as an alternative method to systems using free cells. In the previous book in this series *Biosurfactants and Biotechnology*, Syldatk gave a detailed overview of the production of biosurfactants [9]. In this chapter, the prospects, advantages and limits of using immobilized biocatalysts for the production of rhamnolipids is summarized. As mentioned above, little has been investigated concerning immobilized biocatalysts for the production of biosurfactants. Sidenko et al. [15] investigated the use of immobilized cultures of "destructor microbes" for tertiary treatment of oil-bearing water. Therefore, the main part of this chapter is a summary of our own investigations, which were carried out by Syldatk [16], Matulovic [17, 18], Fischer [19], and Siemann [20] during the last several years.

II. PRODUCTION OF ANIONIC RHAMNOLIPIDS WITH FREE CELLS OF *PSEUDOMONAS* sp. DSM 2874

The anionic rhamnolipids (RLI–RLIV) (Fig. 1) are extracellular metabolites and could be produced under growth-limiting conditions [7, 12]. Experiments on medium optimization for the growth of *Pseudomonas* sp. indicated that the over-production of rhamnolipids is caused either by limitation of nitrogen [14] or by multivalent cations [12]. Both the growth and the product kinetics indicated that rhamnolipids are not formed in substantial amount as long as ammonium ions are present in the medium (Fig. 2). In contrast, at the moment of ammonia limitation the *n*-alkane consumption and the rhamnolipid formation becomes linear up to 140 h. A further increase in rhamnolipid production was obtained by limitation of Fe-, Mg-, or Ca-cations, which are all essential for growth of *Pseudomonas* sp. [10, 12, 14].

Based on these results one can develop a strategy for the production of biosurfactants when separating the growth phase of the bacterium from the production phase. For this purpose, the microbes were grown under optimal growth conditions, separated from the culture broth, and resuspended in a reaction medium for optimal production conditions. This procedure has the advantage that possible by-products in the culture broth are eliminated; also single influence parameters of biosurfactant synthesis can be studied under chemically well-defined conditions.

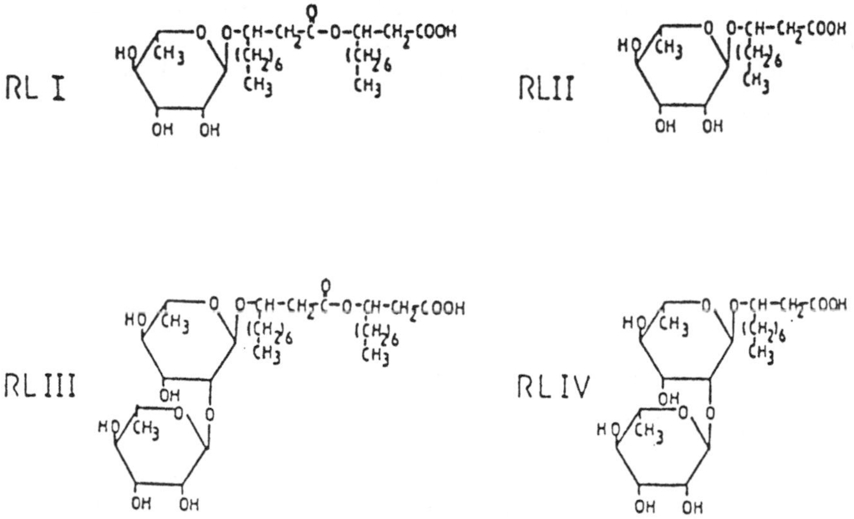

RL I

RLII

RLIII

RLIV

FIG. 1 Rhamnolipids RLI-RLIV produced by *Pseudomonas* sp. DSM 2874 [9].

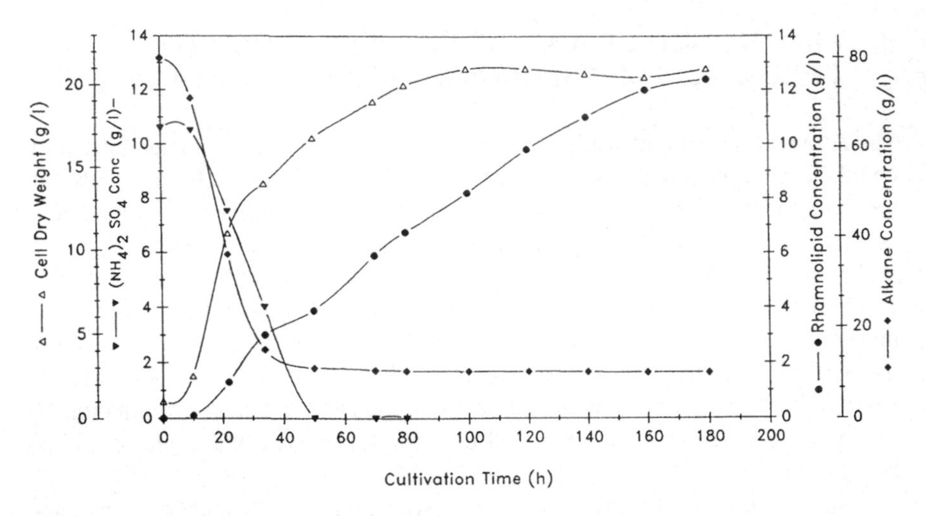

FIG. 2 Rhamnolipid production by *Pseudomonas* sp. DSM 2874 during a 30-L batch cultivation. Conditions: mineral salt medium with 8% *n*-alkanes C-14 to C-15, *T*=30°C, nitrogen (N)-limitation [10, 11, 16].

The production of rhamnolipids with resting cells suspended in 100 mM sodium chloride solution with *n*-alkane as the carbon source is demonstrated in Fig. 3. The biocatalytic stability as well as the specific rhamnolipid productivity were both influenced by nitrogen limitation, when comparing resting cells with growing cells. Whereas growing cells formed both RLI and RLIII, two new rhamnolipids (RLII and RLIV [13]) could be obtained under resting cell conditions (Fig. 1). The influence of different carbon sources and of reaction conditions on the productivity and product composition is summarized in Table 1.

III. PRODUCTION OF RHAMNOLIPIDS WITH IMMOBILIZED CELLS

The main technical problem in the production of extracellular biosurfactants under aerobic conditions either with growing cells or with resting cells is the extensive formation of foam. Rhamnolipids are one of the most effective surfactants known today [8, 10]. Therefore, after reaching a surfactant concentration of approximately 2 g/L in an aerated, mechanically stirred reactor the whole reactor system becomes filled with foam, and there is no possibility of retaining the cells in the reactor system. We have developed a strategy to optimize the production

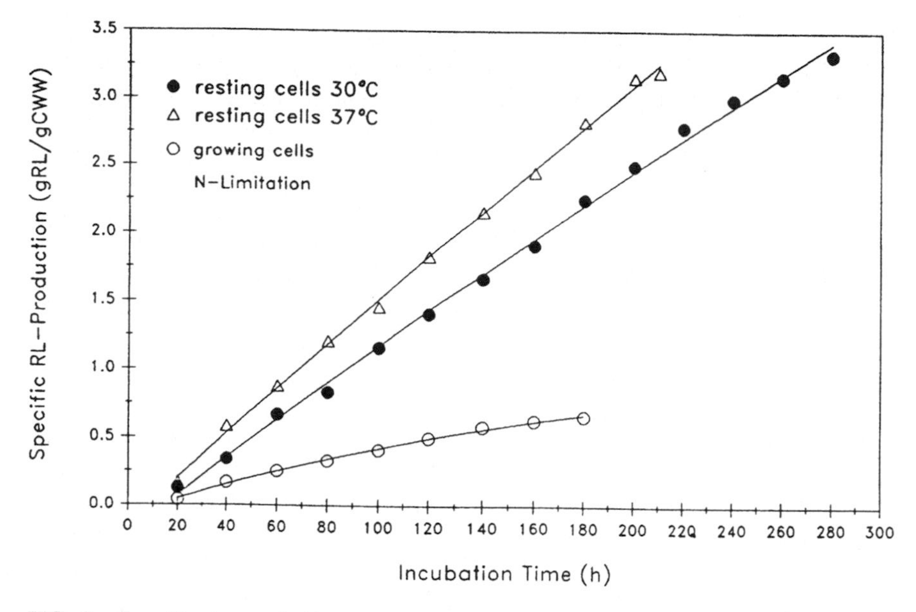

FIG. 3 Specific rhamnolipid production with resting cells in 100 mM sodium chloride solution at T=30°C and T=37°C in comparison with nitrogen (N)-limited growing cells at T=30°C by *Pseudomonas* sp. DSM 2874 [10, 11, 16].

phase using immobilized viable (resting) cells in which the immobilized cell biocatalyst can be easily retained in the fluid reaction phase [18].

The following prerequisites have to be fulfilled when using an immobilization system for the production of biosurfactants.

1. Possibility of a simple and cheap production of the immobilized biocatalyst on large-scale production.
2. Sufficient rhamnolipid productivity compared with free cells.
3. High resistance to stress, which allows continuous processing and reuse of the immobilized biocatalysts.

A. Development of the Biocatalytic System and Screening for an Optimal Support System

1. Entrapment Methods

Several cell entrapment methods [2, 3, 5, 6] in polymers were tested. As demonstrated in Table 2, the highest rhamnolipid production was obtained in an ionotropic network of Ca-alginate [10, 18] using glycerol as the carbon source. It was mentioned above that the specific rhamnolipid production rate could be raised

TABLE 1 Rhamnolipid (RL) Formation by *Pseudomonas* sp. DSM 2874 Influence of Different Reaction Conditions on the Conversion Yield (g Product/g Carbon Substrate), Specific Rhamnolipid Production Rate (mg Rhamnolipid/g Dried Biomass h) and Product Composition [10, 11, 16]

Reaction conditions	Conversion rate (g/g)	Specific production rate (mg/g·h)	Product composition (%)			
			RLI	RLII	RLIII	RLIV
Growing cells						
No limitation						
C source:						
glucose	0.04	9	50	—	50	—
n-alkanes	0.11	10	50	—	50	—
N limitation						
C source:						
n-alkanes	0.18	4	65	—	35	—
Resting cells						
C source:						
n-alkanes, 30°C	0.23	13	42	15	41	2
glycerol, 30°C	0.10	12	22	25	62	1
n-alkanes, 37°C	0.23	22	57	—	43	—

C: carbon; N: nitrogen

efficiently, when using free cells and *n*-alkanes as the single carbon source (see also Table 1). The opposite effect was achieved using immobilized cell systems. The polar properties of all supports used and the resulting diffusion limitation for all nonpolar carbon sources seemed to be the reason for this remarkable loss of activity. Also, the use of an extremely lipophilic polymer, which is termed ENTP-2000 (a derivative of poly(propyleneglycol) 2000 [21] did not lead to increased productivity. This seemed not to be caused by an insufficient influx of the nonpolar *n*-alkanes into the matrix but rather was due to a blocked efflux of the synthesized products, such as rhamnolipids and β-hydroxydecanoic acid in the polar medium. Because of similar limitation effects, some investigators tried to develop several hydrophobic gel matrices [22]. Nevertheless, we tried to prevent these substrate limitations by adsorbing the cells onto chemically inert organic and inorganic supports [18].

2. Adsorption onto Organic and Inorganic Supports

Inorganic supports, based on silica, and organic supports, by means of activated carbon, served as a support for the adsorption of *Pseudomonas* (Table 3). To

TABLE 2 Immobilization of *Pseudomonas* sp. DSM 2874 Using Several Cell Entrapment Methods. Comparison Between the Rhamnolipid Production (%) of Free and Immobilized Resting Cells Using Different Carbon Sources [10, 18]

Immobilization support	Principle of immobilization	C Sources		
		Glycerol	Glucose	*n*-Alkane
Free cells	None	100	100	100
Alginate	Ionic binding	90.00	66.1	21.9
Chitosane	Ionic binding	2.1	2.0	2.5
Agar	Nonspecific gelation	54.9	51.5	20.8
Carrageenan	Nonspecific gelation	44.4	39.2	16.9
Epoxide resin	Covalent binding	2.8	3.8	7.8
Polyacrylamide	Covalent binding	3.5	1.5	6.5

TABLE 3 Adsorption of *Pseudomonas* sp. DSM 2874 Onto Different Supports. The Cell Loading of the Supports (mg CWW/g Support) was Carried Out Under Vacuum [18]. The Remaining Cell Load was Estimated After 4 h of Intensive Shaking (100 rpm) in the Production Medium

Support	Description	Cell Load (mg/g)	Remaining Cell Load (%)
Activated coal	Surface (m^2/g)		
	1000	102.5	
	1300	102.7	
	1500	88.8	
Silica ("Raschig rings")	ϕ of pores (μm)		0.5–1.0
	60–100	137.1	
Fractosil	0.5	61.8	
	1.0	69.0	
	2.5	66.0	

create a more effective adsorption of cells onto the surface of the support, the immobilization procedure was carried out under vacuum [18]. Both, the activated carbon and silica supports allowed a sufficient cell load on the support. Unfortunately, after incubating the immobilized cell system in an aerated and continuously shaken production medium, more than 90% of the cells were removed from the support into the medium (Table 3).

After summarizing these results (poor fixing ability of *Pseudomonas* onto a given support and diffusion limitation for both nonpolar substrates or products), we decided to develop a production process with resting cells of *Pseudomonas* entrapped into the Ca-alginate polymer and with an exclusive use of polar carbon sources such as glycerol.

B. Optimization of Reaction Parameters

To enhance the rhamnolipid production, it is most useful to separate the growth phase from the production phase, as mentioned above in the case of free cells. Therefore, the production of immobilized biocatalists must include two different processes: First, the desired cell mass must be cultivated and harvested. Second, the whole cell mass must be entrapped into spherical Ca-alginate beads, which become directly involved in the production process. To obtain biocatalysts of optimal physiological fitness, it was of utmost importance to determine the phase of maximum ability of rhamnolipid production. Therefore, cells related to different growth phases were harvested, immobilized, and afterward set for production. Regardless of whether resting free cells or immobilized cells were used, the maximum rhamnolipid production was obtained during the transitional phase of growth, as demonstrated in Fig. 4. Moreover, both free and immobilized cells were producing all known rhamnolipids (RLI–IV), which confirms their equal physiological ability for this production.

Obviously, one of the most important differences between these two biocatalytical systems was an enormous decrease in specific rhamnolipid production, when Ca-alginate entrapped cells were used. One of the main objectives of further investigation was to remedy this remarkable loss of activity.

1. Influence of the Cell Load on Rhamnolipid Production by Spherical Ca-Alginate Biocatalysts

The cell load of biocatalysts is defined as the wet cell weight per gram of wet immobilization support (e.g., Ca-alginate). A maximum cell load of approximately 30% could be attained when all of the free room inside a spherical Ca-alginate support is filled with cells [23]. To investigate the dependence of the cell load on the specific rhamnolipid production, spherical Ca-alginate beads containing different amounts of cells were incubated in a production medium. In all tests, the *total* cell concentration was the same. Nevertheless, a simple increase

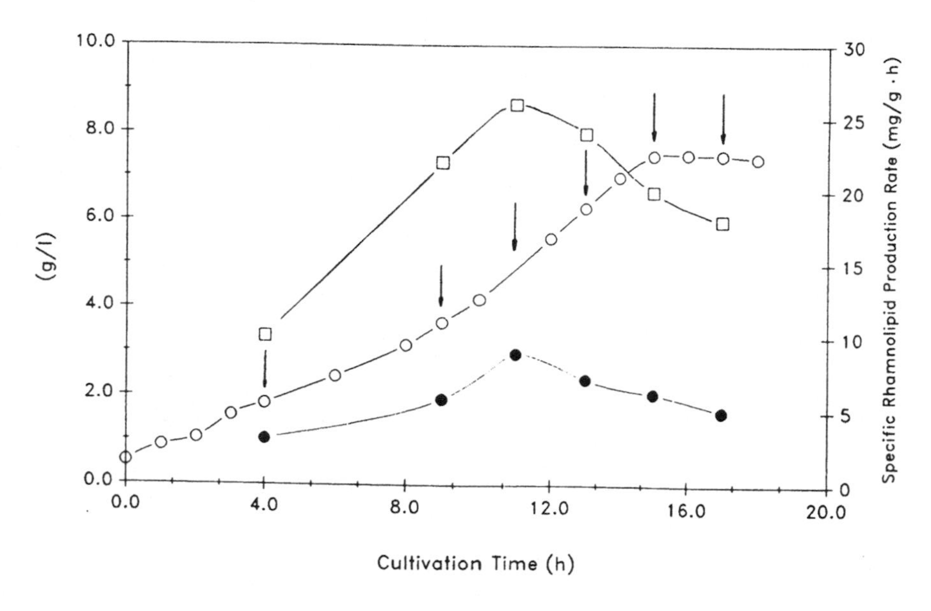

FIG. 4 Production of rhamnolipids with resting free and immobilized cells of *Pseudomonas* sp. DSM 2874 received from different growth phases. The cells were cultivated in a 30-L reactor with glycerol as carbon source [16, 42]. After harvesting and immobilizing, the cells were incubated in production media [18, 20]. The arrows indicate the time of harvest. (○) Cell dry weight; (●) immobilized cells; (□) free cells.

in the biocatalyst cell load did not cause a remarkable increase of the specific rhamnolipid production rate (Fig. 5).

In many cases, this reduction of activity is typical for immobilized biocatalysts, and it is mostly caused by diffusion limitation inside of the support of the biocatalyst [24–27]. The inhibited transport of the substrates, such as glycerol and oxygen, leads to the assumed concentration profile, which is demonstrated in Figure 6 [27].

The immobilization of cells into a Ca-alginate support leads to a homogeneous distribution of the cells inside the gel beads. Following the typical decline of substrate concentration demonstrated in Figure 6, it is possible that all cells positioned somewhere near the center of the beads will be exposed to substrate limitation with the consequence being a loss in activity. A further cell loading of the Ca-alginate beads improved the efficiency of the biocatalysts. This is caused by a better exploitation of the unlimited outer zone of the gel support. The Ca-alginate support seemed to be saturated with cells at a certain cell load (cell load ~16%), resulting in a constant specific rhamnolipid productivity.

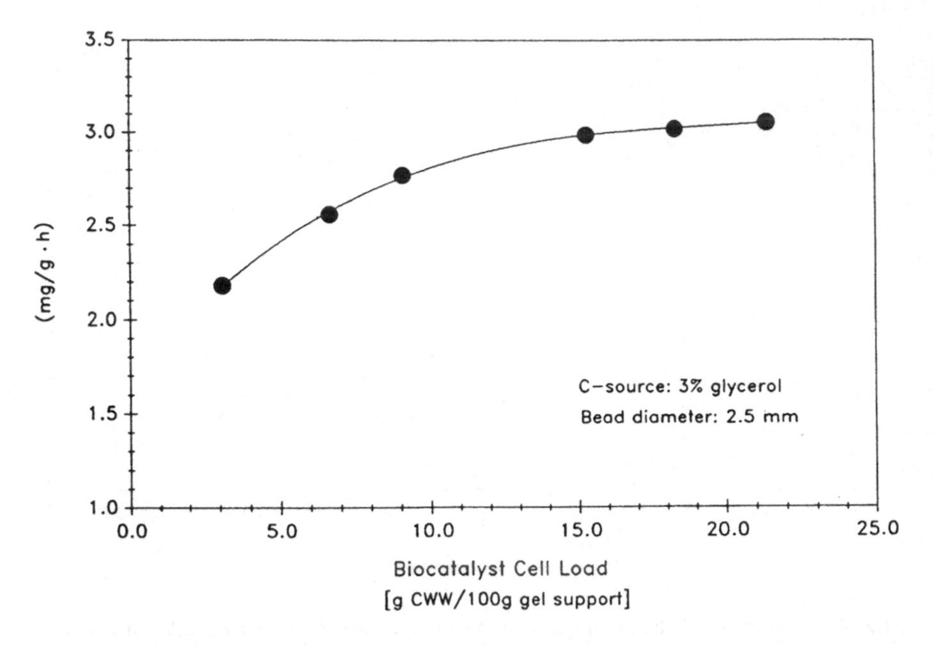

FIG. 5 Dependence of the specific rhamnolipid production rate (P_s) upon the cell loading of Ca-alginate biocatalysts [18, 20]. (●) Specific RL-production rate.

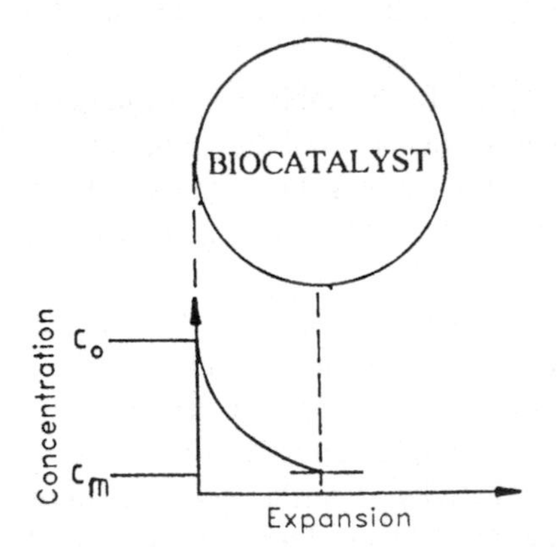

FIG. 6 Concentration profile inside a spherical biocatalyst [18, 27]. c_0 = Concentration of substrate in medium, equal to concentration at the periferic zone of the biocatalyst. c_m = Concentration of substrate in the center of the biocatalyst.

The extent of the diffusion limitation of spherical Ca-alginate beads could be represented by the diffusion index (D_i in cm^2/sec) [28, 29]. The diffusion index could be expressed as the migration velocity of substrates or products into or out of the immobilization support. In all cases of investigation, the diffusion index was found to be very low (see Table 4). It was amazing that, with the tested rhamnolipids, the diffusion velocity of oxygen was the lowest. Moreover, an increase of cell load of the biocatalyst decreases the diffusion velocity of glycerol, whereas the efflux of the formed product rhamnolipid seemed to be slightly enhanced.

2. Reduction of the Diffusion Limitation Aimed at Improving Rhamnolipid Production with Immobilized Cells

Due to the above mentioned correlations, a desired reduction of the diffusion limitation could be attained in two different ways [30]:

1. Further reduction of the biocatalyst size.
2. Inhomogeneous loading of the biocatalyst by means of positioning the main part of catalytic active cells at the outer zones of the biocatalyst.

This could be performed by cultivating the spherical biocatalyst loaded with a small amount of cells into a growing medium. Since diffusion of substrates is

TABLE 4 Diffusion Index *Di* of Different Molecules in Spherical Ca-Alginate Beads [18]

Molecule	Alginate concentration (%)	Cell Load (%)	D_i (cm^2/sec)	Reference
Glycerol	3	0	$7.0 \cdot 10^{-6}$	
	3	9.1	$6.0 \cdot 10^{-6}$	
	3	15.3	$5.5 \cdot 10^{-6}$	18
RLI–IV	3	0	$4.5 \cdot 10^{-7}$	
	3	9.1	$8.2 \cdot 10^{-7}$	
	3	15.3	$8.8 \cdot 10^{-7}$	18
Glycose	2	0	$6.8 \cdot 10^{-6}$	49
Oxygen	3	0	$3.7 \cdot 10^{-6}$	24, 40
L-Tryptophan	2	0	$6.6 \cdot 10^{-6}$	49
α-Lactoalbumin	2	0	$1.0 \cdot 10^{-6}$	49

limited, *Pseudomonas* sp. grows just below the surface of the biocatalyst, whereas the substrate limited center remains free of cells. This growing behavior of the entrapped cells is demonstrated below in Fig. 9.

There are two main problems in producing such inhomogeneously loaded biocatalysts. The first is that the growth of cells inside the porous support will be blocked at a distinct time, because of the continuously reduced space left behind. The other problem is caused by an undesired simultaneous growth of free cells beside the immobilized cells. This formation of free cells is caused by an inevitable damage of the Ca-alginate support (see below).

Nevertheless, comparing both biocatalyst systems, the periferic loaded biocatalysts indicated a considerably enhanced rhamnolipid production, as demonstrated in Figs. 7 and 8. The efficiency of periferic loaded biocatalyst is obviously demonstrated when comparing the rhamnolipid production of both immobilized cell systems at a cell concentration of 2 g/L (9.1% share of the support). In the case of homogeneously loaded cells, only 34% of all cells entrapped in the Ca-alginate support seemed to be catalytically active compared to those of the periferic loaded biocatalyst. How it was expected, an overloading of the support with cells leads to a decreased catalytic activity of the biocatalyst. Nevertheless, even the maximum

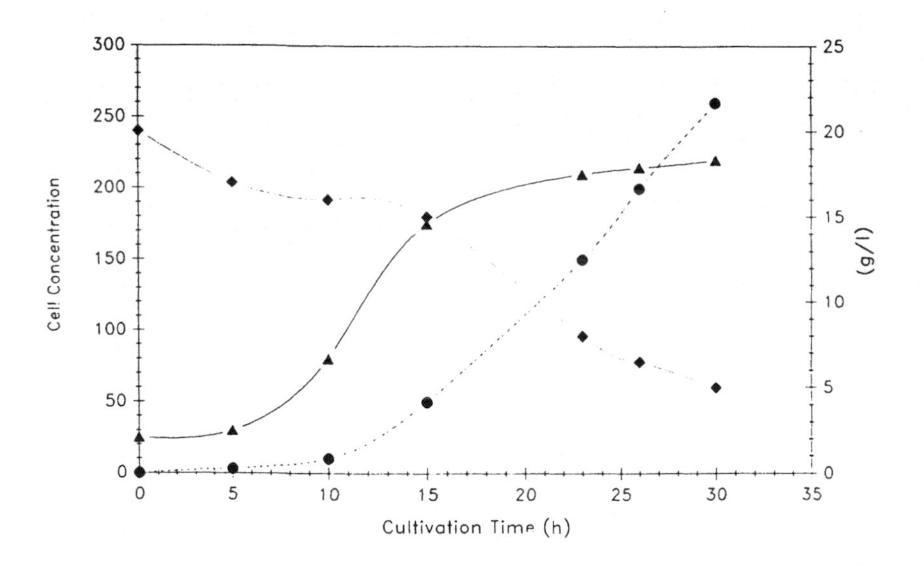

FIG. 7 Cultivation of *Pseudomonas* sp. DSM 2874 in 500-mL shaking flasks. Growing behavior of free and immobilized cells. Conditions: T=30°C, pH 6.8, 100 rpm, size of the beads = 1.5 mm, starting biocatalyst cell load = 1% CWW [18]. (●) CDW (mg/100 mL medium); (▲) CDW (mg/10 g beads); (◆) Glycerol concentration (g/L).

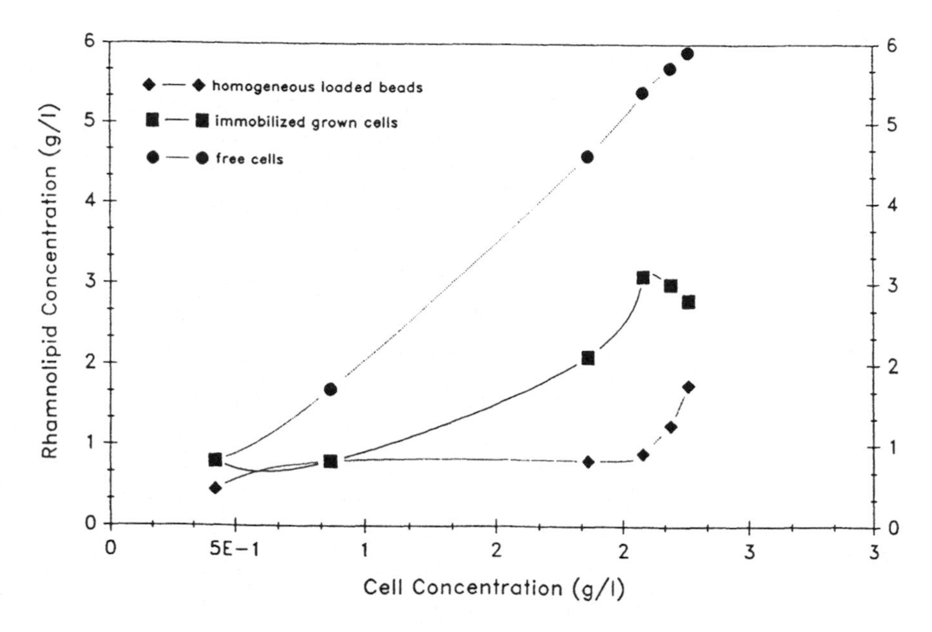

FIG. 8 Rhamnolipid production of resting free, homogeneously loaded, and immobilized (inhomogeneously) grown cells. Conditions: 100-mL production medium in 500-mL shaking flasks, T=30°C, ph=6.8, 100 rpm [18, 44].

rhamnolipid production of the above described inhomogeneously loaded biocatalysts could not attain the productivity achieved with free resting cells.

A further increase in the rhamnolipid production could be achieved, when decreasing the size of the biocatalyst by means of reducing the bead diameter of the spherical Ca-alginate support. Knowing about this correlation (see also Fig. 5), we tried to predict the optimum size of the beads by producing photos of microscopically enlarged thin cuttings through the midplane of calcium alginate biocatalysts of different size (Fig. 9). The biocatalysts were received from several cultivation processes. Since diffusion of substrate is limited, *Pseudomonas* grows just below the surface of the biocatalyst. It could be noticed that *Pseudomonas* forms a layer of approximately 0.125 mm, regardless of bead diameter or cultivation time (see also Table 5). Because of this characteristic growth behavior, the ideal size of the biocatalyst should be approximately 0.25 mm.

The efficiency of biocatalysts could be defined as the proportion of the maximum specific rhamnolipid production (P_s) of the biocatalyst and that of free cells.

$$\text{Efficiency (\%)} = \frac{P_s \text{ immobilized cells}}{P_s \text{ free cells}} \times 100$$

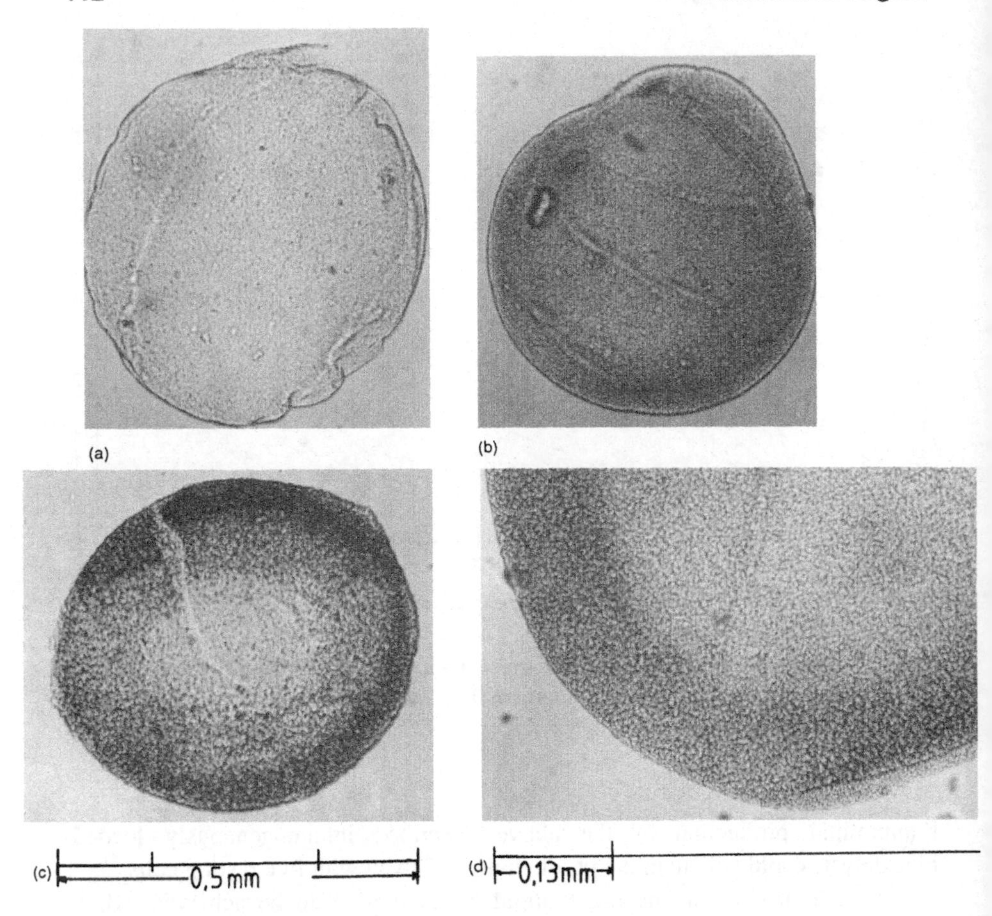

FIG. 9 Photos of microscopically enlarged thin cuttings through the midplane of calcium alginate beads. Before cutting, the biocatalysts were fixed with glutardialdehyde. Thin cuttings were colored with methylene blue. (b)–(d) The biocatalysts were derived from a cultivation in a 23-L fluid bed bioreactor. (a) Shows a thin cutting through a calcium alginate bead, which was inoculated with *Pseudomonas*. (b)–(d) The biocatalysts were derived from a cultivation in a 23-L fluid bed bioreactor. The darker zones indicate a higher concentration of *Pseudomonas*. The bacteria form a layer of 0.125 mm just below the surface, regardless of the bead diameter, biocatalyst cell load (amount of cell wet weight inside the calcium alginate beads; % CWW), or cultivation time [20, 43].

In Figure 10 the efficiency of an ideal biocatalyst is presented. Regardless of whether producing a biocatalyst that was homogeneously loaded with cells or produced by a cultivation procedure (immobilized grown biocatalyst), the maximum specific rhamnolipid production of free cells was almost attained. The losses of activity caused by diffusion limitation were reduced to a minimum by

TABLE 5 Growth Behavior of Immobilized *Pseudomonas* sp. DSM 2874 in a Ca-Alginate Support Depending on Cultivation Time and Bead Diameter [20, 43]

Figure number	Bead diam (mm)	Cultivation time (h)	Biocatalyst cell load (% CWW)	Layer (mm)
9a	0.5	0	1	—
9b	0.5	10	3	~0.1
9c	0.5	30.5	16	0.12
9d	0.8	20	13	0.13

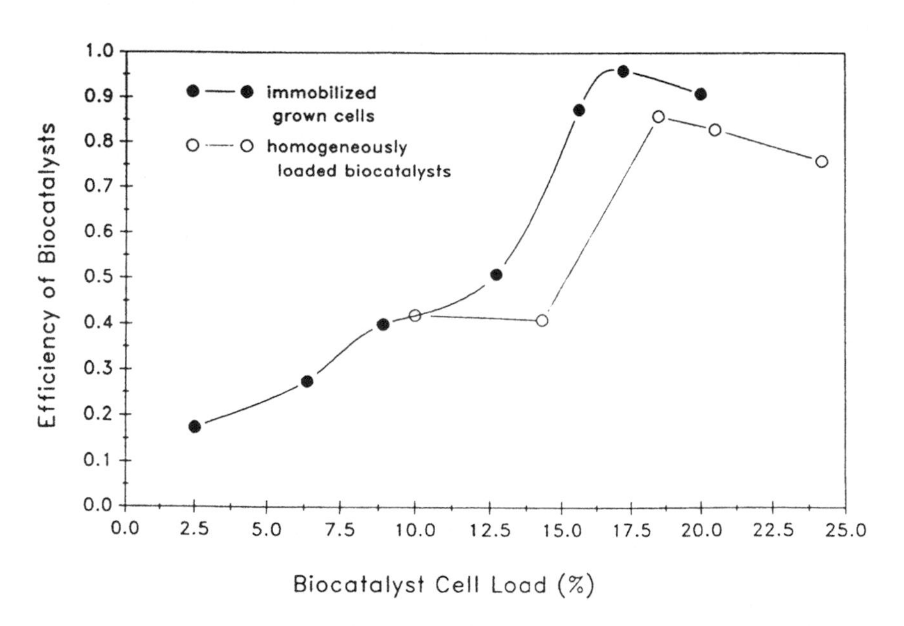

FIG. 10 Efficiency of homogeneously loaded and immobilized grown cells upon the biocatalyst cell load. The immobilized grown biocatalysts were derived from a cultivation procedure in a 23-L fluidized bed reactor. Rhamnolipid production conditions: bead size = 0.25 mm, 100-mL production medium in 500-mL shaking flasks, T=30°C, pH=6.8, duration=100 h [18, 20, 43].

decreasing the biocatalyst size to 0.25 mm (Fig. 11). In Table 6 the efficiency of several types of biocatalysts by means of different size and cell immobilizing procedure is summarized.

C. Development of an Immobilization Procedure Suitable for the Large-Scale Entrapment of Living Cells into a Ca-Alginate Support

Many methods are described in the literature for the scaled-up production of spherical biocatalysts with a diameter larger than 1 mm [5, 6, 31, 32]. In contrast, nothing has been described for the scaled-up production of small sized biocatalysts, for example, smaller than 0.5 mm. Matulovic and Rasch were the first to develop an immobilization procedure involving a rotating nozzle ring [33]. This special rotating ring technique was adapted to the scaled-up production of spherical 0.25-mm beads. The essence is that the diameter as well as the quality of the beads depend on the rotating speed of the nozzle ring as well as on the overpressure. A rotating speed of approximately 5700 rpm is necessary for the production of 0.25-mm beads of high quality (standard deviation less than 10%).

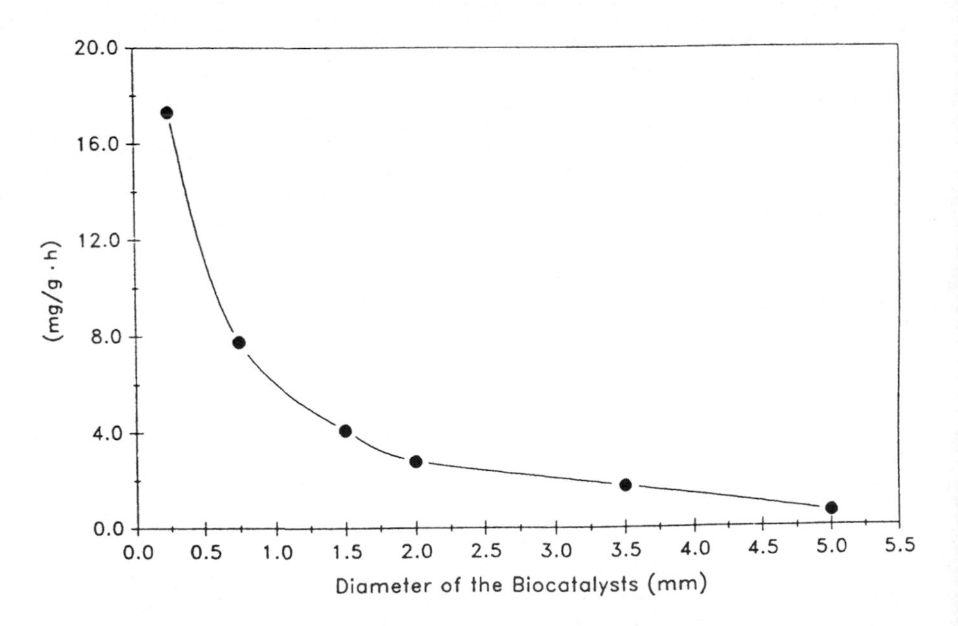

FIG. 11 Dependence of the specific rhamnolipid production rate upon the size of the biocatalyst. Production conditions are same as in Fig. 10 [18–22, 43]. (●) Specific RL-production rate.

TABLE 6 Efficiency of Immobilized Grown Cells and Homogeneously Loaded Biocatalysts Depending on Their Size

Size	Efficiency of biocatalysts: Homogeneously loaded biocat.	Efficiency of biocatalysts: Immobilized grown cells	Reference
3.5	0.1	n.d.	18
1.5	0.3	0.57	18
0.75	n.d.	0.73	19
0.25	0.85	0.95	20
Free cells	1	1	42

At this speed and an overpressure of 50 mbar, we were able to produce 3 kg of Ca-alginate biocatalysts per hour [for more details see Refs. 20 and 34].

The development of an optimal biocatalytic system (see Sec. II) as well as the adaptation of the immobilization procedure to large-scale production of these biocatalysts were both an inevitable prerequisite for further development of the rhamnolipid production process.

IV. DEVELOPMENT OF A PROCESS MODE FOR THE CONTINUOUS PRODUCTION OF RHAMNOLIPIDS

A. Isolation of Rhamnolipids Based on Flotation

Enormous formation of foam could accompany the production of rhamnolipids with resting free or immobilized cells. The extent of this formation depends on the product concentration, aeration rate, and reactor dimensions. When analyzing the foam composition it was important to establish whether biosurfactant ramnolipids were the main component of the foam. This rhamnolipid-caused foam was more or less polluted with proteins (e.g., an exolipase[18]), phospholipids, or free cells, depending on production conditions (Table 7). The increased formation of foam, caused by production of the biosurfactant started at a rhamnolipid concentration of approximately 0.15 g/L in the production medium. During the whole production process, this rhamnolipid concentration remains constant, regardless of continuous draining of the rhamnolipids by flotation.

Nevertheless, such a continuous flotation causes an inevitable decrease of the production medium volume. Due to that, the rhamnolipids have to be continuously

TABLE 7 Composition of the Foam, Which was Formed During the Production Process [18]

	Protein concentration		Phospholipid concentration		Rhamnolipid production
Incubation time (h)	Medium (mg/L)	Foam (mg/L)	Medium (mg/L)	Foam (mg/L)	(gRL/L)
20	65	60	18	10	204
30	50	110	5	45	510
40	50	130	9	52	765
55	50	115	5	55	1122
75	50	200	5	65	1326

isolated from the coflotated production components, whereas the product-free medium has to be recycled to the production process.

B. Isolation of the Rhamnolipids by Adsorption Onto an Amberlite XAD-2 Support

Like other anionic biosurfactants the rhamnolipids could be precipitated by acidifying [9]. Due to their lipophilic ability the rhamnolipids (by means of the collapsed foam) could be adsorbed onto a support, which is termed XAD-2 [18]. This XAD-2 amberlite resin is well known for the fixation of lipophilic compounds like corrinoids, steroids or indole [35–37].

Without further influencing the composition of the production medium 20 g of this support could adsorb 1 g of the rhamnolipids. To achieve complete rhamnolipid isolation from the product containing fluidized phase, the influence of several flow rates on rhamnolipid adsorption behavior was obtained. Therefore, several rhamnolipid solutions were passed through a XAD-2 packed column, altering the flow rate. In all cases of increased flow rates, the adsorption of the rhamnolipid remained constant, even when altering the electric charge of the rhamnolipids (see Table 8).

Nevertheless, after a distinct time of continuous load with rhamnolipids, the XAD-2 support will be exhausted. This has to be recognized immediately to prevent a decreased efficiency of the process recycle system. In the case of an exceeded loading, all precipitated rhamnolipids will pass through the column as a crude and cloudy suspension. This could be recognized immediately by controlling the optical density (546 nm) of the eluted solution on-line using a common laboratory photometer.

TABLE 8 Influence of Various Flow Rates on the Adsorption of Rhamnolipids on an Amberlite XAD-2 Support. 500 ml of a 1g RL/l were Totally Passed Through a XAD-2 Packed Column. (Length: 100 mm, diameter: 300mm, 20 XAD-2) [18]

	Flow rate (mL/min)	Rhamnolipid concentration (g/L)	
		pH 2.0	pH 7.0
Starting solution	—	1.00	1.00
Eluted solution	10	0.02	0.03
	50	0.04	0.04
	100	0.05	0.05
	200	0.05	0.05

C. Construction of a Fluidized Bed Reactor

Conventional reactor systems suitable for a process involving free cells are almost wholly unsuitable for those process modes using immobilized living cells [1, 11, 38]. Mechanical stirring causes extreme *shear force*, which causes damage of the immobilization support and releases the fixed cell material. These effects are minimized if the distribution of the immobilized cells inside a given medium is exclusively caused by aeration of the reactor system [39]. This reactor system, which is described as a fluidized bed reactor system, is based on three different components (Fig. 12):

1. Bottom, manufactured of high-grade steel, including the fittings for several process control instruments, medium influx, and efflux, as well as the aeration gadget.
2. A glass-manufactured, double-coated, and temperature-controlled column.
3. Top, manufactured of high-grade steel, including a biocatalyst support system as well as the flotation grain, respectively.

The combination of glass columns of different length allows a scaled-up production mode in the modular construction system.

In contrast to common types of bioreactors, the aeration of this reactor system does not serve exclusively as an oxygen donor. Additionally, the immobilized biocatalysts have to be homogeneously distributed among the production medium, and a sufficient flotation rate has to be maintained at all times [40, 41].

To prevent coflotation of the immobilized biocatalyst out of the reaction system, a porous platform was constructed at the top of the reactor. A possible

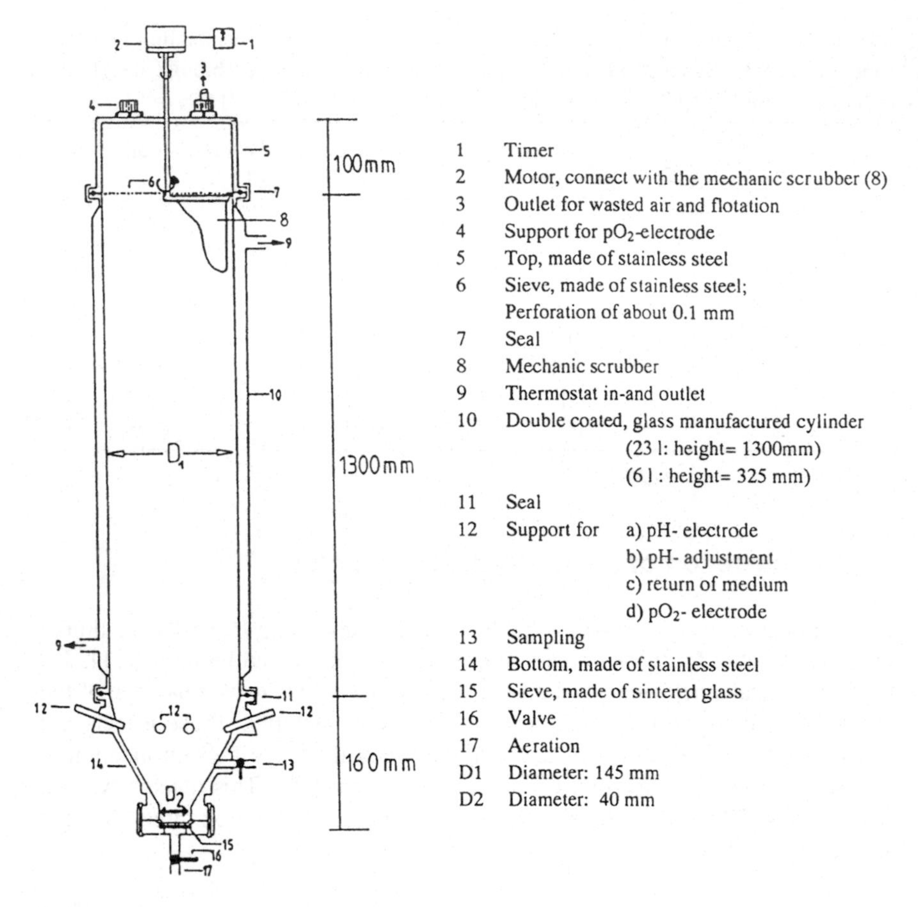

FIG. 12 Fluidized bed reactor [6–23].

blocking of this sieve caused by biocatalyst adhesion was prevented by a mechanical pulsed scrubber [20].

D. Process Mode for Production and Continuous Isolation

Based on the above described findings the following process was developed (Fig. 13). The fluidized bed reactor (1) is filled with the production medium and the immobilized biocatalyst. The temperature inside the reactor is controlled by a thermostat at 30°C (2). The aeration of the reactor is realized by pressing air through an inlet made of sintered glass.

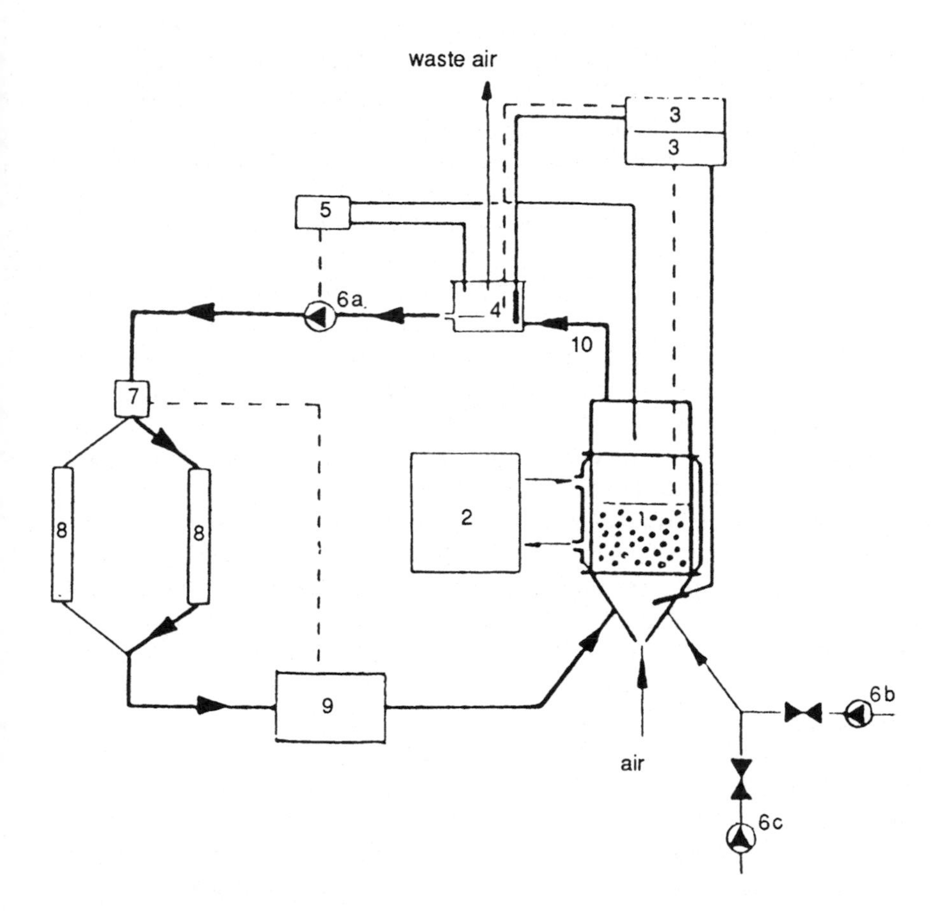

FIG. 13 Continuous production of rhamnolipids in a fluidized bed reactor and on-line isolation of rhamnolipids by foam flotation. (1) Fluidized bed reactor; (2) thermostat; (3) pH adjustment; (4) product precipitation; (5) conductivity measurement; (6) pump for (a) medium circulation, (b) substrate feeding, (c) regeneration of biocatalysts; (7) magnetic valve; (8) XAD-2 adsorption column; (9) photometer; (10) continuous product foam flotation.

As the rhamnolipid production starts, an enormous formation of foam is obtained. The products, which are entitled in the foam, are transferred and floated into the precipitation vessel (4), using the wasted air passage (10). In shifting the pH from 6.8 to 2.8, the rhamnolipids are precipitated into the medium. Both, the pH value of the production medium and the precipitation medium are controlled by pH adjustments (3). A constant volume inside the fluidized bed reactor is realized by an on-line level regulation in the precipitation vessel. Therefore, the vessel is equipped with a conductivity measurement system (5).

In the next step, the rhamnolipid-enriched medium passes the XAD-2 packed adsorption column (8). If the loading capacity of the column is exceeded, which is controlled by the on-line measurement (9) of the optical density (546 nm) of the eluate, a magnetic valve (7) changes the passage from the exceeded column to a freshly prepared one.

Cultivation, as well as a necessary regeneration, of the immobilized biocatalyst could be carried out replacing the production medium with the growth medium. In the case of a continuous production mode, the consumed glycerol could be fed using a pump (6) with off-line detection of glycerol using HPLC [18].

In alteration, two different process schemes were developed for the production of rhamnolipids. The first process includes the production of rhamnolipids using immobilized grown cells. In the alternate, the homogeneously loaded biocatalyst is used in a second process (see Fig. 14). Both processes have several advantages and disadvantages and are compared below.

The optimum biocatalyst cell load of 0.25-mm Ca-alginate beads was estimated to be 18%, as mentioned in Sec. III. At first glance, it seems to be most efficient if this distinct cell load is achieved by simultaneously using the same equipment developed for the production of rhamnolipids. Therefore, Ca-alginate beads, which were slightly loaded with cells (1% CWW) were cultivated in the 23-L fluidized bed reactor. Under normal aeration conditions and caused by a large expansion of free cells, this desired amount of cell load could not be attained in any case. Even if raising the initial cell concentration from 1% to 6% CWW and, in addition, changing the aeration of the growth medium from normal air to pure oxygen, the optimum biocatalyst cell load of 18% could just be achieved (see in Fig. 15). Moreover, after 17 h of cultivation, the cell concentration of the immobilized biocatalysts began to decrease while the concentration of the free cells remained nearly constant. An insufficient oxygen supply in the growing medium is responsible for this critical growing behavior. This could be confirmed when estimating both the partial oxygen concentration in the growth medium (pO_2) as well as the behavior of respiration, expressed by the partial oxygen consumption rate (QO_2) and the carbon dioxide production rate (QCO_2). After only 8 h of cultivation, the pO_2 decreased down to 1 mg/L. Simultaneously, both respiration rates (QO_2 as well as the (QCO_2) diminished. Summarizing all these

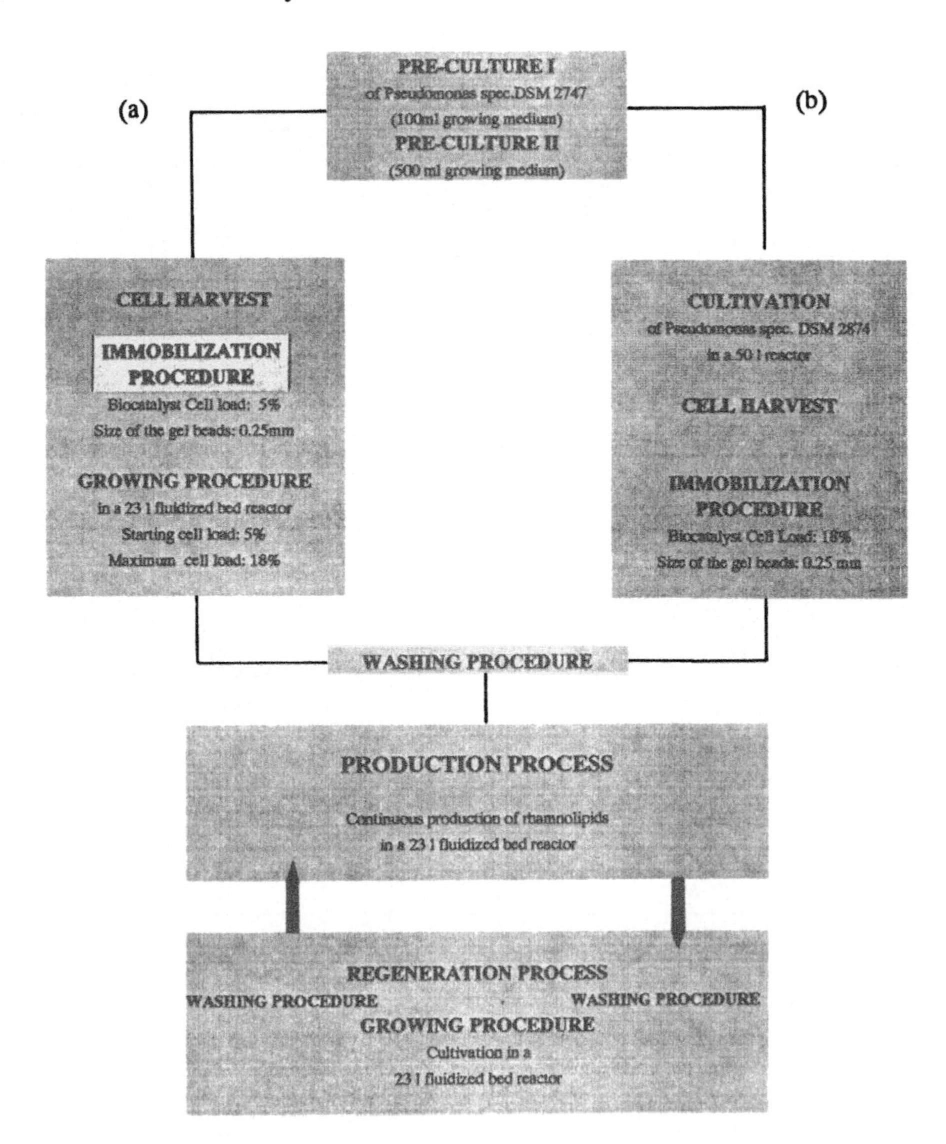

FIG. 14 Rhamnolipid production by the use of (a) immobilized grown cells and (b) homogeneously loaded biocatalysts.

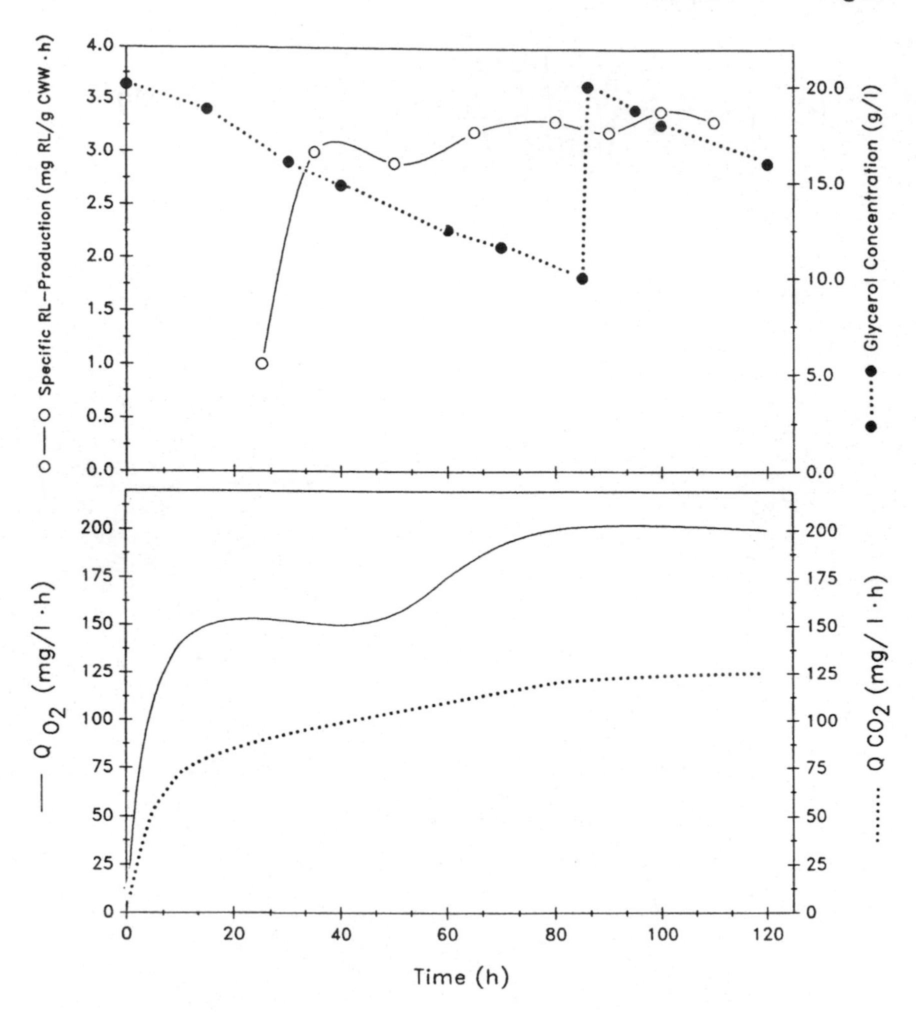

FIG. 15 Growth behavior of immobilized cells and free cells during the cultivation process in a 23-L fluidized bed reactor. The formation of free cells was caused by an inevitable damage of the gel support [18–20]. Conditions: 18 L cultivation medium, 3.75 kg wet Ca-alginate beads, starting at a cell concentration of 6% CWW, aeration with 100% oxygen, aeration rate: 3.5 NL/min, T=30°C, pH 6.8.

effects, it has to be realized that the given fluidized bed reactor set up (which is well optimized for rhamnolipid production) is not suitable for the growth process when immobilized biocatalysts are used.

These problems should be avoidable in further optimization of the fluidized bed reactor. Nevertheless, for the sake of simplicity it seems to be more useful if the production-optimized biocatalysts are produced without using this special growth procedure. Consequently, we produced homogeneously loaded biocatalysts, where the total yield of cells (necessary for the load of such Ca-alginate beads) were received from a separate cultivation procedure in a 50-L reactor [for cultivation details see Ref. 9, 18, and 20]. Immediately after harvesting, the cells were mixed with the Na-alginate support (biocatalyst cell load of 18 g wet cell/100 g alginate support) and immobilized by gelating in a $CaCl_2$ solution [for immobilization details see Refs. 20, 34, 43, 44]. These homogeneously loaded biocatalysts were used in the rhamnolipid production process.

E. Production in a Fluidized Bed Reactor

Matulovic was the first to investigate the production behavior of immobilized biocatalysts in a fluidized bed reactor [18]. The fluid dynamics of this reactor system are influenced by the concentration of the solid phase (immobilized biocatalyst) as well as by the aeration rate [45]. The aeration rate and reactor loading produce a dynamic balance between washout of the biocatalyst, product flotation, and the oxygen transfer rate with respect to an influence of the rhamnolipid production. Thus, examination in a 6-L fluidized bed reactor indicated a maximum aeration rate of approximately 3.5 nL/min [18]. Rhamnolipid production performed in a 6-L fluidized bed reactor, as well as in a reactor with a production volume of 23 L [19], indicated that the concentration of the solid phase (g wet immobilized biocatalyst/100 g fluidized phase) should not exceed more than 20%. Following these distinct process parameters, a continuous rhamnolipid production was carried out in a 23-L fluidized bed reactor using immobilized biocatalyst of production-optimized size.

The rhamnolipid production behavior of this process is demonstrated in Fig. 16. The flotation process started after 16 h of incubation. After a short time of increase, the specific rhamnolipid production rate (P_s) remained constant. The average specific rhamnolipid production rate of approximately 3 mg/g·h obtained during this process was rather bad, consisting that more than 12 mg rhamnolipid/g·h could be produced if optimum process conditions are realized.

An insufficient oxygen supply, caused by an overload of the fluidized bed reactor with immobilized cells, led to a substantial decrease of the rhamnolipid production [40] (see Table 9 and Fig. 17). Thus, if decreasing the loading of

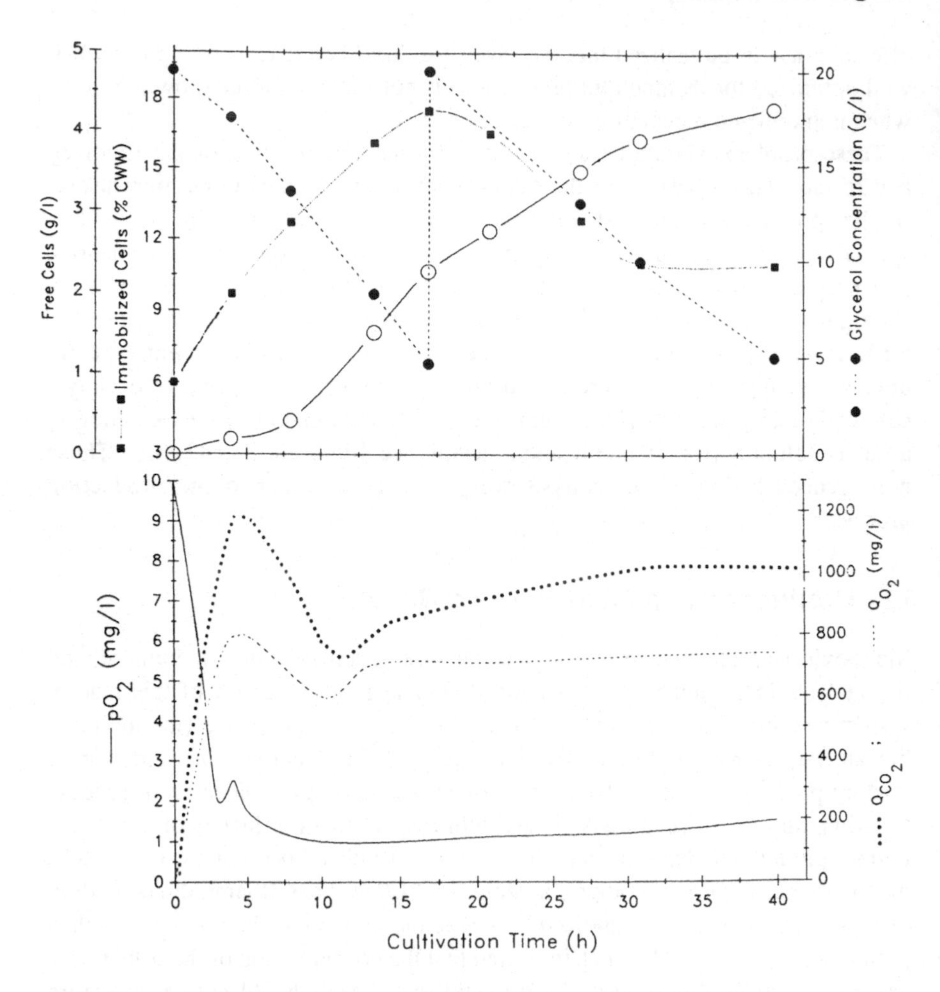

FIG. 16 Continuous production of rhamnolipids in a 23-L fluidized bed reactor using production optimized immobilized biocatalysts (20).

RL-Production conditions	
Reactor	23-L fluidized bed reactor
Medium	18-L Production Medium (9, 20)
Immobilized biocatalyst	3.7 kg wet Ca-alginate biocatalysts
Size of the biocatalysts	0.25 mm
Cell load of the biocatalyst	17.5%
Temperature	30°C
pH	6.8

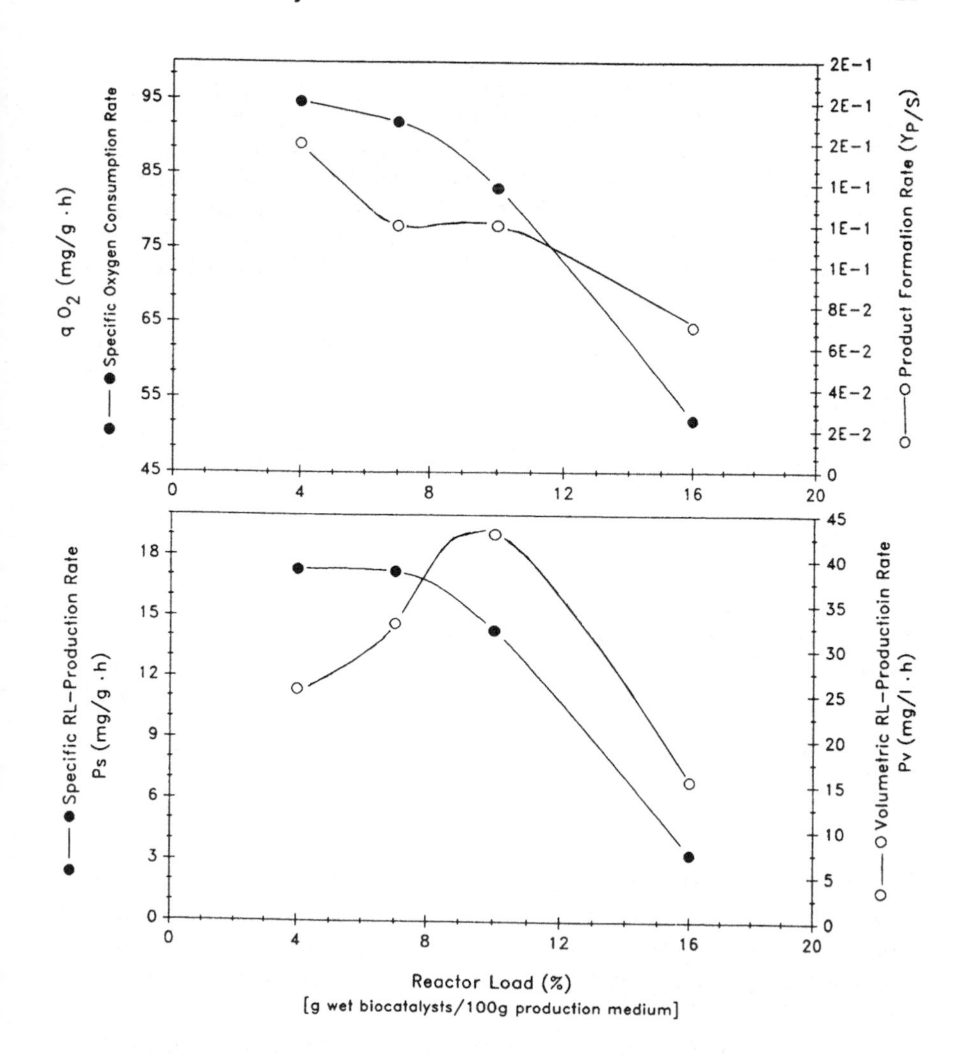

FIG. 17 Dependence of several production parameters upon the load of the fluidized bed reactor with immobilized biocatalysts. Incubation in a 23-L fluidized bed reactor; 0.25 mm Ca-alginate biocatalysts; homogeneously loaded biocatalysts; biocatalyst cell load = 17% CWW; aeration rate = 3.5 NL/min (air), duration of each production process > 20 h.

TABLE 9 Dependence of Several Production Parameters Upon the Load of the Fluidized Bed Reactor with Immobilized Biocatalysts. Incubation in a 23 L Fluidized Bed Reactor; 0.25 mm Ca-Alginate Biocatalysts; Homogeneously Loaded Biocatalysts; Biocatalyst Cell Load = 17% CWW, Aeration rate = 3.5 NL/min (Air), Duration of Each Production Process > 20 h. (P_S) Specific Rhamnolipid Production Rate (mg RL/g CDW·h); (P_V) Volumetric Rhamnolipid Production Rate (mg RL/l·h); ($Y_{P/S}$) Product Formation Rate (g RL/g Glycerol; (qCO_2) Specific Carbon Dioxide Production Rate (mg CO_2/g CDW·h); (qO_2) Specific Oxygen Consumption Rate (mgO_2/gCDW·h)

	Reactor A	Reactor B	Reactor C	Reactor D
Reactor load				
Concentration of biocatalysts (%)	16	10	7	4
Cell concentration (g CDW/L)	4.8	3.05	1.96	1.41
Average of P_s (mg/g·h)	3.3	14.3	17.2	17.3
Average of P_v (mg/L·h)	15.5	42.9	32.9	25.6
Average of $Y_{P/S}$	0.07	0.12	0.12	0.16
qCO_2 (mg/g·h)	46.5	66.3	77.8	75.9
qO_2	51.9	83.1	92.0	94.7

the reactor down to approximately 7% (that corresponds to about 2 g CDW/L production medium) the maximum specific rhamnolipid production rate (P_s) of about 17 mg/g·h was attained. However, the maximum volumetric rhamnolipid production rate (P_v) of approximately 43 mg/L·h) was attained at a reactor load of approximately 10% (3 g CDW/L).

Some of the most important results using immobilized biocatalysts in fluidized bed reactor systems are summarized in Table 10. The maximum specific rhamnolipid production rate of free resting cells employed by an intensor-stirred reactor system [10, 12, 13] was obtained using immobilized biocatalysts, having a bead size of 0.25 mm. In addition, the volumetric rhamnolipid production rate could be raised from 9 mg/l·h to merely 43 mg/l·h. However, at this state of investigation, the maximum volumetric production rate of the reactor system using free resting cells (60 mg/l·h) could not be obtained. Further enhancement of rhamnolipid production in a fluidized bed reactor could be expected if limitations caused by the special reactor design could be prevented [46] (see also Sec. IV.G).

TABLE 10 Comparison of Several Rhamnolipid Production Parameters Between Immobilized Biocatalysts of Different Size and Free Resting Cells Using Different Bioreactor Systems

	Immobilized biocatalysts				Free cells
	23-L fluidized bed reactor			6 L-Fluidized bed reactor	Intensor system (30 L)
Biocatalyst size (mm)	0.25		0.75	1.5	—
Biocatalyst cell load (g CWW/100 g biocat.)	16		8	9.1	—
Reactor load (%)	10		20	20	—
Cell Concentration (g CDW/l)	3.1		3.05	0.9	5
		Maximum values			
P_s (mg RL/g·h)	14.3	17.2 (2 g CDW/L)	7.6	10.0	12
P_v (mg RL/l·h)	42.9	42.9 (3.1 g CDW/L)	23.2	9	60
P_r (g RL/h)	1.0	1.0 (3.1 g CDW/L)	0.51	0.05	1.2
$Y_{p/s}$	0.12	0.16 (2 g CDW/L)	0.10	0.30	0.10
Reference	20		20	19	10, 16

F. Regeneration of the Immobilized Biocatalysts

It is of interest that the specific rhamnolipid production rate of immobilized cells could be preserved over more than 5 days without any loss of activity. Nevertheless, because of gradual leaching of immobilized cells as well as because of continuous starvation, the rhamnolipid production ability started to decline.

For regeneration purposes, the immobilized biocatalysts were separated from the production medium after recognizing a declined rhamnolipid production.

After a short washing procedure the immobilized biocatalysts were recultivated in the fluidized bed reactor. The used up immobilized cells should be relaxed within the cultivation process. Afterward the immobilized cells are separated from the cultivation medium and can be reincubated in a further rhamnolipid production process. This whole continuous rhamnolipid production process, including several reincubation and recultivation processes, is summarized in Tables 11 and 12 [18].

G. Further Development of the Fluidized Bed Reactor

The main limitation in further scaling up rhamnolipid production is the fluidized bed reactor design as described above. Thus, at the Institut für Verfahrens und Kerntechnik of the TU Braunschweig, Potthof developed a new process mode for the production of rhamnolipids with immobilized cells [47]. In contrast to the common fluidized bed reactor system, the new reactor system could be termed a three-phase fluidized bed reactor additionally fitted with a static mixing system [46, 48] (see Fig. 18).

TABLE 11 Continuous Rhamnolipid Production with Immobilized Biocatalysts in a Fluidized Bed Reactor. The Four Production Processes are Disrupted by a Recultivation Process (cultivation time 10 h) in Order to Relax the Immobilized Cells

	Duration of the process (h)	Total CDW in beads (g CDW)	Total RL production (g)	Specific RL production (g RL/g CDW)	Total glycerol consumption (g)	$Y_{P/S}$
First production process	120	4.13	4.58	1.10	31.1	0.15
Second production process	98	5.45	5.34	0.98	37.1	0.14
Third production process	95	6.40	5.12	0.80	38.4	0.13
Fourth production process	95	7.24	3.98	0.55	30.1	0.13

TABLE 12 Regeneration Process of Immobilized Biocatalysts. Recultivation of Production Exceeded Immobilized Cells in a 6 L Fluidized Bed Reactor

	Duration of cultivation (h)	Total CDW in beads (g)	Total glycerol consumption (g)	$Y_{X/S}$
Production of periferic loaded biocatalysts				
Initial biocatalyst cell load	—	2.92	—	—
Cultivation in gel support	11	4.13	3.90	0.31
First recultivation process	10	5.45	4.25	0.31
Second recultivation process	10	6.40	3.52	0.27
Third recultivation process	10	7.24	3.5	0.24

Both the distribution of the biocatalyst and a necessary biocatalyst retention could be realized if the gas phase (air) and the fluid phase are mixed in a well-balanced countercurrent process mode (Fig. 18). According to these conduction principles, the oxygen transfer rate as well as the biocatalyst-retaining system are effectively improved. Both parameters are additionally amplified by equipping the reactor with a special static mixing system. The whole rhamnolipid production process using immobilized biocatalysts is demonstrated in Fig. 18.

Finally, it should be mentioned that this kind of process mode is at a further state of development. To create optimal examination conditions, a two-part cooperation between the department of Verfahrens und Kerntechnik and our department was founded. First, the concentration on bioengineering process parameters, like the determination of the fluid dynamics or some aspects on mass transfer phenomenon, will be examined at the first part of examination. Second, the bioprocessing parameters, like the determination of rhamnolipid productivity or long-time stability of this process, will be investigated.

Only the future will show whether immobilized cells systems producing biosurfactants like the rhamnolipids will find industrial application.

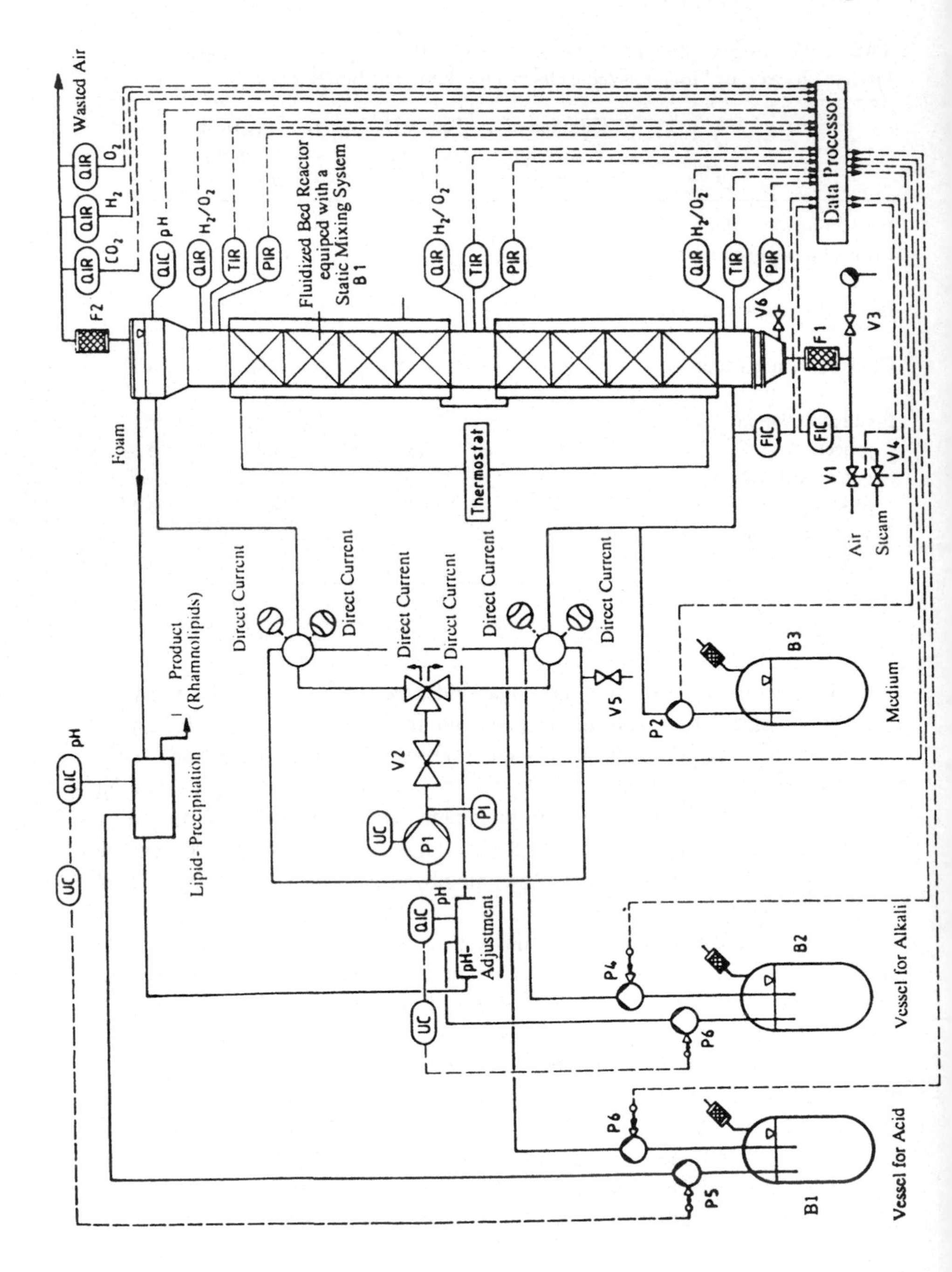
Wasted Air
QIR
QIR
QIR
CO_2
H_2
O_2
F2
QIC
pH
QIR
H_2/O_2
TIR
PIR
Fluidized Bed Reactor
equiped with a
Static Mixing System
B 1
QIR
H_2/O_2
TIR
PIR
QIR
H_2/O_2
TIR
PIR
Data Processor
V6
F1
V3
FIC
FIC
V1
V4
Air
Steam
Foam
Thermostat
Direct Current
Direct Current
Direct Current
Direct Current
Direct Current
Direct Current
Product
(Rhamnolipids)
Lipid- Precipitation
QIC
pH
UC
V5
V2
PI
UC
P1
P2
B3
Medium
QIC
pH
pH-
Adjustment
UC
P4
P6
B2
Vessel for Alkali
P6
P5
B1
Vessel for Acid

REFERENCES

1. J. Tramper, in *Physiology of Immobilized Cells* (J. A. M. de Bont, J. Visser, B. Mattiasson, and J. Tramper, eds.), Elsevier, Amsterdam, 1989, pp. 1–14.
2. J. Klein and F. Wagner, in *Dechema Monographien*, Vol. 82 (H. D. Rehm, ed.), Verlag Chemie, Weinheim, 1978, pp. 142–164.
3. J. Klein and F. Wagner, *Appl. Biochem. Bioeng. 4*: 11 (1983).
4. F. Wagner and S. Lang, in *Dechema Monographien*, Vol. 84 (H. D. Rehm, ed.), Verlag Chemie, Weinheim, 1979, pp. 315–335.
5. J. Klein and F. Wagner, *Appl. Biochem. Bioeng. 4*: 11 (1983).
6. C. D. Scott, *Enzyme Microb. Technol. 9*: 66 (1987).
7. F. Wagner, J.-S. Kim, S. Lang, Z.-Y. Li, G. Marwede, U. Matulovic, E. Ristau, and C. Syldatk, in *European Congress on Biotechnology*, Verlag Chemie, Weinheim, 1984, pp. 3–8.
8. S. Lang and F. Wagner, in *Biosurfactants and Biotechnology*, Surfactant Science Series, Vol. 25, (N. Kosaric, W. L. Cairns, and N. C. C. Gray, eds.), Marcel Dekker, New York, 1985, pp. 21–45.
9. C. Syldatk and F. Wagner, in *Biosurfactants and Biotechnology*, Surfactant Science Series, Vol. 25 (N. Kosaric, W. L. Cairns, and N. C. C. Gray, eds.), Marcel Dekker, New York, 1987, pp. 89–120.
10. C. Syldatk, U. Matulovic, and F. Wagner, *Biotech. Forum 1*: 58 (1984).
11. F. Wagner, *Fat Sci. Technol. 89*: 86 (1987).
12. C. Syldatk, S. Lang, U. Matulovic, and F. Wagner, *Z. Naturforsch. 40c*: 61 (1985).
13. C. Syldatk, S. Lang, F. Wagner, V. Wray, and L. Witte, *Z. Naturforsch. 40c*: 51 (1985).
14. L. Guerra-Santos, O. Käppeli, and A. Feichter, *Appl. Environ. Microbiol. 48*: 301 (1984).
15. C. P. Sidenko, D. I. Mordvinova, N. E. Yarotskaya, N. F. Melyukh, T. M. Klyushnikova, and G. F. Smirnova, *Microbiol. Zh. 48*: 26 (1986).
16. C. Syldatk, Dissertation am Institut für Biochemie und Biotechnologie, TU Braunschweig (1984).
17. U. Matulovic, Diplomarbeit am Institut für Biochemie und Biotechnologie, TU Braunschweig (1984).
18. U. Matulovic, Dissertation am Institut für Biochemie und Biotechnologie, TU Braunschweig (1987).
19. L. Fischer, Diplomarbeit am Institut für Biochemie und Biotechnologie, TU Braunschweig (1988).
20. M. Siemann, Diplomarbeit am Institut für Biochemie und Biotechnologie, TU Braunschweig (1989).

FIG. 18 Process for the continuous rhamnolipid production in a three-phase fluidized bed reactor as it was developed at the Department of Verfahrens und Kerntechnik at the Technical University of Braunschweig [47, 48]. (V) Valve; (F) Filter; (P) Pump; (B) Bottle; (QIR) Quantity indication registration; (QIC) Quantity indication control; (TIR) Temperature indication control; (PIR) Pressure indication control; (FIR) Flow indication control.

21. S. Fukui, S. A. Ahmed, T. Omata, and A. Tanaka, *Eur. J. Appl. Microbiol. Biotechnol. 10*: 289 (1980).
22. H.-J. Steinert, K. D. Vorlop, and J. Klein, in *Biocatalysts in Organic Media* (J. A. M. de Bont, J. Visser, B. Mattiasson, and J. Tramper, eds.), Elsevier, Amsterdam, 1986.
23. J. Klein, *Kontakte 3*: 29 (1980).
24. B. Gosmann and H. Rehm, *Appl. Microbiol. Biotechnol. 23*: 163 (1986).
25. J. M. Radovic, *Biotechnol. Adv. 3*: 1 (1985).
26. M. Dalili and P. C. Chau, *Appl. Microbiol. Biotechnol. 26*: 500 (1987).
27. C. N. Satterfield, MIT Press, Cambridge, MA, 1970.
28. J. M. Radovich, *Enzyme Microbiol. Technol. 7*: 2 (1985).
29. K. Buchholz, *Adv. Biochem. Eng. 24*: 39 (1982).
30. J. Klein, K.-D. Vorlop, and F. Wagner, in *Enzyme Engineering*, Vol. 7 (A. K. Laskin, G. T. Tsao, and L. B. Wingard, ed.), New York Academy of Science, New York, 1985, pp. 437–449.
31. U. Hackel, J. Klein, R. Megnet, and F. Wagner, *Eur. J. Appl. Microbiol. 1*: 291 (1975).
32. J. Klein and K. D. Vorlop, in *Jahrbuch der Biotechnologie* (P. Präve, ed.), Carl Hanser Verlag, München, 1986.
33. U. Matulovic, D. Rasch, and F. Wagner, *Biotech. Letts. 8*: 485 (1986).
34. M. Siemann, R. Müller-Hurtig, and F. Wagner, in *Physiology of Immobilized Cells* (J. A. M. de Bont, J. Visser, B. Mattiasson, and J. Tramper, eds.), Elsevier, Amsterdam, 1990, pp. 275–282.
35. H. Vogelmann and F. Wagner, *J. Chromat. 76*: 359 (1973).
36. C. K. A. Martin and F. Wagner, *Eur. J. Appl. Microbiol. 2*: 243 (1976).
37. W. G. Bang, S. Lang, H. Sahm, and F. Wagner, *Biotechnol. Bioeng. 25*: 999 (1983).
38. J. Klein and F. Wagner, in *Enzyme Engineering*, Vol. 8 (A. I. Laskin, K. Mosbach, D. Thomas, and L. B. Wingard, eds.), New York, Academy of Science, 1987, pp. 306–316.
39. F. Wagner, in Bioreaktoren; Entwicklung einer energiesparenden und umweltfreundlichen Technologie 2. *BMFT Statusseminar Bioverfahrenstechnik*, Jülich, Bundesministerium für Forschung und Entwicklung, 1979, pp. 145–155.
40. J. Lehmann, U. Oels, R. Schügerl, J. Todt, M. Reuss, and F. Wagner, *1st Symposium Mikrobielle Proteingewinnung* (F. Wagner, ed.), Verlag Chemie, Weinheim, FRG, 1975, pp. 133–141.
41. J. Lehmann and F. Wagner, in Bioreaktoren; Entwicklung einer energiesparenden und umweltfreundlichen Technologie. *BMFT-Statusseminar Bioverfahrenstechnik*, Braunschweig, Stockheim. Deutsche Forschungs- und versuchsanstalt für Luft and Raumfahrt e.V. DFVLR, 1977, pp. 173–186.
42. I. Feige, Diplomarbeit am Institut für Biochemie und Biotechnologie, TU Braunschweig (1987).
43. M. Siemann, R. Müller-Hurtig, L. Fischer, and F. Wagner, *Dechema Biotechnology Conferences*, Vol. 3 (H. J. Rehm, ed.), Verlag Chemie, Weinheim, 1989, pp. 649–652.
44. R. Müller-Hurtig, U. Matulovic, I. Feige, and F. Wagner, in *Proceedings of the 4th European Congress on Biotechnology*, Vol. 2, Elsevier, Amsterdam, 1987, pp. 253–256.
45. O. Moebus and M. Teuber, *Forum Mikrobiologie 4*: 187 (1986).

46. G. Wild, M. Sabeian, J. L. Schwartz, and J. C. Karpention, *Int. Cem. Eng. 24*: 639 (1984).
47. M. Potthoff, Dissertation am Institut für Verfahrens—und Kerntechnik, TU Braunschweig (in preparation).
48. J. Tiefke, Dissertation am Institut für Verfahrens—und Kerntechnik, TU Braunschweig (1984).
49. H. Tanaka, M. Matsumura, and I. A. Veliky, *Biotechnol. Bioeng. 26*: 53 (1984).
50. R. I. Scott, *J. Chem. Technol. Biotechnol. 39*: 144 (1987).

4

Lipopeptide Production by *Bacillus licheniformis*

KATHARINA JENNY and V. DELTRIEU Institute of Biotechnology, Swiss Federal Institute of Technology, Zurich, Switzerland

O. KÄPPELI Biosystems Division, BIDECO AG, Dübendorf, Switzerland

I. INTRODUCTION

Many types of surface-active agents are synthesized by a wide variety of microorganisms. Mostly they exhibit the typical amphiphilic character of lipids, but they are generally extracellular. Basically there are six major classes of biosurfactants: glycolipids, lipopeptides/lipoproteins, phospholipids, neutral lipids, substituted fatty acids, and lipopolysaccharides. The structure, function, and physiological role of these biological surface-active agents have been described in several reviews [1–5].

The most often isolated and most thoroughly studied biosurfactants are the structural homogenous glycolipids, for example, sophorose lipids, rhamnolipids, and trehalose lipids. On the contrary, the group of lipopeptides/lipoproteins presents a heterogenous class of biologically active peptides. Table 1 gives a survey of lipopeptides and lipoproteins produced by non-*Bacillus* microorganisms described in the literature. As far as are available the compositions of the active compounds are indicated.

The spectrum of lipopeptides produced by *Bacillus* spp. is broad, but only a few have been reported in relation to their surface activity. Most of them are known because of their antibiotic characteristics. They are grouped into linear and cyclic lipopeptides, which were further divided into subgroups according to the nature of the ring in the molecule [6]. The bifunctional properties exhibited by lipopeptides (surface and antibiotic activity) is equally well known for many other biosurfactants, such as rhamnolipids and viscosin from *Pseudomonas* spp. as well as glycolipids of *Torulopsis apicola* [7–10].

With respect to surface activity, surfactin, a lipopeptide from *B. subtilis*, was characterized most thoroughly. It is a cyclic lipopeptide which turned out to be a highly effective biosurfactant [11]. Already 20 mg L^{-1} of the purified product reduced the surface tension of water from 72 to 27 mN m^{-1}. The biosynthesis of surfactin was studied by Kluge et al. [12] in a cell-free system. Formation of the peptide chart occurred nonribosomally by the thiotemplate mechanism similar to numerous other peptide antibiotics of *Bacillus* spp. Peptide synthesis by the thiotemplate mechanism proceeds in three main steps:

1. Activation of the amino acid by formation of an aminoacyladenylate.
2. Thiolaminoacylation, which is the transfer of the activated amino acid to a specific thiol group of the enzyme complex.
3. Formation of the peptide bond involving a transpeptidation and a transthiolation within the enzyme complex catalyzed by 4′-phosphopantethein.

Compared with the ribosomal peptide and protein synthesis, the thiotemplate mechanism represents a relatively unspecific peptide formation. The incorporation of analogs of amino acids is often observed [13]. An exchange of amino acids within the group of Leu/Ile/Val/Thr or Phe/Tyr/Trp was found in Gramicidin and Tyrocidin.

The lipophilic part of the amphiphilic molecule is formed by fatty acids of different chain lengths. Because branched and hydroxylated fatty acids are characteristic for *Bacillus* spp. [14], they are often found in surface-active lipopeptides produced by these microorganisms. The biosynthesis of branched fatty acids is linked to that of branched amino acids [3, 15–18].

TABLE 1 Survey of Lipopeptides and Lipoproteins Produced by Microorganisms Other than *Bacillus* spp. (from Various Sources)

Organism	Composition
Mycobacterium fortuitum	Lipophilic part: C_{20} or C_{22} fatty acids Hydrophilic part: 9 AA (3 LVal; 2 L Thr; LAla; LPro; 2 MeLeu)
Mycobacterium paratuberculosis	Lipophilic part: C_{20} fatty acid; Hydrophilic part: 9 AA (LPhe, DPhe, LAla, LLeu, LIle)
Nocardia asteroides	Lipophilic part: β-OH C_{20} fatty acids Hydrophilic part: 7 AA (2 LThr, LVal, LPro, LAla, DAla, D-alloIle)
Corynebacterium lepus	Lipophilic part: C_{13}–C_{24} (25%) fatty acid, Corynomycolic acid (75%); Hydrophilic part: 13 different AA
Streptomyces canus *S. violaceus*	Lipophilic part: 3-ai C_{13}, 3-i C_{12} fatty acid; Hydrophilic part: 10 AA
Serratia marescens	Lipophilic part: 2 β-OH C_{10} fatty acid Hydrophilic part: 2 LSer
Pseudomonas fluorescens	β-OH-C_{10}-LLeu-DGlu Dallo Thr DVal
Pseudomonas viscosa	LIle-DSer-LLeu-DSer-LLeu
P. rubescens *Thiobacillus thiooxidans* *Rhodopseudomonas spheroides* *Streptomyces sioyaensis*	Lipophilic part: β-OH fatty acids Hydrophilic part: Ornithin
Agrobacterium tumefaciens	Lipophilic part: β-OH fatty acids Hydrophilic part: Lysin
Gluconobacter cerinus	Lipophilic part: β-OH fatty acids; Hydrophilic part: Taurin and Ornithin
Candida petrophilum	Peptide (Glu, Asp, Ala und Leu) + non-identified fatty acids
Acinetobacter calcoaceticus	Protein lipid
Pseudomonas fluorescens	Carbohydrate-protein-complex with a minor lipid part (M.W. 300 kDa)
Corynebacterium hydrocarboclastus	Protein-lipid-carbohydrate-complex
Pseudomonas aeruginosa	Protein-like activator MW ca. 14,300 (147 AA)
Candida lipolytica	Carbohydrate-protein-complex; (ca. 27.6 kDa)

In the following section, biosurfactant production of *Bacillus licheniformis* is reviewed. The focus is placed on the physiology of production and on the characterization of the surface-active compounds. So far, *B. licheniformis* is used for the production of Bacitracin, a dodecapeptide antibioticum. Hanlon and Hodges [19] investigated its formation with different carbon and nitrogen sources as well as in relation to spore and protease formation. A complex and heterogenous regulation pattern resulted for the various aspects considered. An increased formation of Bacitracin was obtained by the external addition of L-leucine [20]. The role of the amino acid in the activation of the thiotemplate mechanism remained, however, unclear.

II. PHYSIOLOGY OF BIOSURFACTANT PRODUCTION BY *BACILLUS LICHENIFORMIS*

The physiology of biosurfactant production in *Bacillus* spp. has not been investigated intensively. Most work was carried out on surfactin production in batch cultures of *B. subtilis*. The continuous removal of the product by foam separation and increased concentrations of the trace elements iron and manganese enhanced surfactin formation by the cells [11]. Ultraviolet mutants of *B. subtilis* produced under similar conditions approximately three times more surfactin than the wild type [21].

Growth and biosurfactant production by *B. licheniformis* were investigated by Jenneman et al. [22] and by Pfiffner et al. [23]. A saccharose minimal medium was used, and the influence of high NaCl concentrations at different pH values was evaluated. The authors did not notice any significant improvement in biosurfactant formation whatever conditions applied. Even the addition of hexadecane, a known inducer for excretion of surface-active compounds did not change that observation. The active compounds were separated into three components by thin-layer chromatography, but further characterization was not presented.

Javaheri et al. [24] investigated the same strain biosurfactant production under anaerobic conditions with glucose as the carbon source. Yeast extract and $NaNO_3$ were essential for anaerobic growth. However, no significant differences to aerobic biosurfactant production were obvious.

The results available showed that biosurfactant production with *B. licheniformis* was rather difficult to achieve, because the cells exhibited a considerable metabolic flexibility that impeded the definition of conditions for the release of surface-active compounds. In view of this fact a more systematic approach was undertaken in order to correlate environmental conditions to cellular physiology, that is, the formation of biosurfactants and other metabolic products.

A. The Medium

Medium composition plays an important role in the production of surface-active compounds by microorganisms. Minor differences in the amount of critical nutrients may change the limiting nutrient and therefore influence the performance of the cells, as has been shown for rhamnolipid production with *Pseudomonas aeruginosa* [25, 26]. The basis for such investigations is the availability of a defined medium for growth and rudimentary biosurfactant production. Subsequently the composition and the concentration of the components can be optimized. Table 2 gives such a medium for *B. licheniformis*. Besides glucose, saccharose can be used as a carbon source. It is noteworthy, however, that *n*-alkanes are not used as a sole source of carbon and energy by the cells.

TABLE 2 Composition of a Defined Medium for Growth and Biosurfactant Production of *B. licheniformis*

Component	Amount (L^{-1})
Glucose	20 g
$NaNO_3$	4 g
$MgSO_4 \cdot 7H_2O$	0.4 g
KCl	0.2 g
$CaCl_2 \cdot 2H_2O$	0.1 g
H_3PO_4, ($\delta = 1.71$)	0.5 mL
H_3BO_4	1.53 mg
$CuSO_4 \cdot 5H_2O$	0.284 mg
$MnSO_4 \cdot H_2O$	1.71 mg
$Na_2MoO_4 \cdot 2H_2O$	0.7 mg
$ZnSO_4 \cdot 7H_2O$	2.9 mg
$FeSO_4 \cdot 7H_2O$	4.3 mg
$CoCl_2 \cdot 6H_2O$	0.1 mg
EDTA	200 mg
Calcium–Pantothenate	1.176 mg
Biotin	5.88 μg
Folic acid	5.88 μg
Inositol	0.588 mg
Niacin	1.176 mg
p-Aminobenzoic acid	0.588 mg
Pyrodoxine-HCl	1.176 mg
Riboflavin	0.588 mg
Thiamin-HCl	1.176 mg

B. Growth and Product Formation

Growth experiments were carried out with the medium listed in Table 2. The interest was placed on product formation (including biosurfactants) under different environmental conditions, that is, under aerobic, oxygen-limited, and anaerobic conditions.

Continuous culture experiments yielded the basic growth parameters of the *B. licheniformis* strain used (Table 3). Maximum growth rate under anaerobic conditions decreased significantly as compared to those where oxygen was supplied. RQ and yield values indicated, however, that under oxygen-limited incomplete oxidation of the carbon source appeared. Carbon flux analysis confirmed this assumption (Fig. 1). The main products of energy metabolism under oxygen-limited and anaerobic conditions were acetate, acetoin, butanediol, pyruvate, lactate, and ethanol. These are products of the mixed acid formation and of the butanediol cycle. These metabolic pathways seem to be characteristic for *B. licheniformis*, but metabolic regulation is not yet elucidated in detail for this organism [27]. Under fully aerobic conditions, complete oxidation of the carbon source occurred. *B. licheniformis* represents, therefore, an organism where oxygen is a major metabolic regulator.

Biosurfactant formation by the cells under different degrees of oxygen supply was best under oxygen limitation. In an oxygen-limited continuous culture the surface tension was below 30 mN m^{-1} whereas under fully aerobic and anaerobic conditions surface tension remained at 40 mN m^{-1} or above (Fig. 2). These observations indicated that oxygen limitation could form the basis for the production of surface-active compounds by *B. licheniformis*.

C. Toward a Production Process

With *P. aeruginosa*, it was possible to produce rhamnolipids under steady-state conditions in continuous cultures [25]. Medium limitations yielded conditions where product formation was maintained over expanded periods of time.

TABLE 3 Important Biological Parameters of B. *licheniformis* Cultivated at Different Conditions with Respect to Oxygen Supply

Parameter	Aerob.	Oxygen-limited	Anaerob.
D_{max}, (h^{-1})	0.38	0.40	0.16
qO_2 (mmol g^{-1} h^{-1})	13.6	10.0	—
qCO_2 (mmol g^{-1} h^{-1})	14.6	25.2	5.8
$Y_{x/s}$ (g g^{-1})	0.39	0.17	0.4
RQ	1.1	2.25	—

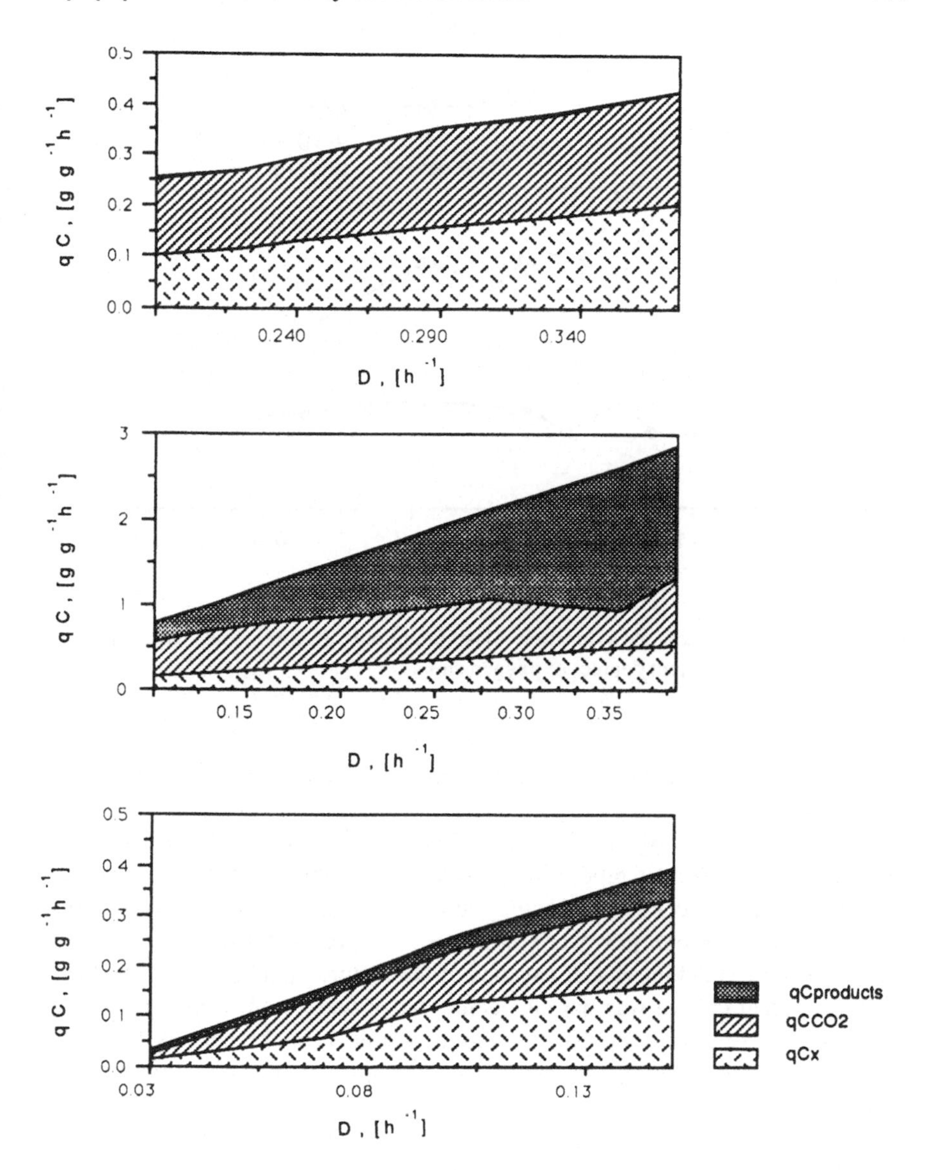

FIG. 1 Carbon flux distribution in continuous cultures of *B. licheniformis* under fully aerobic (a), oxygen-limited (b), and anaerobic (c) conditions. Products observed from the incomplete oxidation of the carbon source were formed under oxygen-limited and under anaerobic conditions. Products were acetate, acetoin, butanediol, pyruvate, lactate, and ethanol.

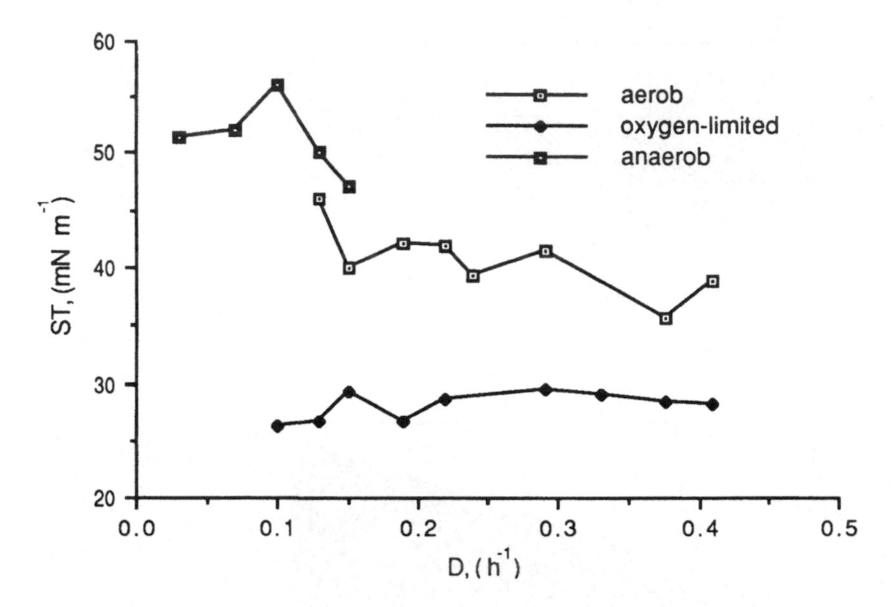

FIG. 2 Surface tension as a function of dilution rate in continuous cultures of *B. licheniformis* with different degrees of oxygen supply.

Obviously, the physiology of the cells remained constant and the productive state was stable.

The cellular physiology of *B. licheniformis* is less stable under steady-state conditions. Biosurfactant production was never possible for longer cultivation periods. Changes in cellular physiology in a biosurfactant-producing continuous culture of *B. licheniformis* were indicated by excess foam formation and increased surface and interfacial tensions of the culture liquid (Fig. 3). The basis for this gradual change was not elucidated, but it was assumed that adaptation processes are very slow or that the genetic stability of the cells is low. The production of biosurfactant appears to be a transitory phenomenon, in this case persisting for about 70 generations.

The importance of transitory states for biosurfactant production was confirmed by pulse experiments in a continuous culture of *B. licheniformis*. Addition of glucose to an inactive culture induced the formation of surface-active compounds (Fig. 4). Consequently, the development of a production process involves the design of a cultivation where transitory states are maintained. Attempts in realizing this by continuously phasing in cultures with *B. subtilis* yielded significantly increased periods of time where surfactin was released by the cells [28].

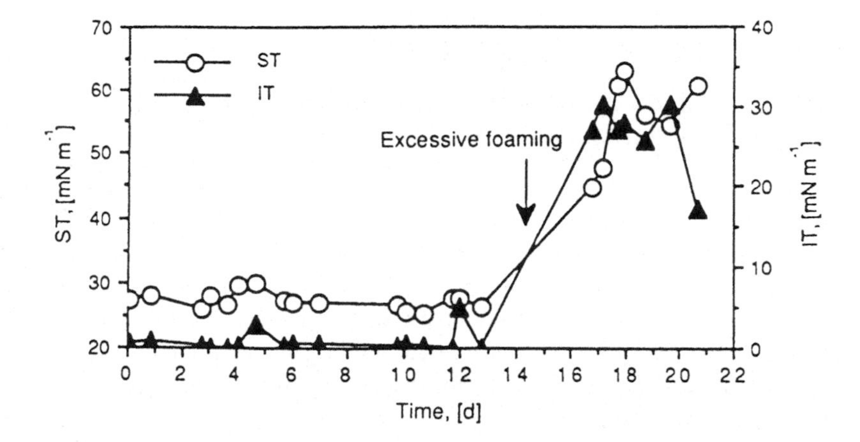

FIG. 3 Influence of cultivation period on surface tension (ST) and interfacial tension (IT) in a continuous culture of *B. licheniformis* (D = 0.15 h^{-1}). After 12 days of cultivation, surface and interfacial tensions of the culture liquid increase, indicating the cessation of biosurfactant production. At the same time, excessive foam formation occurs.

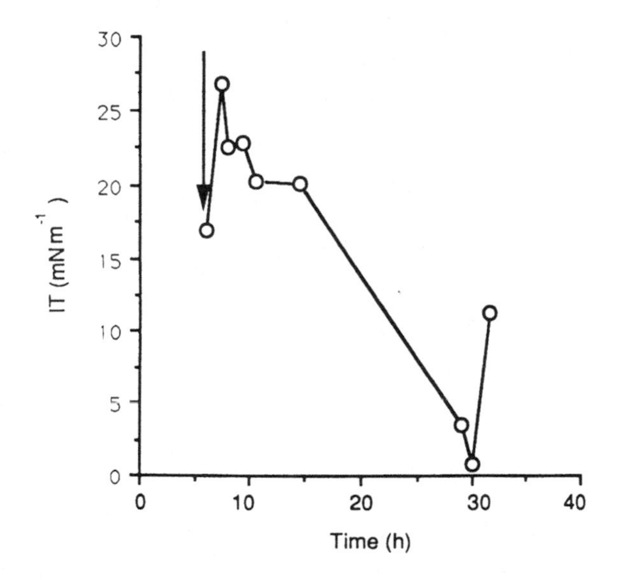

FIG. 4 Induction of biosurfactant production in a continuous culture of *B. licheniformis* (D = 0.04 h^{-1}) by introducing a transitionary state through glucose pulsing (15 g L^{-1}) as indicated by the arrow.

III. PURIFICATION AND IDENTIFICATION OF SURFACE-ACTIVE COMPOUNDS

A. Surface-Active Compounds from the Spent Medium of *B. licheniformis* Cultures

Surface-active compound were isolated from the culture supernatant of continuously growing cells by a step-by-step purification of the raw material (HCl-precipitate) using adsorption chromatography and liquid–liquid partitioning. An enrichment of the active compounds resulted. Adsorption chromatography of the crude product on a silica gel column by stepwise elution using solvents of increasing polarities yielded at least three active product fractions, which exhibited distinctly different retention factor (R_f) values, when submitted to thin layer chromatography (Table 4).

Fraction 1 was directly analyzed on gc/ms without any prior treatment. The resulting ms spectra showed the typical signals of long-chain hydrocarbons with the base peak at m/e 57 ($C_4H_9^+$) and a number of fragment ions at m/e 71, 85, 99, 113, and so on, because of the loss of a methylene group (– 4 resp. —CH_2). The molecular weight represented by the molecular ions (M^+) present ranged from 282 to 310 corresponding C_{20} and C_{22} aliphatic hydrocarbons. Hydrocarbons as extracellular products are rather unusual, because they function more as cellular components than as extracellular products. The ability of microorganisms to synthesize appreciable amounts of hydrocarbons is limited to a few bacterial, algal, and fungal species [29]. Up to the present *Bacillus* spp. are not described as characteristic hydrocarbon producers. Because *B. licheniformis* has been

TABLE 4 Detection of the Separated Active Compounds on TLC (solvent system: $CHCl_3$–CH_3OH–CH_3COOH, 80:15:5)

Fraction	R_f-value	Minimum surface tension ($mN\ m^{-1}$)	Detection with DPH	Amount in the crude extract (%)	cmc ($mg\ L^{-1}$)
$CHCl_3$-eluate	0.96	29.5	+	20	n.d.
Acetone-eluate	0.96	—	+	both together	n.d.
	0.83	32.0	+	3.5	
$CHCl_3$–CH_3OH, CH_3OH-eluate	0.54	28.0	+	67	30

All of the four compounds detected on TLC by their different R_f-values showed surface activity and a lipid-positive reaction with DPH. The highest amount of active fraction was found however in fraction 3, the $CHCl_3$–CH_3OH eluat.

cultivated on a water-soluble substrate, the hydrocarbons do not represent an artifact derived from the extraction procedure as it does for many alkane-utilizing organisms. However, similar substances, described as wax esters, were identified by thin layer chromatography for a *Bacillus* spp. which showed significant emulsifying properties [30].

Fraction 2 was composed of two different compounds. The spot corresponding to a R_f value of 0.96 was due to traces of an unknown glycolipid (detection with orcinol-H_2SO_4 [31]). The second compound (R_f 0.83) of fraction 2 was analyzed by gc and gc/ms. Fatty acid methyl esters of the untreated as well as of the previously hydrolyzed sample showed the same pattern in gc analysis. The chromatogram exhibited two peaks with different retention times. Compared to the standard solutions of fatty acid methyl esters they were identified as saturated, long-chain carboxy acids namely palmitic ($C_{16:0}$) and stearic acid ($C_{18:0}$). The result was verified by gc/ms with the characteristic base peak at m/z 74 typical for saturated fatty acid methyl esters due to the ions formed up on a α–β cleavage with simultaneous migration of one hydrogen atom from the lost fragment, known as the McLafferty rearrangement. Signals at M-29, M-31, and M-43 due to the loss of an ethyl-, a methoxy-, and a propyl- group and the presence of the molecular ions (M^+) at m/z 270 and 298 defined the chain length to be C_{16} and C_{18}.

Fraction 3, the main component of the crude product and the most surface-active one was analyzed by different procedures after further purification. Ion exchange column chromatography of the main surface-active fraction of *B. licheniformis* (fraction 3 from silica gel column chromatography) on DEAE sepharose CL 6B resulted in a single peak in the region where surface tension was low (Fig. 5). The active fractions were collected, dried, and dissolved in a chloroform-methanol (2:1) mixture before submitting to HPLC. HPLC analysis of the pooled fractions resulted in at least two major peaks. They appeared throughout with a ratio of 2.1:1. The isolated peaks were identified as lipid positive spots on thin layer chromatography, both exhibiting the same R_f value as the primary product of the silica gel procedure. This indicated that fraction 3 was made up by two closely related compounds.

B. Identification of Lipopeptides, the Main Surface-Active Compounds

The infrared (IR) spectrum of fraction 3 in KBr (Fig. 6) showed strong bands characteristic for peptides at 3300 cm^{-1} (band A), at 1655 cm^{-1} (band F), and at 1535 cm^{-1} (band G) resulting from the N–H stretching mode, the stretching mode of the >CO—O bond and the deformation mode (combined with the C—N stretching mode) of the N—H bond, respectively. The bands at 2960–2930 cm^{-1}, 2880–2860 cm^{-1} (bands B, C, and D) and at 1470–1430 cm^{-1}, 1390–1370 cm^{-1}

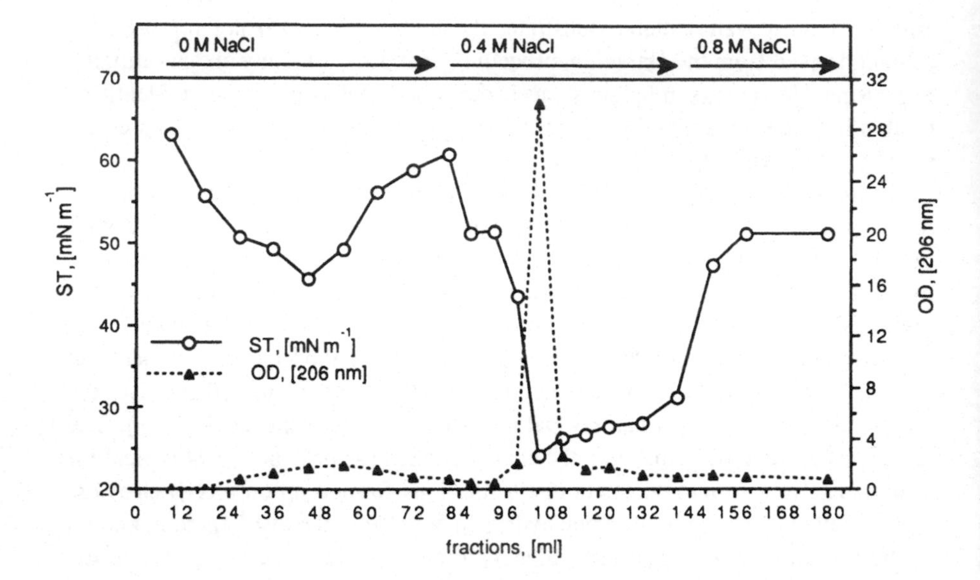

FIG. 5 Ion-exchange chromatography of the surface-active fraction 3 on DEAE-sepharose. Elution was effected by a gradient from 0 M to 0.8 M NaCl in 10 mM Tris-HCl containing 20% EtOH (pH 8). The absorption of the fractions by 206 nm was monitored. The product fraction appeared as a single peak in the region where surface activity was high.

(bands H and I) reflect aliphatic chains ($—CH_3$, $—CH_2—$) of the fraction. Band E was due to lactone carbonyl absorption, indicating that the product contains aliphatic hydrocarbons as well as peptidelike moiety.

The mass of the complete molecules was determined by both combined laser desorption/Fourier transform mass spectrometry and fast atom bombardment mass spectrometry (Fig. 7). Two molecular masses, 1022 and 1036, were determined with both methods as derived from different attachment of H^+, Na^+, and K^+. The mass difference of 14 units characterizes the lipopeptide as a mixture of closely related molecules varying in their fatty acid residues.

The lipophilic part of the biosurfactant was analyzed by combined gc/ms of the fatty acids methyl esters and derivatives it contained. The gc spectrum was dominated by four main peaks, namely β-OH-iso C_{14}, β-OH C_{14}, β-OH-iso C_{15}, and β-OH-anteiso C_{15} with a ratio of 26:45:15:14%. The peaks were identified by mass spectrometry of the fatty acid methyl esters giving a characteristic base peak at m/z 103 [$CH(OH)CH_2COOCH_3$] due to the fragment ion caused by β,γ-fragmentation commonly known for β-OH fatty acid methyl esters and the fragment ion peaks such as M-18, M-50, and M-73 for β-OH acids. Further elucidation was

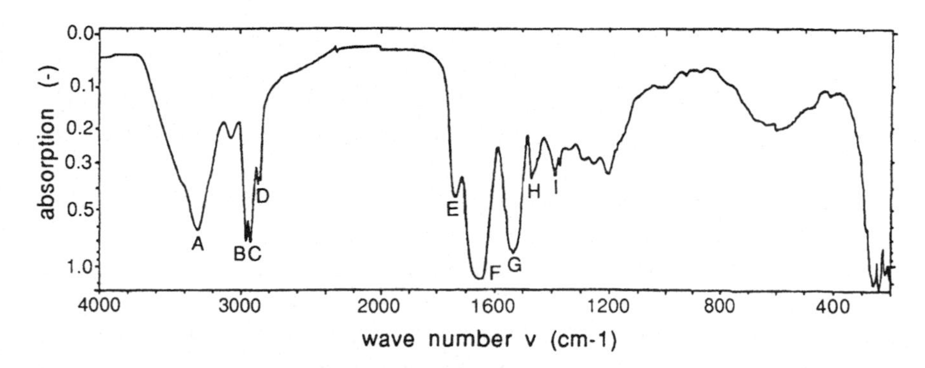

FIG. 6 Infrared spectrum of the surface-active fraction 3.

carried out by mass spectrometry of the dehydrated hydrogenated forms for determining the chain length (yielding m/z, 74, M-31, M-43 for saturated fatty acids) and of the picolinylesters (yielding m/z 92 or 108, +/– M-43 or M-57 characteristic for iso- and anteiso-branching).

For the hydrophilic part, an amino acid composition of Asp, Leu, Val, Glu, and Ile with a ratio of 1:3:1:1:1 was determined. On the basis of the molecular weight and together with the ninhydrin negative reaction of the native

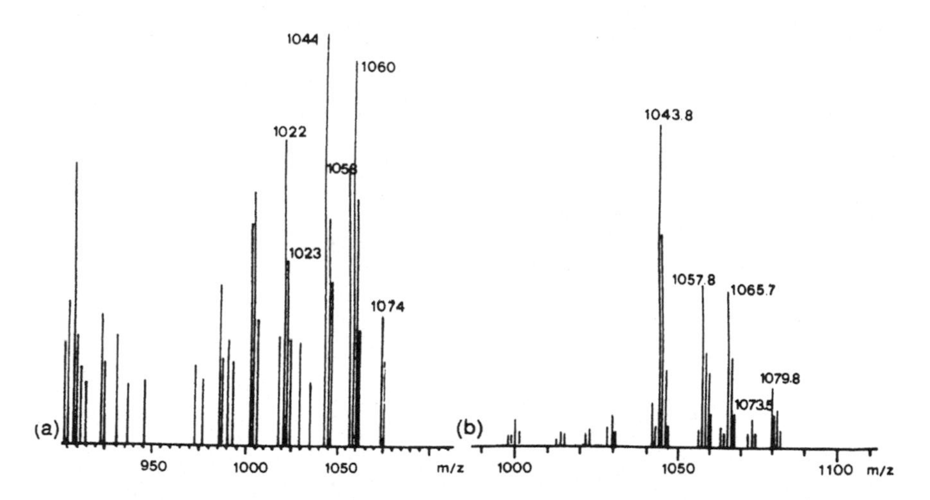

FIG. 7 Partial mass spectrums of fraction 3 showing the region of the molecule masses by the molecular ions (M^+; 1022, 1036) and by abundant ions like ($[M + H]^+$; 1023, 1037), ($[M + Na]^+$; 1044, 1058), ($[M + 2Na]^+$; 1066, 1078) and ($[M + K]^+$; 1060, 1074): (a) Fourier transform mass spectrum. (b) Fast atom bombardment mass spectrum.

compound, a cyclic closed structure of the peptide part was conceivable. For sequence analysis the lipopeptide was hydrolyzed in 12 N HCl—CH_3COOH (2:1) and the subsequent fragments were separated by HPLC. The selected ninhydrin positive fragments were identified yielding the sequences shown in Table 5.

Glutamic acid was not detected in any of the fragments, but the direct linkage of glutamic acid to the fatty acid part and the position of the dipeptide Leu–Ile at the end of the sequence could be demonstrated by the mass spectrum of the permethylated product (Fig. 8; Table 6). The spectrum of Fig. 8 may be divided into groups representing four compounds: Group A shows the permethylated acid itself, which is a mixture of two homologs (C_{14} and C_{15}), and group B shows the same mixture, which has lost one molecule of methanol from the acyl constituent. Therefore in the lower mass region, peaks are forming a characteristic fragment peak group of four signals. The mass spectral fragmentation occurred predominantly at the peptide CO—NMe bonds; the resulting peaks, as outlined in Table 6 and Fig. 8 (for C_{15} constituent only) delineated the sequence.

In the spectrum, the molecular ion peak M^+ is only recognized for the C_{15} constituent at m/z 1207.8. The region of the high mass shows additionally complex information from the attachment with Na, which leads to the formation of quasimolecular ions at m/z 1229.9 $(M + Na)^+$ and at m/z 1197.9 and 1183 $(M—OCH_3 + Na)^+$ and due to the further loss of OMe from the C-terminal amino acid $(M—OCH_3)^+$. The two isomeric amino acids leucine and isoleucine and the different branching of the C_{14} and C_{15} fatty acids are difficult to differentiate,

TABLE 5 Amino Acid Sequences of Peptide Fragments

Peptide fragment	Amino acid sequences					
1		Leu—	Val—	Asp—		
2		Leu—	Val—	Asp—		
3			Val—			
6					Leu—	Ile
9	Leu—	Leu—	Val—			
11		Leu—	Val—			
17		Leu—				
Total sequence	Leu—	Leu—	Val—	Asp—	Leu—	Ile

Six of the seven amino acids of the lipopeptide fraction were found in fragments resulting from HPLC separation of an alkaline hydrolysis.

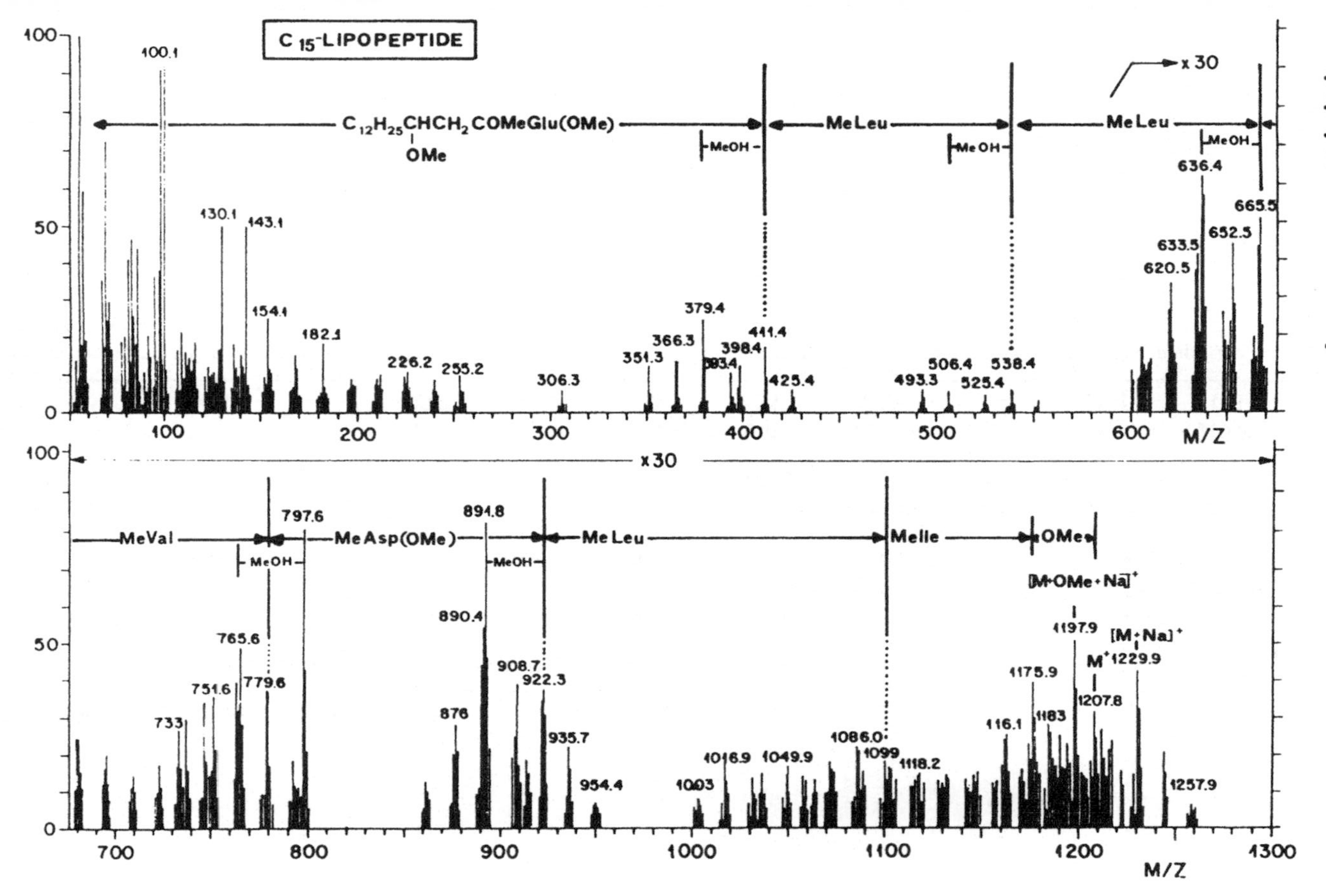

FIG. 8 FAB spectrum of the O,N-permethylated fraction 3 after alkaline hydrolysis in 10% NaOH in EtOH (3-NOBA matrix). The fragmentation ions are outlined by means of the C_{15} constituent.

TABLE 6 Identification of the Characteristic Fragmentation Ions of Permethylated Fraction 3

Group A	$RCHOMeCH_2COMeGluOMe$	MeLeu	MeLeu	MeVal	MeAsp OMe	MeLeu	Melle	OCH_3	Na^+
C_{15}	411.4	538.4	666.4	779.6	922.3	1099	1175.9	1207.8	1229.9
C_{14}	398.4	525.4	652.5	765.6	908.7	1086	1161	—	—
Group B	RCH=CHCOMeGluOMe	MeLeu	MeLeu	MeVal	MeAsp OMe	MeLeu	Melle	OCH_3	Na^+
C_{15}	379.8	506.4	634.4	(746.6)	890.4	1016.9	(1144)	1175.9	1197.9
C_{14}	366.3	493.3	620.5	733	876	1003	—	1161	1183

Group A shows the fragmentation pattern of the permethylated product of each C_{14} and C_{15} constituent. The two homologous constituents in group B are due to a primary loss of MeOH at the acyl end.

because they give rise to the same mass increments. An elimination of M-42/M-28 characteristic of Leu/Ile at the C-terminal end was not obvious due to the complex mass region.

However, the above results led to the formulation of the structure for the acidic form of the main surface-active fraction:

$$R_{1-4}–(CH_2)_8–CHOHCH_2CO–Glu–Leu–Leu–Val–Asp–Leu–Ile–OH$$

The molecular weight and the infrared spectrum of the purified compound, however, indicated a lactone ring bond. The determination of the location of the lactone ring in the compound was carried out by digestion with carboxypeptidase Y, which is known to cleave specifically peptides possessing a free C-terminal residue. The enzyme reaction was controlled on HPLC. The untreated surface-active main fraction did not show any reaction after a 20-h treatment with the enzyme solution, but the free acid form of the lipopeptide derived from alkaline treatment with 10% NaOH in EtOH reacted with the enzyme within 45 min. Here the retention time of the educt was approximately 1 min shorter than the product. These results excluded a lactone ring bond with the free carboxyl group of the aspartic or glutamic acid, and the final structure of the main surface-active fraction was elucidated as it is shown in Fig. 9.

C. Physical Product Characteristics

The critical micelle concentration (cmc) of the purified lipopeptides was found to be approximately 15–20 mg L^{-1} in 0.1 M potassium phosphate buffer (pH 7). The dependence of the surface activity on temperature and pH was investigated in the same buffer solution at concentrations of 1 cmc (Fig. 10). Temperatures in the range 20–60°C did not show any influence on the surface tension or interfacial tension. An optimal activity was detected at a pH range from 5 to 9, below and above these values a slight decrease of the activity was observed.

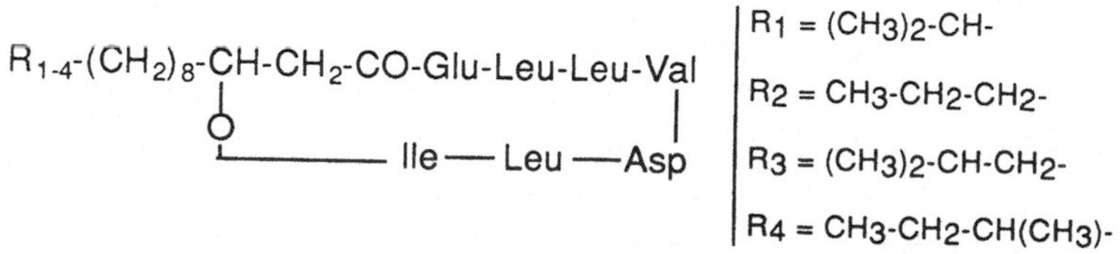

R_{1-4}-$(CH_2)_8$-CH-CH_2-CO-Glu-Leu-Leu-Val
O (lactone bond to Ile) — Ile — Leu — Asp

$R_1 = (CH_3)_2$-CH-
$R_2 = CH_3$-CH_2-CH_2-
$R_3 = (CH_3)_2$-CH-CH_2-
$R_4 = CH_3$-CH_2-CH(CH_3)-

FIG. 9 Proposed structure of the surface-active lipopeptides from *B. licheniformis*.

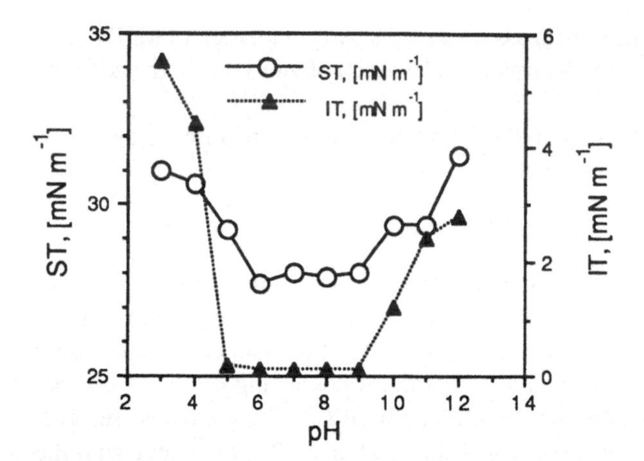

FIG. 10 Dependence of surface tension (ST) and interfacial tension (IT) on pH for the purified lipopeptide fraction 3.

D. Biological Product Characterization

Antibiotic activities of the lipopeptides were tested in concentrations of 0.1–40 mg L^{-1}. A survey of the results is given in Table 7. The lipopeptides showed activity against a variety of yeast strains with older cells being affected by a ten times lower concentration as compared to exponentially growing cells. Growth of the gram negative bacteria *P. aeruginosa* and *E. coli* was inhibited by a minimal concentration of 1 mg mL^{-1}. *Bacillus licheniformis* itself showed high susceptibility to the product. Concentrations as low as 0.5 mg mL^{-1} inhibited growth. Among the various fungi tested (11 strains) only the growth of *Trichoderma reesei* and *Penicillium oxalicum* was negatively influenced by the lipopeptides.

E. Classification of *B. licheniformis* Lipopeptides

The spectrum of lipopeptides produced by *Bacillus* spp. is broad but only a few have been reported in relation to their surface activity. Most of them are known because of their antibiotic characteristics. Within the peptide antibiotic group, the lipopeptides of *B. licheniformis* show analogous structures as representatives of the iturin group (iturin, mycosubtilin, bacillomycin) or of the lactone group (esperin, surfactin, polypeptin) [6]. Most of them show similarities in the hydrophilic part, consisting of a seven-amino acid cyclic peptide which is linked by a hydroxy or an ester peptide linkage to the fatty acid part.

TABLE 7 Antibiotic Activity of the Surface-Active Lipopeptides from *B. licheniformis* against Yeasts, Bacteria, and Fungi in Agar-Diffusion Tests[a]

Microorganisms	Test media[b]	Temperature (°C)	MIC[c] (mg mL^{-1})	MIC[d] (mg mL^{-1})	Inhibition zones[e]
Bacillus licheniformis	LB	37	0.5	—	++
Pseudomonas aeruginosa	LB	30	1	—	+++
Escherichia coli	LB	37	1	0.5	++++
Candida utilis	YEPD	30	1	0.1	+++
Candida tropicalis	YEPD	30	1	0.1	+++
Trichosporon cutaneum	YEPD	30	10	—	+
Saccharomyces cerevisiae	YEPD	30	5	0.1	++
Trichoderma reesei	MEA	30	0.5	0.1	++
Penicillium oxalicum	MEA/PDA	30	0.1	—	++++

[a]No inhibition of growth was obtained with *Aspergillus nidulans, Phanerochaetae chrysosporeum, Bjerkandera adusta, Ustilago maydis, Sordaria fimicola, Botrytis cinerea, Fusarium lycopersicii, Phytophtora infestans, Phytium debarayanum.*

[b]Media composition in g L^{-1}; LB, yeast extract 5, glucose 10, tryptone 10, NaCl 5, agar 15. YEPD, yeast extract 10, glucose 20, peptone 20, 20. MEA, malt extract 20, agar 20. PDA, potato dextrose 20, agar 20.

[c]MIC, minimal inhibitory concentration against microorganisms during the exponential growth phase.

[d]MIC against microorganisms during decreasing and death phase.

[e]Maximium diameter of halos: + = 7–10 mm, ++ = 11–14 mm, +++ = 15–18 mm, ++++ = >18 mm.

Surfactin produced by *B. subtilis* shows structurally the highest analogy to the lipopeptides of *B. licheniformis* [32, 33]. Surfactin is already known as a potent surface-active agent and as an antibiotic [11, 21]. This bifunctional nature of surface-active agents is well known for many other biosurfactants, not only for those originating from representatives of the *Bacillus* genus. The antibiotic spectrum of the lipopeptides from *B. licheniformis* seems to be identical to that of surfactin. High antibiotic activity of surfactin against gram negative bacteria was reported by Bernheimer and Avigad [34] and Takahara et al. [35]. The antibiotic activity was not tested against yeast strains, where in our case growth inhibition was demonstrated.

The structural difference between surfactin and the *B. licheniformis* lipopeptides was established in the C-terminal amino acids (leucine is substituted by isoleucine) and in the composition of the lipophilic part where surfactin shows additionally β-OH-i,ai C_{13} and a higher amount of β-OH-i,ai C_{15} fatty acids. The high similarity between surfactin and the *B. licheniformis* lipopeptides leads to the conclusion that they are synthesized by the thiotemplate mechanism [12]. Minor variations, such as exchange of amino acids in the group of Val/Leu/Ile/Thr or of Phe/Tyr/Trp, are common for peptide antibiotics, for example, Gramicidin S and Tyrocidin. They can be attributed to overlapping specificities of activation reactions within the thiotemplate mechanism. Due to the nonribosomal mechanism, specificities of the amino acid activation reactions are considerably lower than in the ribosomal protein synthesizing system [36, 13].

The structure of the fatty acids in the lipopeptides shows a particular characteristic of *Bacillus* spp. (branched and hydroxy fatty acids), whereas the production of long-chain saturated fatty acids such as $C_{16:0}$ and $C_{18:0}$ is rather exceptional for this genus [15], but high surface activity of free long-chain fatty acids was already demonstrated for other species like *Nocardia erythropolis* and *Corynebacterium lepus* [37, 38]. Since *n*-alkanes were the carbon and energy source in these cases, it was postulated that they derive from the substrate rather than being products of *de novo* synthesis.

IV. OUTLOOK

Microorganisms produce a number of surface-active compounds that could be of interest for the substitution of chemical surfactants. New products with better properties in relation to activity and environmental pollution may finally evolve. The potential for biosurfactants certainly exists if the great number of different compounds isolated and characterized so far is considered.

However, the basic potential of microbiologically produced surface-active agents cannot be exploited unless economically feasible production processes are available. Because of the many fields of possible applications, these production processes need to be adapted in view of the final application. Realizable prices will greatly depend on the final use; therefore, it is of great importance that application studies for biosurfactants are carried out. From such studies, it can be derived which prerequisites production needs to fulfill. Certainly they will be different for biosurfactants produced for detergent than for food ingredients. Biosurfactants could become a valuable biotechnologically manufactured chemical if we succeed in developing production processes that yield not only compounds with superior properties but also are feasible when compared to surfactants derived from chemical synthesis.

ACKNOWLEDGMENT

The authors wish to thank Dr. I. Beretta for her help in arranging the manuscript.

REFERENCES

1. D. G. Cooper and J. E. Zajic, *Adv. Appl. Microbiol. 26*: 229 (1980).
2. J. E. Zajic and W. Seffens, in *CRC in Biotechnology*, Vol. 1 (G. G. Stewart and J. Russell, eds.), Labatt Brewing Company, London, 1983.
3. J. E. Zajic and A. Y. Mahomedy, in *Petroleum Microbiology* (R. M. Atlas, ed.), Macmillan, New York, 1984.
4. D. Haferburg, R. Hommel, R. Claus, and H. P. Kleber, *Adv. Biol. Eng. 33*: 53 (1986).
5. S. Lang and F. Wagner, in *Biosurfactants and Biotechnology*, Surfactant Sciences Series (N. Kosaric, W. L. Cairns, and N. C. C. Gray, eds.), Marcel Dekker, New York, 1987, pp. 21–45.
6. H. Umezawa, T. Takita, and T. Shiba, *Bioactive Peptides Produced by Microorganisms*, Holsted Press, Tokyo, 1978.
7. S. Itoh, H. Honda, F. Tomita, and T. Suzuki, *J. Antibiot. 24*: 855, 1971.
8. M. Hiramoto, K. Okada, and S. Nagai, *Antibiotic. Chemother. 3*: 1239 (1969).
9. T. R. Neu and K. Poralla, *Appl. Microbiol. Technol. 32*: 518 (1990).
10. R. Hommel, O. Stüwer, W. Stuber, D. Haferburg, and H.-P. Kleber, *Appl. Microbiol. Biotechnol. 26*: 199 (1987).
11. D. G. Cooper, C. R. MacDonald, S. J. B. Duff, and N. Kosaric, *Appl. Environm. Microbiol. 42*: 408 (1981).
12. B. Kluge, J. Vater, J. Salnikow, and K. Eckart, *FEBS Lett. 231*: 107 (1988).
13. H. Kleinkauf and H. van Döhren, *Ann. Rev. Microbiol. 41*: 259 (1987).
14. W. M. O'Leary and S.G. Wilkinson, in *Microbial Lipids*, Vol. 1 (C. Ratledge and S. G. Wilkinson, eds.), Academic Press, San Diego, 1989.
15. T. Kaneda, *Bacteriol. Rev. 41*: 391–418 (1977).
16. A. J. Fulco, *Prog. Lipid Res. 22*: 133 (1983).
17. C. A. Boulton and C. Ratledge, in *Biosurfactants and Biotechnology*, Surfactant Science Series (N. Kosaric, W. L. Cairns, and N. C. C. Gray, eds.), Marcel Dekker, New York, 1987, pp. 47–87.
18. W. R. Finnerty, in *Microbial Lipids*, Vol. 2 (C. Ratledge and S. G. Wilkinson, eds.), Academic Press, San Diego, 1989, pp. 525–566.
19. G. W. Hanlon and N. A. Hodges, *J. Bact. 147*: 427 (1981).
20. H. J. Haavik and O. Frøyshov, in *Peptide Antibiotics* (H. Kleinkauf and H. van Döhren, eds.), Walter de Gruyter Verlag, Berlin, 1982, pp. 155–159.
21. C. Mulligan, T. Y.-K. Chow, and B. F. Gibbs, *Appl. Microbiol. Biotechnol. 31*: 486 (1989).
22. G. E. Jenneman, M. J. McInerney, R. M. Knapp, J. B. Clark, and J. M. Feero, *Dev. Ind. Microbiol. 24*: 485 (1983).
23. S. M. Pfiffner, G. E. Jenneman, G. B. Walker, M. Javaheri, M. J. McInerney, and R. M. Knapp, *Int. Bioresources J. 1*: 285–294 (1985).
24. M. Javaheri, G. E. Jenneman, M. J. McInerney, and R. M. Knapp, *Appl. Environ. Microbiol. 50*: 698 (1985).
25. L. Guerra-Santos, O. Käppeli, and A. Fiechter, *Appl. Environm. Microbiol. 48*: 301 (1984).

26. F. Wagner, in *Proceedings of the World Conference on Biotechnology for the Fats and Oils Industry* (T. H. Applewhite, ed.), American Oil Chemistry Society, 1988, pp. 189–194.
27. J. Frankena, H. W. van Verseveld, and A. H. Stouthamer, *Appl. Microbiol. Biotechnol. 22*: 169 (1985).
28. J. D. Sheppard and D. G. Cooper, *Biotechnol. Bioeng. 36*: 539 (1990).
29. D. G. Cooper and B. G. Goldenberg, *Appl. Environm. Microbiol. 53*: 224 (1987).
30. W. W. Christie, in *Lipid Analysis*, Pergamon Press, New York, 1982.
31. K. Biemann and M. A. Stephen, *Mass Spectrom. Rev. 6*: 1 (1987).
32. K. Arima, *Agr. Biol. Chem. 33*: 1669 (1969).
33. K. Hosono and H. Suzuki, *J. Antiobiot. 36*: 194–196, 667–673, 674–678 (1983).
34. A. W. Bernheimer and L. Avigad, *J. Gen. Microbiol. 61*: 361 (1970).
35. Y. Takahara, I. Takeda, and H. Ohsawa, *Agr. Biol. Chem. 40*: 1901 (1976).
36. H. Kleinkauf and H. van Döhren, in *Biotechnology—A Comprehensive Treatise*, H.-J. Rehm and G. Reed, eds.), Verlag Chemie, Weinheim, 1986.
37. C. R. MacDonald, D. G. Cooper, and J. E. Zajic, *Appl. Environm. Microbiol. 41*: 117 (1981).
38. D. G. Cooper, J. E. Zajic, and C. Denis, *J. Am. Oil Chem. Soc.* (SD & C7): 77 (1981).

5

Integrated Process for Continuous Rhamnolipid Biosynthesis

THOMAS GRUBER* **and HORST CHMIEL**[†] Fraunhofer-Institut für Grenzflächen- und Bioverfahrenstechnik, Stuttgart, Germany

O. KÄPPELI Biosystems Division, BIDECO AG, Dübendorf, Switzerland

PATRICK STICHER[‡] **and ARMIN FIECHTER**[§] Institute of Biotechnology, Swiss Federal Institute of Technology, Zurich, Switzerland

I. INTRODUCTION

Up to now, microbial surfactants have not been generally successful in substituting for chemical surfactants. Their use has been restricted to specific applications. One reason for this situation is the higher cost for synthesis and downstream processing. The prerequisites for a more competitive production of biosurfactants are [1]:

1. Highly active compounds with specific properties for particular applications
2. Cheap substrates
3. Cheap production processes (biosynthesis and downstream processing)
4. High biosurfactant yields

Current affiliations:
*Bayer AG, Krefeld, Germany
[†]Institut für Industrielle Reststoff und Abfallwirtschaft GmbH, Saarbrücken, Germany
[‡]EAW AG, Dübendorf, Switzerland
[§]Fraunhofer-Institut für Grenzflächen- und Bioverfahrenstechnik, Stuttgart, Germany

Rhamnolipids (RL) are surfactants that certainly meet the first demand. In the past years, several application areas have been described. Rhamnolipids were proposed as pesticides because of their antibacterial, mycoplasmacidal, and antiviral activities [2, 3], as concrete additives for higher strength and cement saving [4], and as a source of rhamnose [5]. They were also tested in bioremediation in particular in the clean-up of oil spills [6, 7], and in enhanced oil recovery [8, 9].

The realization of the requirement to use cheap substrates is also possible if liquid wastes may serve as substrates. For some waste materials, even a cost bonus can be achieved, because of the simultaneous waste disposal or effluent purification. Principally, a variety of different carbon and energy sources can be envisaged for rhamnolipid synthesis [8–10]. If required, it is possible to extend the list of suitable substrates by tailoring the strains for specific needs, as described in Chapter 8.

More difficult is the design of a feasible process. Most of the known surfactant-producing organisms require aerobic conditions. Using conventional submersal aeration, the formation of very stable foams, due to the high foaminess of some biosurfactants, causes several problems. This is particularly valid for rhamnolipid production. The high foaminess of rhamnolipid solutions is further increased by proteins and microbial cells [11]. This results in high expenditures for foam control and, depending on the control technique, even in decreased productivity. Because operating conditions, especially pH, aeration rate, and surfactant concentration in the bioreactor influence foam volume and foam stability, mechanical foam breakers may fail to reduce the foam [12]. On the other hand, chemical antifoam agents may affect the downstream processing, especially the performance of filtration units [13]. Their possible influence on cell metabolism and the pollution of the reactor effluent have to be considered as well.

A further negative aspect for process costs is the requirement of specific substrate limitations coupled with a low specific growth rate for a maximum specific productivity [8, 9]. Using a chemostat, low dilution rates result in high residence times and therefore in low volumetric productivities increasing bioreactor costs.

As a consequence of the fact that there is an interest in using industrial waste effluents as substrates, huge volumes of aqueous low-concentration product solutions need to be handled, diminishing the efficiency of downstream processing.

In the following, a process involving membrane technology for cell recycling, integrated downstream processing, and alternative aeration is presented. It represents an approach to overcome the problems described above. It has been designed for the production of rhamnolipids from a carbohydrate-containing waste substrate from the food industry. To test the performance of the new process, the production results in laboratory scale were compared to data obtained in previous work with conventional operations [8, 12]. Therefore the experiments were carried out under nearly the same operating conditions as described before [8, 12].

The production strain, the aerobic bacterium *Pseudomonas aeruginosa* DSM 2659 was used for all investigations.

II. PROCESS DEVELOPMENT

Earlier experiments in a chemostat without cell recycle showed for *P. aeruginosa* DSM 2659 a maximal specific productivity at relatively low specific growth rates ($\mu = 0.135\ h^{-1}$) [8]. The product yield increased even further when lower specific growth rates were applied. For such systems, the use of cell recycling or other retention techniques is suitable. By these means, it is possible to increase the concentration of the biocatalyst in the reactor significantly, and the substrate residence time becomes uncoupled from the growth rate of the cells. Both contribute to the design of processes with high volumetric productivities and high product yields.

Several techniques for the retention of biocatalysts in a reaction system are suggested in the literature [14]. They can be divided into two groups: internal or external retention methods. Internal retention techniques include settling, filtration, and immobilization. External separation techniques are in general settling, filtration, centrifugation, and flotation. For the separation of Pseudomonads from their culture liquid, all listed methods are principally suitable. In the literature, mainly centrifugation [12] and immobilization [9, 14, 15] are described. The drawback of the immobilization technique is the low volumetric productivity of such a system, because of the limited diffusional transport of nutrients and product through the beads in which the cells are immobilized [15].

Against the use of centrifuges for cell recycling are the small size of the organisms and the small difference between the densities of the biomass and surrounding liquid (Table 1). In relation to a required separation efficiency of more than 90% and a high biomass concentration in the reactor, the relative throughput in the centrifuge is very low. As a consequence, high capital and energy costs result.

Limited lifetime of filtration membranes, significant decrease of the transmembrane flux, and changes in the retention characteristics during operation are the often discussed disadvantages of a filtration unit. However, flux and retention reach nearly steady-state values after a few hours of operation. Due to the fact that a filtration unit means very short residence times for the microorganisms and a separation efficiency of 100%, this separation method was chosen. Further arguments in favor of the filtration technique are the possibilities for integrated downstream processing in addition to cell separation, which are described in more detail below.

The flow scheme of the designed process for the continuous production of rhamnolipids is shown in Fig. 1. For the biosynthesis, a continuous stirred tank

TABLE 1 Data of *Pseudomonas aeruginosa* DSM 2659 Relating to Gravimetric Separation Methods

Length	1.2–2.6 μm
Diameter	0.6–0.8 μm
Density (at 25°C)	1055 kg m^{-3}

reactor with an external loop was used. Into this loop, two membrane units were integrated. The first unit effects the cell retention and recycle by means of an ultra- or a microfiltration membrane.

The second unit of the loop is a gas exchange system, which allows aeration and exhaust gas removal from the bioreactor avoiding the disadvantages of submerse aeration and subsequent mechanical or chemical foam control. The main part consists of a nonporous solution-diffusion membrane. The setup is designed in such a way that no additional recycle pump is required. The culture liquid flows across one side of the membrane, while the other side is flashed with air. Using an oxygen containing gas for sweeping, both oxygen supply and exhaust gas removal can be achieved simultaneously. The exhaust gas is highly enriched in carbon dioxide. When the sweep gas side of the solution-diffusion membrane was evaporated, carbon dioxide was recovered in a purity of about 98%, which could easily be used for further applications. To minimize the membrane area, oxygen supply was supported by conventional bubble gassing with pure oxygen. However, gaseous carbon dioxide removal was effected exclusively by the membrane.

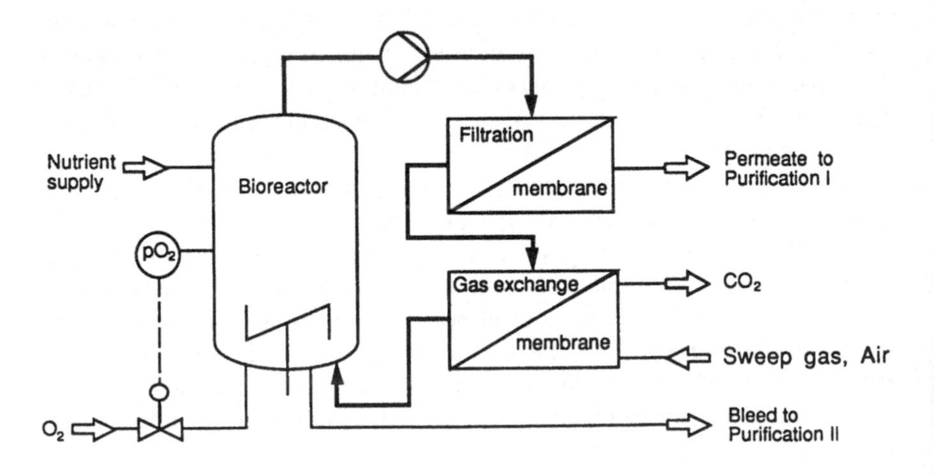

FIG. 1 Flow scheme for continuous production of rhamnolipids.

Additionally, dissolved carbon dioxide is removed in the filtrate of the microfiltration unit and in the bleed stream.

Hydrophobic microporous membranes, which have been used previously for the aeration of cell cultures are not suited [16] in this case, because of the wetting properties of rhamnolipids. These biosurfactants are able to reduce the contact angle of water against hydrophobic surfaces (e.g., polypropylene, polyvinylidenefluoride, and polytetrafluor ethylene) below 90°. Using such porous membranes, the resistance against permeation is reduced allowing the culture liquid to pass the membrane.

The reactor effluent consists of two parts, the bleed, which controls the specific growth rate of the microorganisms, and the cell-free permeate. The fact that rhamnolipids are retained by filtration membranes offers interesting aspects for an increased efficiency of the downstream processing, because of the starting solution is more concentrated.

A. Filtration

The filtration of rhamnolipids (RL) and possibly other biosurfactants is accompanied by an interesting effect. In spite of the low molecular weight of RL (maximum 650 D), there is a significant retention using ultra- or even microfiltration membranes. The retention by microfiltration membranes is nearly independent of the pore diameter of the membrane (Fig. 2).

The retention increased with decreasing cut-off only when ultrafiltration membranes were used with a pore size below 0.01 μm. Mainly two phenomena, which superimpose each other, are responsible for this. One is the membrane fouling and the other is the property of amphiphilic substances to associate and aggregate. The latter effect is being used as a separation technique, which is called micellar enhanced ultrafiltration (MEUF) [17]. The size of rhamnolipid associates has been determined by means of two techniques: by electron microscopy of freeze etched samples and by quasielastic light scattering (QELS). The evaluation of freeze etched samples of pure rhamnolipids solutions by means of electron microscopy revealed that rhamnolipids form vesicles with a diameter of about 80 to 200 nm at the cultivation pH of 6.3 (Fig. 3).

The diameter range of the vesicles estimated by electron microscopy was confirmed by QELS measurements. The average of the intensity of the associate diameters was determined to be 80 to 100 nm for a rhamnolipid concentration between 0.5 and 10 gL^{-1} in a 0.2 M phosphate buffer solution. For this ionic strength, no significant influence of the surfactant concentration on the size of the associates was observed. Through strong shear stress on the vesicles, for example, by vigorous agitation or pumping, the spontaneously formed multilayer structures may be disrupted and the mean diameter may decrease [18]. Experiments where

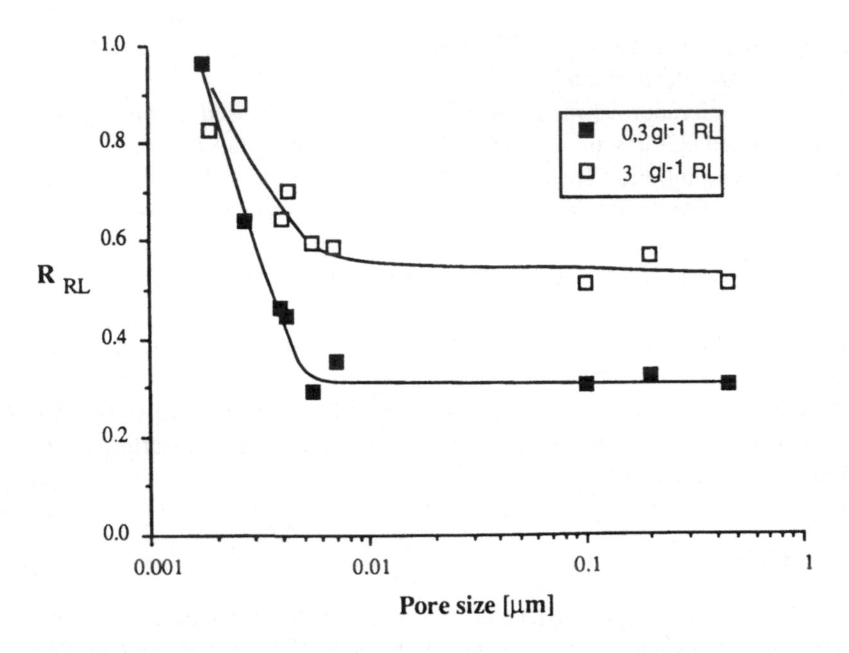

FIG. 2 RL retention as a function of membrane pore size and RL/protein concentration. The ratio of the RL and protein concentrations in both solutions was 8 to 1. The protein was isolated from disrupted cells from *P. aeruginosa* [19].

the samples were treated by ultrasonics showed a reduction of the mean diameter to approximately 50 nm [19].

The level of the rhamnolipid retention by a particular membrane, depends on different parameters, which influence the aggregate size as well as the fouling layer. Besides the concentrations of rhamnolipids and proteins (c.f. Fig. 2) the pH of the solution belongs to these parameters (Fig. 4). Depending on the pH the relative retention of RL by a microfiltration membrane (pore size down to 0.1 μm) varies between zero and one at high and at low pH values, respectively, when pure rhamnolipid solutions are processed.

Very important for the filtration of culture liquid are the interactions between rhamnolipids and proteins, which are released from living as well as lysed cells. The simultaneous presence of proteins in the liquid increases the retention at cultivation pH significantly. In cultivation experiments, retentions of up to more than 0.9 were reached when using microfiltration membranes. Apart from the protein concentration, the concentration ratio between proteins and rhamnolipids is decisive for the level of retention [19]. Figure 5 shows this relation by means of

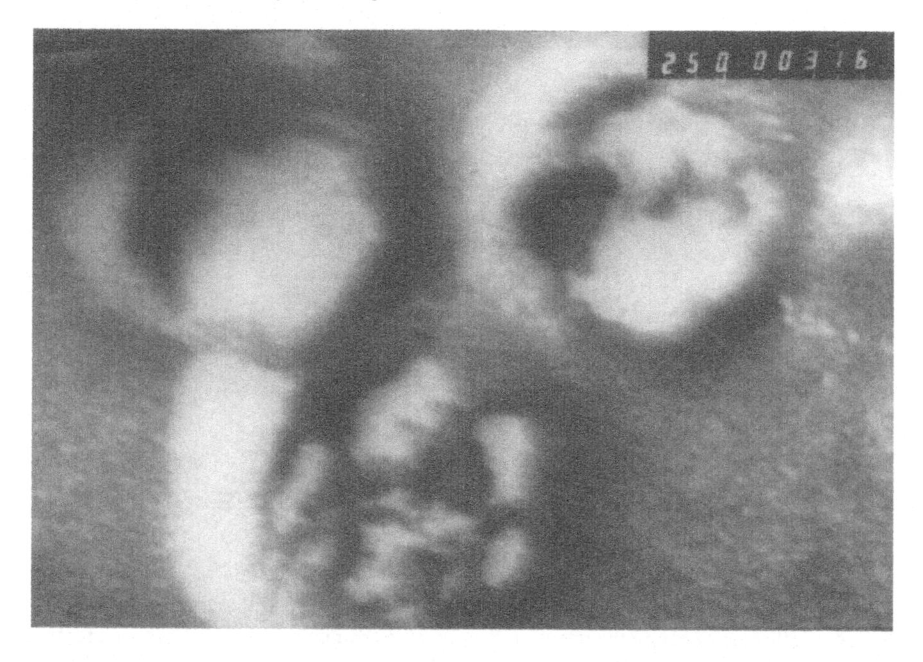

FIG. 3 Electron micrograph of vesicles formed by rhamnolipids. The sample was prepared by freeze etching technique and analyzed by transmission electron microscopy.

the aggregate size measured by QELS as a function of the rhamnolipid concentration for different protein concentrations. In filtration experiments, a similar behavior for the rhamnolipid retention was observed. At very low rhamnolipid/protein concentration ratios, the retention is higher than at ratios where the curves in Fig. 5 show a minimum in the aggregate size [19].

The interactions of the rhamnolipids cause a further effect on the filtration. The part of irreversible fouling of the membrane is significantly reduced, so that through intensive cleaning with water a high degree of purification of the membrane can be achieved. Longterm experiments over more than 1000 h with culture liquids showed that periodical membrane back flushing is sufficient to keep the transmembrane flux at adequate values.

Principally, ultra- and microfiltration membranes are suitable for the recycle of *P. aeruginosa* cells, if the nominal pore size does not exceed 0.2 μm. Above this diameter, the amount of cells that pass the membrane becomes significant. The lower limit of molecular cut-off for the recycle membrane is determined by the retention and accumulation of inhibiting substrates and by-products in the culture liquid.

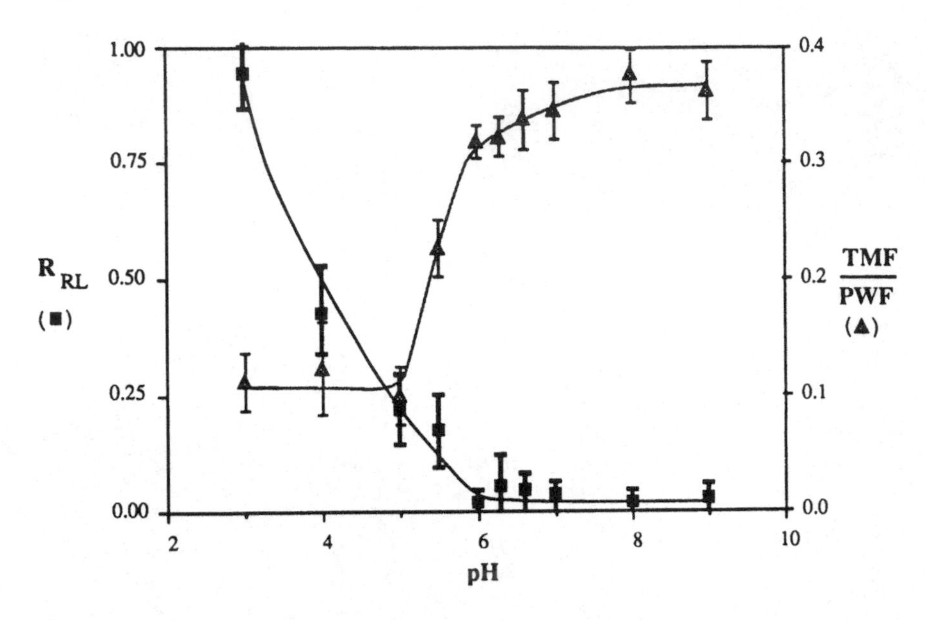

FIG. 4 Influence of pH on the retention R_{RL} of pure rhamnolipid and the transmembrane flux (TMF) which is related on the pure water flux of the clean membrane (PWF) (membrane material polysulfon, pore size 0.1 μm).

B. Downstream Processing

The costs for intermediate storage and for most of the downstream processing methods are primarily correlated with the liquid volume, which has to be processed, and only to a lower degree with the product and contaminant quantities. This necessitates the development of new processes to reduce the volume of fluid in the initial stages of the downstream process, particularly when using dilute substrates. Supplying higher concentrated media to the bioreactor by concentrating the substrate is mostly limited by the occurrence of metabolic inhibition by the substrate.

Therefore, it is more suitable to reduce the volume at the bioreactor outlet. In the designed process, the two parts of the outlet (filtrate and bleed), together with the possibility to accumulate the product using an appropriate membrane can be exploited. The bleed stream contains a correspondingly high rhamnolipid concentration, whereas the filtrate consists of mainly water, salts, and by-products and only small amounts of rhamnolipids. The remaining product may even be extracted by adsorptive downstream processes like foam fractionation or chromatography on XAD 2 resins [12]. The costs of these adsorptive separation

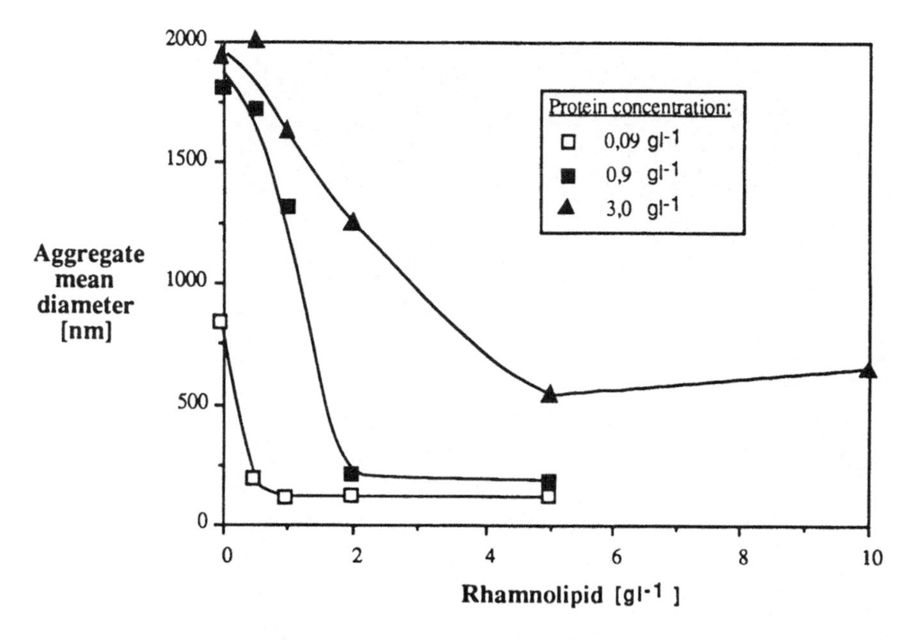

FIG. 5 Mean diameter of RL/protein aggregates for different RL and protein concentrations in 0.2 M phosphate buffer, determined by QELS.

techniques depend primarily on the amount of product. In cultivation experiments, combining rhamnolipid retention and foam fractionation of the permeate, up to 90% of the liquid could be separated with a rhamnolipid yield of over 98% [20].

The process is limited by the occurrence of product inhibition by substances accumulating in the bioreactor. Inhibition concentrations for rhamnolipids and proteins are currently investigated. Considering all factors an optimum for the recycling rate and the retention characteristics of the membranes may be specified for an optimal biosynthesis of RL.

C. Gas Exchange

The mass transfer over gas-exchange membranes depends on the solubility of the gases in and their diffusion through the membrane polymer material. The driving force is the transmembrane partial pressure difference of the molecules. The membrane, which was developed at the Fraunhofer-Institute, consists of a polysulphone supporting layer arranged as hollow fibers. It is coated on the inside with the essential active separating layer, a 2–4 μm thick film of polydimethylsiloxane (Fig. 6).

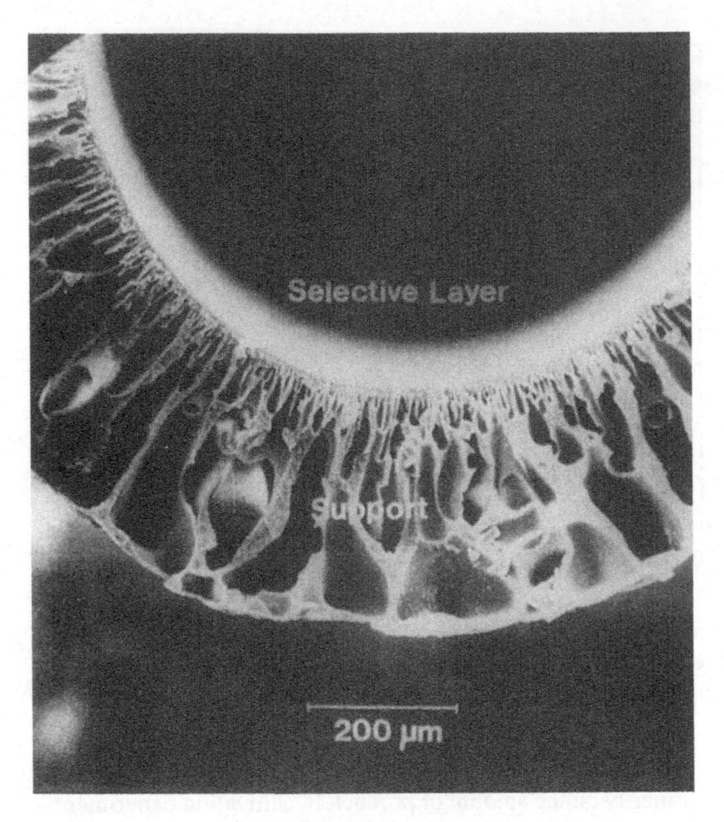

Composite Hollow Fiber

FIG. 6 Electronmicrograph of the cross section of the gas exchange membrane.

The transmembrane partial pressure difference for oxygen and carbon dioxide is related to the corresponding concentration limits in the culture fluid. Beyond these limits an inhibition of product formation by the organisms occurs [19]. A CO_2 partial pressure below approximately 30 kPa did not influence product formation, but an increase above 50 kPa almost completely inhibited RL formation. Similarly, this was true for oxygen. No influence was observed between 30 and 100% of air saturation, but above 200% growth was inhibited.

As a consequence of these limiting conditions and the low solubility of oxygen in aqueous solutions, only a relatively low rate of oxygen transfer across the membrane occurs. In experiments with water as the model solution, oxygen transfer into the liquid is only 0.01 kg $(m^2h\ 0.1\ MPa)^{-1}$ was obtained at 33°C when laminar flow conditions were applied. Because of the presently relatively high prices for membranes (200–350 $ m^{-2}), oxygenation of the culture liquid exclusively by membranes was not considered economical because of the considerable membrane surface area required. Only for organisms with low oxygen requirements, such as animal cell cultures, could membrane gassing be an alternative.

However, oxygen supply is not the decisive problem in excessively foaming systems. The more problematic step is the removal of the exhaust gas without initiating undesired cell and product flotation and blocking of gas filters. If the membrane module is used exclusively for the removal of CO_2, the required membrane surface area can be drastically reduced. The transmembrane flux of carbon dioxide is about 30 times [21] higher than that for oxygen at the same partial pressure difference. Additionally CO_2 interacts with the liquid phase, which also contributes to the removal of CO_2 from the system. In fact, this part is considerably higher than usually anticipated, so that the membrane area can be reduced to only 2% of that required for oxygen transfer [19]. The additionally required oxygen is supplied to the reactor by conventional sparging of pure oxygen, which is put under the control of the respiratory activity of the cells.

The advantages of this combined gas exchange system are manifold. Foaming problems no longer occur, increasing the operational safety for excessively foaming processes. Costs for foam-breaking techniques can be reduced. Production parameters such as pH, temperature, and medium composition are no longer limited because of their possible influence on foam formation but can be adapted to the needs of the microorganism. The working volume of a bioreactor can be reduced up to 50% over systems that work with mechanical foam control [5, 8, 9]. The reasons for this are the reduced gas-phase volume allowing the reduction of the foam drainage zones in the head-space of the bioreactor.

Tests under real working conditions also showed no significant loss in the efficiency of the membranes after 1200 h of operation.

III. RESULTS

By means of the presented process and the bacterium *P. aeruginosa* DSM 2659 four different rhamnolipids were produced. The molecular structures of these biosurfants (Fig. 7) correspond to products of *Pseudomonas* sp. DSM 2874 as described by Syldatk [9]. The rhamnolipids (RL) consist of one or two L-rhamnose units and one or two units of β-hydroxydecanoic acid. The amounts of the different types in the culture liquid was about 90% RL 3 and 10% RL 1. RL 2 and RL 4 occurred only in trace amounts.

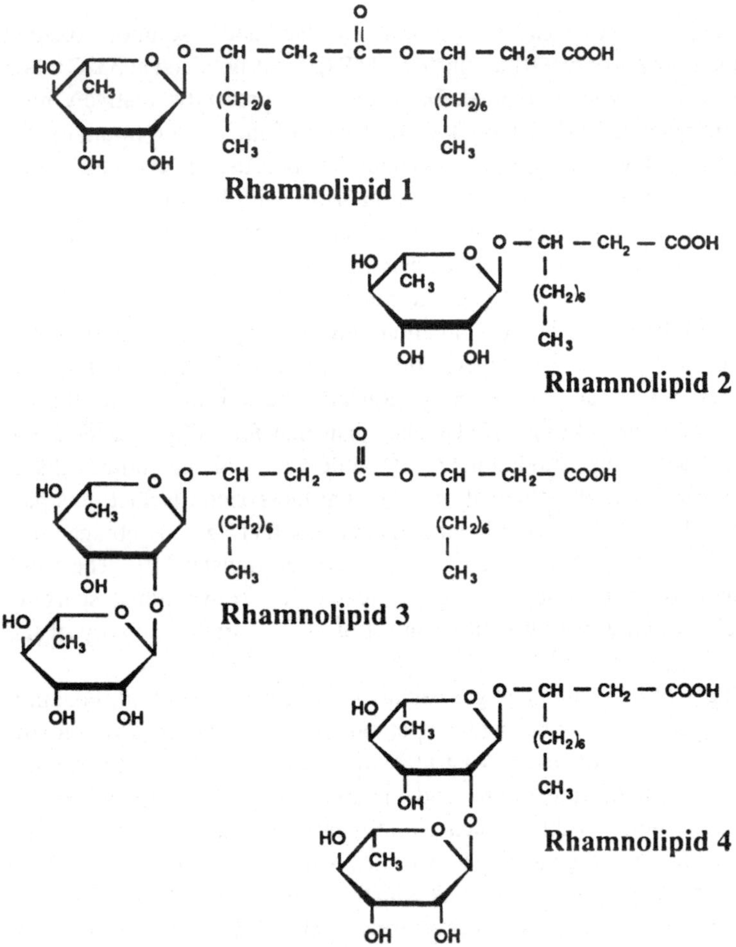

Rhamnolipid 1

Rhamnolipid 2

Rhamnolipid 3

Rhamnolipid 4

FIG. 7 Rhamnolipids identified in the culture liquid of *P. aeruginosa* DSM 2659.

Investigations with high cell recycle ratios revealed that sufficiently high productivities could not be obtained when using the medium [8], which was optimized for RL production in a chemostat without cell recycle. A major reason for this is due to cell lysis and the subsequent accumulation of nitrogenous compounds on the retentate side of the filtration unit. A prerequisite for excess production of rhamnolipids, however, are limiting concentrations of nitrogen and iron. Since proteins are rejected by the filtration membrane, they become a nitrogen source for the organisms disturbing the C/N ratio of the medium. To avoid an excess concentration, the nitrogen concentration of the feed medium needs to be decreased. As a consequence, different C/N-ratios in the medium are required for optimum RL production by the cells (Fig. 8).

For the control of nitrogen addition through the inlet medium a model has been developed, which describes the relation between the protein concentration in the culture liquid, the feed of nitrogen and carbon source by the medium and the specific productivity of the organism. The starting point for the model is the assumption that Pseudomonads are able to take up and utilize proteins from the culture liquid. By using this approach a carbon/nitrogen (c_C/c_N) ratio for the substrate uptake can be defined:

$$\frac{c_C}{c_N} = \frac{D(s_C^0 - s_C) + y\, k_{Np}\, c_{prot} X}{D(s_N^0 - s_N) + k_{Np}\, c_{prot} X}$$

where D is the dilution rate; (s^o-s) is the difference of carbon (index C) and nitrogen (index N) concentrations between bioreactor inlet and outlet; X is the cell dry weight; c_{prot} is the protein concentration, and k_{Np} is an empirically determined coefficient. The term $k_{Np}c_{prot}X$ gives the amount of nitrogen that becomes available from the accumulated proteins per volume and time unit. The weight ratio

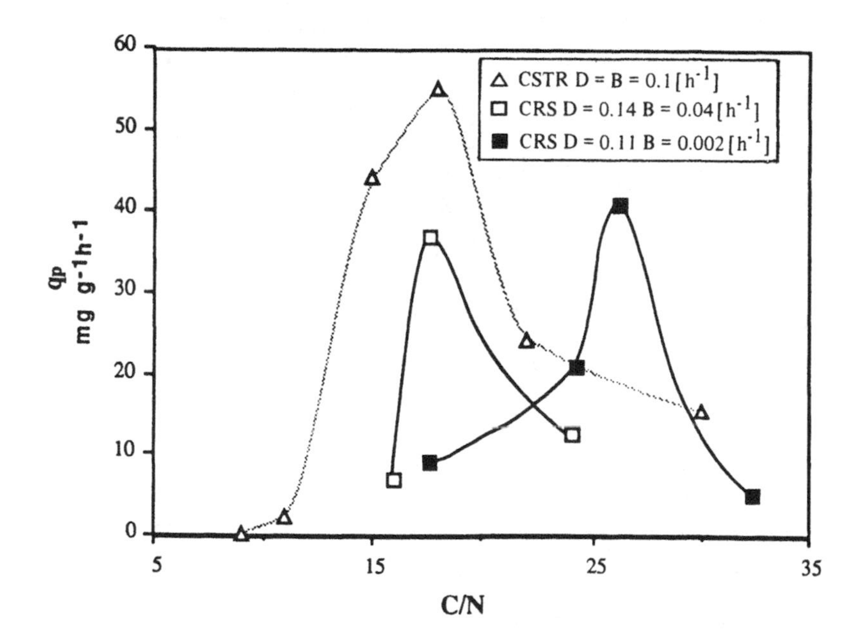

FIG. 8 Influence of carbon to nitrogen ratio in the medium on the specific productivity of *P. aeruginosa* DSM 2659 for different recycle ratios *R*. CRS, cell recycle system.

between carbon and nitrogen in the released nitrogen compounds is described by y. The basis for the estimation of y is the mole ratio for the protein composition with $CH_{1,58}N_{0,29}O_{0,34}$ according to Stouthamer and Verseveld [22]. The time constant k_{Np} was determined numerically for all recycle ratios to $8.9 \cdot 10^{-4}$ L $g^{-1}h^{-1}$. In Fig. 9, the specific productivity is shown as function of the actual C/N-ratio, which was calculated with the formula given above.

According to Fig. 9, only a small range for the C/N-ratio enables maximum specific productivities. The optimum was reached at a ratio of 17.5, which corresponds to the value determined by Guerra-Santos [8] for a chemostat without considering the uptake of nitrogen originating from cell lysis. Because of the low cell and protein concentrations in his experiments, it is justified to ignore its contribution in a chemostat. On the other hand, for a cell recycle process at extremely high cell densities, it has yet to be investigated whether the model has to be extended for other accumulating essential compounds.

Using the cell recycle system and adjusting the nitrogen concentration in the inlet medium as outlined, RL productivity could be improved significantly. Table 2 shows the results of the cultivation experiments in comparison to the data obtained without cell recycle. For the economy of the biosynthesis two factors are

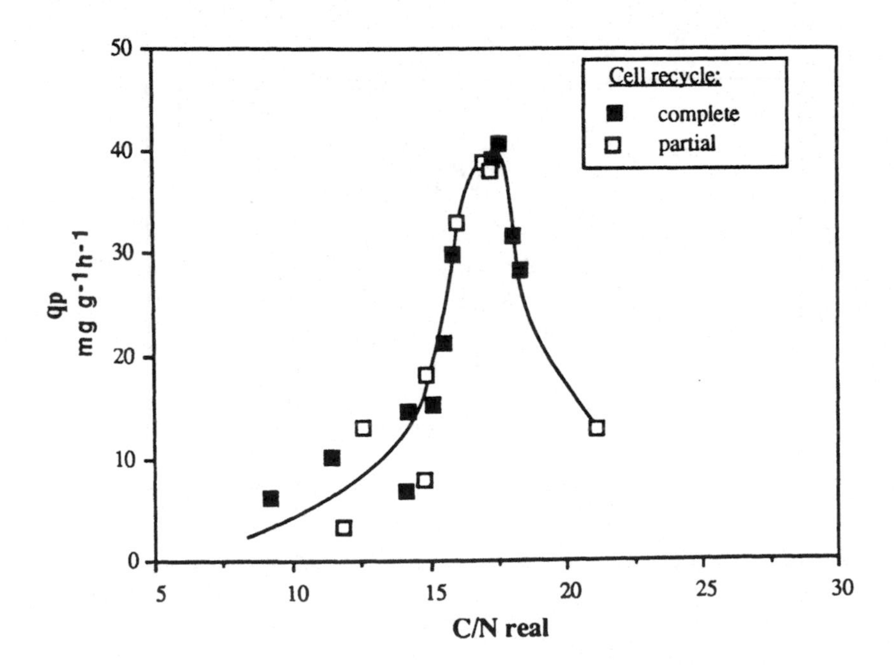

FIG. 9 Influence of the C/N ratio defined for the substrate uptake on the specific productivity of *P. aeruginosa* DSM 2659 for different recycle ratios R.

TABLE 2 Results of Cultivation Experiments with Different Operations

	Dilution rate D [h^{-1}]	Bleed rate B [h^{-1}]	Cell dry weight X [$g\,l^{-1}$]	Volumetric productivity r_P [$mg\,l^{-1}\,h^{-1}$]	Specific productivity q_P [$mg\,g^{-1}h^{-1}$]	Product yield $Y_{P/s}$ [$g\,g^{-1}$]	Cell yield $Y_{x/s}$ [$g\,g^{-1}$]	Volume ratio V_r^L/V_r [$l\,l^{-1}$]
a	0.18	0.002	13.3	545	41	0.15	0.007	1.8/2
	0.10	0.004	7.7	292	38	0.145	0.015	7.2/8
	0.14	0.04	7.9	300	37	0.10	0.11	7.2/8
	0.14	0.07	4.6	179	39	0.07	0.12	1.8/2
b	0.135	0.135	2.4	134	56	0.05	0.12	1.5/5
	0.05	0.05	2.5	77	31	0.08	0.13	1.5/5
c	0.05	0.05	2.5	88	35	0.097	0.14	1.5/5
	0.12	0.12	3.8	309	81	0.070	0.14	1.5/5
	0.065	0.065	2.5	147	59	0.077	0.085	23/50

[a] [19,23] CSTR with cell recycle, *P. aeruginosa* DSM 2659, $S_o = 20\,g\,l^{-1}$ glucose
[b] [8] Chemostat, *P. aeruginosa* DSM 2659, $S_o = 20\,g\,l^{-1}$ glucose
[c] [12] Chemostat, *P. aeruginosa* DSM 2659, $^1S_o = 18{,}2$, $^2S_o = 36{,}5$ $^3S_o = 30\,g\,l^{-1}$ glucose

important: the volumetric productivity and the product yield. By means of the cell recycle system both values could be increased.

The dependence of the product yield from the bleed rate and therefore from the specific growth rate of the microorganisms is obvious. With decreasing dilution and bleed rate, respectively, product yields increase in the chemostat [8] and in the process with cell recycle. The influence of the dilution and bleed rate, respectively, on the specific productivity is, however, different in both systems. Although the productivity in the chemostat has a distinct maximum at approximately 0.135 h^{-1}, no clear dependence could be observed in the recycle experiments.

IV. SUMMARY

A new process for aerobic continuous production of rhamnolipids from low concentrated substrates is presented with *P. aeruginosa* DSM 2659 as production strain. Two membrane units are integrated into this process, one for cell recycle and product stream preparation and the second for oxygenation and carbon dioxide removal. The direct integration of separation units into the bioreactor is a tool to enhance the productivity of the bioreactor and the efficiency of subsequent downstream processing.

Using a combined aeration method and membranes for exhaust gas removal the problem of the formation of stable foams in bioreactors has been solved. It is now possible to handle high surfactant concentrations in an aerobic microbial production without any foam control. The required membrane area is significantly reduced, for example, compared to membrane oxygenizers, by a factor of 50.

ACKNOWLEDGMENT

Industry and the Bundesministerium für Forschung und Technologie are gratefully acknowledged for financial support.

REFERENCES

1. N. Kosaric, N. C. C. Gray, and W. I. Cairns, in *Biosurfactants and Biotechnology*, Surfactant Series, Vol. 25 (N. Kosaric, W. L. Cairns, and N. C. C. Gray), Marcel Dekker, New York, 1987, pp. 1–19.
2. S. Itoh, H. Honda, F. Tomita, and T. Suzuki, *J. Antibiot. 24*: 855 (1971).
3. Syldatk, C., Lang, S., Wagner, F., Wray, V., Witte, L., *Z. Naturforsch. 40c*: 51–60 (1985).
4. D. Haferburg, D. Jass, and H.-P. Kleber, DD 259395 A1 to VEB Wohnbaukombinat Erfurt (1988).
5. R. J. Linhardt, R. Bakhit, L. Daniels, F. Mayerl, and W. Pickenhagen, *Biotechnol. Bioeng. 33*: 365 (1989).
6. S. Lang, B. Frautz, G. A. Henke, J. S. Kim, C. Syldatk, and F. Wagner, Abstracts of the 17th FEBS Meeting, Berlin 1986, in *Biol. Chem. Hoppe-Seyler 367 Suppl*: 212 (1986).
7. K. H. von Bernem, *Senckenbergiana Marit. 16*: 13 (1984).
8. L. H. Guerra-Santos, PhD-thesis no. 7722, ETH Zürich (1985).
9. C. Syldatk, PhD-thesis, TU Braunschweig (1984).
10. M. Robert, M. E. Mercadé, M. P. Bosch, J. L. Parra, M. J. Espuny, M. A. Manresa, and J. Guinea, *Biotechnol. Lett. 11*: 871 (1989).
11. B. Schultze, Diploma-thesis U Stuttgart (1989).
12. H. E. Reiling, U. Thanei-Wyss, L. H. Guerra-Santos, R. Hirt, O. Käppeli, and A. Fiechter, *Appl. Environ. Microbiol. 51*: 985 (1986).
13. J. M. S. Cabral, B. Casale, and C. L. Cooney, *Biotechnol. Lett. 7*: 749 (1985).
14. M. Meier, PhD-thesis, U Stuttgart (1990).
15. Sieman M, Müller-Hurtig, R., Fischer, L., Wagner, F., in *Dechema Biotechnology Conferences*, Vol. 2 (D. Behrens and A. J. Driesel, eds.), VCH-Verlagsgesellschaft, Weinheim, 1989, pp. 649–652.
16. J. Lehmann, G. W. Piehl, and R. Schulz, *Dev. Biol. Stand. 66*: 227 (1987).
17. S. D. Christian and J. F. Scamehorn, in *Surfactant-Based Separation Processes* (J. F. Scamehorn and J. H. Harwell, eds.), Marcel Dekker, New York (1989), pp. 3–28.
18. D. Myers, *Surfactant Science and Technologie*, VCH Verlagsgesellschaft, Weinheim, 1988.
19. T. Gruber, PhD-thesis, U Stuttgart (1991).

20. T. Gruber and H. Chmiel, *DECHEMA Biotechnol. Conf. 4*: 1085 (1990).
21. T. Gruber and H. Chmiel, in *Biochemical Engineering-Stuttgart* (M. Reuss, H.-J. Knackmuss, H. Chmiel, and E.-D. Gilles, eds.), Fischer Verlag, Stuttgart (1991), pp. 212–215.
22. A. H. Stouthamer and H. W. Verseveld, *Trends Biotechnol. 5*: 149 (1987).
23. P. Sticher, Diploma-thesis, ETH Zurich (1990).

6

Production, Properties, and Practical Applications of Fungal Polysaccharides

WALTER STEINER, DIETMAR HALTRICH, and ROBERT M. LAFFERTY Institute for Biotechnology, Technical University of Graz, Graz, Austria

I. INTRODUCTION

Polysaccharides, or gums, have been used by mankind for a number of purposes for many centuries. These traditionally used polysaccharides originated mainly from plant tissues, including seeds, trees, fruits, or seaweeds. Since the 1950s microbial polysaccharides have been investigated and soon afterward were commercially produced. Their technical application is based on their ability to drastically alter the rheological properties of aqueous solutions at low concentrations. Therefore, these biopolymers are used as thickeners and to stabilize emulsions, dispersions, and suspensions in aqueous systems.

Even though most polysaccharides are predominantly hydrophilic substances and have only low surface activity, they are still widely used for stabilization of emulsions and suspensions. The mechanism conventionally regarded as underlying this stabilization is that the alteration of the rheological properties of the continuous phase leads to the thickening or gelling reaction.

This modification of viscosity is not the only way in which polysaccharides act as stabilizers for emulsions or dispersions. A basic property of all polysaccharides, in spite of their obvious chemical differences, is the ability to interact with water, small ions, and other polymers, as well as with groups located at interfaces [1]. Thus the interaction of a polysaccharide with liquid–solid or liquid–liquid interfaces plays an important role in stabilizing emulsions or dispersions. This steric stabilization caused by an adsorbed layer of polysaccharide molecules at the interface may also be of major importance with respect to the formation of emulsions in the presence of emulsifiers. In one proposed model, the stabilization of oil in water emulsion is caused by the combined structure of a primary surfactant layer upon which a secondary layer of polysaccharide is adsorbed [2].

The aim of this review was to focus attention on the polysaccharides of yeasts and fungi, as most of the microbial polysaccharides described in the literature are of bacterial origin and not enough focus has been given to polysaccharides from fungi. A great number of fungi produce such biopolymers, and these biopolymers are mainly uncharged in contrast to most of the bacterial polysaccharides. This property might be an additional advantage in certain fields of application.

II. α-GLUCANS

A. Pullulan

In 1958, Bernier [3] reported on the formation of an extracellular polysaccharide by *Aureobasidium* (formerly *Pullularia*) *pullulans*. This observation was followed by a publication by Bender et al. [4], who characterized the neutral glucan excreted by this organism and determined the predominant existence of α-(1→4)- and α-(1→6)-glucosidic linkages. Wallenfels et al. [5, 6] showed that the ratio of the α-(1→4)- and α-(1→6)-glucosidic linkages is 2:1. Thus, the structure of pullulan was described as a relatively simple linear molecule consisting of maltotriose units, that is, units of three α-(1→4)-linked glucose molecules that are polymerized via α-(1→6)-glucosidic linkages on the terminal glucose residues. Later studies [7, 8] indicated that linear maltotetraose units randomly exist within the repeating polymaltotriose chain in pullulan in amounts of up to 7%. Catley et al. [9] showed that pullulan, when excreted by certain strains, may contain branching points where additional polymaltotriose chains are linked to the main chain of the molecule. The basic structure of pullulan can be seen in Fig. 1.

Polymaltotriose is not the only structure proposed for pullulan that might be used as a generic name for the neutral α-D-glucan elaborated by *A. pullulans*. Other reports on the possible structure of pullulan described the existence of α-(1→3)-linkages [10] together with a markedly different structure consisting of a branched polymer chain. In addition to the neutral glucan, *A. pullulans* also produces acidic glycans, which could be separated into a water-soluble component rich in D-galactose and D-mannose and a water-insoluble component existing mainly of D-glucose [11].

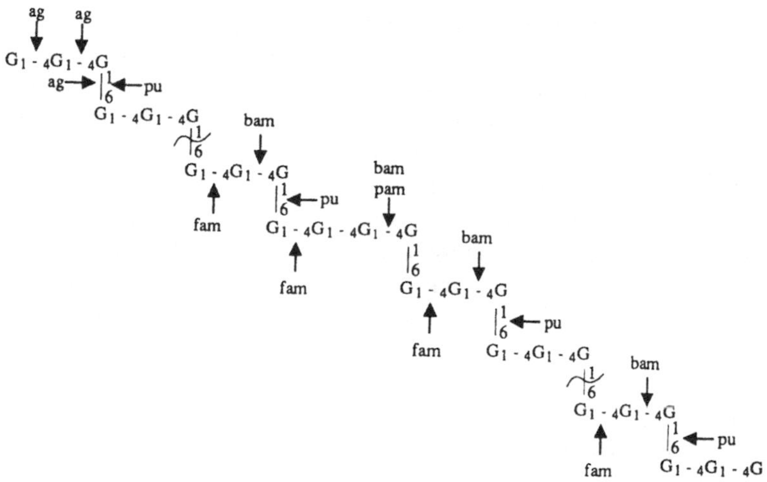

FIG. 1 Basic polymaltotriose structure of pullulan with coexisting maltotetraose unit according to Catley et al. [9] and typical sites of enzymatic attack. Amyloglucosidase (ag) cleaves both (1→4)- and (1→6)-linkages sequentially from the nonreducing end; pullulanase (pu) acts randomly on (1→6)-linkages, porcine alpha-amylase (pam) acts randomly on the terminal (1→4)-linkage of maltotetraose units, bacterial alpha-amylase (bam) and fungal alpha-amylase (fam) cleave (1→4)-linkages adjacent to (1→6)-linkages [16].

Aureobasidium pullulans, which is a polymorphic deuteromycete, is commonly known as belonging to the black yeasts. It occurs ubiquitously, degrading organic matter in soil, rivers, and sewage. Because of its deterioration of paint, blackening of lumber, and attack on plants and plant products, it can cause severe economic loss [12]. It is characterized by a complex polymorphic lifecycle. In its two major vegetative forms, the predominant cell types are either yeast-like or mycelial. However, it also shows a wide range of intermediate morphological types. The factors influencing the morphological form include medium components, dissolved oxygen levels, pH, and temperature [13]. The morphological form of *A. pullulans* is of importance during the production of pullulan.

1. Properties

Pullulan is readily water soluble, forming random coils in the proper solvents. Solutions of pullulan show an enhanced viscosity and this viscosity increasing effect depends on the molecular weight of pullulan. In comparison to other polysaccharides, such as alginate or xanthan, the viscosity of a high-molecular-weight pullulan solution is relatively low. The viscosity of its aqueous solution is stable over a broad range of pH values (pH 2–12) and is hardly affected by high salt concentrations.

The molecular weight of pullulan decreases during the course of fermentation. The polysaccharide initially having a maximum molecular weight of 2×10^6, is degraded during extensive fermentation to one-tenth of its original molecular weight [14]. The molecular weight of pullulan can be adjusted from 5×10^4 to 3×10^6 by controlling the cultivation conditions. The major influences are the pH value and the phosphate concentration of the medium [15].

Pullulan is hydrolyzed by different enzymes. Pullulanase (EC 3.2.1.41) randomly acts on α-(1→6)-linkages. After complete hydrolysis of the α-(1→6)-bonds, only maltotriose and maltotetraose remain. Amyloglucosidase (EC 3.2.1.3) acts on both α-(1→4)- and α-(1→6)-linkages starting from the nonreducing end of the molecule. Porcine α-amylase acts randomly on the α-(1→4)-linkage, that is, on the reducing end of a maltotetraose unit [9]. α-Amylase from *Thermoactinomyces vulgaris* (EC 3.2.1.57) cleaves α-(1→4)-bonds adjacent to the α-(1→6)-glucosidic linkage to produce panose, whereas hydrolysis with α-amylase from *Aspergillus niger* yields isopanose (Fig. 1) [16].

One of the main features of a commercial application of pullulan is related to the formation of films. However, pullulan can also be spun and molded. Pullulan films are colorless and transparent and are formed by casting and drying aqueous solutions of pullulan (5–10% w/v) on a metal plate or roller. Partial esterification or etherification of pullulan reduces its solubility in water, whereas complete esterification or etherification results in water insolubility [17]. Carboxylation enhances the solubility of pullulan in cold water.

2. Production

Aureobasidium pullulans utilizes a wide range of carbon sources for both growth and pullulan formation. These carbon sources include simple sugars such as glucose, xylose, fructose, sucrose, and maltose [18]. Commercial production uses partially hydrolyzed starch (dextrose equivalent ~50%) as a carbon source. Maximum yields of 76% from a 10% starch hydrolysate [17] can be attained. More complex carbon sources have been investigated as alternatives for pullulan production. These include inulin [19], spent sulfite liquor [20], and peat hydrolysate [21].

Earlier reports revealed two main factors that affected pullulan synthesis: the concentration of ammonium ions and the cell morphology [13, 22, 23]. When ammonium salts are used as a nitrogen source, a marked effect of the initial concentration on polysaccharide formation is evident [24, 25]. In batch culture experiments, pullulan synthesis only commenced when NH_4^+ was exhausted. After this lag period, the polysaccharide concentration increased parallel to that of biomass. Increasing the initial concentration of ammonium ions lengthened this lag period for polysaccharide formation, thus leading to reduced pullulan elaboration and to a higher biomass concentration [24].

The yeastlike form of *A. pullulans* has been regarded as the primary producer of pullulan [26]. One of the main factors influencing cell morphology is the pH value; at low pH levels the filamentous form is predominant, whereas at higher pH levels the proportion of the yeastlike form also increases, thus leading to an increase of pullulan production [22]. McNeil et al. [27] concluded that in continuous culture, the filamentous form contributes significantly to the formation of the polysaccharide. Different strains of *A. pullulans* were compared with respect to pullulan formation in submerged culture. These strains possessed distinct differences with regard to pullulan yield and its properties [28].

Auer and Seviour [29] reported on the influence of nitrogen sources other than NH_4^+ on pullulan production. Their results question some of the earlier published conclusions; they found that, depending on the nitrogen source, pullulan synthesis was carried out by a predominantly mycelial population.

Pullulan has been commercially produced by Hayashibara Co., Ltd. since 1972. Production has reached about 200 tons per year. The selling price is from 2000–3500 Japanese yen/kilo. Production strains were selected on the basis of low pigment formation and productivity was enhanced by mutation. After fermentation, cells are removed by centrifugation or filtration. The resulting fluid is heated to inactivate enzymes and decolorized with activated carbon to remove the black pigment. Salts can be removed by fractionation with alcohol or by membrane filtration. Thereafter, the pullulan solution is concentrated, dried, and pulverized.

Pullulan is sold as a powder, either as a desalinated pharmaceutical grade or as a food grade, or as a pullulan film.

3. Application

One of the unique properties of pullulan is its film-forming ability. Pullulan films have a very low oxygen permeability in comparison to commonly used packaging films. It is edible, oil resistant, and thermo sealable; thus, it is a good wrapping or sealing material not only for high fat- or oil-content foods, but also for other easily oxidizable products. The films are also expected to be used for Japanese snack food and to form a composite film containing granular food [30]. Water-resistant or water-insoluble films can be made by etherification or esterification of pullulan to different degrees. Plasticizers such as glycerol, sorbitol, or maltitol enhance flexibility of the film. Powdered pullulan mixed with a small amount of water can also be compression molded into shaped articles [15]. As a molding plastic, it resembles styrene in transparency, gloss, and hardness, but is more elastic. It is biodegradable and heat resistant. However, decomposition starts above a temperature of 200°C.

The use of a modified pullulan as a water-soluble contrast enhancing material has been investigated [31]. Pullulan can be used for the synthesis of oligosaccharides using enzymatic and/or chemical reactions [32]. Pullulan forms an aqueous two-phase system with polyethylene glycol. This two-phase system can be used for the recovery and purification of biochemical products such as enzymes. The low cost and low viscosity of pullulan make it a suitable replacement for dextran, a component widely used in aqueous two-phase formulations [33]. Pullulan is an efficient flocculating agent, being comparable to synthetic materials. Pullulan can be used as a binder for different applications such as nonwoven fabrics or fertilizers. Derivatives of pullulan can be useful in the production of adhesives for a variety of boards and other substances or as an adhesive ingredient for coating materials used to make high-quality printing paper [18].

Certain toxins can be combined with pullulan through conjugation. Thus, a vaccine for human immunization has been developed that shows a very good antibody response [34]. Pullulan can even be used as a blood plasma substitute [35]. Acetylated pullulan has been used for microencapsulating drugs that are orally administered [36].

B. Nigeran

Nigeran, also called mycodextran, was first isolated from mycelia of *Penicillium expansum* and *Aspergillus niger* by Dox and Niedig [37, 38]. Nigeran is a linear α-D-glucan with alternating (1→3)- and (1→4)-linkages (Fig. 2).

-α-D-Glc*p*-(1→3)-α-D-Glc*p*-(1→4)-

FIG. 2 Repeating unit of nigeran.

In certain *Aspergillus* and *Penicillium* spp. nigeran is part of the hyphal cell wall, where it can contribute up to 40% of cell dry weight [39]. Deposition of nigeran is primarily on the outer surface of the hyphal wall and commences with depletion of available nitrogen in the culture medium [40]. By creating conditions of environmental stress, such as under nitrogen or carbon limitation, or by increasing concentrations of certain metal ions (copper, iron), the nigeran content of cell walls can be increased [41]. X-ray diffraction studies indicate that at least part of nigeran in the cell wall is in a highly crystalline state [42]. The folded lamellar configuration in this crystalline region may fulfill a protective function, because it protects the polymer (and thus the wall) from enzymatic attack.

Nigeran is readily soluble in hot water, but is quite insoluble at room temperature. This property is used for isolation of nigeran from mycelia. Nigeran is cleaved by mycodextranase, an α-(1→4)-glucanase.

Under certain culture conditions, mainly characterized by nitrogen depletion and an excess of an available carbon source, *Aspergillus awamori* excretes a nigeran-protein complex into the medium [39]. This complex consists of 94% nigeran and 2%–4% protein, which is not covalently linked to the polysaccharide. This extracellular nigeran possesses crystalline regions and is supposed to be excreted by a displacement mechanism of a portion of the hyphal wall into the medium.

C. Elsinan

Takaya et al. [43] isolated a fungus from the white scab of tea leaves that produced a mucous layer on plate cultures. The fungus was identified as *Elsinoe leucospila*. The structure of elsinan, as this polysaccharide was designated, was elucidated by Tsumuraya et al. [44]. The main chain of elsinan consists of maltotriose units joined by single α-(1→3)-linkages (Fig. 3). Within the repeating maltotriose chain, linear maltotetraose units exist. The ratio of α-(1→4)- to α-(1→3)-linkages is 2.3–2.5:1.0. Elsinan is essentially linear; however, there may be few (1→6)-linked glucose branches. Elsinan has a structure similar to that of pullulan in which the maltotriose and maltotetraose units are joined by α-(1→6)-linkages. Lichenan, a water-soluble β-D-glucan extracted from Iceland moss (*Cetraria islandica*), is a linear polysaccharide with a similar structural sequence except for the opposite anomeric configurations; cellotriose and cellotetraose units are

-α-D-Glc*p*-(1→4)-α-D-Glc*p*-(1→4)-α-D-Glc*p*-(1→3)-

FIG. 3 Repeating unit of elsinan.

connected by β-(1→3)-glucosidic linkages. Other plant pathogenic fungi belonging to the *Elsinoe* species, such as *E. fawetti*, which causes citrus scab, were found to produce a similar α-D-glucan.

Production of elsinan has been studied using small laboratory fermenters. *Elsinoe leucospila* utilizes simple sugars, such as sucrose, fructose, mannose, or glucose; the maximum yield of the extracellular polysaccharide reaches more than 50% on the basis of the carbon source employed [45].

The molecular weight of the native glucan was found to be in the range of $2–6 \times 10^6$, as estimated by gel exclusion chromatography. Elsinan is readily soluble in warm water, yielding highly viscous, pseudoplastic solutions. Viscosity is stable over a broad range of pH values (pH 3–11) and salt concentrations (0–30% NaCl). At concentrations of 5% or higher, elsinan tends to form a gel. Elsinan is susceptible to enzymatic degradation. Because it consists of glucose residues linked by α-(1→3)- and α-(1→4)-linkages it can be hydrolyzed by certain α-amylases such as salivary α-amylase. The main oligosaccharide following enzymatic hydrolysis is 4-*O*-α-nigerosyl-D-glucose [45].

Films of elsinan can be made by evaporation of an aqueous solution. Like pullulan films they are almost impermeable for oxygen and they possess high tensile strength.

III. β-GLUCANS

β-Glucans are important structural elements of the fungal cell wall. A common pattern for the composition of fungal cell walls is a fibrillar array of polymers interwoven to form a net. These fibrillar polymers are mainly formed by chitin and to a lesser extent by cellulose and chitosan. The spaces within the net are filled by a matrix of more flexible polymers. The most important matrix components are β-(1→3)-glucans, which also often contain β-(1→6)-side chains, α-(1→3)-glucans and mannans [46].

Certain yeasts and fungi contain not only polysaccharides in the cell wall but also excrete large amounts into the extracellular environment where these polymers cannot be considered as part of the supportive cell-wall structure. They can be loosely attached to the cell or totally free from it. The structure of these extracellular polysaccharides can be similar to cell-wall components or can be quite different. Very often excretion of polysaccharides is dependent on the chemical and physical conditions of the environment in which the organism lives.

A. Pachyman

Pachyman is extracted from the sclerotia or hyphae of *Poria cocos* Wolf, a fungus that grows on the roots of pine trees [47]. This polysaccharide has a molecular weight of approximately 370.000. It is a linear β-(1→3)-glucan with only few β-(1→6)-linked glucopyranosyl groups and an internal β-(1→6)-linkage. Among the fungal β-(1→3),(1→6)-glucans, it most closely resembles the bacterial, linear β-(1→3)-glucan curdlan [48].

B. Schizophyllan

According to Essig [49] *Schizophyllum commune*, the organism producing schizophyllan, was first described by Dillenius in 1719. In 1964, Fujimoto et al. reported on schizophyllan [50]. Schizophyllan is a neutral homoglucan that is excreted into the extracellular environment. It consists of a linear main chain of β-(1→3)-linked D-glucopyranosyl units with a single β-(1→6)-D-glucopyranosyl residue attached to every third glucose molecule in the main chain [51], the repeating unit of schizophyllan can be seen in Fig. 4. β-Glucans play an important role in the structure of the cell wall of *S. commune*, which has been very well examined [52, 53]. Approximately 70% of its dry weight consists of three types of polysaccharides: glucosamineglycans (chitin, chitosan, and heteropolysaccharides containing both *N*-acetylglucosamine and glucosamine), the alkali-soluble *S*-glucan (α-(1→3)-glucan) and the alkali-insoluble or alkali-resistant *R*-glucan (β-(1→3),β-(1→6)-linked glucan). The *R*-glucan is highly branched with (1→3)- and (1→6)-linkages, however the proportion of linkage types, the chain length, and the degree of crosslinking vary with different strains. Different structures may even be found from the same strain. A fraction of *R*-glucan which contains many (1→6)-linked branches of single β-D-glucosyl residues on the (1→3)-linked chain resembles the water-soluble, extracellular glucan schizophyllan excreted by the fungus. Glucosamineglycans and the *R*-glucan are covalently linked. This complex is insoluble in water and alkali whereas

```
                              β-D-Glcp
                                  | 1
                                  ↓ 6
-β-D-Glcp-(1→3)-β-D-Glcp-(1→3)-β-D-Glcp-(1→3)-
```

FIG. 4 Repeating unit of schizophyllan and scleroglucan.

enzymatic or chemical depolymerization of the glucosamineglycan renders all of the β-glucan soluble in water or alkali.

Schizophyllum commune Fries is a cosmopolitan organism that grows under extremely variable conditions throughout the temperate and tropic zones [54]. It decomposes hardwood and also roots and stems of herbaceous plants. Being able to degrade complex carbon sources, it excretes significant amounts of cellulolytic, xylanolytic, and amylolytic enzymes in laboratory cultures [55–57]. In submerged culture, it produces L-malate, the yield of which, depending on the strain and on the components of the medium, possibly amounts to 90% when glucose is employed [58].

1. Chemical and Physical Properties

Schizophyllan dissolves in water and dilute sodium hydroxide solutions to form highly viscous solutions. In aqueous solutions it forms a trimer with a triple helical structure. This structure is stabilized by interchain hydrogen bonds with the β-(1→6)-glucose residues protruding outside the helix backbone [59]. When dissolved in dimethyl sulfoxide (DMSO), the triple helix dissociates and the individual chains assume a random coil configuration. In DMSO-water mixtures of increasing DMSO concentrations, dissociation occurs when the DMSO content exceeds 87%. This abrupt change from the triple helical trimer to the random coil monomer is accompanied by an almost discontinuous decrease in intrinsic viscosity. Once the triple helical structure has been denatured in DMSO, it cannot be reformed [60]. Yanaki et al. [61] and Kashigawi et al. [62] reported on the pitch per glucose residue, the hydrodynamic diameter, and the persistence length of the schizophyllan triple helix, which were found to be 0.30 ± 0.01, 2.6 ± 0.4, and 180 ± 30 nm, respectively. Up to an average molecular weight (M_w) of 5×10^5 g mol^{-1} the triple helix is almost perfectly rigid, but becomes semiflexible at larger values of M_w. Schizophyllan exhibits a highly cooperative order–disorder transition in aqueous solutions. This transition is not associated with the helical structure of the main chains but with the conformation of the side chains. It is assumed that, at lower temperatures, side-chain glucose residues form a well-organized structure together with surrounding water molecules, thus forming a helical chain on the outside of the main triple helix, which remains intact. This ordered structure surrounding the helical core increases the rigidity of the molecule [63–65]. The schematic structure of the schizophyllan triple helix with the glucose side chains sticking out of the main core, drawn roughly to scale, can be seen in Fig. 5.

Aqueous solutions of schizophyllan form a cholesteric mesophase above a certain critical concentration, and there is a concentration range where the isotropic and cholesteric phases coexist in equilibrium [66].

Schizophyllan in aqueous solutions has an extreme thermal stability. At a temperature of approximately 135°C, the triple helical structure of schizophyllan

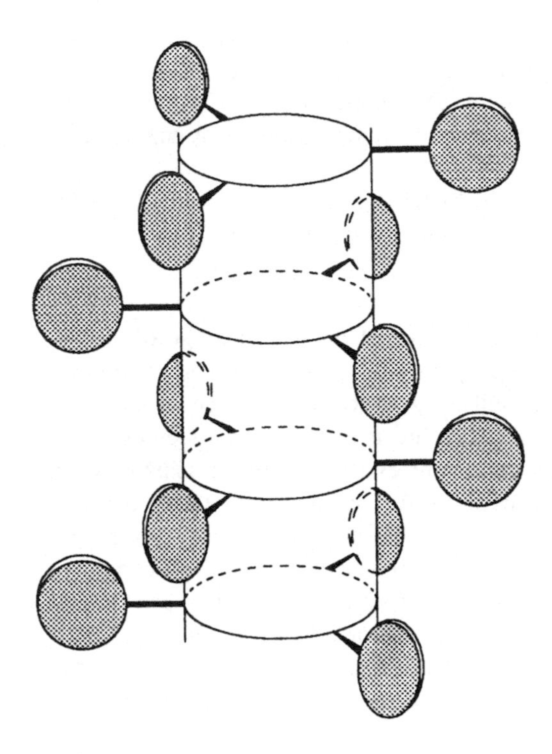

FIG. 5 Schematic structure of the schizophyllan triple helix, with the central cylinder representing the triple helical core of the molecule and the disks representing the side chain glucose residues. From Ref. 64.

is destroyed, and the molecule then forms random coils. On cooling, the original triple helix is not restored, but random aggregates are formed. Above a certain concentration of schizophyllan, these aggregates form a gel, thus leading to a viscosity higher than that expected for the rodlike triple helix structure [67].

Aqueous solutions of schizophyllan have a very complex rheological behavior. They are highly viscous and pseudoplastic, which means that viscosity depends on the shear rate. Schizophyllan solutions also have thixotropic properties, that is, sheared polymer solutions reach their zero shear viscosity only after a certain time shift, which can be explained by entanglements of the polymer molecules [68].

The high viscosity of polysaccharide solutions is not influenced by high salt concentrations. It is very stable over a large range of pH values (pH 2–12). However, when the pH value exceeds 12, a decrease of viscosity is observed, which is caused by a destruction of the triple helical structure.

Ultrasonic treatment of native schizophyllan solutions produce a polysaccharide with a lower molecular weight but with the same rodlike triple helical structure and the same biological activity. The shorter helices of ultrasonicated schizophyllan shows a very narrow molecular weight distribution as compared to the native sample [69].

Schizophyllan is also depolymerized by high shear rates, which were caused by forcing the polymer solution through a capillary by high pressure [70].

2. Production

Both growth and polymer formation are sustained by a wide range of carbon sources. Simple mono- or disaccharides, such as glucose, xylose, mannose, sucrose, or maltose, are utilized as well as polysaccharides, such as starch or xylans, as would be expected from the high levels of polysaccharide degrading enzymes [71]. However, complex carbon sources result in better schizophyllan yields than when using simple monosaccharides as substrates. Other medium components for the cultivation of *S. commune* are organic nitrogen sources and mineral salts. Good yields of schizophyllan are obtained when the carbon to nitrogen ratio is quite high; under conditions of nitrogen limitation, growth and biomass formation are reduced in favor of schizophyllan formation. In batch fermentation, the highest glucan concentrations are obtained after 4–7 days depending on the substrate used. Formation of the extracellular polysaccharide is associated with growth. After depletion of the carbon source the organism excretes glucanases, which degrade the polysaccharide, thus it can be utilized for further metabolic activities. The molecular weight of schizophyllan formed during batch fermentation reaches a broad maximum of 1.3×10^7 g/mol and decreases again because of the activity of degrading enzymes [72].

3. Application

The first reports of an antitumor action of schizophyllan were reported in Japan [73]. Dissatisfaction with cancer chemotherapy experiments, in which chemical compounds used as cytostatic agents often produce undesirable side effects, raised interest in noncytotoxic antitumor substances. Some polysaccharides were shown to be very successful in this respect. Schizophyllan was tested against various forms of tumors and significant results were obtained against solid forms of sarcoma 37, sarcoma 180, Ehrlich carcinoma, and Yoshida sarcoma, but not against ascites forms of these tumors with the exception of sarcoma 180 [73]. Schizophyllan is active in an indirect, host-mediated way, which means that a host reaction is initiated. Macrophage activation is in part responsible for this antitumor effect [74]. One feature necessary for the antitumor activity of schizophyllan is the triple-stranded helical structure. Random coil single strands of schizophyllan lose their antitumor potency [59]. Intensive studies on antitumor polysaccharides have only been carried out in Japan. Schizophyllan is now in clinical use in Japan with patients having various forms of cancer. It can be used alone or in combination with other forms of therapy [74].

An antiviral effect of schizophyllan has also been reported. It is claimed to totally control the tobacco mosaic virus in infected tobacco leaves [75]. Schizophyllan sulfates were tested for action against the HIV virus and exhibited mitogenic activity, which was not observed with schizophyllan itself. Schizophyllan sulfates showed strong anti-HIV viral activities, distinctly suppressing proliferation of the virus. Antiviral activity was evaluated in terms of cytopathic effect, by fluorescence antibody technique and by reverse transcriptase assay [76].

Schizophyllan was tested as an immunomodulator-mediated protection agent against acute radiation injury, and significant radioprotection was observed for schizophyllan after exposure to ^{60}Co γ-rays, based on enhanced survival rates after irradiation [77].

Since schizophyllan and scleroglucan have very similar structures and comparable properties, schizophyllan can substitute for scleroglucan in applications conceived or experimentally investigated for that polysaccharide.

C. Scleroglucan

Scleroglucan is an extracellular polysaccharide secreted by species of the genus *Sclerotium. S. rolfsii* was first described by Henry Rolfs almost 100 years ago as a plant pathogen causing tomato blight. Other names suggested for the basidial stage of this genus were *Corticium* or *Athelia* [78].

In 1967, Halleck [79] reported on the formation of this polysaccharide by different fungi of closely related genera, *Sclerotium delphinii, S. glucanicum, S. rolfsii, S. coffeicolum, Corticium rolfsii, Sclerotinia gladioli*, and *Stromatinia narcissi.*

The structure of scleroglucan is very similar to that of schizophyllan. Scleroglucan has repeating units of three β-(1→3)-linked D-glucopyranosyl residues to one of which a single D-glucopyranosyl molecule is attached through a β-(1→6)-linkage (Fig. 4). This structure of scleroglucan has been confirmed by NMR-investigation [80]. Although very similar in structure, there might be some slight structural differences between these two polysaccharides [81].

Sclerotium rolfsii is an aggressive plant pathogen of many crop plants in the tropics and subtropics [82]. During its attack on plant material, it produces large amounts of different enzymes, which rapidly destroy host tissue and cell walls, thus enabling it to enter the host organism. Some of these enzymes are polygalacturonase, galactanase, mannanase, xylanase, and arabanase [83–85]. Most of the earlier studies on the production and activities of various hydrolytic enzymes were related to pathogenesis and degradation of cell-wall material during host infection.

In the case of *S. rolfsii,* Bateman [86] described cellulase production excreted by this organism into the medium during fungal growth. This cellulolytic system as well as xylanolytic enzymes of *S. rolfsii* have been well described [87–89]. *Sclerotium rolfsii* also produces an enzyme saccharifying raw starch. This enzyme only hydrolyzes raw starch, that is starch that has not been gelatinized [90].

During pathogenesis *S. rolfsii* produces high amounts of oxalic acid, which binds calcium ions from calcium pectate of the host cell wall. Together with acidification of the host tissue, this reaction enables the fungal polygalacturonase to hydrolyze pectates of the middle lamella. Oxalic acid is also formed during cultivation of *S. rolfsii* in submerged culture [83].

Sclerotium rolfsii is soilborne, principally occurring in light sandy soils. It colonizes organic matter from where the fungus may parasitize certain plants. It can also form sclerotia and by means of these resistant structures the fungus survives in the ground [91]. Sclerotia are also formed on agar plates.

1. Properties

Like schizophyllan, scleroglucan is also soluble in water and in dilute sodium hydroxide solutions, yielding highly viscous solutions. Both polymers, being structurally very much alike, possess very similar properties. Scleroglucan solutions, even at low concentrations, show non-Newtonian character of pseudoplasticity, which might be due to the anisotropy of the rigid triple helical structure of the polymer [68]. In water or dilute NaOH, scleroglucan exists as a rigid, semiflexible trimer with a helical structure. This structure is dissociated in solutions containing 87% or more DMSO in which scleroglucan is dispersed as a random coil monomer. The same structures have also been observed for schizophyllan. However, when the dependence of the radius of gyration and the intrinsic viscosity on the average molecular weight of the random coils of these two polymers in DMSO is compared, there seems to be a distinct difference. Scleroglucan as a monomeric random coil appears to be almost in an unperturbed state, whereas schizophyllan is expanded appreciably by volume exclusion [92]. This finding suggests, that there must be a minor difference in the molecular structure of these two polysaccharides.

Concentrated solutions of scleroglucan (14%–19% according to molecular weight of scleroglucan) form two phases: an isotropic and a liquid crystalline phase. When the concentration is further increased to approximately 20%–30% according to molecular weight of scleroglucan, the entire solution forms a cholesteric mesophase [93]. Scleroglucan again resembles schizophyllan in this property; however, another distinct difference between these two polymers has been detected, namely the interlayer interaction of the liquid crystals.

When cooling a diluted scleroglucan solution below a critical temperature (~6°C) a thermoreversible gelation occurs. As judged from the behavior of the optical rotation, this sol-to-gel transition is caused by a crosslinking mechanism [94].

When the pH value increases to approximately 12, the triple helix structure of scleroglucan is destroyed, and the trimer dissociates to the random coil monomer form [95].

The viscosity of scleroglucan solutions is not influenced within a broad range of pH values [81], is insensitive to high salt concentrations, and is stable over a long period of time even at elevated temperatures [96]. The scleroglucan molecule is very shear-insensitive and even at high shear rates, the molecular chains do not break.

2. Production

A wide range of carbon sources can be used for scleroglucan formation. The best results are achieved using glucose, mannose, cellobiose, maltose, saccharose, starch hydrolyzate, or xylan. The other components of the medium are an organic nitrogen source, such as yeast extract, and mineral salts. High yields of polymer can be achieved by employing a high carbon-to-nitrogen ratio of the medium. Limiting the concentration of the organic nitrogen source leads to a maximum yield of 79% from glucose (30 g/L) (unpublished results). The pH value of the fermentation medium, which is initially in the range of 3–5, drops to approximately 2 due to oxalic acid formation. This low pH value diminishes the risk of infections. Thus, it should be possible to produce scleroglucan to raise the viscosity of oil drilling fluids and for purpose of tertiary oil recovery in the direct vicinity of oil fields under less strictly controlled conditions than would otherwise be necessary.

During fermentation, the molecular weight of scleroglucan goes through a broad maximum of 7.4×10^6. After depletion of the carbon source, scleroglucan is cleaved by β-(1→3)- and β-(1→6)-glucanases excreted by the fungus, which can reutilize scleroglucan as a carbon source. Approximately 90% of the polymer formed during fungal growth can be degraded by these extracellular β-glucanases [97].

To purify the polymer, the fermentation broth is heated to 80°C to inactivate these hydrolytic enzymes. The following purification steps include removal of mycelium by centrifugation or filtration after appropriate dilution, concentration of the filtrate, purification of the polymer either by precipitation or by ultrafiltration, and drying the final product [98].

Scleroglucan is commercially produced by Mero-Rousselot-Satia (France) under the trade name Actigum CS. Further experiments are currently being carried out as to the commercial development of this polymer.

3. Application

Because of the similar structures of scleroglucan and schizophyllan, these two polymers have the same possible fields of application. The antitumor activity of scleroglucan has also been proven [99, 100]; however, it has not been as intensively investigated as schizophyllan.

Scleroglucan can be used as a matrix for tablets that do not disintegrate but slowly release drugs. By using a variety of different additives, this rate of release can be modified [101].

The most promising field for commercial application of scleroglucan is enhanced oil recovery (EOR) or tertiary oil recovery. By using these EOR methods, the percentage of oil that can be obtained from a crude oil reservoir can be appreciably increased. One of these EOR methods is termed polymer flooding. In this procedure, viscous polymer solutions are injected into an oil reservoir to maintain reservoir pressure. The polymers employed must possess good viscosifying properties and a good flow performance through porous media under conditions of high temperature, high salinity, and high pressure [102]. In a comparative study of more than 140 polymers (polyacrylamides, polyvinylpyrrolidones, cellulose derivatives, and biopolymers) scleroglucan was classified as giving the best results on the basis of the criteria mentioned above. The thermal stability of scleroglucan (no loss of viscosity of a scleroglucan solution at 90°C over 500 days) and its salinity tolerance, even in the presence of divalent metal ions, are especially remarkable. Scleroglucan can also be utilized as a drag reducing agent in cleaning fluids used in pipelines for gas or petroleum or added to solutions used in treatments of petroleum wells, such as in acidic solutions or completion fluids.

Other proposed applications for scleroglucan are its use in the production of porcelain and ceramic glazes, paints, inks, pesticide sprays, and extruded refractory products [103]. Since scleroglucan has rheological properties similar to those of xanthan, it could substitute for this commercially successful polysaccharide in certain fields. It has certain advantages due to its nonionic character. Scleroglucan has not yet been approved by the FDA for use as a food additive. In tests on rats, scleroglucan was assimilated with a caloric equivalent of starch. Dogs and chicks exhibited a reduced level of cholesterol when a standard diet was supplemented with scleroglucan [98]. Another possible use of scleroglucan could be as a viscosifying agent for cosmetic preparations.

D. Other β-(1→3),β-(1→6)-Glucans

Neutral glucans with the same basic structure consisting of a main chain of β-(1→3)-linked D-glucopyranosyl units substituted at O6 with β-D-glucopyranosyl side chains are found in many different species of fungi. These neutral glucans can either be located in the mycelium or in the fruiting body or are excreted into the culture medium.

Different members of the genera *Acremonium* and *Cephalosporium* (a synonym of *Acremonium*) were found to produce extracellular β-glucans. Although their exact chemical structure has not been elucidated, they are believed to be β-(1→3),β-(1→6)-glucans. *Acremonium persicinum* produces an exopolysaccharide with a yield of approximately 25% from glucose. The overall yield of the exopolymer is dependent upon the type and concentration of the nitrogen source present in the medium. A high carbon-to-nitrogen ratio favors polymer formation [104].

Claviceps purpurea forms an extracellular glucan that has single β-(1→6)-side chains on approximately every fourth unit of the main chain [105].

A more branched exopolysaccharide is elaborated by *Glomerella cingulata*, which was termed *glomerellan*. The main chain of this glucan, which contains β-(1→3)-with a minor portion of β-(1→2)-linkages, carries side chains attached via β-(1→6)-linkages to approximately every second unit of the main chain. The side chains contain β-(1→4)-and β-(1→6)-linkages. This glucan shows viscosifying properties as well as pseudoplastic and thixotropic behavior in solutions [106].

The extracellular glucan from *Phanerochaete chrysosporium*, a lignin degrading fungus, was found statistically to carry single (1→6)-linked β-D-glucopyranosyl groups on every second residue of the main chain. This glucan is secreted when the fungus is grown under nitrogen-limited conditions [107]. The extracellular polysaccharide might regulate the concentration of extracellular glucose, because glucose represses enzymes necessary for the breakdown of lignin [108].

A glucan structurally similar to schizophyllan or scleroglucan where single side chains are attached to every third unit of the main chain is formed extracellularly by *Nomuraea rileyi*. This fungus is a pathogen of various insects that damage several agricultural products [109]. A similar glucan, called *pendulan*, is excreted by *Porodisculus pendulus*. It seems to have a similar molecular conformation, that is, a multistranded helix, as schizophyllan has. The average molecular weight M_w is approximately 2×10^6; however, during fermentation, the molecular weight is reduced, probably due to the action of an endoglucanase produced by the organism [110].

The extracellular glucan of *Pestalotia* sp. 815 was found to produce a highly branched polysaccharide, designated *pestalotan*, with a molecular weight above 2×10^6. In the case of this polymer three out of the five units of the main chain are substituted at O6, mainly with single glucosyl groups and a few short β-(1→6)-linked oligosaccharide units [111].

The extracellular polysaccharide from *Ulocladium atrum* has one β-(1→6-side chain on every second unit of the β-(1→3)-main chain [112], a structure also found in the glucan from *Monilinia fructigena* [113].

A glucan isolated from *Pleurotus ostreatus*, an edible mushroom known commonly as "oyster mushroom," has a single side chain for every four glucose units of the main chain. This polymer was found to be highly active against tumors [114]. Other members of the same genus were also found to form extracellular glucans [115].

Especially in Japan, polysaccharides have been extracted from fungi that are used either as food or as a crude drug. Many of these were found to be β-glucans containing a β-(1→3)-linked main chain with β-(1→6)-linked side chains. However, the number and length of these side residues vary. A number of these glucans also show antitumor activity when administered by intramuscular injections. It is interesting to note that hot water extracts of different mushrooms traditionally have been used for cancer therapy in Japan and China for a long time [116].

Lentinan, the polysaccharide from *Lentinus edodes*, is extracted from the edible fruiting bodies known as shiitake and have long been used as a home remedy in Japan. Lentinan has a linear β-(1→3)-linked main chain. Two out of every five glucose residues contain branch points at O6. These side chains consist of β-(1→6)- and β-(1→3)-linked glucose residues. The main chain appears as well to have some β-(1→6)-linkages [117]. Lentinan shows antitumor activity in very low doses [118] and is presently in clinical use in Japan being used either alone or in combination with chemotherapy or radiotherapy mainly after surgical removal of a primary tumor [74].

Another alkali-soluble glucan has been extracted from *Omphalia lapidescens*. This polymer possesses approximately two single side chains for every three main glucosyl units [119].

Three different glucans were extracted from the fruit bodies of *Dictyophora indusiata*. A water-soluble polysaccharide has a molecular weight (M_w) of approximately 5.1×10^5 and has two side groups attached to every five sugar residues of the main chain. It can be simply extracted using hot water. After extraction with 2% sodium carbonate, a glucan of higher molecular weight (5.5×10^6) with similar structure was isolated. However, the glucosyl side chains were found to be mainly localized in the neighborhood of the nonreducing end of the main chain. Extraction with 1 M sodium hydroxide yielded a less branched glucan with two side chains for each seven glucosyl residues of the main chain. A molecular weight of 3.3×10^5 was found for this polymer. It was suggested that these polysaccharides have a triple helical structure [120–122].

A glucan with short branches (one or two glucose residues) on every second or third residue of the main chain was extracted from the fruiting bodies of *Cyttaria harioti* [123]. A similar structure is to be found for the glucan of *Cochliobolus miyabeanus*, which was prepared from the mycelial cell walls. Every third unit of the basic linear molecule possesses short chains of β-(1→3)-D-glucosyl residues [124].

A slightly different pattern for the side chains was observed with the glucan obtained from *Phytophtora parasitica*. Every fourth unit of the main chain carried a β-(1→6)-linked side chain, which could have a length of up to three β-(1→3)-linked glucosyl units. However, single side groups were dominant. This polysaccharide also showed an antitumor effect [125, 126].

The fruiting bodies of *Grifola frondosa*, which is a commercially available, edible mushroom, contain a glucan with single side chains on approximately every third main chain unit. However, the distribution of these branches may be irregular. The polysaccharide named grifolan is extracted from both the fruiting bodies and from the mycelium. It has a potent activity against sarcoma 180 solid tumors. It has been suggested that grifolan forms a weakly rigid, helical structure [116, 127].

Fewer side branches were also found in the case of the glucans of *Coryceps cicadae.* Single glucose residues were found only on every 25th residue of the main chain on an average. The glucan obtained from *Ganoderma japonicum* has, for example, single side chains only on every 30th unit of the backbone. Due to the lack of a branching structure, these polysaccharides are insoluble in water but are soluble in dilute alkaline solutions [128, 129].

From the fruiting bodies of an *Auricularia* spp. (Chinese name: Yu er), which is used as a food and as a crude drug in China, a β-glucan can be extracted that has a single glucose residue as side chains. These account for a quarter of the total sugar residues. The distribution of the single glucose entities along the main chain is irregular. However, these β-(1→6)-linked side groups are mainly located near the nonreducing end of the main chain [130].

IV. YEAST CELL WALL GLYCAN

Although β-glucan comprises a large portion of the yeast cell wall material, one cannot speak of a homogeneous polysaccharide. The cell wall actually consists for the most part of two different classes of polymers—β-glucan and α-mannan—together with a small portion of chitin. The most-investigated structure is that of the cell wall of *Saccharomyces cerevisiae*. Following an alkaline extraction procedure, two different glucan fractions can be distinguished. These two fractions comprise approximately 12%–14% of the cell dry weight. The major component of the alkali-insoluble glucan fraction consists of a slightly branched β-(1→3)-glucan having approximately 3% β-(1→6)-glucosidic interchain linkages. The minor component of the alkali-insoluble glucan fraction is highly branched and mainly consists of β-(1→6)-linkages with 14% of β-(1→3)-bonds. It makes up

15%–20% of the total glucan. The β-(1→3)-glucan forms triple helices, thus giving the cell wall its rigidity and stiffness. The β-(1→6)-glucan possibly serves as a plasticizer to retain some flexibility of the cell wall [131]. The β-(1→6)-glucan is soluble under nonalkaline conditions. The insolubility of the β-(1→3)-glucan under alkaline conditions is not a property of the glucan chains themselves, but is a result of their close association with chitin. The small amount of chitin present in the cell wall of *S. cerevisiae* is responsible for maintaining the β-glucan in an alkaline-insoluble form [53].

The second fraction obtained from an alkaline extraction is alkaline-soluble glucan. It consists of a β-(1→3)-glucan chain with occasional β-(1→6)-linked residues. Various side chains containing both β-(1→3)- and β-(1→6)-linkages are attached to the basic structure. This fraction comprises approximately 20% of the dry weight of the cell wall of *S. cerevisiae*. Since this alkaline-soluble glucan always contains a mannan fragment, it seems likely that this glucan and a mannan fraction are closely related or even covalently bound [131]. The mannan found in *S. cerevisiae* is part of a phosphoglycoprotein. The mannan moiety contains an α-(1→6)-D-mannopyranosyl main chain substituted at O2 or O3 with side chains of various length and structure containing α-(1→2)- and α-(1→3)-linkages. This mannan fraction is linked to protein by means of an *N*-acetylglucosamine-containing segment [132].

Other *Saccharomyces* spp. possess related cell wall structures. However, variations do exist with respect to the relative amounts of different glucans and mannans present, the proportions of linkages present, the molecular size, and the degree of branching [133].

By changing the composition of the growth medium or by limiting some nutrient, both the polysaccharide content and the ratio of glucan to mannan in the cell wall of *S. cerevisiae* can be changed. When *S. cerevisiae* is grown in a medium limited in either inositol or biotin, the cells have a much higher glucan-to-mannan ratio than cells grown in a complete medium. A direct correlation between the amount of glucose in the medium and the amount of glucan in the cell wall exists [131].

A different approach to alter the properties of the cell wall polysaccharide of *S. cerevisiae* was chosen by Jamas et al. [134]. Based on the observation that glucans containing a higher degree of β-(1→6)-cross links are more resistant to digestion with laminarinase which is an endo-β-(1→3)-glucanase, mutants have been isolated that contained an altered cell wall glucan structure. In a rheological investigation, only the cell wall particle containing β-glucan was considered, a vigorous alkaline extraction procedure permitted the separation of intracellular components as well as other components of the cell wall, such as the mannan and chitin fractions, leaving a cell wall glucan particle that retains the intact morphology of the yeast cell. As judged from rheological analysis, the amount of β-(1→6)-D-glycosidic linkages distinctly affects the morphology of yeast cell walls and the rheological properties of the glucan by increasing its rigidity.

Yeast glycan has been produced by the Anheuser-Busch company. Yeast cells were ruptured, then separated into a soluble and a cell wall fraction. The cell wall fraction was washed with water, concentrated, and then dried. Yeast glycan from *S. cerevisiae* contained 84.8% carbohydrates; the ratio of glucose to mannose was 3:2 [135].

A yeast β-glucan—Fibercel™—is produced by Alpha-Beta Technology (in Worcester, MA) and has reached the stage of precommercial production. This yeast polysaccharide, containing approximately 86% β-glucan, does not greatly enhance the viscosity of aqueous solutions but does result in a fatty or oily mouth feel. Thus, it could replace fats or oils in low calorie foods. Proposed applications as a fat substitute range in use from frozen to processed foods. In animal experiments using hamsters, it could reduce total cholesterol and significantly LDL cholesterol when added to a high fat diet. This particular yeast polysaccharide could also serve as a flavor entrapment agent.

Alpha-Beta Technology has also developed a yeast glucan from genetically engineered yeast that was found to stimulate the immune system by targeting the β-glucan receptor on macrophages. This polysaccharide is undergoing preclinical evaluation and the company intends to file an application to begin human trials in the United States. The same company developed an adjuvant to enhance the effectiveness of vaccines by incorporating a vaccine antigen into microparticles made of yeast glucan.

Yeast glucan can be also used in cosmetics as an agent to increase viscosity that enhances moisture retention as well [136]. In animals, the topical application of yeast glucan accelerated wound healing [137]. Yeast glucan also has an immunomodulating activity and a stimulation of host defence reactions against infections could be observed [138]. Yeast polysaccharides, especially yeast mannans, have antiviral properties. These properties were tested against plant viruses and infections by tobacco mosaic virus could be inhibited [139].

V. CHITIN, CHITOSAN, AND OTHER POLYHEXOSAMINES

Chitin is a linear β-(1→4)-linked polymer of *N*-acetyl-glucosamine (2-acetamido-2-deoxy-D-glucose), however, a varying number of free amino groups may also be present. The term chitosan is used to designate the long-chain polymer of β-(1→4)-linked glucosamine (2-amino-2-deoxy-D-glucose) in which few *N*-acetyl groups may be present.

Chitin exists in three polymorphic forms with various degrees of crystallinity. In α-chitin, the form found in fungal cell walls, the neighboring chains are antiparallel. Each chain is stabilized by intramolecular hydrogen bonds.

Successive chains are linked by intermolecular hydrogen bonds to form sheets, which results in a very stable structure. Chitin is a characteristic component of the fungal cell wall, and although it is a common feature of the wall of a vast majority of fungi the amount can vary greatly. In some cases, chitin may comprise up to 60% of the dry weight of the fungal cell wall [140].

The structure of chitosan has until now not been fully determined. Six different polymorphs have been proposed for different preparations of chitosan [141]. It is a major part of the hyphal walls of members of the Zygomycotina such as *Mucor* species.

In fungal cells, chitin synthetase is packed into microvesicles, the so-called chitosomes. Chitin synthetase is activated by periplasmic proteases and chitin microfibrils assemble only *in situ*, outside of the cytoplasm. Chitosan results from the deacetylation of chitin and its synthesis is closely coupled with that of chitin. Only growing chitin chains are attacked by chitin deacetylase [142].

Commercially available chitin is produced from waste shells of crustaceans, and chitosan is prepared by deacetylation of chitin using alkaline conditions. However, waste fungal biomass from the fermentation industry could be a significant source of these two polymers, which could then be produced from this material under controlled conditions. Citric acid and antibiotic production processes lead to the formation of 790,000 tons of mycelial waste that could potentially yield 32,000 tons of chitin per year [143].

A possible method to produce chitosan, for which a greater demand exists, could be by means of a fermentation process. In a screening program, 125 *Mucoraceae* strains of the genera *Absidia, Actinomucor, Circinella, Mucor, Phycomyces, Rhizopus*, and *Zygorhynchus* were tested for chitosan production. Chitosan was extracted from the total biomass using 2% acetic acid after an alkaline extraction procedure. The best results with respect to a high chitosan productivity were found when using the genera *Absidia* and *Rhizopus*. *Absidia butleri* produced up to 1 g/L chitosan in shake flask experiments. The molecular weights, as determined by viscosimetry and by gel permeation chromatography, were 7.0×10^5 and 1.2×10^6, respectively. The degree of acetylation was found to be 13% as was determined by infrared spectroscopy [144].

The cell wall of *Mucor rouxii*, a dimorphic fungus, ranges from 16% to 22% of the total dry weight of the mycelium and contains approximately 35% to 40% of glucosamine. The actual yield of chitosan extracted from cell walls of *M. rouxii* ranged from 4% to 8% of the total cell dry weights (9–14 g/L in batch fermentations), which yields approximately 1 g/L chitosan [145].

During a fermentation, the average molecular weight of chitosan that can be extracted from cell walls of *M. rouxii* increases rapidly at the beginning, reaching maximum values of about 1×10^6, then declines as the period of incubation continues [146].

Paecilomyces spp. were found to produce an extracellular polyhexosamine that consists of a α-(1→4)-linked galactosamine chain. The degree of acetylation is approximately 8% and no glucosamine groups are present. The molecular weight is greater than 3×10^5. This polygalactosamine has a similar conformation and properties resembling those of chitosan in spite of the different chemical compositions [147, 148]. A partially acetylated galactosamine is also secreted by *Aspergillus parasiticus* [149].

During a screening program, approximately 1200 strains of fungi were tested for their ability to synthesize polyhexosamines. Most strains that were found to be producers of polyhexosamines belonged to the genus *Aspergillus*. Preliminary studies showed that these polymers are heteropolysaccharides with varying contents of glucosamine and galactosamine [150].

Chitosan has "the potential of being one of the major new commercial polysaccharides" [151]. Chitin, chitosan, and derivatives of these substances have a very broad range of useful properties. Possible applications would be for the production of pharmaceuticals. Chitin, chitosan, and chitosan derivatives accelerate the healing of wounds when used as wound dressings [152]. Water-soluble chitooligosaccharides show antitumor activity against sarcoma 180 solid tumors [153] and chitosan beads have been introduced as carriers for anticancer agents. Carboxymethyl chitin can serve as a controlled-release drug carrier or as a hapten carrier. Deacetylated chitin (70% deacetylated) has a relatively long-acting immunoadjuvant activity [154].

For cosmetic applications, chitosan can serve as a moisturizer in skin care products. Chitosan is also a very effective flocculant due to its cationic properties. It is used for the clarification of drinking water, fruit juices, and beverages, for waste water in paper mills and sewage effluents. It can also chelate metal ions, so it can be used to remove toxic heavy metal ions such as copper, cadmium, mercury, chromium, or uranium from waste water [155].

Further possible applications for chitosan have been described in an excellent review [156] in which uses for chitosan were classified in the following categories: clarification and purification of water and beverages, pharmaceutical products, cosmetics and personal care preparations, agriculture, and biotechnology.

From this relatively extensive spectrum of possible uses for chitosan, it can be seen that the attractiveness of this substance, and its potential economic value, will further stimulate research and the development of effective biotechnological production processes.

VI. CONCLUSIONS

It seems that nearly all fungi are capable of producing polysaccharides that can have several functions:

1. Protection against negative environmental conditions
2. Carbon and energy storage
3. Adhesion to surfaces and biofilm formation
4. Trapping and concentration of nutrients
5. Structural element

As summarized in Table 1 most of the more fully investigated extracellular fungal polysaccharides are homopolymers consisting of D-glucose units, that is, polyglucans. In contrast to most bacterial polysaccharides, they are uncharged,

TABLE 1 Various Fungal Polysaccharides with Their Basic Structural Elements

Polymer	Organism	Linkages (ratio)	Monomeric unit	Properties
Pullulan	*Aureobasidium pullulans*	α-(1→4),α-(1→6)- (2:1)	D-glucose	M_w 5 × 10^4–3 × 10^6, water soluble, pH-stable (pH 2–12)
Nigeran	*Aspergillus* spp. *Penicillium* spp.	α-(1→3),α-(1→4)- (1:1)	D-glucose	Soluble in hot water, insoluble in cold water
Elsinan	*Elsinoe* spp.	α-(1→4),α-(1→3)- (2.3–2.5:1)	D-glucose	M_w 2–6 × 10^6 water soluble, pH-stable (pH 3–11)
Schizo-phyllan	*Schizophyllum commune*	β-(1→3),β-(1→6)- (3:1)	D-glucose	M_w up to 1.3 × 10^7, water soluble temperature-stable (up to 135°C) pH-stable (pH 2–12)
Sclero-glucan	*Sclerotium* spp.	β-(1→3),β-(1→6)- (3:1)	D-glucose	M_w up to 7.4 × 10^6, water soluble temperature-stable (up to 135°C) pH-stable (pH 2–12)
Yeast glycan	*Saccharomyces* spp.	α-(1→6),α-(1→3)- α-(1→2)- β-(1→3),β-(1→6)-	D-mannose D-glucose	Water insoluble
Chitosan	*Mucoraceae*	β-(1→4)-	glucosamine	M_w 1.2 × 10^6, water insoluble, charged

neutral molecules. The vast number of different organisms excreting such polymers should be stimulating for further investigations in this exciting field for future applications of biopolymers in diverse categories such as pharmaceutical applications, personal and health care, as food additives, for agricultural purposes, and for different industrial applications including enhanced oil recovery.

REFERENCES

1. E. Dickinson, in *Gums and Stabilisers for the Food Industry*, Vol. 4 (G. O. Phillips, P. A. Williams, and D. J. Wedlock, eds.), IRL Press, Oxford, 1988, pp. 249–263.
2. B. Bergenståhl, in *Gums and Stabilisers for the Food Industry*, Vol. 4 (G. O. Phillips, P. A. Williams, and D. J. Wedlock, eds.), IRL Press, Oxford, 1988, pp. 363–369.
3. B. Bernier, *Can. J. Microbiol. 4*: 195 (1958).
4. H. Bender, J. Lehmann, and K. Wallenfels, *Biochem. Biophys. Acta 36*: 309 (1959).
5. K. Wallenfels, H. Bender, G. Keilich, and G. Bechtler, *Angew. Chem. 73*: 245 (1961).
6. K. Wallenfels, G. Keilich, G. Bechtler, and D. Freudenberger, *Biochem. Z. 341*: 433 (1965).
7. B. J. Catley and W. J. Whelan, *Arch. Biochem. Biophys. 143*: 138 (1971).
8. G. Carolan, B. J. Catley, and F. J. McDougal, *Carbohydr. Res. 114*: 237 (1983).
9. B. J. Catley, A. Ramsay, and C. Servis, *Carbohydr. Res. 153*: 79 (1986).
10. N. P. Elinov, V. A. Marikhin, A. N. Dranishnikov, L. P., Myasnikova, and Y. B. Maryukhta, *Dokl. Akad. Nauk. S.S.S.R. 221*: 213 (1975).
11. H. O. Bouveng, H. Kessling, B. Lindberg, and J. McKay, *Acta Chem. Scand. 16*: 615 (1962).
12. A. Jeanes, in *Extracellular Microbial Polysaccharides* (P. A. Sandford and A. Laskin, eds.), ACS Symposium Series, Vol. 45, American Chemical Society, Washington, D.C., 1977, pp. 284–298.
13. B. McNeil and B. Kristiansen, *Enzyme Microb. Technol. 12*: 521 (1990).
14. B. J. Catley, *FEBS Lett. 10*: 190 (1970).
15. K. Sugimoto, *J. Ferm. Assoc. Jap. 36*: 98 (1977).
16. Y. Sakano, M. Kogure, T. Kobayashi, M. Tamura, and M. Suekane, *Carbohydr. Res. 61*: 175 (1978).
17. S. Yuen, *Process Biochem. 9*: 7 (1974).
18. A. LeDuy, L. Choplin, J. E. Zajic, and J. H. T. Luong, in *Encyclopedia of Polymer Science and Engineering*, Vol. 13, Wiley, New York, 1988, pp. 650–660.
19. Y. C. Shin, Y. H. Kim, H. S. Lee, S. J. Cho, and S. M. Byun, *Biotechnol. Bioeng. 33*: 129 (1989).
20. J. E. Zajic, K. K. Ho, and N. Kosaric, *Dev. Ind. Microbiol. 20*: 631 (1979).
21. J. M. Boa and A. Leduy, *Appl. Environ. Microbiol. 48*: 26 (1984).
22. P. J. Heald and B. Kristiansen, *Biotechnol. Bioeng. 27*: 1516 (1985).
23. K. Ono, N. Yasuda, and S. Ueda, *Agric. Biol. Chem. 41*: 2113 (1977).
24. R. J. Seviour and B. Kristiansen, *Eur. J. Appl. Microbiol. Biotechnol. 17*: 178 (1983).
25. M. A. Bulmer, B. J. Catley, and P. J. Kelly, *Appl. Microbiol. Biotechnol. 25*: 362 (1987).
26. B. J. Catley, in *Microbial Polysaccharides and Polysaccharases* (R. C. W. Berkely, G. W. Gooday, and D. C Ellwood, eds.), Academic Press, London, 1979, pp. 69–84.

27. B. McNeil, B. Kristiansen, and R. J. Seviour, *Biotechnol. Bioeng. 33*: 1210 (1989).
28. R. W. Silman, W. L. Bryan, and T. D. Leathers, *FEMS Microbiol. Lett. 77*: 65 (1990).
29. D. P. F. Auer and R. J. Seviour, *Appl. Microbiol. Biotechnol. 32*: 637 (1990).
30. K. Nishinari, in *Gums and Stabilisers for the Food Industry*, Vol. 4 (G. O. Phillips, P. A. Williams, and D. J. Wedlock, eds.), IRL Press, Oxford, 1988, pp. 373–390.
31. M. Endo, M. Sasago, Y. Hirai, K. Ogawa, and T. Ishihara, *J. Vac. Sci. Technol. B6*: 1600 (1988).
32. N. Sakairi, M. Hayashida, and H. Kuzuhara, *Carbohydr. Res. 185*: 91 (1989).
33. A.-L. Nguyen, S. Grothe, and J. H. T. Luong, *Appl. Microbiol. Biotechnol. 27*: 341 (1988).
34. S. Yamaya, A. Yamamoto, T. Komiya, J. Mizuguchi, and T. Mituhasi, *Vaccine 8*: 65 (1990).
35. G. S. Alekseeva, T. N. Telkova, S. M. Yarovaya, M. A. Chlenov, V. A. Yanin, and V. A. Dombrovskii, *Khim-Farm. Zh. 23*: 789 (1989).
36. T. Uchida, I. Fujimoto, and S. Goto, *Yakuzaigaku 49*: 53 (1989).
37. A. W. Dox and R. E. Niedig, *J. Biol. Chem. 18*: 167 (1914).
38. A. W. Dox, *J. Biol. Chem. 20*: 83 (1915).
39. T. F. Bobbitt and J. H. Nordin, *J. Bacteriol. 150*: 365 (1982).
40. M. H. Gold, D. L. Mitzel, and I. H. Segel, *J. Bacteriol. 113*: 856 (1973).
41. R. Gupta and K. G. Mukerji. Folia Microbiol. *27*: 38 (1982).
42. J. H. Nordin, T. F. Bobbitt, and R. H. Marchessault, in *Fungal Polysaccharides* (P. A. Sandford and K. Matsuda, eds.), ACS Symposium Series, Vol. 126, American Chemical Society, Washington, D.C., 1980, pp. 143–157.
43. S. Takaya, T. Fukuda, and Y. Oihe, *Chagyo-Gijutsu-Kenkyu* (Study of Tea) *49*: 79 (1975).
44. Y. Tsumuraya, A. Misaki, S. Takaya, and M. Torii, *Carbohydr. Res. 66*: 53 (1978).
45. A. Misaki and Y. Tsumuraya, in *Fungal Polysaccharides* (P. A. Sandford and K. Matsuda, eds.), ACS Symposium Series, Vol. 126, American Chemical Society, Washington, D.C., 1980, pp. 197–220.
46. B. J. Catley, in *Microbial Polysaccharides* (M. E. Bushell, ed.), Progress in Industrial Microbiology, Vol. 18, Elsevier, Amsterdam, 1983, pp. 129–200.
47. T. Narui, K. Takahashi, M. Kobayashi, and S. Shibata, *Carbohydr. Res. 87*: 161 (1980).
48. P. A. J. Gorin and Barreto-Bergter, in *The Polysaccharides* (G. O. Aspinall, ed.), Vol. 2, Academic Press, New York, 1983, pp. 365–409.
49. F. M. Essig, *Univ. Cal. Publ. Bot. 7*: 447 (1922).
50. S. Fujimoto, S. Kikumoto and T. Kaneuchi, *Taito Kenkyusho Hokoku 22*: 77 (1964).
51. S. Kikumoto, T. Miyajima, K. Kimura, S. Okubo, and N. Komatsu, *Nippon Nogeikagaku Kaishi 45*: 162 (1971).
52. J. G. H. Wessels and J. H. Sietsma, in *Fungal Walls and Hyphal Growth* (J. H. Burnett and A. P. J. Trinci, eds.), Cambridge University Press, 1979, pp. 27–48.
53. J. G. H. Wessels, *Lignin Enzymic and Microbial Degradation*, INRA, Paris (Les Colloques de l'INRA, n° 40), 1987, pp. 19–42.
54. J. G. H. Wessels, *Wentia 13*: 1 (1965).
55. M. Desrochers, L. Jurasek, and M. G. Paice, *Dev. Ind. Microbiol. 22*: 675 (1981).
56. W. Steiner, R. M. Lafferty, I. Gomes, and H. Esterbauer, *Biotechnol. Bioeng. 30*: 169 (1987).

57. P. L. Manachini, *Ann. Microbiol. 32*: 23 (1982).
58. S. Tachibana, J. Siode, and T. Hanai, *J. Ferment. Technol. 45*: 1130 (1967).
59. T. Norisuye, *Makromol. Chem. S14*: 105 (1985).
60. T. Norisuye, T. Yanaki, and H. Fujita, *J. Polym. Sci. Polym. Phys. 18*: 547 (1980).
61. T. Yanaki, T. Norisuye, and H. Fujita, *Macromolecules 13*: 1462 (1980).
62. Y. Kashiwagi, T. Norisuye, and H. Fujita, *Macromolecules 14*: 1220 (1981).
63. T. Itou, A. Teramoto, T. Matsuo, and H. Suga, *Macromolecules 19*: 1234 (1986).
64. T. Itou, A. Teramoto, T. Matsuo, and H. Suga, *Carbohydr. Res. 160*: 243 (1987).
65. T. Hirao, T. Sato, A. Teramoto, T. Matsuo, and H. Suga, *Biopolymers 29*: 1867 (1990).
66. T. Itou and A. Teramoto, *Polymer J. 16*: 779 (1984).
67. T. Yanaki, K. Tabata, and T. Kojima, *Carbohydr. Polymer 5*: 275 (1985).
68. E. Steiner, H. Divjak, W. Steiner, R. M. Lafferty, and H. Esterbauer, *Progr. Colloid Polymer Sci. 77*: 217 (1988).
69. T. Yanaki, K. Nishii, K. Tabata, and T. Kojima, *J. Appl. Polym. Sci. 28*: 873 (1983).
70. T. Kojima, K. Tabata, T. Ikumoto, and T. Yanaki, *Agric. Biol. Chem. 48*: 915 (1984).
71. T. Miyajima, S. Yoshizumi, S. Kikumoto, and H. Takahashi, *Seito Gijutsu Keukyukaishi 22*: 35 (1970).
72. U. Rau, R.-J. Müller, K. Cordes, and J. Klein, *Bioprocess Eng. 5*: 89 (1990).
73. N. Komatsu, S. Okubo, S. Kikumoto, K. Kimura, G. Saito, and S. Sakai, *Gann 60*: 137 (1969).
74. G. Franz, *Farm. Tijdschr. Belg. 64*: 301 (1987).
75. M. Aoki, A. Fukushima, Y. Mikami, M. Tan, and S. Kubo, Japanese Patent 01,272,509, to Japan Tobacco, Inc. (1989).
76. W. Itoh, I. Sugawara, S. Kimura, K. Tabata, A. Hirata, T. Kojima, S. Mori, and K. Shimada, *Int. J. Immunopharmacol. 12*: 225 (1990).
77. M. L. Patchen, M. A. Chirigos, and I. Brook, *Comments Toxicol. 2*: 217 (1988).
78. Z. K. Punja, R. G. Grogan, and G. C. Adams, *Mycologia 74*: 917 (1982).
79. F. E. Halleck, U.S. Patent 3,301,848 to Pillsbury Co. (1967).
80. M. Rinaudo and M. Vincendon, *Carbohydr. Polymers 2*: 135 (1982).
81. W. Steiner, S. Leonhartsberger, S. Zettl, H. Divjak, and R. M. Lafferty, in *Interbiotech '89. Mathematical Modelling in Biotechnology* (A. Blazej and A. Ottová, eds.), Progress in Biotechnology, Vol. 6, Elsevier, Amsterdam, 1990, pp. 337–344.
82. T.-C. Tseng and D. F. Bateman, *Phytopathology 59*: 359 (1969).
83. D. F. Bateman and S. V. Beer, *Phytopathology 55*: 204 (1965).
84. H. D. Van Etten and D. F. Bateman, *Phytopathology 59*: 968 (1969).
85. A. L. J. Cole and D. F. Bateman, *Phytopathology 59*: 1750 (1969).
86. D. F. Bateman, *Phytopathology 59*: 37 (1969).
87. A. H. Lachke and M. V. Deshpande, *FEMS Microbiol. Rev. 54*: 177 (1988).
88. J. G. Shewale and J. C. Sadana, *Arch. Biochem. Biophys. 207*: 185 (1981).
89. J. C. Sadana, A. H. Lachke, and R. V. Patil, *Carbohydr. Res. 133*: 297 (1984).
90. S. Takao, H. Sasaki, K. Kurosawa, M. Tanida, and Y. Kamagata, *Agric. Biol. Chem. 50*: 1979 (1986).
91. Z. K. Punja and R. G. Grogan, *Plant Dis. 67*: 875 (1983).
92. T. Yanaki, T. Kojima, and T. Norisuye, *Polymer J. 13*: 1135 (1981).

93. T. Yanaki, T. Norisuye, and A. Teramoto, *Polymer J. 16*: 165 (1984).
94. C. Biver, J. Lesec, C. Allain, L. Salomé, and J. Lecourtier, *Polymer Commun. 27*: 351 (1986).
95. T. L. Bluhm, Y. Deslandes, R. H. Marchessault, S. Pérez, and M. Rinaudo, *Carbohydr. Res. 100*: 117 (1982).
96. C. Noik, J. Lecourtier, G. Chauveteau, and A. Bonche, *Polymer Mat. Sci. Eng. 57*: 380 (1987).
97. P. Rapp, *J. Gen. Microbiol. 135*: 2847 (1989).
98. N. E. Rogers, in *Industrial Gums* (R. L. Whistler and J. N. Bemiller, eds.), Academic Press, New York, 1973, pp. 499–511.
99. P. P. Singh, R. L. Whistler, R. Tokuzen, and W. Nakahara, *Carbohydr. Res. 37*: 245 (1974).
100. N. Ohno, I. Suzuki, and T. Yadomae, *Chem. Pharm. Bull. 34*: 1362 (1986).
101. F. Alhaique, F. M. Riccieri, G. Riccioni, E. Santucci, and E. Touitou, in *Biomedical and Biotechnological Advances in Industrial Polysaccharides* (V. Crescenzi, I. C. M. Dea, S. Paoletti, S. S. Stivala, and I. W. Sutherland, eds.), Gordon and Breach, New York, 1989, pp. 157–165.
102. P. Davison and E. Mentzer, *Soc. Pet. Eng. J.* 353 (1982).
103. I. W. Cottrell, in *Fungal Polysaccharides* ACS Symposium Series, Vol. 126 (P. A. Sandford and K. Matsuda, eds.), American Chemical Society, Washington, D.C., 1980, pp. 251–270.
104. S. J. Stasinopoulos and R. J. Seviour, *Mycol. Res. 92*: 55 (1989).
105. A. S. Perlin and W. A. Taber, *Can. J. Chem. 41*: 2278 (1963).
106. J. M. Sarkar, G. L. Hennebert, and J. Mayaudon, *Biotechnol. Lett. 7*: 631 (1985).
107. A. J. Buchala and M. Leisola, *Carbohydr. Res. 165*: 146 (1987).
108. K.-E. Eriksson, P. Ander, and B. Pettersson, *Proceedings of the International Conference on Biotechnology, Pulp Paper Industry*, Stockholm, 1986, pp. 24–27.
109. J.-P. Latgé, D. G. Boucias, and B. Fournet, *Carbohydr. Res. 181*: 282 (1988).
110. Y. Iwamuro, M. Aoki, Y. Mikami, Y. Obi, and T. Kisaki, *J. Ferment. Technol. 60*: 405 (1982).
111. A. Misaki, K. Kawaguchi, H. Miyaji, H. Nagae, S. Hokkoku, M. Kakuta, and T. Sasaki, *Carbohydr. Res. 129*: 209 (1984).
112. A. T. Martínez and M. J. Martínez, *Soil Biol. Biochem. 18*: 469 (1986).
113. F. Santamaría, F. Reyes, and R. Lahoz, *J. Gen. Microbiol. 109*: 287 (1978).
114. Y. Yoshioka, R. Tabeta, H. Saito, N. Uehara, and F. Fukuoka, *Carbohydr. Res. 140*: 93 (1985).
115. A. L. Compere, W. L. Griffith, and S. V. Greene, *Dev. Ind. Microbiol. 21*: 461 (1980).
116. K. Iino, N. Ohno, I. Suzuki, T. Miyazaki, T. Yadomae, S. Oikawa, and K. Sato, *Carbohydr. Res. 141*: 111 (1985).
117. T. Sasaki and N. Takasuka, *Carbohydr. Res. 47*: 99 (1976).
118. T. Sasaki, N. Takasuka, G. Chihara, and Y. Y. Maeda, *Gann 67*: 191 (1976).
119. K. Saito, M. Nishijima, and T. Miyazaki, *Chem. Pharm. Bull. 38*: 1745 (1990).
120. C. Hara, T. Kiho, and S. Ukai, *Carbohydr. Res. 117*: 201 (1983).
121. S. Ukai, C. Hara, and T. Kiho, *Chem. Pharm. Bull. 30*: 2147 (1982).

122. C. Hara, T. Kiho, and S. Ukai, *Carbohydr. Res. 145*: 237 (1986).
123. A. F. Cirelli, J. A. Covian, N. Ohno, Y. Adachi, and T. Yaomae, *Carbohydr. Res. 190*: 329 (1989).
124. H. Nanba and H. Kuroda, *Chem. Pharm. Bull. 19*: 448 (1971).
125. I. Fabre, M. Bruneteau, P. Ricci, and G. Michel, *Eur. J. Biochem. 142*: 99 (1984).
126. M. Bruneteau, I. Fabre, J. Perret, G. Michel, P. Ricci, J.-P. Joseleau, J. Kraus, M. Schneider, W. Blaschek, and G. Franz, *Carbohydr. Res. 175*: 137 (1988).
127. N. Ohno, K. Iino, T. Takeyama, I. Suzuki, K. Sato, S. Oikawa, T. Miyazaki, and T. Yadomae, *Chem. Pharm. Bull. 33*: 3395 (1985).
128. T. Kiho, M. Ito, I. Yoshida, K. Nagai, C. Hara, and S. Ukai, *Chem. Pharm. Bull. 37*: 2770 (1989).
129. S. Ukai, S. Yokoyama, C. Hara, and T. Kiho, *Carbohydr. Res. 105*: 237 (1982).
130. T. Kiho, M. Ito, K. Nagai, C. Hara, and S. Ukai, *Chem. Pharm. Bull. 35*: 4286 (1987).
131. J. H. Duffus, C. Levi, and D. J. Manners, *Adv. Microb. Physiol. 23*: 151 (1982).
132. P. N. Lipke and C. E. Ballou, *J. Bacteriol. 141*: 1170 (1980).
133. D. J. Manners, A. J. Masson, and J. C. Patterson, *J. Gen. Microbiol. 80*: 411 (1974).
134. S. Jamas, C. K. Rha, and A. J. Sinskey, *Biotechnol. Bioeng. 28*: 769 (1986).
135. R. W. Sucher, E. A. Robbins, D. R. Sidoti, E. H. Schuldt, and R. D. Seeley, U.S. Patent 3,867,554 (1975).
136. H. Konishi, H. Kato, and K. Itaya, Japanese Patent 63,083,012 (1988).
137. M. Wolk and D. Danon, *Med. Biol. 63*: 73 (1985).
138. G. Kogan, L. Masler, J. Sanula, J. Navarová, and T. Trnovec, in *Biomedical and Biotechnological Advances in Industrial Polysaccharides* (V. Crescenzi, I. C. M. Dea, S. Paoletti, S. S. Stivala, and I. W. Sutherland, eds.), Gordon and Breach, New York, 1989, pp. 251–258.
139. A. G. Kovalenko, *Zentralbl. Mikrobiol. 142*: 301 (1987).
140. R. C. W. Berkeley, in *Microbial Polysaccharides and Polysaccharases* (R. C. W. Berkeley, G. W. Gooday, D. C. Ellwood, eds.), Academic Press, London, 1979, pp. 205–236.
141. H. Saito, R. Tabeta, and K. Ogawa, in *Industrial Polysaccharides*, Progress in Biotechnology, Vol. 3 (M. Yalpani, ed.), Elsevier Science Publishers, Amsterdam, 1987, pp. 267–280.
142. S. Bartnicki-Garcia, in *Chitin and Chitosan* (G. Skjåk-Bræk, T. Anthonsen, and P. Sandford, eds.), Elsevier, London, 1988, pp. 23–35.
143. R. A. A. Muzzarelli, in *The Polysaccharides*, Vol. 3 (G. O. Aspinall, ed.), Academic Press, Orlando, 1985, pp. 417–450.
144. K. Shimahara, Y. Takiguchi, T. Kobayashi, K. Uda, and T. Sannan, in *Chitin and Chitosan* (G. Skjåk-Bræk, T. Anthonsen, and P. Sandford, eds.), Elsevier, London, 1988, pp. 171–178.
145. S. A. White, P. R. Farina, and I. Fulton, *Appl. Environ. Microbiol. 38*: 323 (1979).
146. S. Arcidiacono, S. J. Lombardi, and D. L. Kaplan, in *Chitin and Chitosan* (G. Skjåk-Bræk, T. Anthonsen, and P. Sandford, eds.), Elsevier, London, 1988, pp. 319–332.
147. H. Takagi and K. Kadowaki, *Agric. Biol. Chem. 49*: 3151 (1985).
148. H. Takagi and K. Kadowaki, *Agric. Biol. Chem. 49*: 3158 (1985).

149. O. Hayashi, H. Yamada, and T. Miyazaki, *Agric. Biol. Chem. 40*: 1643 (1976).
150. T. Yokoyama, E. Murakami, K. Hasegawa, S. Tukada, H. Takagi, K. Kadowaki, and K. Oishi, in *Chitin and Chitosan* (G. Skjåk-Bræk, T. Anthonsen, and P. Sandford, eds.), Elsevier, London, 1988, pp. 333–341.
151. M. Yalpani and P. Sandford, in *Industrial Polysaccharides*, Progress in Biotechnology, Vol. 3 (M. Yalpani, ed.), Elsevier, Amsterdam, 1987, pp. 311–335.
152. R. Muzzarelli, G. Biagini, A. Damadei, A. Pugnaloni, and J. Da Lio, in *Biomedical and Biotechnological Advances in Industrial Polysaccharides* (V. Crescenzi, I. C. M. Dea, S. Paoletti, S. S. Stivala, and I. W. Sutherland, eds.), Gordon and Breach, New York, 1989, pp. 77–88.
153. K. Suzuki, T. Mikami, Y. Okawa, A. Tokoro, S. Suzuki, and M. Suzuki, *Carbohydr. Res. 151*: 403 (1986).
154. S. Tokura, N. Nishi, and I. Azuma, in *Industrial Polysaccharides*, Progress in Biotechnology, Vol. 3, (M. Yalpani, ed.), Elsevier, Amsterdam, 1987, pp. 347–362.
155. K. Kurita, in *Industrial Polysaccharides*, Progress in Biotechnology, Vol. 3 (M. Yalpani, ed.), Elsevier, Amsterdam, 1987, pp. 337–346.
156. P. A. Sandford and G. P. Hutchings, in *Industrial Polysaccharides*, Progress in Biotechnology, Vol. 3 (M. Yalpani, ed.), Elsevier, Amsterdam, 1987, pp. 363–376.

7

Bioconversion of Oils and Sugars to Glycolipids

SIEGMUND LANG and FRITZ WAGNER Institute of Biochemistry and Biotechnology, Technical University of Braunschweig, Braunschweig, Germany

I. INTRODUCTION

The most important renewable resources are polysaccharides, sugars, and oils and fats. They are attractive alternatives to mineral oils as feedstocks for chemicals and could make a contribution to

Utilization of agricultural surpluses in some countries
Broadening of the raw material basis for industrial purposes
Development of new products
Improvement of the world around us

One of their advantages is that they are renewable by agricultural methods within a short time.

The intention of this chapter is to show the efforts underway to refine the above-mentioned raw materials into more valuable products. Biotechnological processes are important tools for attaining this target. Among the various possibilities are biomodification of oils, fats, and sugars with microbes or enzymes leading to natural glycolipids.

Reviews on the occurrence and types of microbial surfactants have been published during the last decade by Zajic et al. [1], Haferburg et al. [2], Kosaric et al. [3], and Hommel [4]. These publications report on the predominant use of *n*-alkanes as substrates for the formation of biosurfactants, as well as on the occurrence of only traces of these compounds when other carbon sources were utilized. In contrast to these surveys, this chapter focuses on potential developments concerning

1. Overproduction of glycolipids by cultivation under growth limitation and by reactions with resting cells using oils and sugars
2. Directed synthesis of glycolipids from oils, fatty acids/alcohols, and sugars by aid of lipases and glycosidases

II. MICROBIAL CELLS UNDER GROWTH-LIMITING CONDITIONS

Some yeasts, fungi, and bacteria are able to utilize triglycerides (usually containing fatty acids in the range of C8 to C22, saturated and unsaturated) or carbohydrates for growth and synthesis of glycolipids predominantly during the stationary phase. A common scheme of possible biosynthetic routes is presented in Fig. 1.

A substantial increase in biosurfactant production after the exponential growth phase was observed in the case of

Nitrogen limitation
Multivalent cation limitation
Temperature shift, pH shift

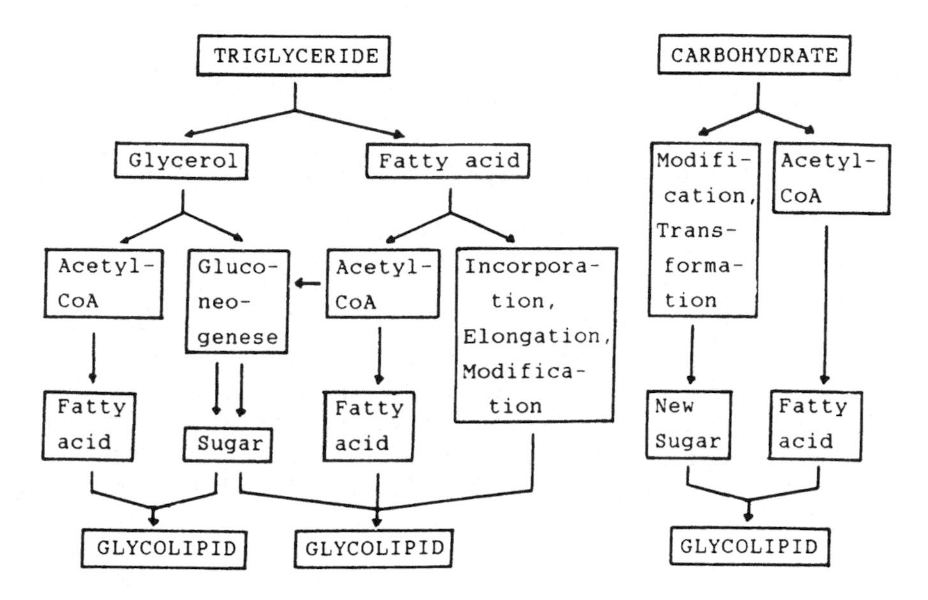

FIG. 1 General scheme of possible routes for the microbial glycolipid synthesis from triglycerides and/or glucose.

Interesting examples for these observations together with the overproduction of mannosylerythritol, sophorose, cellobiose, and rhamnose lipids on the above substrates are presented below.
The following statements could be made about the biosynthesizing enzymes.

1. The corresponding enzymes are produced during the exponential growth phase; they are in a nonactive stage.
2. Upon change of one (or more) of the environmental parameters (nutrient requirements, temperature, pH), growth is inhibited and the enzyme activities for biosurfactant synthesis are switched on.
3. As only carbon, hydrogen, and oxygen are important elements for the development of the molecular structure, the glycolipid production does not need additional nitrogen-containing salts and continues as long as the carbon source and oxygen are sufficient and the cell energy allows this work, respectively.

With regard to the regulation of glycolipid production, Mulligan and Gibbs [5] found that in a proteose peptone/glucose medium with *Pseudomonas aeruginosa* ATCC 9027 (wild type) and the corresponding chloramphenicol-tolerant strain, ATCC 9027 var. RCII, the glutamine synthetase and rhamnolipid formation were regulated by glutamate, glutamine, and ammonia levels. As growth slowed, glutamate and ammonia were utilized, and glutamine synthetase activity increased. Glutamine inhibited this process. After this the cell metabolism switched from nitrogen as it became limiting (glutamate) to glucose (reverse diauxie), resulting in rhamnolipid production.

A. Production of Mannosylerythritol Lipids

Following previous studies with *Candida* sp. B-7 on *n*-alkanes and vegetable oils leading to an apparently growth-associated mannosylerythritol lipid [6, 7], recently Kitamoto et al. [8, 9] reported on the extracellular accumulation of similar products by a strain of *Candida antarctica* T-34. The biosurfactants were found to be a mixture of four compounds, including two new mannosylerythritol lipids as major components (Fig. 2, MEL-A, MEL-B).

This strain produced the lipids from different vegetable oils (soybean, safflower, coconut, cottonseed, corn, and palm oil), but failed to produce them from *n*-alkanes or carbohydrates. Under optimum conditions in shaken cultures, the

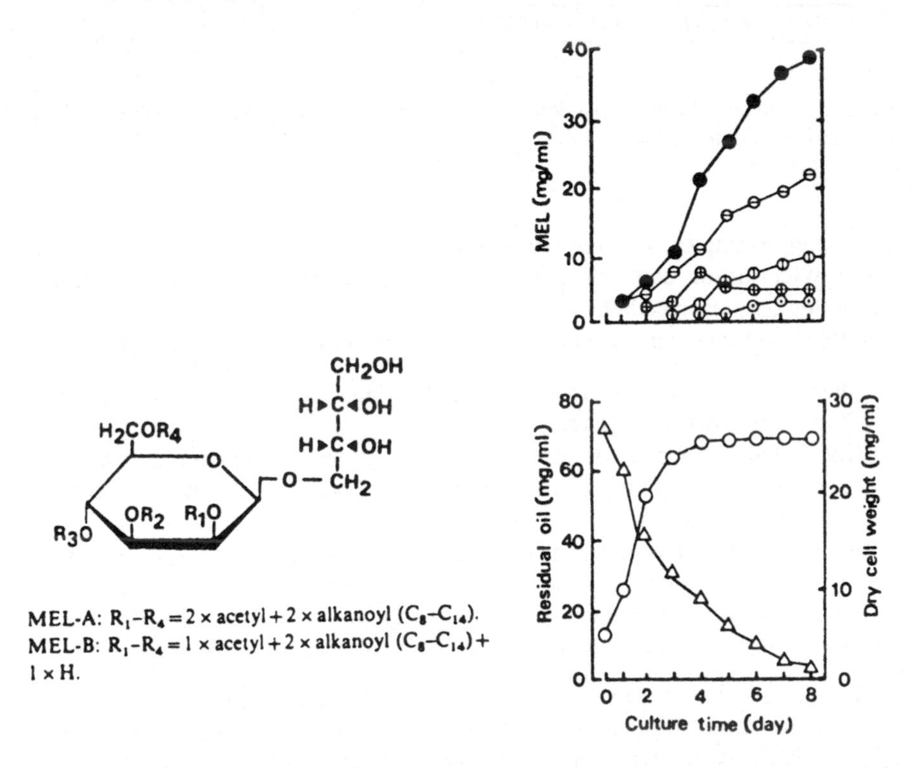

FIG. 2 Structures of mannosylerythritol-lipids (left). Time course of their production from *Candida antarctica* (right). Source: Refs. 8 and 9. –Δ– soybean oil, –O– biomass, –●– total MEL, –⊖– MEL-A, –⦶– MEL-B, –⊕– MEL-C, –⊙– MEL-D.

concentration of total lipids amounted to approximately 40 g/L after 8 days when 8% soybean oil was used as substrate; that means a specific production of 0.5 g/g carbon source. Figure 2 clearly shows the dramatic increase of products at the end of growth and little change in the composition of the lipid mixture.

B. Production of Sophorose Lipids

Since 1961, it has been known that yeasts of the genus *Torulopsis*, e.g., *T. magnoliae, T. bombicola* (now belonging to the genus *Candida*) excrete a mixture of sophorose lipids during growth on 20% glucose, 1.25% yeast extract and 0.2% of urea [10]. In further studies the biosurfactant yield could be improved by the stepwise addition of lipophilic long-chain fatty acid esters or *n*-alkanes [11] and of triglycerides [12, 13]. Using 20% carbon source (10% glucose and 10% sunflower oil), 67 g/L of crude sophorolipid mixture corresponding to 0.335 g/g carbon source could be isolated after cultivation of *T. bombicola* ATCC 22214; most of the glycolipids appeared in the late logarithmic phase of growth [13].

Using *T. bombicola* KSM-36, Inoue [14] succeeded in a high overproduction in the range of 100–150 g/L product from 10% glucose and 15% palm oil or safflower oil (0.4 to 0.6 g/g substrates, respectively). The analyzed molecular structures indicated a series of lactonic and acidic sophorose lipids.

Our own studies on sophorose lipid production by *T. bombicola* using both 100-mL and 30-L cultures showed the following results:

1. The basic medium of Tulloch et al. [15], containing 10% glucose, 1% yeast extract, and 0.1% urea, was the best. All components had to be sterilized together.
2. The pH was adjusted to 6.0 before sterilization. During cultivation pH should not be adjusted.
3. Incubation temperatures between 23°C and 30°C promoted growth and glycolipid production.
4. Stepwise or continuous addition of the second carbon source, such as triglycerides, fatty acids, esters, alcohols, and *n*-alkanes, after 24 h to the basic medium enhanced the sophorolipid production.
5. A ratio of glucose/second carbon source of about 3:1 favored the overproduction of a single sophorolipid. Lower ratios led to more hydrophilic sophorolipids.

In Fig. 3, a typical 30-L cultivation with a total concentration of 136 g/L of carbon sources including soybean oil is shown. The yield amounted to 33 g/L or 0.24 g/g substrates. In comparison, glucose alone (136 g/L) was converted to 15 g/L of glycolipids, that is, less than 50% of the illustrated example. The concentration of the principal sophorolipid 1′,4″-lactone 6′6″-diacetate (mainly containing

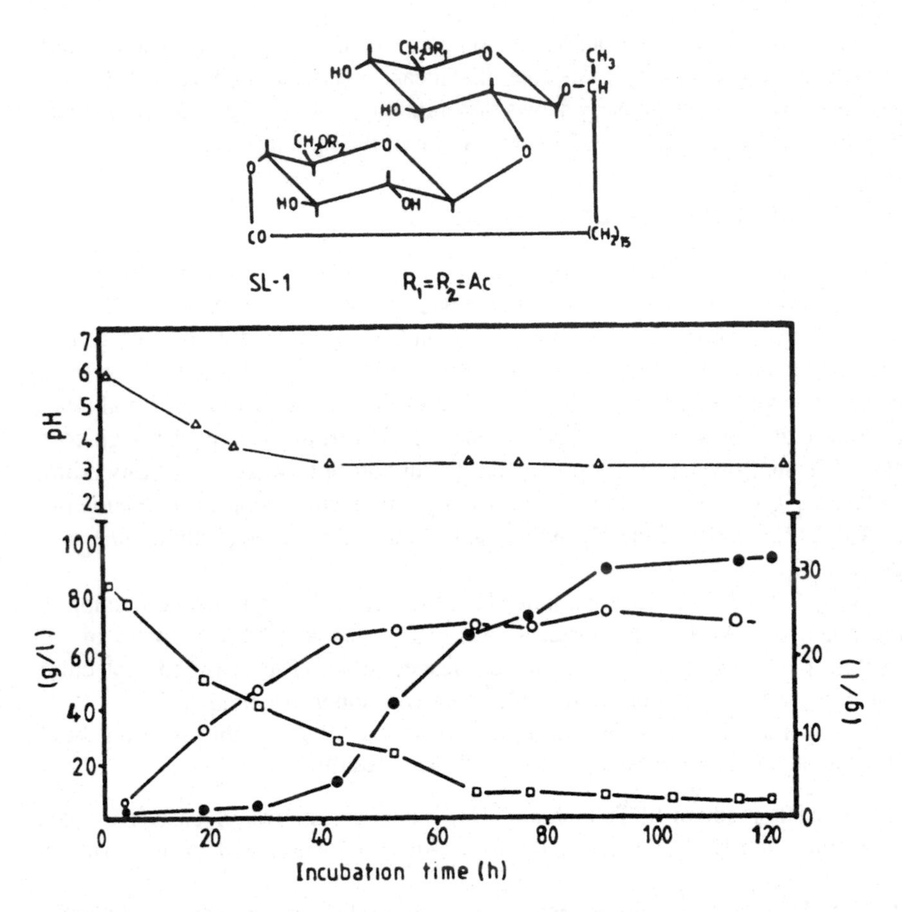

FIG. 3 Structure of the main component of sophorolipids (above). Growth and sophorolipid production of *Torulopsis bombicola* on 136 g/L of glucose/soybean oil mixture (below). Conditions: see text. (○) biomass (g/l); (●) sophorose lipids (g/l); (□) glucose (g/l); (Δ) pH.

17-OH-octadecanoic acid) within the crude product amounted to 40%–60%. On the other hand after feeding the yeast with oleic acid under the same conditions, the yield increased to 77 g/L (0.57 g/g substrate); now a double bond was incorporated in the lipophilic moiety [16].

Stüwer et al. [17] obtained a crystalline glycolipid after culturing *T. apicola* IMET 43734 on a mixture of glucose and alkanes or on vegetable oils. The excretion of product started with the beginning of the stationary growth phase and continued at a constant rate.

C. Production of Cellobiose Lipids

From the literature, it seems that among the great number of fungi only the *Ustilaginales* are able to form glycolipids. At first the group of Haskins and Lemieux reported that anionic cellobiose lipids were produced by *Ustilago zeae* PRL-119 when growing on glucose [18, 19]. Already in 1954, a process for the production of the so-called ustilagic acids was scaled up to 200-gal fermentors by Roxburgh et al. [20]. Under nitrogen-limiting conditions, the maximum efficiency, calculated on the basis of the ratio of carbon recovered as ustilagic acid to carbon added as glucose, was shown to be about 32% at 10% glucose. This is equivalent to a yield of 0.23 g/g substrate.

Using *Ustilago maydis* ATCC 14826, our group found differences with regard to the structures proposed by the above mentioned authors [18, 19], when coconut oil served as carbon source. C12 and C14 fatty acids (27% and 16%, respectively) were the major lipophilic fractions beside the C6 acid (37%) in position 2″ of cellobiose, whereas glucose mainly produced the C6 acid (69%). Polar differences in the substitution pattern concerning position 6′ (acylated and deacylated) and position 2″ (H or OH) indicate a mixture of glycolipids (Fig. 4). The results from a 20-L cultivation using a mineral salts medium supplemented with 0.03% urea and 2% coconut oil, are also presented in Fig. 4. The glycolipid synthesis began after utilization of urea, that is, under nitrogen-limiting conditions. Further increase in biomass in the absence of nitrogen was not due to cell multiplication but to cell-located product accumulation. The specific cellobiose lipid formation amounted to 0.59 g/g substrate. In comparison, other vegetable oils like palm, rapeseed, or safflower oil, gave yields in the range of 0.21 to 0.27 g/g [21, 22].

D. Production of Rhamnose Lipids

Both *Pseudomonas fluorescens* and *P. aeruginosa* are known to synthesize different extracellular rhamnolipids when grown on *n*-alkanes, glucose, or glycerol [23–25]. These lipids play an important role during growth on hydrocarbons [26, 27] and are produced in considerable amounts (0.18 g/g *n*-C-14,15-alkanes) during the stationary phase [28]. Concerning water soluble substrates, Fiechter's group developed a continuous process by culturing *P. aeruginosa* DSM 2659 on glucose [29, 30]. The main characteristics of the media were carbon and phosphorus excess as well as nitrogen and iron limitation. The biosurfactant yield on glucose was 0.077 g/g h, and the productivity was 0.147 g/L h.

Recently, our group [31], expanding the spectrum of carbon sources, obtained high yields using C14 to C18 fatty alcohols (32 g/L) or soybean oil (36 g/L), when culturing *Pseudomonas* sp. DSM 2874 for 5 days on 4% of these oils under nitrogen-limiting conditions [31]. Figure 5 shows the structures of both rhamnolipids as well as the time course of a soybean oil cultivation with respect to

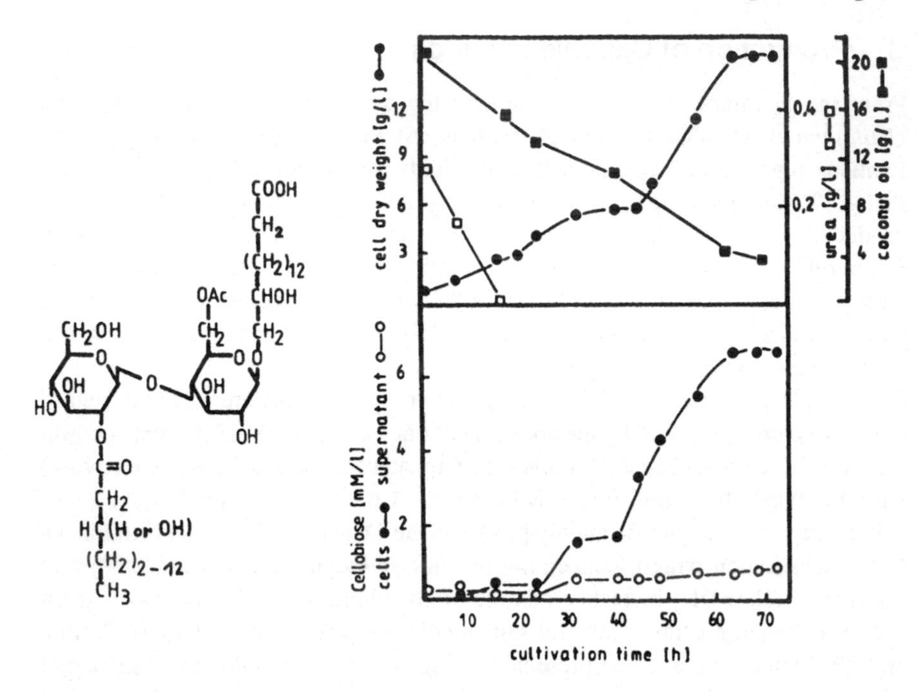

FIG. 4 Structures of cellobiose lipids (left). Time course of their production from *Ustilago maydis* on 2% of coconut oil (right). 1 mM/L cellobiose = 1.5 g/L cellobiose lipids. From Ref. 22.

growth and production of total rhamnolipids (R1 and R3). The yield was 0.88 g/g substrate.

E. Production of Other Glycolipids

From *Rhodococcus erythropolis* DSM 43215, anionic 2-*O*-succinoyl-3,4,2′-tri-*O*-alkanoyl-α,α-trehalose is known in connection with the cultivation on *n*-alkanes under nitrogen limitation [32, 33]. The use of sugars did not lead to this compound, and vegetable oils produced low amounts.

Also in the case of *R. erythropolis* SD-74 the overproduction of similar lipids, 2,3,4,2′-di-*O*-succinoyl-di-*O*-alkanoyl-α,α-trehalose and 2,3,4-mono-*O*-succinoyl-di-*O*-alkanoyl-α,α-trehalose was favored by hydrocarbons as carbon sources [34, 35].

Nocardia corynebacteroides SM1 synthesizes a pentasaccharide lipid in remarkable amounts when grown on *n*-alkanes, especially after nitrogen exhaustion; other substrates were unsuitable [36].

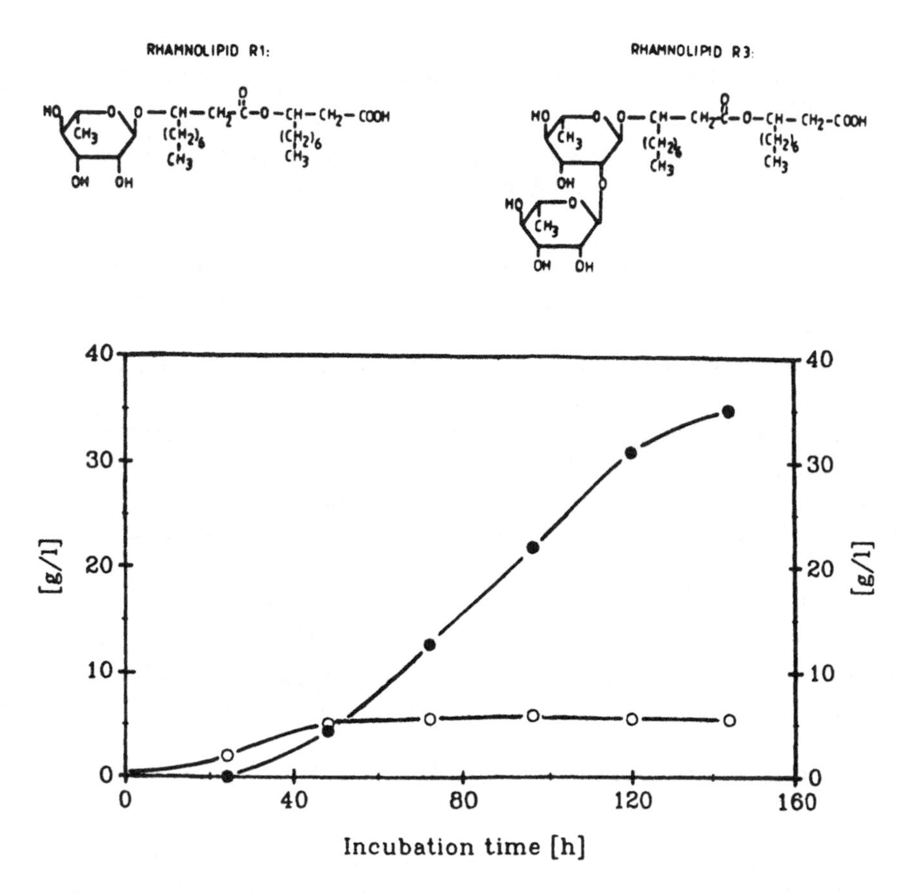

FIG. 5 Structures of rhamnolipids (above). Time course of their production from *Pseudomonas* sp. DSM 2874 under N-limitation on 4% of soybean oil (below) (31); 30°C, pH 6.8 (○) biomass (g/*l*); (●) rhamnolipids (g/*l*).

III. MICROBIAL RESTING CELLS

In summing up the results on glycolipid overproduction (as discussed in Sec. II), we find that these special processes seem to belong to the microbial production of secondary metabolites. Such compounds are usually produced at the end of the logarithmic phase and during the stationary phase. Therefore it seems possible to separate the growth phase from the biosurfactant production phase:

At first, the microorganisms are cultivated under optimum conditions.

After logarithmic growth, the cells are harvested (e.g., by centrifugation) and the wet biomass is washed in a buffer solution.

These cells can now be used as *free* or, after corresponding procedures, as *immobilized* biocatalysts for glycolipid biosynthesis under specific conditions.

On condition that they possess a sufficient energy level for a *multistep* biosynthesis, the *resting cells* have some advantages and allow important studies, such as the following:

The influence of single factors (e.g., different cations or concentration of salts) on the synthesis of the desired product can be examined more exactly because of more definite environmental reaction conditions.
The inhibition of glycolipid production by inorganic salts, necessary for growth but not for secondary metabolites, could be examined.
Feeding with various sugars or lipophilic compounds could allow another (more desired) composition of the glycolipid.
The effect of possibly disturbing by-products can be eliminated.
Immobilization of resting cells allows a continuous process and a more facilitated product recovery; certainly water-soluble carbon sources are necessary for an economic production.
Repeating batch reactions with immobilized cells is possible.

Examples of these biotechnological methods, while considering utilization of agricultural renewal resources, are reported in the following section.

A. Production of Sophorose Lipids

Preliminary studies with resting *T. bombicola* cells (formerly grown on glucose) in a 0.1 M McIlvaine buffer indicated that optimum pH and temperature were different from optimum growth conditions. Thus the results in view of the most suitable parameters (see Sec. II.B) during the cultivation under nitrogen-limiting conditions were confirmed [37]. A further intention was to replace either the sophorose or the fatty acid unit by another carbohydrate or lipophilic moiety. However, the precursor experiments always indicated the presence of the same molecular structure: a sophorose, acetylated in positions 6′ and 6″ and associated with 17-hydroxy-octadec*en*oic acid. The corresponding specific productions (g/g) related to metabolized sugars and oils after 3 days are summarized in Table 1. Additionally, by extending the incubation time for glucose conversion, approximately 15 g/L product of 0.75 g/g substrate or 1.5 g/g biomass could be isolated.

B. Production of Cellobiose Lipids

In the case of *U. maydis* a simple buffer solution used as reaction medium was unsuitable, but a modified mineral salts medium without nitrogen source

TABLE 1 Influence of Various Carbon Sources on the Specific Sophorolipid Production by Resting Cells of *Torulopsis bombicola*

	Spec. sophorolipid production	
Carbon source	(g/g substrate)	(g/g biomass)
Glucose	0.325	0.65
Fructose	0.205	0.41
Mannose	0.245	0.49
Sucrose	0.160	0.32
Maltose	0.100	0.20
Raffinose	0.205	0.41
Soybean oil	0.340	0.68
Oleic acid	0.350	0.70

Conditions: 100 mL of 0.1 M McIlvaine buffer, pH 3.5; 10 g/L of biomass; 20 g/L of carbon source; 30°C; 100 rpm; 3 days.
Source: Ref. 37.

supported the conversion of renewable resources to a mixture of cellobiose lipids. Table 2 shows the influence of various carbon substrates on glycolipid production [22]. Because of its lowest initial concentration (2%), coconut oil was more metabolized than other compounds. The specific production amounted to 0.71 g/g substrate using cells formerly precultured on glucose. Coconut precultured cells led to a value of 0.81 g/g.

TABLE 2 Influence of Various Carbon Sources on the Specific Cellobiose Lipid Production by Resting Cells of Glucose-Precultured *Ustilago maydis*

	Spec. cellobiose lipid production	
Carbon source	(g/g substrate)	(g/g biomass)
Glucose (5%)	0.234	2.785
Fructose (5%)	0.315	3.750
Sucrose (5%)	0.279	3.321
Glycerol (5%)	0.204	2.428
Coconut oil (2%)	0.712	3.392

Conditions: 100 mL of mineral salts medium (without nitrogen source), pH 5.5; 4.2 g/L of biomass; 33°C; 100 rpm; 6 days.
Source: Ref. 22.

C. Production of Rhamnose Lipids

Syldatk et al. [38] reported that the specific rhamnolipid production of nitrogen-limited *Pseudomonas* sp. cells was slightly improved by resting cells. In NaCl or 0.1 M phosphate buffer solutions, the energy gain by degradation of hydrocarbons or vegetable oils was high enough to permit synthesis of rhamnolipids over a long period of time leading to yields of 0.18 g/g of carbon source (soybean oil; see Table 3). In addition, two more rhamnolipids were also synthesized. The productivity and the yield based on glycerol (a hydrolysis product of triglycerides) as substrate, could be increased by cells immobilized in Ca-alginate beads [28, 39].

Using resting cells of *P. aeruginosa* CFTR-6, Ramana and Karanth [40] showed that the elimination of phosphate in the production medium is advantageous for biosurfactant production from glucose. A maximum of 720 mg/L of glycolipid was produced within 24 h compared to 430 mg/L in 89 h in phosphate-supplemented medium in conventional batch fermentation, taking into account both growth and surfactant production.

D. Formation of Various Sugar Corynomycolates

This section shows the *directed* biochemical synthesis of carbohydrate esters from sugar and fatty acid. The biotechnological route is preferred, in contrast to the competitive chemical methods, as it fulfills the following marginal conditions:

1. Sugar acylation only in definite positions, for example, acylation of *primary* hydroxy groups or of *secondary* hydroxy groups.
2. No introduction of protecting groups for the sugar.

TABLE 3 Influence of Various Carbon Sources on the Specific Rhamnolipid Production by Resting Cells of *Pseudomonas* sp. Precultured on Glucose

	Spec. rhamnolipid production	
Carbon source	(g/g substrate)	(g/g biomass)
n-Alkanes (C_{14}, C_{15})	0.150	1.200
Soybean oil	0.187	1.500
Glycerol	0.212	1.700
Glucose	0.112	0.900
Sucrose	0.025	0.200

Conditions: 100 mL of 0.1 M phosphate buffer, pH 7; 5 g/L of biomass; 40 g/L of carbon source; 30°C; 100 rpm; 7 days.
Source: Ref. 38.

3. Use of harmless solvents like water or *n*-hexane instead of pyridine or dimethylformamide, especially if food application of surfactant is desired.

It can be shown that the carbohydrate substrate indeed determined the type of carbohydrate moiety, whereas the lipophilic part remained constant. After growth on glucose and formation of cell-associated glucose corynomycolates, the harvested *Arthrobacter* sp. DSM 2567 cells transferred the α-branched-β hydroxy fatty acids from glucose to other mono- di-, or trisaccharides fed for bioconversion in a simple phosphate buffer [41]. Continuing these studies the following was observed at yields in the range of 0.1 g/g biomass [42]. The bacterial transesterification reaction proved to be regioselective:

1. Only primary hydroxy groups at C6 (mono- or disaccharides) or C6′ (disaccharides) were acylated.
2. Axial hydroxy groups in the C4 position prevented the acylation (e.g., galactose, lactose).
3. In the case of disaccharides with only one nonreducing end the C6 and C6′ positions were acylated one after another (e.g., maltose). If there are two nonreducing parts, the substitution is simultaneous (e.g., maltitol).

Figure 6 presents the bacterial transfer of carbohydrate/fatty acid.

Similar results were found by extending the studies with *Corynebacterium* sp. M9b [42 43]. Corresponding data compared to the values obtained with *Arthrobacter* sp. are summarized in Table 4. Qualitative experiments on further transfer of the corynomycolic acid to a third and after this to a fourth sugar (cascadelike) indicated positive results.

Our efforts to exchange the fatty acid moiety by feeding with lipophilic precursors instead of carbohydrate moiety, failed until now. Looking for the localization of the transferring enzyme system after cell disruption, we discovered the activity to be bound to the cell debris. In conversion reactions comparable to resting whole cell experiments, the cell particles were also able to produce disaccharide-corynomycolates from glucose corynomycolates. Further isolation of the enzyme system including treatment with octylglucoside (especially recommended for membrane proteins) did not succeed up to now. The occurrence of a simple lipase should be excluded.

IV. LIPASE CATALYSIS FOR THE ACYLATION OF SUGARS AND SUGAR DERIVATIVES

The costs for biochemical acylation of sugars, in competition to chemical processes, are estimated to be high, therefore, considerable advantages must result from use of lipases to make them economic. As mentioned in the foregoing

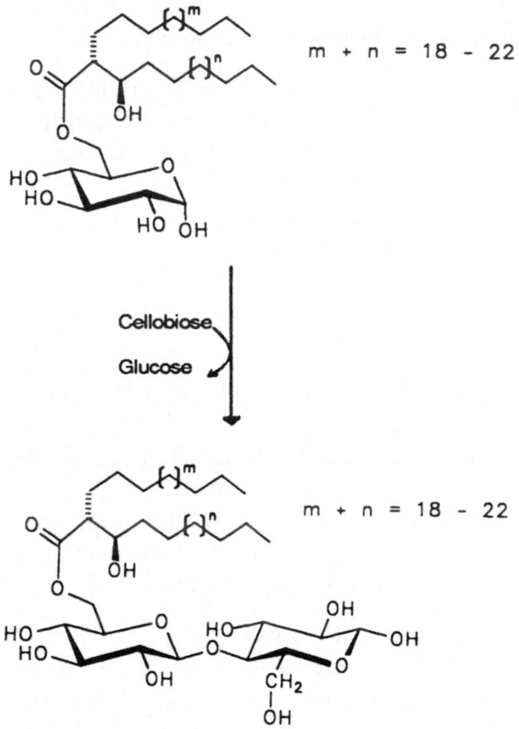

m + n = 18 - 22

Cellobiose

Glucose

m + n = 18 - 22

FIG. 6 Scheme of the regioselective enzyme-catalyzed transfer of the corynomycolic acid onto a new carbohydrate by aid of *Arthrobacter* sp. or *Corynebacterium* sp.

paragraphs, the marginal conditions with regard to regioselectivity of acylation, exclusion of protecting group chemistry, and use of harmless solvents (food or cosmetic applications) have to be considered.

The following examples from the literature indicate that these strong conditions have not been fulfilled so far.

A. Lipases in Aqueous Solutions

Using commercially available crude lipase preparations, Seino et al. [44, 45] succeeded in 1984 in a one-step synthesis of glycolipids from sugar and fatty acid. They observed that in a simple phosphate buffer solution and at 40°C, a lipase of *Candida cylindracea* (e.g., Lipase MY, see Ref. 46) was the most active enzyme in linking together sucrose, glucose, or fructose on the one hand and stearic, oleic,

TABLE 4 Bacterial Acylation of Carbohydrates by Resting Cells Carrying Glucose-Corynomycolate

	Yield on carbohydrate corynomycolates (g/g biomass)	
Carbohydrate	*Arthrobacter* sp.	*Corynebacterium* sp.
Sucrose	0.300	0.020
Palatinose	0.180	0.015
Maltose	0.300	0.015
Isomaltose	0.120	0.010
Gentiobiose	0.100	0.010
Cellobiose	0.080	0.010
Melizitose	0.220	0.015
Trehalose	n.d.[a]	0.020
Maltotriose	0.020	n.d.
Maltitol	0.160	n.d.
Palatinol	0.170	n.d.
Amygdalin	0.110	n.d.

Conditions: 100 mL of 0.08 M phosphate buffer, pH 6.3–6.5; 10–50 g/L of biomass (dry weight); 30 or 40 g/L of carbohydrate; concentration of glucose corynomycolate: 0.2 g/g biomass (*Arthrobacter* sp.); 0.02 g/g biomass (*Corynebacterium* sp.); 30°C; 100 rpm; 24 or 48 h, respectively.
[a]n.d.: not determined.
Source: Refs. 42 and 43.

or linoleic acid on the other hand. Thin-layer chromatography showed that the products were mixtures of both single and multiple acylated sugars. In a patent, the same group extended the spectrum of substrates to polysaccharides and to other fatty acids [46]. In general, 10%–200% by weight of lipase relative to the sugar must be employed. In Table 5, some results of Seino et al. [46] and of Klibanov's group [53] about lipase catalyzed acylations of carbohydrates in organic solvents are presented.

Diverse patents on this subject, especially from Japanese companies followed but are not reviewed here in detail. For example, two patents describe the production of maltose- and lactoselike sugar fatty acid esters using acetylmaltose (-lactose), a C_8–C_{22} fatty acid and a *Rhizopus delemar* lipase in a phosphate buffer [47, 48].

Sugar alcohols, such as the hydrophilic moiety of a surfactant, e.g., sorbitol, sorbitan, mannitol, or xylitol, were connected with higher fatty acids including C_8–C_{22} saturated as well as unsaturated common compounds. Also in this case,

TABLE 5 Crude Lipase-Catalyzed Acylations of Sugars/Sugar Derivatives with Fatty Acids/Fatty Acid Derivatives[a]

Substrates				Surfactant production		
Sugar or sugar derivative (g/L)		Fatty acid or derivative		(g/L)	(g.g)[b]	Ref.
(a) *Candida cylindracea* lipase (4 or 2 g/L)						
Glucose	(3.6)	Oleic acid	(22.0 g/L)	10.7	2.67	46
Fructose	(3.6)	Oleic acid	(22.0 g/L)	12.2	3.05	46
Sucrose	(4.0)	Oleic acid	(11.3 g/L)	7.7	3.84	46
Sorbitol	(3.6)	Oleic acid	22.5 g/L)	7.5	3.74	46
(b) Porcine pancreatic lipase (200 g/L)						
Sorbitol	(25.5)	Corn oil	(0.15 L/L)	8.4	0.04	53
Sorbitol	(25.5)	Olive oil	(0.15 L/L)	6.4	0.03	53
Sorbitol	(25.5)	Soybean oil	(0.15 L/L)	5.1	0.02	53
Sorbitol	(25.5)	Triolein	(0.15 L/L)	11.0	0.06	53
Ribitol	(30.4)	Triolein	(0.15 L/L)	13.8	0.07	53
(c) *Candida antarctica* lipase (2.5%, w/w)[c,d]						
Ethyl		Decanoic[e]	(55.7 g)	76.9[f]	29.13	55
D-glycopyranoside						55
(50.0 g)		Tetradecanoic[e]	(73.9 g)	89.5[f]	28.89	55
		Octadecanoic[e]	(92.0 g)	131.8[f]	37.12	55
		Octadecenoic[e]	(91.4 g)	126.2[f]	35.70	55

[a]Reaction times: (a) 3 days; (b) 3 days; (c) 1 day.
[b]g biosurfactant per g biocatalyst.
[c]% immobilized lipase per weight of substrates.
[d]Reaction *without* solvent.
[e]Acid.
[f]g of 6-*O*-monoester.

Seino et al. [45] observed mixtures of esters because of a probable nonregioselectivity of the crude lipase.

Recently Janssen et al. [49] reported on the esterification of decanoic acid with D-sorbitol in water using *Candida rugosa* lipase. In a two-phase membrane reactor, the initial esterification rate was 6.8 mmole/g h^{-1}. After 570 h, this reaction rate was reduced by 15%. A mixture of mono-, di-, and triesters was detected.

B. Lipases in Organic Solvents or Without Solvents

The enzymatic catalysis in nonaqueous media significantly extends the conventional aqueous-based biocatalysis. Dordick [50] presented an excellent overview on the potential advantages of employing enzymes like lipases, esterases, proteases, or dehydrogenases in monophasic organic solvents. The insolubility in these solvents, however, appears to preclude the enzymes from undergoing severe inactivation. Of course, a small amount of water (below 3%) is necessary to stabilize the conformation of the enzyme.

Therisod and Klibanov [51] discovered that lipases suspended in anhydrous pyridine can be used for the stereospecific and regioselective acylation of hydroxyl groups of multifunctional alcohols or sugars. Concerning various unprotected monosaccharides, the acylation of primary hydroxy positions was favored when dried porcine pancreatic lipase and an activated fatty acid was used. By employing trichloroethyl as a suitable leaving group on the acyl donor, high yields of ester synthesis through transesterification have been obtained; preparative and regioselective primary acylations of glucose, galactose, mannose, or fructose were carried out. Regioselective synthesis of secondary monosaccharide esters were synthesized by a similar approach, although protection of the primary hydroxyl groups was required prior to secondary acylations [52]. However, the activated esters (trichloroethyl) are expensive and their use would not be commercially feasible.

Sugar-alcohol acylations using triglyceride donors for the production of biosurfactants have been presented as alternatives [53]. In the presence of natural vegetable oils in pyridine, the above-mentioned lipase as well as *Chromobacterium viscosum* lipase catalyzed the monoacylation of sorbitol in the primary position (Fig. 7). For all these reactions, generally, it should be noted that the amount of enzyme necessary for sufficient turnover of substrates is very high (see Table 5).

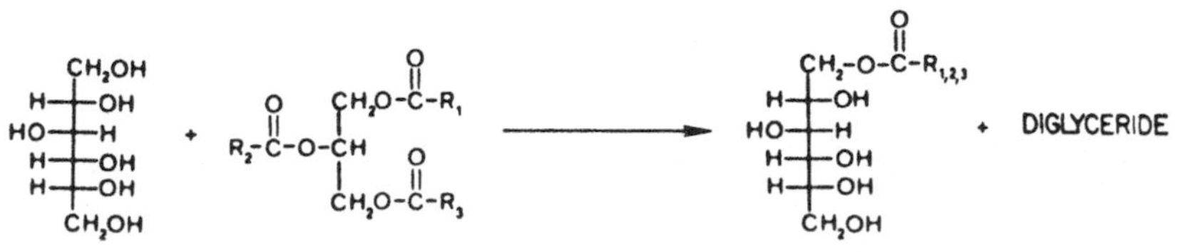

+ DIGLYCERIDE

FIG. 7 Porcine pancreatic lipase-catalyzed regioselective acylation of sorbitol with triglycerides in pyridine. From Refs. 50 and 53.

Recently Ciuffreda et al. [54] reported the stereospecific acylation of methyl-alpha-6-deoxy-L and of methyl-alpha-D-hexosides through lipase-catalyzed transesterification in, for example, tetrahydrofuran. Although the D-sugar derivatives always gave the 2-butyric acid ester as the main product, in the L series the rather unreactive 4-OH function was preferentially acylated.

Developing a solvent-free process at 70°C and at 0.01 bar using alkyl-glucosides and long-chain fatty acids, Björkling et al. [55] and Kirk et al. [56] succeeded in producing relatively high amounts of biosurfactants. In the presence of a substantially lower content of lipase (*Candida antarctica*) than in the above-mentioned studies, they were able to acylate ethyl D-glucopyranoside with C8–C18 fatty acids in yields of 85%–95% of the 6-*O*-monoesters (Fig. 8, Table 5). Methyl substitution at the hydroxy group at the terminal anomeric carbon atom reduced the conversion to 50%. Propyl- or butyl-substitution minimized the regioselectivity.

Surprisingly crude preparations of proteases also catalyzed the ester formation: subtilisin in the case of transesterification between sucrose and trichloroethyl or trifluoroethyl esters in anhydrous dimethylformamide [57]; porcine pancreatic trypsin (after reductive alkylation) in the case of esterification of oleic acid with glucose in the same solvent [58].

V. GLYCOSIDASE CATALYSIS FOR THE ALKYL-GLYCOSIDATION OF SUGARS

Beside chemical syntheses, the possibility of preparing glycosides via the catalytic action of diverse glycosidases has been studied intensively. However, the methods did not find any practical exploitation because the free enzymes in aqueous

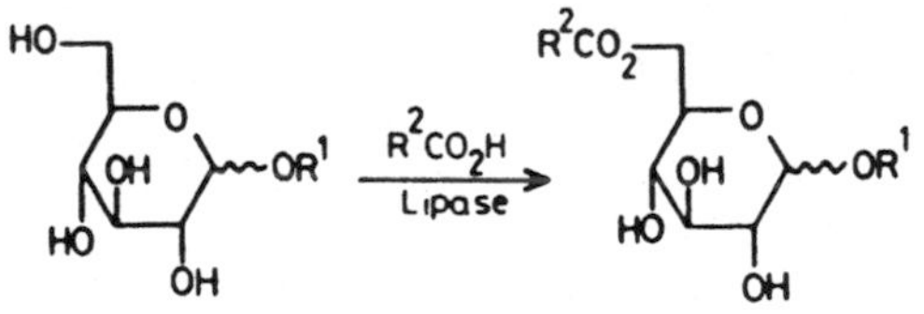

R^1 = H, Me, Et, Pr^n, Pr^i, or Bu^n. R^2 = C_7H_{15}, C_9H_{19}, $C_{11}H_{23}$, $C_{13}H_{27}$, $C_{15}H_{31}$, or $C_{17}H_{35}$.

FIG. 8 *Candida antarctica* lipase-catalyzed regioselective acylation of alkyl glucosides with fatty acids.
Source: Ref. 55.

solution markedly shifted the reaction equilibrium toward the hydrolysis of the glycosidic bond.

Nevertheless, in the case of other hydrolases (e.g., lipases), it could be demonstrated in the last years that it is possible to shift the equilibrium toward a thermodynamically disadvantageous synthesis using a mixture of an organic solvent immiscible with water and a small amount of water.

In a similar fashion to the lipase-catalyzed sugar acylation, the main restriction for the use of glycosidases in the production of alkyl-glucosides is poor solubility of carbohydrates in the most appropriate organic solvents. Vulfson et al. [59, 60] attempted to overcome this problem. Almond β-glucosidase was adsorbed on XAD-4 and alkyl-β-glucosides were formed in a water/organic two-phase system using a concentrated aqueous solution of either methyl-β-glucoside or glucose and a water-immiscible primary alcohol. The reactions proceeded quickly at elevated temperatures (37°C and 60°C). A decrease in reaction rate and an optimum water content of the system was observed when the alcohol chain length was increased from C5 to C12. Increasing glucose concentration in the aqueous phase led to a corresponding increase in alkyl-β-glucoside concentration in the organic phase, whereas the overall product yield dropped. The product yield was nearly doubled by doubling the volume of the organic phase along with higher enzyme loadings and/or longer incubation time. A reaction yield of approximately 10 g/L was achieved. Analytic studies showed that the product contained more than 96% of 1-*O*-alkyl-β-glucoside.

Taking PEG-modified β-galactosidase of *Aspergillus oryzae*, Beecher et al. [61] exchanged the methyl group of methyl-β-galactoside for a hexyl group using hexanol in a mixture of chlorinated hydrocarbons and dimethyl sulfoxide.

VI. PHYSICOCHEMICAL PROPERTIES OF GLYCOLIPIDS

One of the most important characteristics of a surfactant is its ability to reduce the surface and interfacial tension of oil/water interfaces as measured by the classical du Nouy ring method. Because of summaries documented in other publications [43, 62, 63] a listing of those properties of microbial glycolipids is not necessary. Only the best results are noted here.

Glycolipids with hydrophilic lipophilic balance (HLB) values near 10 are able to reduce the surface tension down to <30 mN/m. The interfacial tension between water and *n*-hexadecane was lowered to values <1 mN/m; the critical micelles concentrations (CMC) were in the range of 5 to 200 mg/L. Examples of these excellent properties are some single metabolites after chromatographic purification (rhamnolipids R1–R4, trehalose-2,2′,3,4-tetraester, sucrose-monocorynomycolate) or glycolipid mixtures themselves (cellobiose lipids, rhamnolipids).

The chemically synthesized compounds (e.g., alkyl-sulfonates, alkyl-phenyl-oxyethylates, or polyglycolesters) seem to have less suitable surface-active properties than biosurfactants, especially with respect to CMC. Only some synthetic anionic and nonionic fluorinated compounds are able to compensate for this disadvantage by lowering the surface tension to <20 mN/m.

Chopineau et al. [53] ascertained that the enzymatically formed sugar alcohol monoesters possess superior surface-active properties compared to chemically produced sorbitan monoesters. The authors showed that the interfacial tension between xylenes and water rapidly dropped upon addition of biosurfactants (0.01%) to about 1 mN/m.

The surfactant properties of alkyl glucoside esters found by Björkling et al. [55] are similar to those of common synthetic nonionic surfactants. With C8–C12 fatty acid esters of ethyl D-glycopyranoside the surface tension was reduced to 31 mN/m at CMC of 5–40 × 10^{-5} mol/L, whereas higher fatty acids increased these values.

Another method to describe the interfacial behavior of a surfactant is the measurement of the film pressure using a Langmuir film balance. Figure 9 presents schematically the (outer) pressure-induced packing of amphiphiles in a monolayer. At room temperature the monolayers of fatty alcohols, fatty acids, and

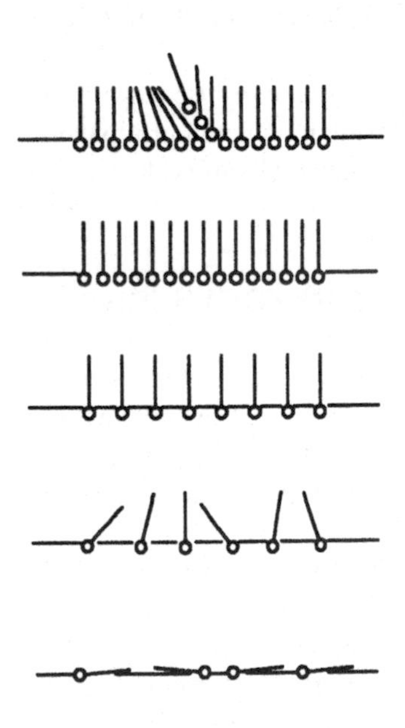

FIG. 9 Scheme of the pressure-induced packing of lipophilic compounds (fatty acids, esters, alcohols) at water–air interfaces (below: outer pressure = 0).

hydroxy wax esters usually indicated good stability in response to outer compression, expressed as high film pressures (50 to 60 mN/m) at different molecular areas (20 A/molecule and more).

As for the behavior of biochemical glycolipids, only two compounds showed comparable properties. With sucrose-monocorynomycolate and trehalose-2,2′,3,4-tetraester, large molecular areas were derived, because they contain broader polar groups (carbohydrates). However, the collapse point at approximately 40 mN/m and the requirement for low temperature (9°C) are disadvantageous [64].

Summing up, in reducing surface and interfacial tension most of the reviewed glycolipids are potential surfactants and seem to be suitable for special emulsification purposes. Based on their behavior in the Langmuir film balance, they seem to be insufficient for wetting or coating.

VII. CONCLUSION

Agricultural renewable resources like sugars and oils are suitable carbon sources for biocatalysts to produce glycolipids with good surfactant properties. Potential biotechnological processes contain either

Multi-step biosynthesis routes performed by microorganisms (cultivation under growth limitation; resting cells conditions) or
One-step reactions using lipases or glycosidases

The cultivation of whole yeasts, fungi, and bacteria, under nitrogen-limitation especially should be mentioned with respect to specific production values of more than 50% conversion of vegetable oils into glycolipids. Concerning lipases, a highly selective esterification of alkyl glucosides accompanied by enormous yields was observed in a special case; but also the target of a direct acylation of a sugar with a fatty acid could not be attained. As for glycosidase-catalyzed connections between sugar and fatty alcohol, studies should be intensified.

REFERENCES

1. J. E. Zajic and W. Seffens, *CRC Crit. Rev. Biotechnol. 1*: 87 (1984).
2. D. Haferburg, R. Hommel, R. Claus, and H.-P. Kleber, *Adv. Biochem. Eng./Biotechnol. 33*: 53 (1986).
3. N. Kosaric, W. L. Cairns, and N. C. C. Gray (eds.), *Biosurfactants and Biotechnology*, Surfactant Science Series, Vol. 25, Marcel Dekker, New York, 1987.
4. R. K. Hommel, *Biodegradation 1*: 107 (1990).
5. C. N. Mulligan and B. F. Gibbs, *Appl. Environ. Microbiol. 55*: 3016 (1989).
6. T. Nakahara, H. Kawashima, T. Sugisawa, U. Takamori, and T. Tabuchi, *J. Ferment. Technol. 61*: 19 (1983).

7. H. Kawashima, T. Nakahara, M. Oogaki, and T. Tabuchi, *J. Ferment. Technol. 61*: 143 (1983).
8. D. Kitamoto, S. Akiba, C. Hioki, and T. Tabuchi, *Agr. Biol. Chem. 54*: 31 (1990).
9. D. Kitamoto, K. Haneishi, T. Nakahara, and T. Tabuchi, *Agr. Biol. Chem. 54*: 37 (1990).
10. P. A. J. Gorin, J. F. T. Spencer, and A. P. Tulloch, *Can. J. Chem. 39*: 846 (1961).
11. A. P. Tulloch, J. F. T. Spencer, and P. A. Gorin, *Can. J. Chem. 40*: 1326 (1962).
12. S. Inoue and S. Ito, *Biotechnol. Lett. 4*: 3 (1982).
13. D. G. Cooper and D. A. Paddock, *Appl. Environ. Microbiol. 47*: 173 (1984).
14. S. Inoue, *Proceedings of the World Conference on Biotechnology for the Fats and Oils Industry* (T. H. Applewhite, ed.), AOCS (1988), pp. 206–210.
15. A. P. Tulloch, A. Hill, and J. F. T. Spencer, *Can. J. Chem. 46*: 3337 (1968).
16. H. J. Asmer, S. Lang, F. Wagner, and V. Wray, *J. Am. Oil Chem. Soc. 65*: 1460 (1988).
17. O. Stüwer, R. Hommel, D. Haferburg, and H. P. Kleber, *J. Biotechnol. 6*: 259 (1987).
18. R. U. Lemieux, R. H. Haskins, J. A. Thorn, and C. Brice, *Can. J. Chem. 29*: 409 (1951).
19. S. S. Bahattacharjee, R. H. Haskins, and P. A. J. Gorin, *Carbohydrate Res. 13*: 235 (1970).
20. J. M. Roxburgh, J. F. T. Spencer, and H. R. Sallans, *J. Agr. Food Chem. 2*: 1121 (1954).
21. B. Frautz, S. Lang, and F. Wagner, *Proceedings of the 3rd European Congress of Biotechnology*, Vol. 1, Verlag Chemie, München, 1984, pp. 79–83.
22. B. Frautz, S. Lang, and F. Wagner, *Biotechnol. Lett. 8*: 757 (1986).
23. G. Hauser and M. L. Karnowsky, *J. Biol. Chem. 229*: 91 (1957).
24. S. Itoh, H. Honda, F. Tomita, and T. Suzuki, *J. Antibiot. 24*: 855 (1971).
25. M. Yamaguchi, A. Sato, and A. Yukuyama, *Chem. Ind. 17*: 741 (1976).
26. K. Hisatsuka, T. Nakahara, N. Sano, and K. Yamada, *Agr. Biol. Chem. 35*: 686 (1971).
27. S. Itoh and T. Suzuki, *Agr. Biol. Chem. 36*: 2233 (1972).
28. C. Syldatk, U. Matulovic, and F. Wagner, *Biotech. For. 1*: 58 (1984).
29. L. Guerra-Santos, O. Käppeli, and A. Fiechter, *Appl. Environ. Microbiol. 48*: 301 (1984).
30. H. E. Reiling, U. Thanei-Wyss, L. H. Guerra-Santos, R. Hirt, O. Käppeli, and A. Fiechter, *Appl. Environ. Microbiol. 51*: 985 (1986).
31. C. Gross and F. Wagner, unpublished results.
32. E. Ristau and F. Wagner, *Biotechnol. Lett. 5*: 95 (1983).
33. J. S. Kim, M. Powalla, S. Lang, F. Wagner, H. Lünsdorf, and V. Wray, *J. Biotechnol. 13*: 257 (1990).
34. Y. Uchida, R. Tsuchiya, M. Chino, J. Hirano, and T. Tabuchi, *Agr. Biol. Chem. 53*: 757 (1989).
35. Y. Uchida, S. Misawa, T. Nakahara, and T. Tabuchi, *Agr. Biol. Chem. 53*: 765 (1989).
36. M. Powalla, S. Lang, and V. Wray, *Appl. Microbiol. Biotechnol. 31*: 473 (1989).
37. U. Göbbert, S. Lang, and F. Wagner, *Biotechnol. Lett. 6*: 225 (1984).
38. C. Syldatk, S. Lang, U. Matulovic, and F. Wagner, *Z. Naturforsch. 40c*: 61 (1985).
39. R. Müller-Hurtig, U. Matulovic, I. Feige, and F. Wagner, *Proceedings of the 4th European Congress of Biotechnology*, Vol. 2 (O. M. Neijssel, R. R. van der Meer, and K. C. A. M. Luyben, eds.), Elsevier, Amsterdam, 1987, pp. 257–260.
40. K. V. Ramana and N. G. Karanth, *Biotechnol. Lett. 11*: 437 (1989).

41. Z.-Y. Li, S. Lang, F. Wagner, L. Witte, and V. Wray, *Appl. Environ. Microbiol. 48*: 610 (1984).
42. U. Göbbert, A. Schmeichel, S. Lang, and F. Wagner, *J. Am. Oil Chem. Soc. 65*: 1519 (1988).
43. A. Schmeichel, S. Lang, and F. Wagner, *DECHEMA Biotechnology Conference*, Vol. 3 (D. Behrens and A. J. Driesel, eds.), VCH Verlagsgesellschaft, 1989, pp. 267–270.
44. H. Seino, T. Uchibori, T. Nishitani, and S. Inamasu, *J. Am. Oil Chem. Soc. 61*: 672 (1984).
45. H. Seino, T. Uchibori, T. Nishitana, and I. Morita, *J. Am. Oil Chem. Soc. 61*: 1761 (1984).
46. H. Seino, T. Uchibori, S. Inamasu, and T. Nishitani, US Patent 4,614,718 (1986).
47. Asahi-Electrochem., Japanese Patent J6 3222-697 (1988).
48. Asahi-Electrochem., Japanese Patent J6 3222-698 (1988).
49. A. E. M. Janssen, A. G. Lefferts, and K. van't Riet, *Biotechnol. Lett. 12*: 711 (1990).
50. J. S. Dordick, *Enzyme Microb. Technol. 11*: 194 (1989).
51. M. Therisod and A. M. Klibanov, *J. Am. Chem. Soc. 108*: 5638 (1986).
52. M. Therisod and A. M. Klibanov, *J. Am. Chem. Soc. 109*: 3977 (1987).
53. J. Chopineau, F. D. McCafferty, M. Therisod, and A. M. Klibanov, *Biotechnol. Bioeng. 31*: 208 (1988).
54. P. Ciuffreda, D. Colombo, F. Ronchetti, and L. Toma, *J. Org. Chem. 55*: 4187 (1990).
55. F. Björkling, S. E. Godtfredsen, and O. Kirk, *J. Chem. Soc. Chem. Commun.* 934 (1989).
56. O. Kirk, F. Björkling, and S. E. Godtfredsen, International Patent WO 89/01480 (1989).
57. G. Carrea, S. Riva, F. Secundo, and B. Danieli, *J. Chem. Soc. Perkin Trans. I, 5*: 1057 (1989).
58. K. Ampon, A. B. Salleh, A. Treoh, W. M. Z. Wan Yunus, C. N. A. Razak, and M. Basri, *Biotechnol. Lett. 13*: 25 (1991).
59. E. N. Vulfson, R. Patel, J. E. Beecher, A. T. Andrews, and B. A. Law, *Enzyme Microb. Technol. 12*: 950 (1990).
60. E. N. Vulfson, R. Patel, and B. A. Law, *Biotechnol. Let. 12*: 397 (1990).
61. J. E. Beecher, A. T. Andrews, and E. N. Vulfson, *Enzyme Microb. Technol. 12*: 95 (1990).
62. S. Lang and F. Wagner, in *Biosurfactants and Biotechnology*, Surfactant Science Series, Vol. 25, (N. Kosaric, W. L. Cairns, and N. C. C. Gray, eds.), Marcel Dekker, New York, 1987, pp. 21–45.
63. F. Wagner and S. Lang, *Proceedings of the 2nd World Surfactants Congress*, Vol. II (organized by ASPA), Paris, 1988, pp. 71–80.
64. S. Lang, R. Multzsch, A. Passeri, A. Schmeichel, B. Steffen, F. Wagner, D. Hamann, and H. K. Cammenga, *Acta Biotechnol. 11*: 379 (1991).

II

Properties

8

Genetics of Surface-Active Compounds

JAKOB REISER, ANDREAS K. KOCH, URS A. OCHSNER, and ARMIN FIECHTER* Institute of Biotechnology, Swiss Federal Institute of Technology, Zurich, Switzerland

**Current affiliation*: Fraunhofer-Institut für Grenzflächen- und Bioverfahrenstechnik, Stuttgart, Germany

I. INTRODUCTION

Many microorganisms, both prokaryotic and eukaryotic, are able to grow on water-insoluble substrates like *n*-alkanes, due to alkane-specific oxidation systems, to substrate-induced alterations in their cell surfaces, and to the production of emulsifying agents, some of them acting as biosurfactants. Many of these extracellular and cell-wall-associated compounds have the potential to promote cell attachment to hydrophobic surfaces, to emulsify water-insoluble substances, and to mediate the transport of these unconventional substrates into the cell. Thus, they are important in determining the degree of hydrophobicity that a cell achieves, the wettability of substrates, the adhesion of a cell to the substrate, and the distribution of cells between oil and water. There is an increased interest in developing industrial processes in which highly active biosurfactants with specific properties for specific applications are produced in large amounts. Therefore, the overexpression of genes involved in the biosynthesis of surfactants and emulsifiers and the control of these genes is an important goal to be achieved.

This chapter reviews the current knowledge concerning the genetics of factors affecting the synthesis and/or composition of amphipathic surface-active compounds such as lipopolysaccharides, glycolipids, lipopeptides, and amphiphilic peptides. We also present molecular genetic strategies that have been useful in unraveling the structures of biosurfactants present in the lung.

II. GENETICS OF SURFACE-ACTIVE COMPOUNDS IN GRAM-NEGATIVES

A number of publications have dealt with molecular investigations concerning extracellular or cell-associated compounds of Gram-negative bacteria that, as emulsion-forming agents, enable such microorganisms to grow on water-insoluble substrates [1–8]. Only a few of those reports present results concerning the genetic analysis of biosurfactants and bioemulsifiers, but mutant strains of several Gram-negative species affected in the production of such factors have been described. Such mutant strains may facilitate the isolation of genes involved in the biosynthesis of biosurfactants and bioemulsifiers in the future.

A. *Acinetobacter calcoaceticus*

1. Involvement of Plasmids in Growth on and Dispersion of Crude Oil

A crude-oil-degrading *Acinetobacter* species, *A. calcoaceticus* RA57, was isolated by standard enrichment culture techniques on the basis of its ability to utilize oil sludge [6]. Strain RA57 was found to contain four plasmids. Among those, a 20-kb plasmid called pSR4 was studied in detail. Colonies were isolated at random after growth in the presence of acridine orange and found to fall into two categories: (1) those that had lost the ability to grow on and disperse crude oil in liquid culture and concurrently were cured of pSR4 and (2) those that retained the ability to both grow on and disperse crude oil and contained pSR4. Strains from the first class continued to grow on hydrocarbon vapors, indicating that the defect associated with the curing of pSR4 was related to the physical interaction of the cells with the hydrocarbon substrate rather than to its metabolism. No differences in either adherence to hydrocarbons or production of extracellular emulsifying activity were found between the two classes of mutants. In growth experiments on crude oil in mixed culture with strains which either contained or lacked pSR4, no sparing of the growth defect was observed. The results are consistent with the possibility that pSR4 encodes a factor(s) that is tightly associated with the cell surface.

2. Mutants of *A. calcoaceticus* Affected in Emulsan Production

Emulsan is an extracellular lipoheteropolysaccharide polyanionic bioemulsifier. Pines and Gutnick [4] have demonstrated the capacity of emulsan to allow wild type *A. calcoaceticus* RAG-1 cells to grow on water-insoluble substrates. This heteropolysaccharide is both a cell-associated capsule and a cell-free product. Only the cell-associated form was biologically active and was required for growth on crude oil. A crude oil-containing medium supplemented with emulsan did not stimulate the growth of the emulsan-deficient mutant strain TR3. In addition, only wild-type cells grew well in a mixed culture experiment where wild-type cells and mutant TR3 cells were co-inoculated in equal numbers in a seawater medium supplemented with 2% crude oil. From the TR3 mutant strain, a revertant strain still affected in emulsan production but capable of growth on crude oil could be isolated. This indicates that besides emulsan other extracellular factors may facilitate the growth of *A. calcoaceticus* on crude oil.

3. Overproduction of Emulsan by Mutant Strains of *A. calcoaceticus*

Nitrosoguanidine-induced, cetyltrimethylammonium bromide (CTAB) tolerant mutant derivatives of the *A. calcoaceticus* RAG-1 strain were shown to grow faster and to produce emulsan at an earlier growth phase as compared to the wild-type strain yielding threefold higher product concentrations [5]. Experiments with resting cells induced upon the addition of chloramphenicol to exponentially growing cultures showed that the CTR-10-49 mutant strain released emulsan at a rate of 30 U/mg h^{-1} and produced almost double the amount of the wild-type within the same time and by the same number of cells, indicating that the effect seen was not simply the result of faster growth. Similar results were obtained with the emulsan overproducing lysine auxotrophic mutant

CTRL-100-1, a CTAB-resistant derivative of RAG-92, which in turn is a lysine auxotrophic derivative of RAG-1. Upon lysine starvation, the method applied to induce resting cells, the CTRL-100-1 mutant strain synthesized emulsan at a rate of 70 U/mg h^{-1} whereas RAG-92 synthesized the product of interest at 20 U/mg h^{-1}. These results indicate that the genetic modification caused an alteration in an emulsifier synthesis-specific step leading to an enhanced capsule production. The fact that the emulsan-overproducing mutants showed an enhanced resistance against CTAB points to a protecting function of emulsan toward the toxicity of this cationic detergent.

B. Pseudomonads

1. Assimilation of Alkanes by *Pseudomonas oleovorans*: The *alkBAC* Operon

The *P. oleovornas* alkane-utilizing system is encoded by the *alkBAC* genes. This operon is located on the so-called OCT plasmid and contains the essential genes for alkane terminal hydroxylation and alkanol dehydrogenation, as well as alkanal dehydrogenation [9]. The OCT plasmid enables *P. oleovorans* to use C_6–C_{12} *n*-alkanes [10]. The question remains whether host-encoded factors are involved in alkane uptake as well. Witholt et al. [11] have proposed that cell wall lipopolysaccharides may be needed to emulsify the hydrophobic substrate prior to its uptake by the cell. Together with the regulatory *alkR* locus, the enzymes of this operon could be expressed successfully in *Pseudomonas putida* and in *Escherichia coli fadR* mutant strains. In those mutants, the fatty acid degradation enzymes are expressed constitutively thus allowing them to grow on *n*-octane as the sole source of carbon and energy [12].

2. Mutational Changes in Physicochemical Cell Surface Properties of Plant-Growth-Stimulating *Pseudomonas* spp.

In an attempt to reach a better understanding of the factors involved in colonizing plant surfaces by bacteria, Weger et al. [13] have investigated bacteriophage-resistant mutant strains of the root-colonizing *Pseudomonas* strains WCS358 and WCS374 lacking the *O*-antigenic side chain of the lipopolysaccharide. These strains were found to differ from their parent strains in cell surface hydrophobicity and in cell surface charge. The observed variations in these physicochemical characteristics were explained by the differences in sugar composition. However, the mutant strains had no altered properties of adherence to sterile potato roots compared with their parental strains, nor were differences observed in the firm adhesion to hydrophilic, lipophilic, negatively charged, or positively charged artificial surfaces. These results showed that neither physicochemical cell surface properties nor the presence of the *O*-antigenic side chain play major roles in the firm adhesion of these bacterial cells to solid surfaces, including potato roots.

3. Effect of Surface-Active *Pseudomonas* spp. on Leaf Wettability

Leaf surfaces represent a biological interface called the phyllosphere, which is colonized by bacteria, yeasts, and filamentous fungi. These surfaces are covered by a hydrophobic cuticular wax layer, so that water distributes as discrete droplets. It has been proposed that the wettability of the leaf surface may be increased by surface-active compounds produced by bacteria. To more directly assess the role(s) of biosurfactants in leaf wettability, Bunster et al. [14] have isolated *Pseudomonas fluorescens* and *P. putida* strains from the rhizosphere and phyllosphere and tested them for surface activity in droplet cultures on polystyrene. Some of the strains spread over the surface during incubation, and these strains were considered surface-active; strains not showing this reaction were considered nonsurface-active. Similar reactions were observed on pieces of wheat leaves. Supernatants from centrifuged broth cultures behaved like droplets of suspensions in broth, and exposure to 100°C destroyed the activity, indicating that biosurfactants were released into the medium. The average contact angles of the supernatants of surface-active and nonsurface-active strains were 24 deg and 72 deg, respectively, and the minimal surface tensions were 46 mN/m and 64 mN/m, respectively, as estimated from Zinsman plots. After 6 d incubation, wheat flag leaves sprayed with a dilute suspension of a surface-active strain of *P. putida* (WCS 358 RR) showed a significant increase in leaf wettability, which was determined by contact angle measurements. Interestingly, however, a closely related strain (WCS 358 U), with no surface activity on polystyrene did not affect leaf wettability, although it was present in densities similar to those of the surface-active strain. Evidently leaf wettability was caused by *Pseudomonas* biosurfactants. Leaf wettability may affect both the availability of water to microorganisms as well as the redistribution of microorganisms and nutrients on the plant surface. Also, leaf wetness duration, which is an important parameter in the epidemiology of bacterial and fungal diseases may be increased by biosurfactants.

4. Biosynthesis of Rhamnose-Containing Exolipids in *Pseudomonas aeruginosa*

Rhamnolipids are produced by *P. aeruginosa* cells during the late growth phase [15]. The synthetic pathway was elucidated [16–19] and the optimal conditions for rhamnolipid production by this organism from various radioactive precursors, such as acetate, glycerol, glucose, and fructose were established. Burger et al. [18, 19] described the complete enzymatic synthesis of a rhamnolipid by extracts of *P. aeruginosa*. They were able to show that the synthesis of rhamnolipids proceeds by sequential glycosyl transfer reactions, each catalyzed by a specific rhamnosyl transferase and that TDP-rhamnose is an efficient rhamnosyl donor in the synthesis of the rhamnolipid according to the following reactions:

$$\begin{array}{c}\text{TDP-L-rhamnose} + \beta\text{-hydroxydecanoyl-}\beta\text{-hydroxydecanoate} \\ \downarrow \text{Transferase 1} \\ \text{TDP} + \text{L-rhamnosyl-}\beta\text{-hydroxydecanoyl-}\beta\text{-hydroxydecanoate}\end{array} \quad (1)$$

$$\begin{array}{c}\text{TDP-L-rhamnose} + \text{L-rhamnosyl-}\beta\text{-hydroxydecanoyl-}\beta\text{-hydroxydecanoate} \\ \downarrow \text{Transferase 2} \\ \text{TDP} + \text{L-rhamnosyl-L-rhamnosyl-}\beta\text{-hydroxydecanoyl-}\beta\text{-hydroxydecanoate}\end{array} \quad (2)$$

L-Rhamnosyl-β-hydroxydecanoyl-β-hydroxydecanoate has been designated as rhamnolipid 1 (RL1) and L-rhamnosyl-L-rhamnosyl-β-hydroxydecanoyl-β-hydroxydecanoate as rhamnolipid 2 (RL2), respectively. Whereas RL1 and RL2 are the principle rhamnolipids produced in liquid cultures, RL3 and RL4, containing two sugar moieties and one fatty acid moiety and one sugar and one fatty acid moiety, respectively, appear to be produced exclusively by resting cells [20].

Previously, it has been shown that rhamnolipid formation by *P. aeruginosa* in a mineral salt medium with 2% alkanes as a carbon source was increased after NO_3^- limitation during the stationary growth phase [21]. Evidently, the control of rhamnolipid production is linked, in some way, to the control of nitrogen metabolism. A direct relationship between increased glutamine synthetase activity and enhanced biosurfactant production was recently found in *P. aeruginosa* cells grown in nitrate and proteose peptone media and increased ammonium and glutamine concentrations repressed both phenomena [22]. It is conceivable that a screen for mutants of *P. aeruginosa* that have increased levels of glutamine synthetase could lead to an enhanced production of biosurfactants. Alternatively, the transfer to and expression in *P. aeruginosa* of a cloned glutamine synthetase gene under the control of a strong promoter may yield the same effect.

A chloramphenicol tolerant strain of *P. aeruginosa* was investigated in terms of its capacity to produce biosurfactants in an inorganic phosphate-limited medium supplemented with chloramphenicol [23]. Several intracellular processes were monitored to correlate biosurfactant production with metabolic changes. In particular, biosurfactant production was preceded by phosphate depletion, followed by increased secretion of alkaline phosphatase and glutamate, and induction of transhydrogenase and glucose-6-phosphate dehydrogenase activity. Cosecretion of alkaline phosphatase and biosurfactant occurred to a greater extent in the chloramphenicol-tolerant strain as compared to the wild type, and extracellular alkaline phosphatase activity increased fourfold while intracellular activity decreased by 50%. Evidently, phosphate metabolism also plays an important role in surfactant production, alkaline phosphatase induction, and glucose metabolism.

5. Genetic Manipulation in the Production of *P. aeruginosa* Biosurfactants and Bioemulsifiers

Rhamnolipids were shown to act as emulsifiers and to stimulate growth of *P. aeruginosa* S_7B_1 on hexadecane [24, 25]. *Pseudomonas aeruginosa* PU-1, a chemically induced mutant derivative of strain KY-4025, is affected in growth on *n*-paraffin and in surfactant production [25]. This strain produces more than 10 times less rhamnolipid than the wild type when grown on *n*-paraffin and is stimulated for growth on the water-insoluble substrate upon exogenous rhamnolipid addition, demonstrating that the mutation must be located in a step involved in rhamnolipid biosynthesis. By taking advantage of the fact that purified rhamnolipids stimulate the growth of *P. aeruginosa* on alkanes, Koch et al. [26] isolated transposon Tn5-GM induced mutants of *P. aeruginosa* PG201, which were unable to grow in minimal media containing hexadecane as the sole carbon source.

Some of these mutants turned out to be affected in rhamnolipid production in the following way. The mutant derivative 65E12 was found to be unable to produce extracellular rhamnolipids under any of the conditions tested and lacked the capacity to take up ^{14}C-labeled hexadecane nor did it grow in media containing individual alkanes with chain lengths ranging from C_{12} to C_{19}. However, growth on these alkanes and uptake of ^{14}C-hexadecane could be restored provided that small amounts of purified rhamnolipids were added to the cultures. Mutant 59C7 did not produce rhamnolipids and was unable to grow in media containing hexadecane, nor was it able to take up ^{14}C-hexadecane in significant amounts, but the addition of small quantities of rhamnolipids restored growth on alkanes and ^{14}C-hexadecane uptake. Interestingly, the rhamnolipid production capacity of mutant 59C7 on hexadecane was restored by this rhamnolipid boost. In glucose-containing media, however, mutant 59C7 produced rhamnolipids at levels about twice as high as those of the wild-type strain. Using rhamnosyl transferase assays, mutant 65E12 turned out to be affected in the production of this key enzyme, whereas strain 59C7 showed transferase activities indistinguishable from the wild-type cell extracts. It is thus possible that this strain carries a mutation affecting a gene whose product is involved in the control of rhamnolipid biosynthesis. These mutants will be instrumental in isolating the corresponding wild-type genes. The isolation and characterization of such genes will allow us to unravel rhamnolipid biosynthesis at the molecular level and may, in the long run, yield strains with increased production capacities.

By chemostatic selection with paraffins isolated from high-paraffin oil, a strain of *P. aeruginosa*, SB1, that can use even-numbered straight-chain alkanes from C_{10} to C_{36} and beyond was isolated [3]. This strain was reported to grow rapidly with oil-well paraffin or candle wax as a sole source of carbon and energy, producing large amounts of surface-active agents. The production of emulsifying agents was particularly apparent when liquid hydrocarbons such as decane, dodecane, tetradecane, and hexadecane were used as growth substrates. A mutant derivative, SB3, that could not produce the emulsifier did not grow on any of the liquid hydrocarbons, unless the emulsifier was added to the culture medium. Interestingly, however, mutant SB3 could grow on solid hydrocarbons as rapidly as the wild type. During growth on solid hydrocarbons, such as tetracosane ($C_{24}H_{50}$) or candle wax, both SB1 and SB3 cells were reported to produce surface-active agents that wet these hydrophobic substrates so as to disperse them throughout the media. Thus, different surface-active compounds were produced when the cells were grown on different hydrocarbon substrates. The *P. aeruginosa* SB30 strain, which is a derepressed mutant derivative of the original SB1 strain was shown to produce large amounts of an emulsifier while growing on either hexadecane, glucose, or a cheap substrate such as chicken fat [3]. Normally, the wild-type strain produces large amounts of emulsifier only during growth on hexadecane but not on glucose or chicken fat. SB30, however, produced five to six times more emulsifier while growing on such substrates. This mutant, therefore, allows rapid production of high amounts of emulsifier, with considerable savings in time and expense.

6. Genetic Construction of Lactose-Utilizing Strains of *P. aeruginosa* and Their Application in Biosurfactant Production

With a view toward using whey as a cheap substrate for the production of biosurfactants, Koch et al. [27] have constructed lactose-utilizing strains of *P. aeruginosa*. For this purpose the *E. coli lac*ZY genes were inserted into the chromosomes of *P. aeruginosa* strains PAO-1 and PG-201 using a bicomponent transposition system, yielding transconjugant strains with one set of *lac*ZY genes integrated into the chromosomes at unique locations. The transconjugants grew well in lactose-based media (minimal medium and whey), albeit with reduced initial rates as compared to growth in glucose-based minimal media. *Pseudomonas* rhamnolipids were produced during stationary growth in lactose-based minimal media and whey, showing that waste products can be effectively used for important biotechnological processes.

C. *Serratia marcescens*

1. Mutants of *S. marcescens* Affected in Cell-Surface Hydrophobicity

Distinct extracellular and cell-bound factors were shown to affect the cell-surface hydrophobicity of certain *Serratia* strains [2]. Enrichment for spontaneously occurring nonyhydrophobic mutants of *S. marcescens* yielded two types: (1) a hydrophobic mutant that exhibited partial hydrophobic characteristics compared to the wild type, as determined by adherence to hexadecane and polystyrene, and (2) a pigmented, nonhydrophobic mutant whose colonies were translucent with respect to those of the wild type. These data suggest that the pronounced cell-surface hydrophobicity of the wild type is mediated by a combination of several surface factors.

2. Serraphobin and Serratamolide as Modulators of *Serratia* Cell-Surface Hydrophobicity

Serraphobin, a 70 kDa protein recovered from the cell surface and from the culture supernatants of wild-type *S. marcescens* cells has been demonstrated to be capable of binding to hexadecane droplets [8]. Serraphobin was either totally absent or present only in minor amounts in hydrophobicity-deficient mutants and in wild-type cultures grown at elevated temperatures (39°C instead of 30°C). Besides this hydrophobin, *S. marcescens* produces a hydrophilin called serratamolide, an aminolipid [28] that plays an important role as a wetting agent [7]. Mutant NS38-9, a derivative of the *S. marcescens* NS38 strain, is deficient in the production of this compound. This was shown by direct colony thin-layer chromatography [29] and by spreading small droplets of washed cell suspensions over a polystyrene surface. In contrast to the wild type, mutant cells did not show spreading activity. The production of this spreading activity was shown to be temperature dependent in that wild-type colonies lacked the capacity to produce serratamolide at 38°C. The data suggest that the presence of serratamolide on *S. marcescens* cells results in a reduction in hydrophobicity possibly by blocking hydrophobic sites on the cell surface and, together with serraphobin, serratamolide seems to act as a modulator of cell-surface hydrophobicity. On the one hand, the production of amphiphilic wetting agents such as serratamolide or the rubiwettins [30] lower the surface tensions of aqueous media and may, therefore, act as emulsion-forming molecules. On the other hand, increased cell-surface hydrophobicity results in a better adherence to hydrophobic substrates and, consequently, in faster growth rates on carbon sources like hexadecane.

III. GENETICS OF BIOSURFACTANT PRODUCTION IN *Bacillus subtilis*

A. Biosurfactants Produced by *Bacillus* spp.

Bacillus subtilis produces surfactin, which is a cyclic lipopeptide with exceptional surface activity [31, 32]. Surfactin lowers the surface tension of water from 72 mN/m to 27 mN/m, inhibits fibrin clotting, and lyses erythrocytes. The lipopeptide contains a carboxylic acid (3-hydroxy-13-methyl tetradecanoic acid) and seven amino acids (Glu-Leu-Leu-Val-Asp-Leu-Leu) [33]. The biosynthesis of surfactin has previously been studied in intact *B. subtilis* cells by incorporating ^{14}C-labeled precursor amino acids directly into the product [34]. [^{14}C]-Acetate appeared in the fatty acid portion of surfactin and was also partially converted into leucine. An enzyme was subsequently isolated and partially purified from a cell-free extract, which catalyzed ATP-P_i-exchange reactions mediated by the amino acid components of surfactin. This activation pattern is consistent with a peptide-synthesizing multienzyme complex that activates its substrate amino acids simultaneously as reactive aminoacyl phosphates. The large-scale production of this surfactant has been investigated [35] and a yield of surfactin of up to 0.8 g/L was obtained in a batch cultivation by continuously removing the product by foam fractionation. A similar aminolipid has been isolated from *B. licheniformis*, and its properties are described in a separate chapter in this volume [36].

B. Identification of Genetic Loci Responsible for Surfactin Production

A study of peptide antibiotic synthesis in *B. subtilis* focusing on the isolation of genes required for the production of surfactin has recently been initiated [37]. Three genes, *sfp*, *srfA*, and *comA* (previously called *srfB*) have been identified and subsequently isolated [38, 39]. The first, *sfp*, is found in surfactin-producing strains of *B. subtilis* and all of the surfactin-nonproducing strains of *B. subtilis* examined so far carry the genes required for surfactin production, with the exception of a functional *sfp* gene. When transferred by genetic transformation, *sfp* is necessary and sufficient to render cells of a nonproducing strain surfactin positive [39]. Mutations of *comA*, in addition to blocking competence development (a stationary phase-induced phenomenon) at an early stage, also render *sfp*-bearing cells surfactin negative [39], and the *comA* gene product has been shown to be required for the expression of a *srfA*::Tn*917lac* fusion as well as other *com* genes that function in later stages of competence development [39]. The *comA* product is homologous to the response regulator class of two-component regulatory proteins and is likely to be a DNA-binding protein [40]. This observation suggests that surfactin production and competence development are regulated by a common signal transduction pathway. The *srfA* gene has been shown to be a large operon of over 25 kb encoding functions responsible for surfactin production, competence development, and sporulation [41].

C. Overproduction of Surfactin by Stable Mutants of *B. subtilis*

Ultraviolet (UV) mutagenesis of *B. subtilis* ATCC 21332 cells yielded a stable mutant strain that produced over three times more surfactin than the parent strain [42]. Approximately 1000 colonies of UV-treated cells were examined for enhanced hemolytic activity on blood agar plates. One mutant (*Suf-1*) produced a significantly larger hemolytic zone and was not an auxotroph. The mutation was mapped by protoplast fusion. This information provides a target site for future genetic manipulation.

IV. MOLECULAR GENETICS OF LUNG SURFACTANTS

A. Physiological Roles and Composition of Lung Surfactant

Lung surfactant is a complex mixture containing primarily phospholipids (phosphatidylcholine and phosphatidylglycerol) with small amounts of proteins, carbohydrates, and neutral lipids. It is found at the air–liquid interface of the alveoli and is essential for normal respiration. In premature infants, an insufficiency of surfactant can cause alveolar collapse, leading to hyaline membrane disease. One promising treatment for the disease is the use of bovine-based surfactant replacement therapy using preparations derived from organic solvent extracts of bovine surfactant. The biochemistry and physiology of the pulmonary surfactant system has been reviewed [see, e.g., Refs. 43 and 44].

B. Structure of Surfactant-Associated Proteins

Hydrophobic proteins with molecular weights between 5 and 18 kD have been identified as the only protein components in the bovine-based preparations now under investigation in treating hyaline membrane disease. An additional protein with a molecular weight of approximately 35 kDa (termed SP-A or SAP-35) is also present in natural surfactant [45]. It is a more hydrophilic glycoprotein, which has been shown to be unrelated to the small hydrophobic proteins and is not found in appreciable amounts in organic extracts from mammalian surfactants being investigated for clinical use [46, 47]. SP-A is insoluble in ether/ethanol or chloroform/methanol and contains an approximately 70-amino-acid collagenlike amino-terminal domain which is rich in glycine and hydroxyproline. The proteins SP-A and Clq, a subunit of the first component (C1) of the classical complement pathway, are structurally homologous molecules, each having an extended collagenlike domain. Evidence has been presented indicating that SP-A can substitute for Clq in enhancing FcR-mediated phagocytosis by monocytes and macrophages and CR1-mediated phagocytosis by macrophages *in vitro* [48].

Two classes of small, organic solvent-soluble proteins have been identified as important components of lung surfactant. These proteins are distinct gene products with unique amino acid sequences. In humans, one of the surfactant-associated proteins with an N-terminal phenylalanine, SP-B [SPL(Phe)] has a molecular weight of approximately 6 to 8 kDa (reduced) or 18 kDa (unreduced).

The other, with an N-terminal glycine, contains a stretch of polyvaline residues and is now termed SP-C [also referred to as SPL (pVal) or SP5]. It has a molecular weight of 4.5 kDa. Both of these proteins have been identified in human surfactant and their primary amino acid sequences have been deduced from cDNAs derived from the mRNAs encoding them [49–51]. The amino acid sequences of bovine and porcine SP-B were recently reported [52, 53]. The porcine peptide was shown to be 79 amino acids long, but the precise C-terminal end of the human protein is not known.

Reconstitution of hydrophobic surfactant peptides with synthetic phospholipids imparts virtually complete surfactantlike properties to synthetic phospholipids, including rapid surface adsorption and lowering of surface tension during dynamic compression. Surfactant lipid extracts containing the two hydrophobic peptides as the sole apoproteins have been used for therapy in hyaline membrane disease in newborn babies. Sarin et al. [54] have synthesized pulmonary surfactant apoprotein SP-B peptides by solid phase chemistry and demonstrated their ability to enhance the surface-active properties of synthetic lipid mixtures. The synthetic peptides were reactive with antiserum generated against the native bovine surfactant peptide. The peptides conferred surfactantlike properties to synthetic lipid mixtures as assessed by a Wilhelmy balance and pulsating bubble surfactometer. Likewise, mixtures of synthetic SP-B peptides and lipids restored compliance of isolated surfactant-deficient rat lungs. This work demonstrates the utility of SP-B as a functional component of pulmonary surfactant mixtures for treatment of respiratory distress syndrome or other disorders characterized by surfactant deficiency. Curstedt et al. [53] have presented unambiguous evidence that native SP-C is a lipopeptide with two palmitoyl groups covalently linked to the polypeptide chain. The deacylation conditions involving treatment with KOH, trimethylamine, or dithioerythrol, the presence of two cysteine residues in the polypeptide, and the absence of other possible attachment sites establish that the palmitoyl groups are thioester-linked to the two adjacent cysteine residues (Fig. 1). In contrast, the major form of porcine SP-B is a dimer without fatty acid components. Long-chain acylation may constitute a means for association of proteins with membranes and could conceivably modulate the stability and biological activity of surfactant films.

C. Molecular Biology of Surfactant Protein Genes

Molecular genetic tools are currently being applied by a number of groups in order to understand the biophysical properties of the surfactant-associated proteins and their mechanism(s) of formation and to provide substrates for reconstitution studies. A review describing the function and regulation of expression of pulmonary surfactant-associated proteins has recently appeared [55].

```
                H3C          CH3
                 |            |
               (H2C)14      (CH2)14
                 |            |
                O=C          C=O
                   \        /
                    S    S
1                   |    |                 10
Phe-Gly-Ile-Pro-Cys-Cys-Pro-Val-His-Leu-Lys-Arg-Leu-Leu-Ile-
Leu Arg                         Asn                     Val


                    20                                  30
Val-Val-Val-Val-Val-Val-Leu-Ile-Val-Val-Val-Ile-Val-Gly-Ala-
                            Val


Leu-Leu-Met-Gly-Leu
```

FIG. 1 Structure of intact SP-C. The amino acid sequence of human SP-C is shown with residues that differ in the porcine molecule indicated below the human sequence. The sulfurs implicit in the Cys symbols are shown to emphasize the thioester nature of the linkages. Adapted from Ref. 53, with permission.

1. The Gene Encoding SP-A

The cloning of the human SP-A gene and cDNA was first reported in 1985 [45]. For this purpose, a full-length cDNA encoding the canine 32-kDa pulmonary surfactant apoprotein [56] was used in reduced stringency hybridization conditions to screen bacteriophage lambda-based genomic and adult lung cDNA libraries. The 248-amino-acid sequence deduced by the single open reading frame was also found to be encoded by the genomic isolate. In a parallel study, Floros et al. [57] have prepared and sequenced tryptic fragments of the 35-kDa SP-A protein and oligonucleotide probes were synthesized based on the amino acid sequences. A human lung cDNA library was then screened using the oligonucleotide probes, and clones encoding these proteins were identified and characterized. By *in vitro* transcription–translation experiments individual clones were associated with particular proteins of 29 and 31 kDa. The data obtained suggest that cotranslational modifications of two primary translation products account for many of the isoforms observed.

2. The Gene Encoding SP-B

Antiserum generated against small hydrophobic surfactant-associated proteins with molecular weights between 6 and 14 kDa was used to screen a bacteriophage expression library constructed from adult human lung polyA$^+$ RNA [49]. This resulted in the identification of a 1.4 kb cDNA clone that was shown to encode the N-terminus of the surfactant polypeptide SP-B. Hybrid-arrested translation with this cDNA and immunoprecipitation of ^{35}S-methionine-labeled *in vitro* translation products of human polyA$^+$ RNA with a surfactant polyclonal antibody resulted in identification of a 42-kDa precursor protein. Blot hybridization analysis of electrophoretically fractionated RNA from human lung detected a 2.0 kb RNA that was more abundant in adult lung than in fetal lung. The larger RNA and translation product indicates that SP-B is derived by proteolysis of a large polypeptide precursor. In a parallel study, Jacobs et al. [58] have partially sequenced one of the low-molecular-weight proteins and specific oligonucleotides deduced from the sequence were used as probes. The cDNA clone selected by hybridization a human lung mRNA, which, upon *in vitro* translation, yielded a 42-kDa protein. The same protein was observed when the cDNA clone was expressed following transfection into monkey COS cells. The SP-B precursor is not homologous to SP-A and has no collagenlike regions, nor do other parts of SP-A share homology with SP-B. A cDNA sequence encoding a related protein from canine lungs termed SP-18 has recently been described [50].

A complete SP-B cDNA was used to isolate the gene encoding the SP-B precursor from a genomic library of human embryonic kidney DNA [59]. The entire SP-B gene was sequenced and is approximately 9.5 kb long, with 11 exons and 10 introns. Southern blotting of human genomic DNA probed with SP-B cDNA indicated the presence of only one SP-B gene in the human genome, and the gene was localized on chromosome 2.

3. The Gene Encoding SP-C

An oligonucleotide probe based on the valine-rich amino-terminal amino acid sequence of SP-C was utilized to isolate cDNA and genomic DNA encoding the human proteolipid SP-C [51]. The primary structure of a precursor protein of approximately 20 kDa, containing the SP-C peptide, was deduced from the nucleotide sequence of the cDNAs. Hybrid-arrested translation and immunoprecipitation of labeled translation products of human mRNA demonstrated a 22-kDa precursor protein, the active hydrophobic peptide being produced by proteolytic processing to 5 to 6 kDa. The precursor contains an extremely hydrophobic region of 34 amino acids that comprises most of the mature SP-C [60]. This hydrophobicity explains the unusual solubility characteristics of SP-C and the fact that it is lipid-associated when isolated from lung. Two distinct human genes encoding SP-C were identified and sequenced [61]. In both genes, the active hydrophobic region of the polypeptide was located in the second exon that encodes a peptide of 53 amino acids. The entire nucleotide sequence of the two classes of SP-C genes differed by only 1%. The SP-C gene locus was assigned to chromosome 8.

V. SURFACE ACTIVITY OF PROTEINS AND AMPHIPHILIC α-HELICES

A. Surface Activity of Proteins

The various structural properties of proteins, namely, surface hydrophobicity, net charge, molecular size, and the presence of regions involved in protein–protein interaction, have all been considered as factors influencing their surface properties. In addition, the stability of proteins seems to be an important factor. The conformation of an unstable protein can easily change to become hydrophilic toward an aqueous phase and hydrophobic toward an air or oil boundary, so that a pronounced reduction of surface or interfacial tensions facilitates foaming and emulsification.

To elucidate the role of structural stability in determining the surface properties of proteins, Kato and Yutani [62] have compared the surface properties of wild-type and six mutant α-subunits of tryptophan synthase substituted at position 49. The surface tension, foaming power, emulsifying activity, and foam stability of the seven proteins tested were found to correlate linearly with the free energy of denaturation of the proteins. These results indicated that the surface properties of the α-subunits of tryptophan synthase change in proportion to their conformational stabilities.

The observations of Kato and Yutani [62] on tryptophan synthase mutants are consistent with differences in surface properties of hemoglobin (Hb) mutants in previous studies [63]. The Hb studies were stimulated by the observation of decreased mechanical stability of deoxy sickle cell Hb (HbS). *Mechanical stability* is a term used to describe the tendency of proteins to precipitate from solution upon shaking. Deoxy-HbS precipitates much more readily during shaking in comparison to the normal HbA variant and thus has a lower mechanical stability. When other hemoglobin variants were tested, a relatively large range of mechanical stability was observed. Furthermore, the surface tension of solutions of Hb mutants was also found to vary. Among a variety of Hb mutants studied, there was a generally good correlation between surface properties and mechanical stability. In addition, the mechanical stability of oxy-HbS is much greater than deoxy-HbS. These differences appear to be due to the greater denaturability of deoxy-HbS, which makes it more ready to denature and spread at the air–water interface.

Fundamental understanding of the behavior of proteins at interfaces can contribute to the solution of many problems, including (1) the utilization of adsorbed enzymes as sensors, (2) the enhanced retention of albumin preadsorbed on biomaterials to reduce thrombosis, (3) the use of more stable variants of cell attachment proteins to encourage cellular growth in culture and vascular healing in implants, and (4) improved separation processes for proteins. Many other problems involving proteins at interfaces will undoubtedly benefit from improvements in our knowledge in this area. Current and future advances in the genetic engineering of proteins should make possible the production of proteins with novel surface properties.

There is potential for improving the physical and functional properties of certain food proteins to develop better quality foods or novel food products [64, 65]. Some natural mutations of milk proteins were found to have a marked effect on their functional properties. For example the B variant of β-lactoglobulin, a milk protein with a known three-dimensional structure, has Ala 118 in place of Val 118 and Gly 64 in place of Asp 64. It is more soluble than the A variant, and its rate of denaturation in heated milk is different. Both of these effects are caused by the changed surface of the protein: Val 118 is much more hydrophobic than Ala, whereas Asp carries a charge. It should be possible, through genetic engineering, to engineer regions or domains in proteins that will enhance their foaming or emulsifying abilities.

B. Amphiphilic α-Helices

The amphiphilic/amphipathic α-helix is an often-encountered secondary structural motif in biologically active peptides and proteins [for a recent review see Ref. 66]. An amphiphilic/amphipathic α-helix is defined as an α-helix with opposing polar and nonpolar faces oriented along the axis of the helix. The venom of the European honey bee *Apis mellifera* contains as its major protein component a 26-amino-acid peptide, called melittin, which has a powerful hemolytic activity. In addition to its hemolytic activity, melittin induces voltage-dependent ion-conductance across planar lipid bilayers and causes selective micellation of bilayers as well as membrane fusion. The actions of melittin on membranes have been reviewed [67]. In common with other membrane-binding peptides and membrane proteins, melittin is predominantly hydrophobic and a skewed distribution of polar and nonpolar amino acids is apparent if the melittin peptide is placed in an α-helical configuration [68]. The polar amino acids of the first 21 residues lie on one face of the helix with the nonpolar amino acids segregated almost completely on an opposite side of the helix. Furthermore, melittin has been shown to have surface properties including the ability to form surface monolayers at water–air interfaces, dramatically lowering the surface tension of water, and to penetrate lipid surface monolayers [69].

ACKNOWLEDGMENTS

Our work dealing with the molecular genetics of *Pseudomonas* biosurfactants and bioemulsifiers is being supported by the Swiss National Science Foundation, grant No. 31-28763.90.

REFERENCES

1. M. Rosenberg and E. Rosenberg, *J. Bacteriol. 148*: 51 (1981).
2. M. Rosenberg, *J. Bacteriol. 160*:480 (1984).
3. A. M. Chakrabarty, *Trends Biotechnol. 3*: 32 (1985).
4. O. Pines and D. Gutnick, *Appl. Environ. Microbiol. 51*: 661 (1986).
5. Y. Shabtai and D. L. Gutnick, *Appl. Environ. Microbiol. 52*: 146 (1986).
6. S. Rusansky, R. Avigad, S. Michaeli, and D. L. Gutnick, *Appl. Environ. Microbiol. 53*: 1918 (1987).
7. R. Bar-Ness, N. Avrahamy, T. Matsuyama, and M. Rosenberg, *J. Bacteriol. 170*: 4361 (1988).
8. R. Bar-Ness and M. Rosenberg, *J. Gen. Microbiol. 135*: 2277 (1989).
9. M. Kok, R. Oldenhuis, M. P. G. van der Linden, C. H. C. Meulenberg, J. Kingma, and B. Witholt, *J. Biol. Chem. 264*: 5442 (1989).
10. J. N. Baptist, R. K. Gholson, and M. J. Coon, *Biochim. Biophys. Acta 69*: 40 (1963).
11. B. Witholt, M.-J. de Smet, J. Kingma, J. B. van Beilen, M. Kok, R. G. Langeveen, and G. Eggink, *Trends Biotechnol. 8*: 46 (1990).
12. G. Eggink, R. G. Lageveen, B. Altenburg, and B. Witholt, *J. Biol. Chem. 262*: 17712 (1987).
13. L. A. Weger, M. C. M. van Loosdrecht, H. E. Klaassen, and B. Lugtenberg, *J. Bacteriol. 171*: 2756 (1989).
14. L. Bunster, N. J. Fokkema, and B. Schippers, *Appl. Environ. Microbiol. 55*: 1340 (1989).
15. G. Hauser and M. Karnovsky, *J. Bacteriol. 68*: 645 (1954).
16. G. Hauser and M. Karnovsky, *J. Biol. Chem. 224*: 91 (1957).
17. G. Hauser and M. Karnovsky, *J. Biol. Chem. 233*: 287 (1958).
18. M. M. Burger, L. Glaser, and R. M. Burton, *J. Biol. Chem. 238*: 2595 (1963).
19. M. M. Burger, L. Glaser, and R. M. Burton, in *Methods in Enzymology* (E. F. Neufeld and V. Ginsburg, eds.), Academic Press, New York, 1966, pp. 441–445.
20. C. Syldatk, S. Lang, and F. Wagner, *Z. Naturforsch. 40c*: 51 (1985).
21. F. Wagner, H. Bock, A. Kretschmer, S. Lang, and C. Syldatk, in *Microbial Enhanced Oil Recovery* (J. E. Zajic, D. G. Cooper, T. R. Jack, and N. Kosaric, eds.), Penn Well Publishing, Tucson, 1983, pp. 55–60.
22. C. N. Mulligan and B. F. Gibbs, *Appl. Environ. Microbiol. 55*: 3016 (1989).
23. C. N. Mulligan, G. Mahmourides, and B. F. Gibbs, *J. Biotechnol. 12*: 37 (1989).
24. K. Hisatsuka, T. Nakahara, N. Sano, and K. Yamada, *Agr. Biol. Chem. 35*: 686 (1971).

25. S. Itoh and T. Suzuki, *Agr. Biol. Chem. 36*: 2233 (1972).
26. A. K. Koch, O. Käppeli, A. Fiechter, and J. Reiser, *J. Bacteriol. 173*: 4212 (1991).
27. A. K. Koch, J. Reiser, O. Käppeli, and A. Fiechter, *Bio/Technology 6*: 1335 (1988).
28. H. H. Wassermann, J. J. Keggi, and J. E. McKeon, *J. Am. Chem. Soc. 84*:2978 (1962).
29. T. Matsuyama, M. Sogawa, and I. Yano, *Appl. Environ. Microbiol. 53*: 1186 (1987).
30. T. Matsuyama, K. Kaneda, I. Ishizuka, T. Toida, and I. Yano, *J. Bacteriol. 172*: 3015 (1990).
31. K. Arima, A. Kakinuma, and G. Tamura, *Biochem. Biophys. Res. Commun. 31*: 488 (1968).
32. A. W. Bernheimer and L. S. Avigad, *J. Gen. Microbiol. 61*: 361 (1970).
33. A. Kakinuma, A. Oachida, T. Shima, H. Sugino, M. Isono, G. Tamura, and K. Arima, *Agr. Biol. Chem. 33*: 1669 (1969).
34. B. Kluge, J. Vater, S. Salnikow, and K. Eckart, *FEBS Lett. 231*: 107 (1988).
35. D. G. Cooper, C. R. Macdonald, S. J. B. Duff, and N. Kosaric, *Appl. Environ. Microbiol. 42*: 408 (1981).
36. K. Jenny et al., this volume.
37. M. M. Nakano and P. Zuber, *Crit. Rev. Biotechnol. 10*: 223 (1990).
38. M. M. Nakano, M. A. Marahiel, and P. Zuber, *J. Bacteriol. 170*: 5662 (1988).
39. M. M. Nakano and P. Zuber, *J. Bacteriol. 171*: 5347 (1989).
40. Y. Weinrauch, N. Guillen, and D. A. Dubnau, *J. Bacteriol. 171*: 5362 (1989).
41. M. M. Nakano, R. Magnuson, A. Myers, J. Curry, A. D. Grossman, and P. Zuber, *J. Bacteriol. 173*: 1770 (1991).
42. C. N. Mulligan, T. Y. K. Chow, and B. F. Gibbs, *Appl. Microbiol. Biotechnol. 31*: 486 (1989).
43. E. V. Cosmi and E. M. Scarpelli (eds.), *Pulmonary Surfactant Systems*, Elsevier, Amsterdam, 1983.
44. B. Robertson, L. G. M. Van Golde, and J. J. Batenburg (eds.), *Pulmonary Surfactant*, Elsevier, Amsterdam, 1986.
45. R. T. White, D. Damm, J. Miller, K. Spratt, J. Schilling, S. Hawgood, B. Benson, and B. Cordell, *Nature 317*: 361 (1985).
46. J. A. Whitsett, W. M. Hull, B. Ohning, G. Ross, and T. E. Weaver, *Pediatr. Res. 20*: 744 (1986).
47. J. A. Whitsett, B. L. Ohning, G. Ross, J. Meuth, T. Weaver, B. A. Holm, D. L. Shapiro, and R. N. Notter, *Pediatr. Res. 20*: 460 (1986).
48. A. J. Tenner, S. L. Robinson, J. Borchelt, and J. R. Wright, *J. Biol. Chem. 264*: 13923 (1989).
49. S. W. Glasser, T. R. Korfhagen, T. Weaver, T. Pilot-Matias, J. L. Fox, and J. A. Whitsett, *Proc. Natl. Acad. Sci. USA 84*: 4007 (1987).
50. S. Hawgood, B. J. Benson, J. Schilling, D. Damm, J. A. Clements, and R. T. White, *Proc. Natl. Acad. Sci. USA 84*: 66 (1987).
51. S. W. Glasser, T. R. Korfhagen, T. E. Weaver, J. C. Clark, T. Pilot-Matias, J. Meuth, J. L. Fox, and J. A. Whitsett, *J. Biol. Chem. 263*: 9 (1988).
52. R. W. Olafson, U. Rink, S. Kielland, S.-H. Yu, J. Chung, P. G. R. Harding, and F. Possmeyer, *Biochem. Biophys. Res. Commun. 148*: 1406 (1987).

53. T. Curstedt, J. Johansson, P. Persson, A. Eklund, B. Robertson, B. Löwenadler, and H. Jörnvall, *Proc. Natl. Acad. Sci. 87*: 2985 (1990).
54. V. K. Sarin, G. Gupta, T. K. Leung, V. E. Taylor, B. L. Ohning, J. A. Whitsett, and J. L. Fox, *Proc. Natl. Acad. Sci. USA 87*: 2633 (1990).
55. T. E. Weaver and J. A. Whitsett, *Biochem. J. 273*: 249 (1991).
56. B. Benson, S. Hawgood, J. Schilling, J. Clements, D. Damm, B. Cordell, and R. T. White, *Proc. Natl. Acad. Sci. USA 82*: 6379 (1985).
57. J. Floros, R. Steinbrink, K. Jacobs, D. Phelps, R. Kriz, M. Recny, L. Sultzman, S. Jones, H. W. Taeusch, H. A. Frank, and E. F. Fritsch, *J. Biol. Chem. 261*: 9029 (1986).
58. K. A. Jacobs, D. S. Phelps, R. Steinbrink, J. Fisch, R. Kriz, L. Mitsock, J. P. Dougherty, H. W. Taeusch, and J. Floros, *J. Biol. Chem. 262*: 9808 (1987).
59. T. J. Pilot-Matias, S. E. Kister, J. L. Fox, K. Kropp, S. W. Glasser, and J. A. Whitsett, *DNA 8*: 75 (1989).
60. R. G. Warr, S. Hawgood, D. I. Buckley, T. M. Crisp, J. Schilling, B. Benson, P. L. Ballard, J. A. Clements, and R. T. White, *Proc. Natl. Acad. Sci. USA 84*: 7915 (1987).
61. S. W. Glasser, T. R. Korfhagen, C. M. Perme, T. J. Pilot-Matias, S. E. Kister, and J. A. Whitsett, *J. Biol. Chem. 263*: 10326 (1988).
62. A. Kato and K. Yutani, *Protein Eng. 2*: 153 (1988).
63. D. Elbaum, J. Harrington, E. F. Roth, Jr., and R. L. Nagel, *Biochim. Biophys. Acta 427*: 57 (1976).
64. S. Nakai, *J. Agric. Food Chem. 31*: 676–683 (1983).
65. L. K. Creamer, R. Jimenez-Flores, and T. Richardson, *Trends Biotechnol. 6*: 163 (1988).
66. J. P. Segrest, H. De Loof, J. G. Dohlmann, C. G. Brouillett, and G. M. Anatharamaiah, *Proteins: Structure, Function, and Genetics 8*: 103 (1990).
67. C. E. Dempsey, *Biochim. Biophys. Acta 1031*: 143 (1990).
68. F. Podo, R. Strom, C. Crifo, and M. Zulauf, *Int. J. Pept. Protein Res. 19*: 514 (1982).
69. G. Sessa, J. H. Freer, G. Colacicco, and G. Weissmann, *J. Biol. Chem. 244*: 3575 (1969).

9

Biological Activities of Biosurfactants

SIEGMUND LANG and FRITZ WAGNER Institute of Biochemistry and Biotechnology, Technical University of Braunschweig, Braunschweig, Germany

I. INTRODUCTION

For several decades, surfactants have been used in the care of the body and teeth, in drugs and cosmetics, in washing agents, and in many other applications. Their widespread use and ecological considerations have necessitated the study of the biological and toxic properties of these substances. The same is true for biosurfactants, which have been attracting attention during the last 15 years and have a potential to partially or fully replace synthetic detergents.

The effects of synthetic surfactants generally influence the permeability of biological membranes [1–5]. Cationic and hybridionic surfactants are known for their antibacterial activities [6] by effecting the release of intracellular material or by inhibiting respiration [7]. In general, anionic and nonionic surfactants were shown to be less toxic and more directed toward Gram-positive than toward Gram-negative bacteria because of different cell wall structures and osmolarities. The protein and lipopolysaccharide moieties of the cell wall of Gram-negative bacteria are able to protect the cell membrane from surfactant attack.

Many studies of antimicrobial responses to synthetic surfactants have been published [8–11]. Some publications have focused on the phytotoxicity of these amphiphiles, because they are used to improve the effectiveness of herbicides, pesticides, and defoliants [3, 12]. According to their type and concentration, surfactants are also able to positively influence plant growth.

Strong effects were observed in the case of fish. Sublethal concentrations led to conspicuous behavioral changes in fish, indicating water contamination [13–15]. Studies on the spermicidal properties of surfactants were performed with liposomes containing spermatozoal lipids [16].

Haferburg et al. [17] has presented an excellent review of the biological activities of microbially produced biosurfactants. Therefore the objectives of our contribution are as follows:

To repeat shortly the most important properties described in the literature before 1986.
To summarize the results of the following years.

Our objective is not to present a survey on the physiological importance of biosurfactants (e.g., for hydrocarbon assimilation by microbial cells) but to inform about effects (positive or negative) on microorganisms, plants, enzymes, and so on. The most-studied surface-active agents of biological origin are lipopeptides and glycolipids.

II. LIPOPEPTIDES

In amino acids containing lipids, Haferburg et al. [17] reviewed in detail the biocidic effects of siolipin and surfactin. Surfactin, a cyclic lipopeptide, was studied with respect to both its physicochemical and its biological activities [18–20]. It is an effective surfactant and lowers the surface tension to as low as 27 mN/m and the interfacial tension against hexadecane to less than 1 mN/m [20]. Surfactin inhibits the formation of fibrin clots and lyses erythrocytes and several bacterial spheroplasts and protoplasts [19, 21]. Also this biosurfactant represses the growth of microorganisms belonging to the genus *Mycobacterium* [minimum inhibitory concentration, 5 μg/L] (MIC). In combination with pyrrolnitrin, surfactin increased the antifungal activity against *Trichophyton mentagrophytes* (responsible for dermatophytosis) [22].

In the case of viscosin, which was first isolated from *Pseudomonas viscosa* and characterized as an antimycobacterial and an antiviral agent [23, 24], Neu et al. [25] detected surface-active properties. Surface-tension measurements revealed a value of 26.5 mN/m and a critical micelle concentration of 0.15 mg/mL. Also for iturine, a peptidolipid antibiotic from *Bacillus subtilis* [26, 27], Neu and Poralla [28] proved surface activity.

Because of these recent results, similar surface-active properties should be expected from other structural analog lipopeptides [reviewed by Kleinkauf and von Döhren, 29]. Besides molecular structures and biosynthetic pathways these authors summarized the most important biological activities of the following:

1. Classical cyclopeptides (e.g., gramicidin S) acting as nucleotide binding complexones with phase-transfer activity.
2. Antitumor peptides (e.g., actinomycins; high toxicity LD50 = 1 mg/kg in mice).
3. Inhibitors of cell-wall formation (e.g., nisin).
4. Membrane-active peptides (e.g., polymyxin).
5. Iron carriers (e.g., ferrichromes).
6. Proteinase inhibitors and immunomodifiers (e.g., pepstatin, cyclosporins).

Disorganization of the membrane by membrane-active peptides (item 4) can be achieved by a variety of compounds with detergentlike properties. All effects of these lipopeptides are dependent on concentration, and resistance effects are related to the primary target, the outer membrane.

This review does not repeat Kleinkauf and von Döhren's survey [29]; therefore only some more recent examples are presented. In most cases, the surface activity of such compounds was not noticed. However, because of the overall combination of polar amino acids and apolar fatty acid components as structural elements, these compounds have the best conditions for potential surface activity.

A. Lactonic Lipopeptides

Using *Bacillus cereus*, Umezawa's group isolated Plipastatins A1, A2, B1, and B2 [30–33]. These lactonic lipopeptides are a family of acylated decapeptides that differ from each other by amino acid composition and the nature of the fatty acid side chain [3(R)-hydroxyhexadecanoic acid and 14(S)-methyl-3(R)-hydroxydecanoic acid]. They inhibit pig pancreas phospholipase A2, which is involved in inflammatory reactions.

The acidic lipopeptide antibiotic complex A21978C (13 amino acids, methyl-substituted C10 and C12 acids) from *Streptomyces roseosporus*, and its chemically modified derivatives are useful as antibacterial agents and as feed-additive for poultry [34–37]. Another new cyclic substance mixture of this class, A54145, consisting of eight factors (Fig. 1), was produced by *Streptomyces fradiae*. These compounds are suitable for treatment of bacterial infections and as feed additives [38–43]. They are active against Gram-positive aerobic organisms (Table 1).

B. Cyclic Lipopeptides with β- and α-Amino Fatty Acid Moieties

Known representatives of lipopeptides with β-amino fatty acid moieties include different compounds of the iturin group: iturin A, bacillomycin, mycosubtilin [44–47]. They are synthesized by *B. subtilis* strains and are antifungal agents [29]. For example, bacillomycin-F inhibits *Aspergillus niger* and *Saccharomyces cerevisiae* by 40 and 10 μg/mL, respectively [44].

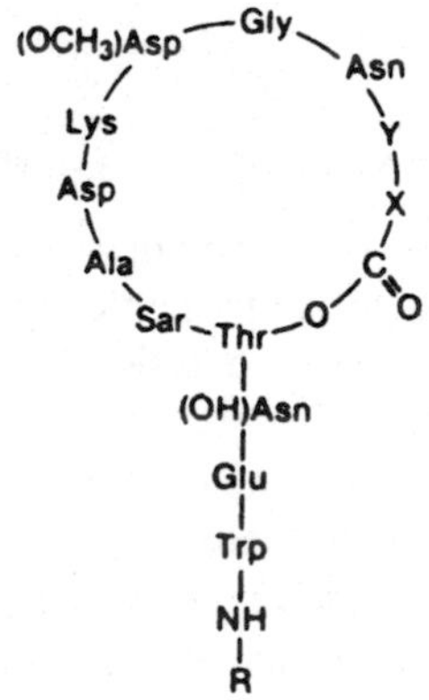

(OCH_3)Asp—Gly—Asn—Y—X—C(=O)—O—Thr—Sar—Ala—Asp—Lys—(OCH_3)Asp (ring)
Thr—(OH)Asn—Glu—Trp—NH—R

Factor	X	Y	R
A	Ile	Glu	8-Methylnonanoyl (iC_{10})
A_1	Ile	Glu	*n*-Decanoyl (nC_{10})
B	Ile	3-MethylGlu	*n*-Decanoyl (nC_{10})
B_1	Ile	3-MethylGlu	8-Methylnonanoyl (iC_{10})
C	Val	3-MethylGlu	8-Methyldecanoyl (aC_{11})
D	Ile	Glu	8-Methyldecanoyl (aC_{11})
E	Ile	3-MethylGlu	8-Methyldecanoyl (aC_{11})
F	Val	Glu	8-Methylnonanoyl (iC_{10})

FIG. 1 Molecular structures of A54145 complex antibiotics from *Streptomyces fradiae*. *Source*: Ref. 40, with kind permission of *J. Antibiot.*

TABLE 1 Antimicrobial Activity of A54145 Factors

	Agar dilution MICs (μg/ml)				
Antibiotic	*Staphylococcus aureus* V 41	*Staphylococcus epidermidis* 222	*Streptococcus pyogenes* C203	*Streptococcus pneumoniae* Park	*Enterococcus* 2041
A54145A	1.0	1.0	0.5	1.0	16
A54145A$_1$	8.0	4.0	1.0	2.0	32
A54145B	1.0	1.0	0.5	1.0	16
A54145B$_1$	2.0	1.0	0.5	2.0	32
A54145C	4.0	2.0	2.0	4.0	8.0
A54145D	2.0	1.0	0.5	4.0	32
A54145E	1.0	1.0	0.25	2.0	4.0
A54145F	16	8.0	2.0	32	128

Source: Ref. 43, with kind permission of *J. Antibiot.*

New cytostatically operating substances containing very similar structural elements with surfactant properties were reported by Ajinomoto [48] also using *B. subtilis*. All of the eight products have antibiotic and cytostatic activity; in particular they show strong growth-inhibitory activity in mouse fibroblast M-MSV BALB 3T3 transformed with the Moloney strain of mouse sarcoma virus.

Concerning lipopeptides with α-amino fatty acids, new cyclotetrapeptides, trapoxins A and B (Fig. 2), recently were isolated from the culture broth of *Helicoma ambiens* [49]. These compounds exhibit detransformation activities against *v-sis* oncogene-transformed NIH3T3 cells (*sis*/NIH3T3) as antitumor agents.

C. Cyclic Lipopeptides with β-Hydroxy Fatty Acid Moieties

Besides the often-cited surfactin, examples of this class, also called cyclodepsipeptides, are empedopeptin from *Empedobacter halabium* [50] and another antibioticum from *Fusarium roseum*. Empedopeptin inhibits a variety of aerobic and anaerobic Gram-positive bacteria both *in vivo* and *in vitro*. The second one (Fig. 3) swelled *Penicillium digitatum* cells to 10 times their normal diameter and inhibited their germination [51–53].

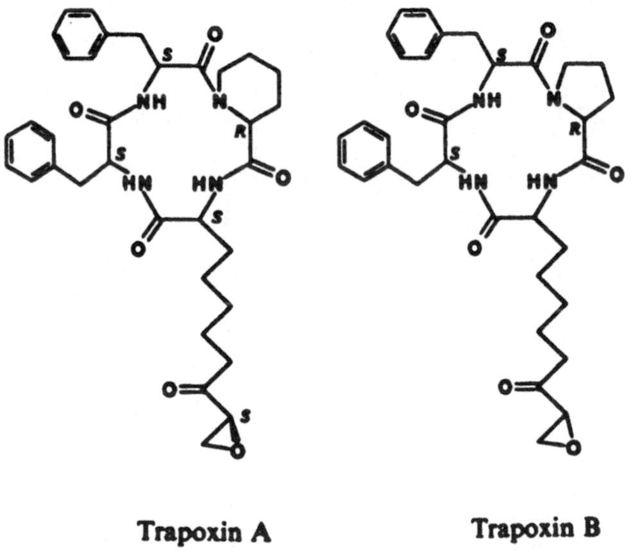

Trapoxin A **Trapoxin B**

FIG. 2 Molecular structures of trapoxins A and B from *Helicoma ambiens*. *Source*: Ref. 49, with kind permission of *J. Antibiot.*

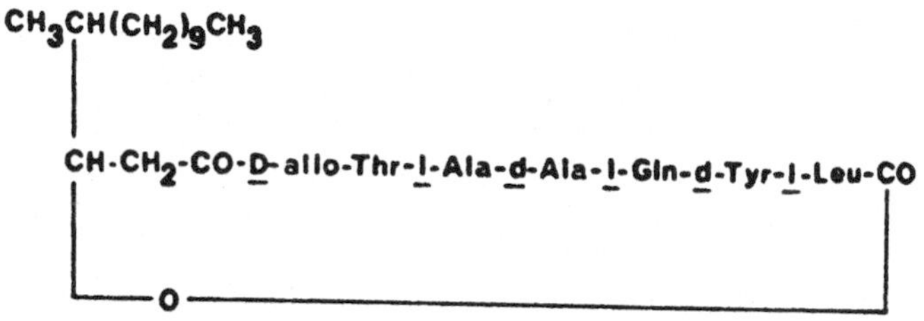

$CH_3CH(CH_2)_9CH_3$

CH-CH_2-CO-D-allo-Thr-l-Ala-d-Ala-l-Gln-d-Tyr-l-Leu-CO

O

FIG. 3 Molecular structure of a cyclodepsipeptide from *Fusarium roseum*. *Source*: Ref. 53, with kind permission of *J. Org. Chem.*

D. Cyclic Lipopeptides of the Echinocandin Class

Mulundocandin, a new lipopeptide of the echinocandin class (Fig. 4), was isolated from the culture liquid of *Aspergillus sydowi* [54, 55]. It has been found to possess antiyeast and antifungal properties. Table 2 shows the minimum inhibitory concentrations required to inhibit different yeast and fungal strains.

E. Glycolipodepsipeptides

Ramoplanin (also called A16686), a novel glycolipodepsipeptide antibiotic mixture, was produced by *Actinoplanes* sp. [56–59]. All three components have

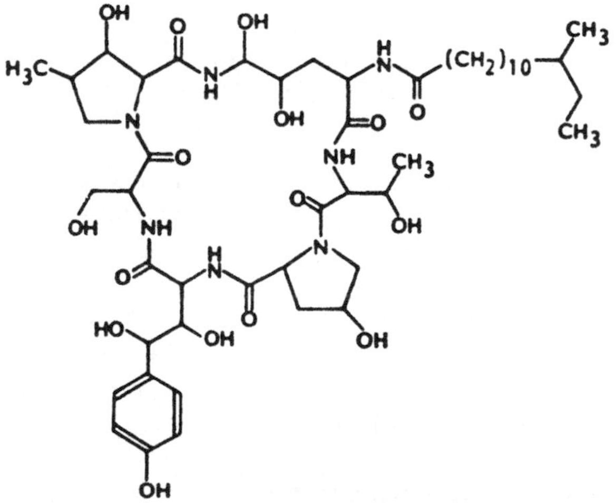

FIG. 4 Molecular structure of Mulundocandin from *Aspergillus sydowi*. *Source*: Ref. 55, with kind permission of *J. Antibiot.*

TABLE 2 Antifungal Activity of Mulundocandin

Test organism	MIC (µg/ml)
Candida albicans 200/175	0.97
Saccharomyces cerevisiae	10
Penicillium italicum	>1,000
Aspergillus niger 500/284	31.25
Cercospora beticola	4
Fusarium nivale	>1,000
Botrytis cinerea	125
Trichophyton mentagrophytes 100/25	>125
Trichophyton rubrum 100/58	>125
Microsporum gypseum	4
Microsporum canis 150/353	>125

Source: Ref. 54, with kind permission of *J. Antibiot.*

structures formed by a common depsipeptide skeleton carrying a dimannosyl group, and are differentiated by the presence of various acylamide moieties, derived from C8-, C9-, and C10-fatty acids. The compounds were very active against Gram-positive and anaerobic bacteria (Table 3); MIC values ranged from 0.016 to 2 µg/mL. There was no cross-resistance with clinically used antibiotics.

III. GLYCOLIPIDS

As mentioned before, Haferburg et al. [17] presented an extensive summary of publications on biological activities of biosurfactants. The main topics on glycolipid effects were as follows:

Mycolic acid type glycolipids and similar biosurfactants from *Actinomycetes* (e.g., trehalose-6,6′-dimycolate cord factor, negative effects on migration of leucocytes, pyridine-nucleotide-dependent dehydrogenases, mitochondrial membranes)

Glycolipids from non-*Actinomycetes* (e.g., antibacterial activities of mannosylerythritol lipids or rhamnolipids, antiviral activities of rhamnolipids against vaccinia virus at low concentrations, hemolytic activities of rhamnolipids, stimulation of leucocyte chemotaxis and chemokinesis).

A. Effects of Glycolipids on Microorganisms

In 1987, Hommel et al. [60] found that the biosurfactant (later determined as sophorose lipids) isolated from the culture medium of *Torulopsis apicola*,

TABLE 3 In vitro Activity of A16686 Against Selected Organisms

Organism	MIC (μg/ml)
Staphylococcus aureus ATCC 6538	0.5
S. aureus TOUR	1.0
Staphylococcus epidermidis ATCC 12228	0.5
Streptococcus pyogenes C 203 SKF 13400	0.063
Streptococcus pneumoniae UC 41	0.032
Streptococcus faecalis ATCC 7080	0.25
Streptococcus faecium ATCC 10541	0.125
Streptococcus bovis ATCC 9809	0.125
Streptococcus agalactiae ATCC 7077	0.063
Streptococcus dysgalactiae ATCC 9926	0.125
Streptococcus mutans ATCC 27531	0.5
Corynebacterium diphteriae type mitis ATCC 11051	1.0
Clostridium perfringens ISS	0.25
Clostridium difficile ATCC 9689	1.0
Propionibacterium acnes ATCC 6919	0.25
Actinomyces viscosus ATCC 19246	0.125
Actinomyces naeslundi ATCC 12104	0.5

Source: Ref. 57, with kind permission of *J. Antibiot.*

exhibited antibiotic effects especially on Gram-positive bacteria (Table 4). Furthermore, an antiphagal effect was observed in the system *Plectonema boryanum/Cyanophage* LPP-1 at concentrations below those required for inhibition of *P. boryanum.*

Lang et al. [61] tested the effect of a variety of microbially produced glycolipids on fungi as well as Gram-positive and Gram-negative bacteria; included were also those bacterial frequently found in the obstructed sebaceous glands of the skin causing reduced respiration. The results were as following:

1. Among three trehalose lipids (from *Rhodococcus erythropolis*) only the most hydrophobic trehalose-dicorynomycolate prevented the conidia germination of the fungus *Glomerella cingulata* at an MIC of 300 μg/mL. All other test organisms did not show growth restriction.
2. Cellobiose lipids from *Ustilago maydis* did not effect growth inhibition.
3. The purified sophorose lipids of *T. bombicola* showed different behavior depending on the amphiphile character of the glycolipids. Table 5 indicates that the lowest MIC values were derived by the most hydrophobic sophorolipid SL-1.
4. Rhamnolipid (RL) mixtures of RL-1/RL-3 prevented the growth of *B. subtilis* at MIC values of 35 μg/mL and of *Staphylococcus epidermidis* at 350 μg/mL.

TABLE 4 Effect of *Torulopsis apicola* IMET 43747 Biosurfactant on Microbial Growth in Plate Dilution Test

Microorganism	Minimal inhibitorious concentration (μg anthronereactive compounds ml^{-1})
Acinetobacter calcoaceticus 69 V	>31.3
Azotobacter chroococcum	1.95
Bacillus subtilis	0.12
Corynebacterium fascians IMET 10513	1.95
Micrococcus luteus	0.48
Mycobacterium rubrum B4a	0.12
Mycobacterium lacticola	0.12
Pseudomonas aeruginosa 196 Aa	7.8
Proteus vulgaris	>31.3
Escherichia coli	7.8
Serratia marcescens	1.95
Plectonema boryanum	0.12

Source: Ref. 60, with kind permission of *Appl. Microbiol. Biotechnol.*

In a later experiment Hommel et al. [62] studied the effect of 0.2 g/L of sophorolipid mixture on the growth of *Acinetobacter calcoaceticus* on *n*-hexadecane. Besides of prolongation of the lag-phase, they detected an increased release of cytoplasmic malate dehydrogenase (high excretions) and of glucose dehydrogenase.

Studying also effects of biosurfactants on the enzymatic level of *n*-alkane cultured *A. calcoaceticus*, Münstermann et al. [63] observed that the activity of

TABLE 5 Minimum Inhibitory Concentrations (50% of Growth) of the Sophorolipids

	MIC (μg/mL) of		
Microorganisms	SL-1	SL-2	SL-3
Bacillus subtilis	6	25	500
Staphylococcus epidermidis	6	25	n.i.[a]
Streptococcus faecium	15	29	n.i.[a]
Glomerella cingulata	n.i.[a]	50	n.i.[a]

[a]n.i. = no inhibition.
Source: Ref. 61, with kind permission of *Fat Sci. Technol.*

the NADP-depending fatty aldehyde dehydrogenase was enhanced if biosurfactants were added during growth (Fig. 5). For instance, rhamnolipids RL increased this activity by a factor of two using concentrations lower, equal, or higher compared to the critical micelle concentration. Synthetic detergents on ethylene oxide base (EO9) or sucrose lipid base (DK50) did not influence activity, but, on lipopeptide base (EAP), the aldehyde dehydrogenase activity decreased. Possible reasons for different responses of surfactant treatment could be changes of the enzyme conformation, changes of the membrane region surrounding the enzyme, or changes of the octanal solubility.

B. Effects of Glycolipids on Viruses

Succinyl-trehalose lipid containing 1–2 moles of succinic acid and 1–2 moles of fatty acid, produced by *R. erythropolis* [64, 65], seems to be useful against viruses, for example, against

1. Herpes simplex virus, *in vitro* LD_{50} value of 11 μg/mL.
2. Influenza virus, *in vitro* with LD_{50} value of 33 μg/mL.

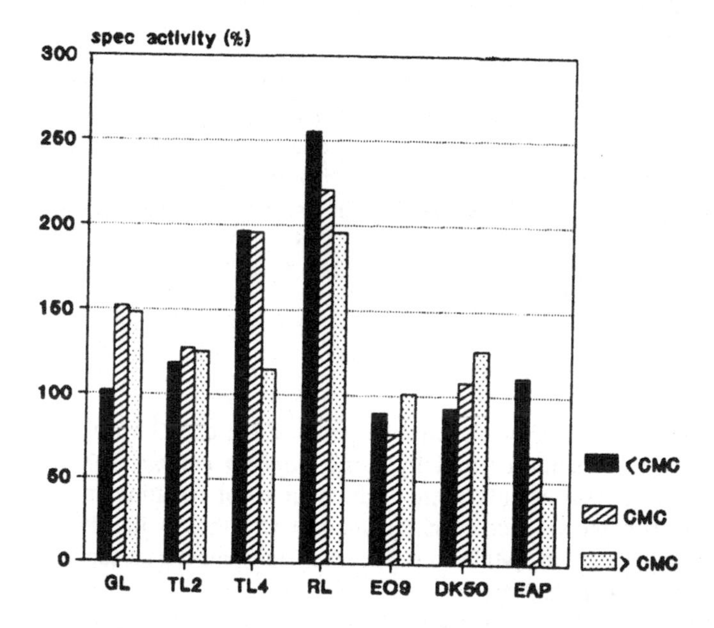

FIG. 5 Influence of surfactant addition to growing cells of *Acinetobacter calcoaceticus* with respect to their fatty aldehyde dehydrogenase activity. *Source*: Ref. 63.

Low toxicity was reported [66].

Haferburg et al. [67] investigated the application of rhamnolipid from *Pseudomonas* with regard to antiphytoviral activity. The use of a 1% emulsion of rhamnolipid for the treatment of leaves of *Nicotiana glutinosa* infected with tobacco mosaic virus (TMV) led to a 90% reduction in the number of lesions. In another experiment, 1% of rhamnolipid decreased the potato virus X (PVX) content of systematically infected *N. tabacum* L. Sansun by 46% and secondarily infected leaves by 43%. Also the red-clover-mottle virus content of the systemic host *Pisum sativum convar.* sp. was reduced.

C. Effects of Glycolipids on Fusion Rates of Plant Protoplasts

Rhamnolipid was found to allow mild fusion of protoplasts from herbaceous plants [68]. The addition of a sophorose lipid mixture (90% lactonic form) increased the protoplast fusion rate 14-fold in the case of pea (*Pisum sativum*) leaf tissue [69].

IV. MISCELLANEOUS BIOSURFACTANTS

A. Effects on Freshwater and Marine Inhabitants

The crude oil contamination of rivers and marine ecosystems is a world-wide problem. To overcome these pollutants biosurfactants could be useful in the near future. In comparison to chemically synthesized detergents, a better biodegradability and a lower toxicity could be expected from microbially surface-active substances because of their biogenetic origin. Nevertheless toxicity testing for biosurfactants is also necessary. Known biological tests for these purposes are as follows:

1. Survival of fishes, daphnia, or algae.
2. Germination tests.
3. Enzymatic tests.
4. Bioluminescence of bacteria.

Some applications of these methods should be given. Zajic and Gerson [70] tested crude microbial surfactants for toxicity using *Daphnia magna*. Table 6 shows that some microbial surfactants seem to be sources of nutrition to the *Daphnia* and death rates in the absence of surfactant exceeded those with biosurfactant (concentrations all near the CMC). Results for several synthetic surfactants are included for comparative purposes.

Studying the effects of surfactants on special marine organisms during tests to combat oil pollution in the European North Sea, van Bernem [71–73] and as Lang et al. [74] found that some biosurfactants were superior to some synthetic

TABLE 6 Toxicity of Surfactants to *Daphnia magna*

Surfactant	Source	Concentration[a] (ppm, w/v)	Mortality[b]
Microbial			
OSGB1	*Corynebacterium*	100	−12
ASPH-A1	*Pseudomonas*	200	−4
5A	*Arthrobacter*	300	17
POSE	*Corynebacterium*	500	25
CD1	*Corynebacterium*	2000	65
Synthetic			
Standamid		1050	100
		105	55
Unamide JJ-35		100	100
		10	100
Petrostep A50		900	100
		90	100
Span 20	Atlas	30	100
		3	70

[a]1.0 or 0.1 × CMC.
[b]% mortality over controls in 48 h; negative numbers indicate greater survival than controls.
Source: Ref. 70, with kind permission of Verlag Chemie.

detergents, stored for such purposes. In field tests, the brine shrimp *Corophium volutator* was more sensitive to synthetic detergents. In laboratory tests using larva of the brine shrimp *Artemia*, the results were similar (Table 7).

Recently Poremba et al. [75] reported several toxicity testing series with marine microorganisms and their response to biosurfactant treatment. Bacterial growth (*A. calcoaceticus, S. marinorubra, Photobacterium phosphoreum*) was slightly effected or stimulated, whereas that of algae (e.g., *Dunaliella tertiolecta*) and flagellates was reduced. Table 8 indicates a decreased multiplication of marine heterotrophic flagellates, especially in the case of some synthetic surfactants.

The highest sensitivity toward surfactants was found with the bioluminescence inhibition test (Table 9). With the exception of GL, a new glucose lipid from the marine bacterial strain MMI [76], all surfactants decreased the bioluminescence of *P. phosphoreum*. Most EC values of synthetic surfactants were higher than those of biosurfactants (exception: the synthetic sucrose esters DK50 and DK160). Only special sophorose- and rhamnolipids (SL, RL) showed similar toxic effects. A subsequent ranking demonstrated that most biogenetic surfactants were less toxic than the synthetic ones.

TABLE 7 Influence of Surfactants on Survival of Marine Organisms

Crude oil + surfactant	Survival of *Corophium volutator* %	*Artemia*[a] LC_{50}/48 h (ppm)
Microbial		
Trehalose-tetraester	75	7,500
Rhamnolipids	72	3,000
Synthetic		
Finasol	0	64
Corexit	5	440
Without surfactant	35	n.d.[b]

[a]Without crude oil.
[b]n.d. = not determined.
Source: From Refs. 71–74.

TABLE 8 Growth Inhibition of Marine Heterotrophic Flagellates by Surfactants ($EC_{fla\text{-}tox}$ = surfactant concentration, in which no mass development over $10^5/cm^3$ occurred within 7 days)

Surfactant	$EC_{fla\text{-}tox}$ (mg/L)
Biosurfactants	
TL-4	>1000
TL-2	500–1000
GL	>1000
LP	>1000
Suc	>1000
SL	100–500
SS	>1000
RL	25–50
Emu	>1000
Synthetic surfactants	
Finasol	13–50
Corexit	50–100
Pril	25–50
CTAB	3–5
E04,5	15–20
E09	60–80
DK 50	>1000
DK 160	>1000

Source: Ref. 75, with kind permission of *Z. Naturforsch.*

TABLE 9 Inhibition of the Bioluminescence of *Photobacterium phosphoreum* by Surfactants; EC_{50}, EC_{20} = Effective Concentration that Inhibits 50(20)% of Luminescence; EC_{max} = Maximal Measured Reduction of Luminescence

Surfactant	EC_{50} (mg/L)	EC_{20} (mg/L)	EC_{max} (%)
Biosurfactants			
TL-4	286	33	24
TL-2	49	7	43
GL		>3000	5
LP	>3000	386	18
Suc	84	25	45
SS	141	12	54
SL	12	1	87
RL	50	6	100
Emu	202	10	50
Synthetic surfactants			
Finasol	7	1	100
Corexit	5	1	96
Pril	35	4	88
CTAB	0.5	0.3	100
E04,5	79	38	45
E09	78	7	57
DK 50	67	27	20
DK 160	334	88	17

Source: Ref. 75, with kind permission of *Z. Naturforsch*.

V. CONCLUSION

Among biosurfactants, lipopeptides and glycolipids have been studied intensively with regard to their biological activities. In this review, the response of a variety of micro- and macroorganisms on the treatment of such substances is summarized. As with chemically synthesized detergents, the point of attack is the biological membrane. Different cell-wall structures of the organisms and the different amphiphile character of the biosurfactants (e.g., ionic charge, hydrophilic/hydrophobic moieties) are responsible for different effects.

Fortunately some of the surface-active agents of biological origin show antimicrobial and/or antiviral activities at very low concentrations; because of low toxicity others seem suitable for reducing environmental pollutants.

REFERENCES

1. R. D. Hotchkiss, in *Surface Active Agents* (R. W. Miner and E. I. Valko, eds.), New York Academy of Science, New York, 1946, pp. 479–492.
2. P. L. Healy, R. Ernst, and J. Arditty, *New Phytol. 70*: 477 (1977).
3. R. Ernst and J. Arditty, *New Phytol. 96*: 197 (1984).
4. A. M. James, in *Surface Activity and the Microbial Cell*, Monograph 19, The Society of Chemical Industry, London, 1965, p. 3–22.
5. W. B. Hugo, in *Surface Activity and the Microbial Cell*, Monograph 19, Society of Chemical Industry, London, 1965, pp. 67–80.
6. F. Devinsky, I. Lacko, D. Mlynarcik, V. Racansky, and L. Krasnec, *Tenside Detergents, 22*: 10 (1985).
7. A. Kopecka-Leitmanova, F. Devinsky, D. Mlynarcik, and I. Lacko, *Drug Metab. Drug Interact. 7*: 29 (1989).
8. A. T. King, K. C. Lowe, and B. J. Mulligan, *Biotechnol. Lett. 10*: 873 (1988).
9. A. T. King, M. R. Davey, I. R. Mellor, and B. J. Mulligan, *Enzyme Microb. Technol. 13*: 148 (1991).
10. Y. Igarashi, K. Yagami, R. Imai, and S. Watanabe, *J. Industrial Microbiol. 6*: 223 (1990).
11. T. Cserhati, Z. Illes, and I. Nemes, *Appl. Microbiol. Biotechnol. 35*: 115 (1991).
12. M. Singh and J. R. Orsenigo, *Bull. Environ. Contam. Toxicol. 32*: 119 (1984).
13. P. D. Abel, *J. Fish Biol. 6*: 279 (1974).
14. N. R. Bromage and A. Fuchs, *J. Fish Biol. 8*: 529 (1976).
15. H. Lal, V. Misra, P. N. Viswanathan, and C. R. Krishna Murti, *Bull. Environ. Contam. Toxicol. 32*: 109 (1984).
16. J. Sunamoto, K. Iwamoto, T. Uesugi, K. Kojima, and K. Furuse, *Chem. Pharmaceut. Bull. 32*: 2891 (1984).
17. D. Haferburg, R. Hommel, R. Claus, and H. P. Kleber, *Adv. Biochem. Eng./Biotechnol. 33*: 53 (1986).
18. K. Arima, A. Kakinuma, and G. Tamura, *Biochem. Biophys. Res. Commun. 31*: 488 (1968).
19. A. W. Bernheimer and L. S. Avigad, *J. Gen. Microbiol. 61*: 361 (1970).
20. D. G. Cooper, C. R. MacDonald, S. J. B. Duff, and N. Kosaric, *Appl. Environ. Microbiol. 65*: 408 (1986).
21. A. Kakinuma, M. Hori, M. Isono, G. Tamura, and K. Arima, *Agric. Biol. Chem. 33*: 971 (1969).
22. K. Arima, G. Tamura, and A. Kasinuma, DE-OS 1803 987 (1968).
23. M. Kochi, D. W. Weiss, L. H. Pugh, and V. Groupe, *Bacteriol. Proc. 29*: 30 (1951).
24. V. Groupe, L. Pugh, D. Weiss, and M. Kochi, *Proc. Soc. Exp. Biol. Med. 78*: 354 (1951).
25. T. R. Neu, T. Härtner, and K. Poralla, *Appl. Microbiol. Biotechnol. 32*: 518 (1990).
26. F. Besson, F. Peypoux, G. Michel, and L. Delchambe, *J. Antibiot. (Tokyo) 31*: 283 (1978).
27. F. Peypoux, M. Guinand, G. Michel, L. Delchambe, B. C. Das, and E. Lederer, *Biochemistry 17*: 3992 (1978).
28. T. R. Neu and K. Poralla, *Appl. Microbiol. Biotechnol. 32*: 521 (1990).

29. H. Kleinkauf and H. von Döhren, in *Biotechnology*, Vol. 4 (H. Pape and H. J. Rehm, eds.), VCH-Verlag, Weinheim, 1986, pp. 283–307.
30. H. Umezawa, T. Aoyagi, T. Nishikiori, A. Okuyama, Y. Yamagishi, and M. Hamada, *J. Antiobiot. 39*: 737 (1986).
31. T. Nishikiori, H. Naganawa, Y. Muraoka, T. Aoyagi, and H. Umezawa, *J. Antibiot. 39*: 745 (1986).
32. T. Nishikiori, H. Naganawa, Y. Muraoka, T. Aoyagi, and H. Umezawa, *J. Antibiot. 39*: 755 (1986).
33. T. Nishikiori, H. Naganawa, Y. Muraoka, T. Aoyagi, and H. Umezawa, *J. Antibiot. 39*: 860 (1986).
34. Lilly, U. S. Patent No. RE 32-333 (1985).
35. M. Debono, M. Bernhart, C. B. Carrell, J. A. Hoffmann, J. L. Occolowitz, B. J. Abbot, D. S. Fukuda, R. L. Hamill, K. Biemann, and W. C. Herlihy, *J. Antibiot. 40*: 761 (1987).
36. L. V. D. Boeck, D. S. Fukuda, B. J. Abbot, and M. Debono, *J. Antibiot. 41*: 1085 (1988).
37. M. D. Debono, B. J. Abbot, R. M. Molloy, D. S. Fukuda, A. H. Hunt, V. M. Daupert, F. T. Counter, J. L. Ott, C. B. Carrell, L. C. Howard, L. D. Boeck, and R. L. Hamill, *J. Antibiot. 41*: 1093 (1988).
38. Lilly, U.S. Patent No. EP 337-731 (1988).
39. L. D. Boeck, H. R. Papiska, R. W. Wetzel, J. S. Mynderse, D. S. Fukuda, F. P. Mertz, and D. M. Berry, *J. Antibiot. 43*: 587 (1990).
40. D. S. Fukuda, R. H. Du Bus, P. J. Baker, D. M. Berry, and J. S. Mynderse, *J. Antibiot. 43*: 594 (1990).
41. D. S. Fukuda, M. Debono, R. M. Molloy, and J. S. Mynderse, *J. Antibiot. 43*: 601 (1990).
42. L. D. Boeck and R. W. Wetzel, *J. Antibiot. 43*: 607 (1990).
43. F. T. Counter, N. E. Allen, D. S. Fukuda, J. N. Hobbs, J. Ott, P. W. Enzminger, J. S. Mynderse, D. A. Preston, and C. Y. E. Wu, *J. Antibiot. 43*: 616 (1990).
44. A. Mhammedi, F. Peypoux, F. Besson, and G. Michel, *J. Antibiot. 35*: 306 (1982).
45. F. Peypoux, D. Marion, R. Maget-Dana, M. Ptak, B. C. Das, and G. Michel, *Eur. J. Biochem. 153*: 335 (1985).
46. F. Peypoux, M. T. Pommier, D. Marion, M. Ptak, B. C. Das, and G. Michel, *J. Antibiot. 39*: 636 (1986).
47. J. Vater, B. Kluge, and H. Kleinkauf, in *Proceedings of the 4th European Congress on Biotechnology*, Vol. 3 (O. M. Neijssel, R. R. van der Meer, and K. C. A. M. Luyben, eds.), Elsevier, Amsterdam, 1987, pp. 266–269.
48. Ajinomoto, J6 1093-125 (1984).
49. H. Itazaki, K. Nagashima, K. Sugita, H. Yoshida, Y. Kawamura, Y. Yasuda, K. Matsumoto, K. Ishii, N. Uotani, H. Nakai, A. Terui, S. Yoshimatsu, Y. Ikenishi, and Y. Nakagawa, *J. Antibiot. 43*: 1524 (1990).
50. K. Sugawara, K. Numata, M. Konishi, and H. Kawaguchi, *J. Antibiot. 37*: 958 (1984).
51. H. R. Burmeister, R. F. Vesonder, and C. W. Hesseltine, *Mycpathlogia 62*: 53 (1977).
52. H. R. Burmeister, J. Ellis, and R. F. Vesonder, *Mycopathologia 74*: 29 (1980).
53. S. A. Carr, E. Block, and C. E. Costello, *J. Org. Chem. 50*: 2854 (1985).
54. K. Roy, T. Mukhopadhyay, G. C. S. Reddy, K. R. Desikan, and B. N. Ganguli, *J. Antibiot. 40*: 275 (1987).

55. T. Mukhopadhyay, B. N. Ganguli, H. W. Fehlhaber, H. Kogler, and L. Vertesy, *J. Antibiot. 40*: 281 (1987).
56. B. Cavalleri, H. Pagani, G. Volpe, E. Selva, and F. Parenti, *J. Antibiot. 37*: 309 (1984).
57. R. Pallanza, M. Berti, R. Scotti, E. Randisi, and V. Arioli, *J. Antibiot. 37*: 318 (1984).
58. R. Ciabatti, J. K. Kettenring, G. Winters, G. Tuan, L. Zerilli, and B. Cavalleri, *J. Antibiot. 42*: 254 (1989).
59. J. K. Kettenring, R. Ciabatti, G. Winters, G. Tamborini, and B. Cavalleri, *J. Antibiot. 42*: 268 (1989).
60. R. Hommel, O. Stüwer, W. Stuber, D. Haferburg, and H. P. Kleber, *Appl. Microbiol. Biotechnol. 26*: 199 (1987).
61. S. Lang, E. Katsiwela, and F. Wagner, *Fat Sci. Technol. 91*: 363 (1989).
62. R. Hommel, M. Götzrath, and H. P. Kleber, *Acta Biotechnol. 9*: 461 (1989).
63. B. Münstermann, S. Lang, and F. Wagner, unpublished results (1991).
64. Y. Uchida, R. Tsuchiya, M. Chino, J. Hirano, and T. Tabuchi, *Agric. Biol. Chem. 53*: 757 (1989).
65. Y. Uchida, S. Misawa, T. Nakahara, and T. Tabuchi, *Agric. Biol. Chem. 53*: 765 (1989).
66. Nippon Oils and Fat, Japan Patent No. J6 3126-493 (1986).
67. D. Haferburg, R. Hommel, H. P. Kleber, S. Kluge, G. Schuster, and H. J. Zschiegner, *Acta Biotechnol. 7*: 353 (1987).
68. Karl Marx University, Doctoral Dissertation DD-254-316 (1986).
69. Karl Marx University, Doctoral Dissertation DD 280-551 (1989).
70. J. E. Zajic and D. F. Gerson, in *Oil Sand and Oil Shale Chemistry* (O. P. Strausz and E. M. Lown, eds.), Verlag Chemie, Weinheim, 1978, pp. 145–161.
71. K. H. van Bernem, *Senckenbergiana Marit. 16*: 13 (1984).
72. K. H. van Bernem, in *Texte 6/87, Meereskundliche Untersuchung von Ölunfällen*, Umweltbundesamt Deutschland, Berlin, 1987, pp. 22–33.
73. K. H. van Bernem, in *Texte 6/87, Meereskundliche Untersuchung von Ölunfällen*, Umweltbundesamt Deutschland, Berlin, 1987, pp. 64–74.
74. S. Lang, B. Frautz, G. A. Henke, J. S. Kim, C. Syldatk, and F. Wagner, *Biol. Chem. Hoppe Seyler 367* (Supplement 17th FEBS Meeting, Berlin, 1986): 212 (1986).
75. K. Poremba, W. Gunkel, S. Lang, and F. Wagner, *Z. Naturforsch. 46c*: 210 (1991).
76. D. Schulz, A. Passeri, M. Schmidt, S. Lang, F. Wagner, V. Wray, and W. Gunkel, *Z. Naturforsch. 46c*, 197 (1991).

10

The Biophysics of Microbial Surfactants: Growth on Insoluble Substrates

DONALD F. GERSON* Process Development, Connaught Laboratories Ltd., Willowdale, Ontario, Canada

**Current affiliation*: Research and Development, Apotex Fermentation, Inc., Winnipeg, Manitoba, Canada

I. INTRODUCTION

Microbes that degrade hydrophobic, water-insoluble substrates such as hydrocarbon liquids or solids, fats, oils or waxes, or elemental sulfur usually produce surfactant substances or biosurfactants [1]. Insolubility, having a very low saturation concentration in water, limits the available aqueous concentration of these substrates. Surfactant improves the availability of the substrate to the microbial cells by allowing emulsion formation. This expands the interfacial area at the aqueous-substrate interface, increasing the rate of substrate dissolution and thus supporting a greater rate of substrate utilization by the microbial population.

Surfactants cause increased wetting of the insoluble substrate, affecting intimate contact among the substrate, the aqueous medium and the cells. The exact nature of these interactions is complicated and difficult to predict but is amenable to analysis by the methods of surface physics.

Effective adhesion is a necessary prerequisite to direct microbial attack of solid substrates [2]. Biodeterioration and microbial corrosion may be significantly reduced if microbial adhesion to the substrate is prevented. Microbes that degrade solid insoluble substrates also produce biosurfactants to facilitate substrate breakdown and dissolution.

The interface between a liquid and either an immiscible fluid or an insoluble solid has special physical and biological characteristics [3]. At an interface, the intermolecular forces in the bulk material are unbalanced, leaving excess free energy, known as the surface energy, in the exposed layer of molecules. For liquids, this is measured as a macroscopic property known as the surface tension, and, for solids, this is inferred from macroscopic wetting behavior and generally referred to as surface energy. Biological systems produce many hydrophobic or amphipathic molecules and particles that tend to accumulate at interfaces and affect the surface tension between liquids and air or surface energies at solid–liquid interfaces.

Examination of the surface energetics involved in microbial growth on insoluble substrates yields an understanding of the effects of biosurfactants on the physical processes involved in the biodegradation of insoluble or hydrophobic substrates. Insoluble substrates can be rich sources for microbial growth; however, the microbial utilization of insoluble substrates imposes a great restriction on the availability of substrate to the cell. By producing biosurfactants and altering the physical nature of the cell-substrate interaction, microbes significantly increase the bioavailability of the substrate and their growth potential [2].

The production of biosurfactants by microbes that utilize insoluble substrates is ubiquitous. Some microbial genera known to produce surfactants are classified by surfactant type and listed in Table 1.

TABLE 1 Microbial Surfactants

Surfactants	Organisms (genus)
Fatty acids	*Acinetobacter, Aspergillus, Candida, Corynebacterium, Mycococcus, Nocardia, Penicillium, Pseudomonas*
Neutral lipids	*Acinetobacter, Arthrobacter, Mycobacterium, Thiobacillus*
Phospholipids	*Candida, Corynebacterium, Micrococcus, Thiobacillus*
Ornithine lipids	*Agrobacterium, Gluconobacter, Pseudomonas, Thiobacillus*
Lipopeptides	*Bacillus, Candida, Corynebacterium, Mycobacterium, Nocardia, Streptomyces*
Trehalose lipids	*Arthrobacter, Brevibacterium, Corynebacterium, Mycobacterium, Nocardia*
Rhamnose lipids	*Arthrobacter, Corynebacterium, Nocardia, Pseudomonas*
Sophorose lipids	*Torulopsis*
Glycosyldiglycerides	*Lactobacillus*

II. BIOSURFACTANTS AND THE UPTAKE OF INSOLUBLE LIQUIDS

The mechanism of uptake of liquid hydrocarbon substrates by microbial cells involves both chemical and interfacial phenomena [1, 2, 4, 5]. By placing emphasis on the importance of interfacial phenomena in developing an understanding of hydrocarbon uptake by microbial cells during fermentation processes, or during hydrocarbon degradation in the environment, it is possible to separate general physical interactions from specific biochemical or receptor-mediated interactions.

Figure 1 shows the range of physical interactions that could occur among cells, hydrocarbon droplets, and air bubbles during fermentations to produce biosurfactants or during hydrocarbon degradation in an aqueous environment. The surface physics of these situations is amenable to analysis using new techniques for the calculation of interfacial tensions, and provides an approach to the understanding of many of the situations that are observed. These include various degrees of adhesion of microbial cells to air bubbles or hydrocarbon droplets and the formation of flocs composed of cells, bubbles, and droplets in a multitude of configurations.

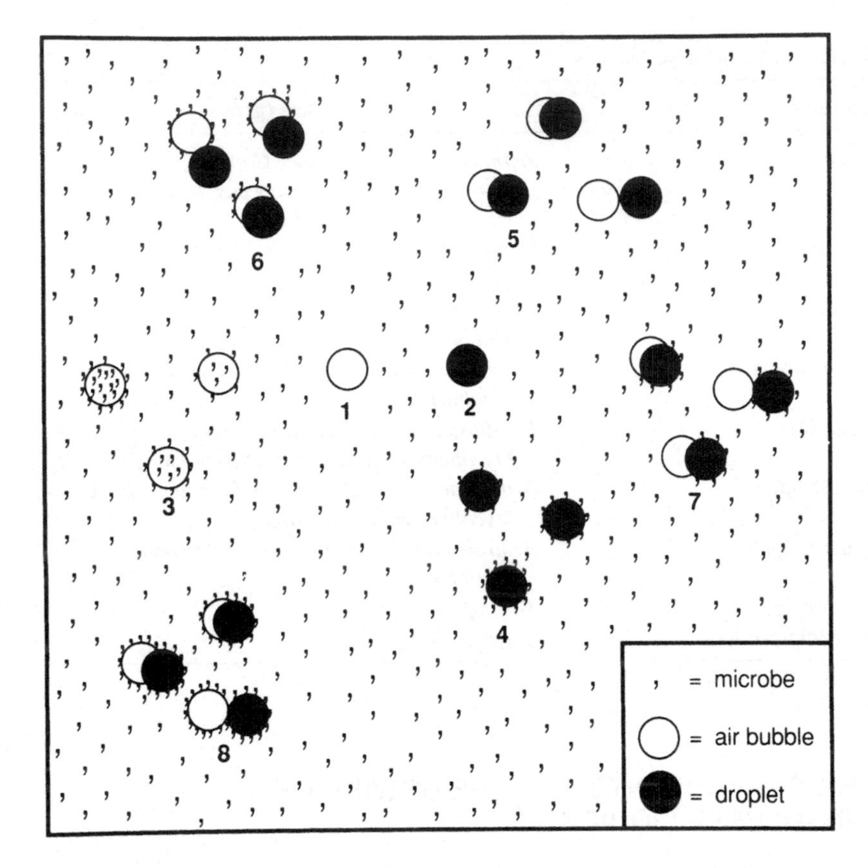

FIG. 1 Components of the hydrocarbon biodegradation system. (1) Air bubble, (2) hydrocarbon drop, (3) microbial cells attached to air bubbles, (4) microbial cells attached to hydrocarbon droplets, (5) air–hydrocarbon doublets, (6) air–hydrocarbon doublets with microbes attached to the air bubbles, (7) air–hydrocarbon doublets with microbes attached to the hydrocarbon droplets, and (8) air–hydrocarbon doublets with microbes attached to both bubbles and droplets.

Biosurfactants increase the bioavailability of insoluble substrates by several mechanisms. In the presence of biosurfactants, liquid substrates are emulsified, incorporated into micelles, and spread over the cell surface. Solid substrates are wetted by biosurfactants, cracks are more easily penetrated, and the surface becomes crazed, increasing the surface area available for dissolution or colonization. In addition, there is the possibility of the dissolution of the solid directly into the lipophilic interior or micelles or cell membranes.

III. BIOSURFACTANT CHARACTERISTICS

To understand the interfacial tensions that determine the adhesion of cells to air bubbles, liquid droplets, and solid surfaces, it is necessary to develop an understanding of surface tension—the interfacial tension at the air–liquid interface.

At room temperature, the surface tension of water is approximately 72 mN/m; this is the highest surface tension of any biologically relevant liquid, and it results primarily from the effects of hydrogen bonding. Many surfactants can reduce surface tension to approximately 25 ± 5 mN/m depending on concentration and surfactant type. Few surfactants are capable of reducing surface tension below this range under normal circumstances. Surfactants achieve this effect by acting as a bridge between the two materials meeting at the interface. Surfactant molecules typically have a hydrophilic moiety and a hydrophobic moiety; the hydrophilic portion of the molecule is imbedded in the aqueous phase, whereas the hydrophobic portion faces the air.

Adsorption of surfactants occurs at the air–aqueous interface. This makes the surfactant-coated air–water interface hydrophobic if approached from the air side and hydrophilic if approached from the aqueous side. The degree of adsorption varies tremendously from one surfactant to another, and it is difficult to generalize about the effects of adsorbed layers of surfactant at the air–aqueous interface. Surfactant adsorption to the air–aqueous interface in fermentations reduces the surface tension, can alter oxygen diffusion from air bubbles by directly impeding diffusion, and can alter the degree of bubble dispersion and coalescence. The measure of surfactant adsorption to the surface is the *Gibbs surface excess* (Eq. 1), which is easily calculated from a plot of surface tension against surfactant concentration.

$$\Gamma = -\frac{1}{RT}\frac{d\gamma}{d\ln C} \tag{1}$$

where Γ is the Gibbs surface excess (mole cm^2), R is the gas constant (erg/mol degree^{-1}), T is the temperature (°K), γ is the surface tension (mN/m), and C is the surfactant concentration (moles/L).

Biosurfactants include classes of molecular structures quite different from synthetic surfactants while having effects on surface and interfacial tensions within the same range. Biosurfactants as a class contribute significantly to the repertoire of available surfactant structures but do not present radically different effects on surface or interfacial tensions. The contributions of biosurfactants to expanding the surfactant repertoire are in extending the possibilities for selection of the hydrophilic–lipophilic balance (HLB), emulsion or foam stability, production technology, and, especially, biological and environmental compatibility.

As with the selection of synthetic surfactants, each application has special considerations that ultimately lead to the selection of the most appropriate surfactant. Selection of biosurfactants for a particular application proceeds along similar lines, but allows greater flexibility while maintaining lower toxicity and greater biodegradability [6–8].

Most biosurfactants are lipids, and some of the major types are listed in Table 1. These surfactants generally are produced by microbes grown on hydrocarbons and not produced by the same microbes when grown on a soluble substrate. Some biosurfactants, however, are produced only from soluble substrates and serve purposes other than those considered here. The surfactants range in chemical structure from simple fatty acids (soaps) to complex cyclic peptides, glycolipids, and proteins.

IV. BIOSURFACTANTS AND THE HYDROCARBON–AQUEOUS INTERFACE

The interfacial tension between water and hexadecane is about 50 mN/m; for kerosene and a typical growth medium, the interfacial tension can be as low as 35–40 mN/m. Biosurfactants produced from hydrocarbons by fermentation typically reduce the hydrocarbon–aqueous interfacial tension to between 0.1 and 1.0 mN/m [1, 4, 5, 7, 8].

In continuously stirred bioreactors, reduced interfacial tension increases emulsification of the hydrocarbon because droplet dispersion will be increased and droplet coalescence will be reduced. Emulsion stability and the quantitative nature of the emulsion depend on the details of a particular situation and are difficult to generalize.

At fluid–fluid interfaces, the interfacial tension is easily measured by a variety of techniques [3]. Interfacial tension is important in bubble formation and droplet dispersion or coalescence and in other processes involving changes in the interfacial area. The tendency to form emulsions is related to the free energy of formation of a unit area of new interface, and the lower the interfacial tension, the greater the surface area that will form under given mechanical conditions.

V. THE MICROBE–AQUEOUS INTERFACE

The solid–solid interface at the cell surface involves difficult aspects of surface physics. The surface is chemically diverse and mechanically complicated. The surface energies of cells are an aspect of the surface energies of solids. This is a developing area of surface physics; advances made by Zisman [9], Neumann [10],

Van Oss [11], and Gerson [12] bring the measurement of the surface energies of solids and biological surfaces into the range of experimental study.

The surface energies of microbes have been measured by van Oss et al. [13] and Gerson [14–16]. Typically, microbial cells range in surface energy (standardized to the surface energy against air) from around 70 mN/m for hydrophilic microbes to 15 mN/m for highly hydrophobic organisms such as *Mycobacterium butyricum*. Table 2 gives the surface energies for a variety of microbial cells.

The significance of cell-surface energy in hydrocarbon fermentations is as a quantitative measure that may be used to analyze the physical interactions among cells, air bubbles, and hydrocarbon droplets. Production of biosurfactants during growth alters the interfacial free energies of the cells, the aqueous medium, and the hydrocarbon substrate.

TABLE 2 Cell Surface Energy[a]

	Against	
Species	Air	Water
Thiobacillus thiooxidans	72	—
Staphylococcus epidermidis	67.9	0.27
Staphylococcus aureus Smith	67.3	0.36
Staphylococcus pneumoniae 1	67.1	0.40
Escherichia coli 0111	67.0	0.41
Haemophilus influenzae β	66.9	0.43
Staphylococcus aureus	66.5	0.51
Salmonella arizonae	66.4	0.53
Serratia marcescens	66.3	0.55
Salmonella typhimurium	66.0	0.61
Escherichia coli 07	64.9	0.87
Listeria monocytogenes	63.4	1.31
Neisseria gonorrhoeae	63.3	1.34
Brucella abortus	63.2	1.38
Candida tropicalis	62.0	1.81
Flexibacter sp.	60.0	2.67
Pseudomonas asphaltenicus	51.7	7.95
Acinetobacter calcoaceticus	50.0	9.31
Corynebacterium lepus	48.7	10.4
Nocardia erythropolis	15.9	41.1
Mycobacterium butyricum	15.6	41.4

[a]Surface energy in mN/m, calculated using Eq. 3.

VI. INTERFACES IN A HYDROCARBON FERMENTATION

The three-phase contact between dispersed hydrocarbon droplets and air bubbles in the aqueous medium is characterized by the interfacial tensions between pairs of phases and the contact angles between the hydrocarbon and aqueous phases. Several types of interaction of an air bubble with a hydrocarbon droplet are possible (Fig. 1): (1) touching at a point with no discernible contact angle, (2) touching with a finite contact angle, and (3) complete spreading of the hydrocarbon over the bubble surface.

Spreading can be predicted from the spreading tension (Eq. 2). If the spreading tension is positive, spreading of the hydrocarbon a the air–aqueous interface will occur, forming an oil-coated bubble.

$$\gamma_{sp} = \gamma_{aq} - \gamma_{hc} - \gamma_{i} \tag{2}$$

where γ_{sp} is the spreading tension (mN/m), γ_{aq} is the aqueous surface tension (mN/m), γ_{hc} is the hydrocarbon surface tension (mN/m), and γ_{i} is the interfacial tension between the hydrocarbon and aqueous phases (mN/m).

The attachment of microbial cells to the air–aqueous and the hydrocarbon–aqueous interfaces can be considered in terms of the interfacial energies. Mass transport of oxygen or the hydrocarbon to the cell can be enhanced by direct contact of the cell with an air bubble or a hydrocarbon drop. The partition of cells from the bulk phases to the interface is dependent on the relative values of the cell-surface energy against air, the aqueous phase, or the hydrocarbon.

All of these interfacial tensions change as biosurfactant is elaborated by the microbes during growth [17]. When growth is initiated, surface tension will be high; then it will fall toward its minimum value as the concentration of biosurfactant reaches the critical micelle concentration (CMC).

We have developed a method to determine the effective concentration of microbial surfactant by titration back to the CMC [1, 4]. This method acknowledges the chemical diversity of biosurfactant while providing a useful measure of surfactant quantity.

Interfacial tension against the hydrocarbon will also decrease to its minimum value at the CMC. As surfactant concentration increases above the CMC, micelles and other surfactant aggregates form. The surface energy of the cell will also change during fermentation, as surfactant accumulates at the cell surface. Cell-surface energies at various times during fermentation can be determined by measuring the contact angle [2] and calculating the surface energy from Eq. 3 below. From these measurements, a complete picture of the surface phenomena occurring throughout the fermentation can be developed.

Interfacial free energies at solid surfaces have been difficult to determine. The best technique available is the contact angle technique, in which a drop of liquid is placed on the solid and the angle through the drop to its tangent is measured [3, 12], as in Fig. 2. The contact angle and the surface tension or interfacial tension of the drop, γ_{12}, can be entered into Eq. 3 to calculate the interfacial free energies between the cell surface and the bulk fluid (γ_{13}) and between the cell surface and the drop (γ_{23}).

Equation 3 gives the relation between contact angles, liquid surface energies, solid surface energies, and solid–liquid interfacial free energies. Equation 3 must be solved numerically, and this can be done easily with any programmable calculator having a "solve $f(X) = 0$" function (e.g., a Hewlett Packard 15C

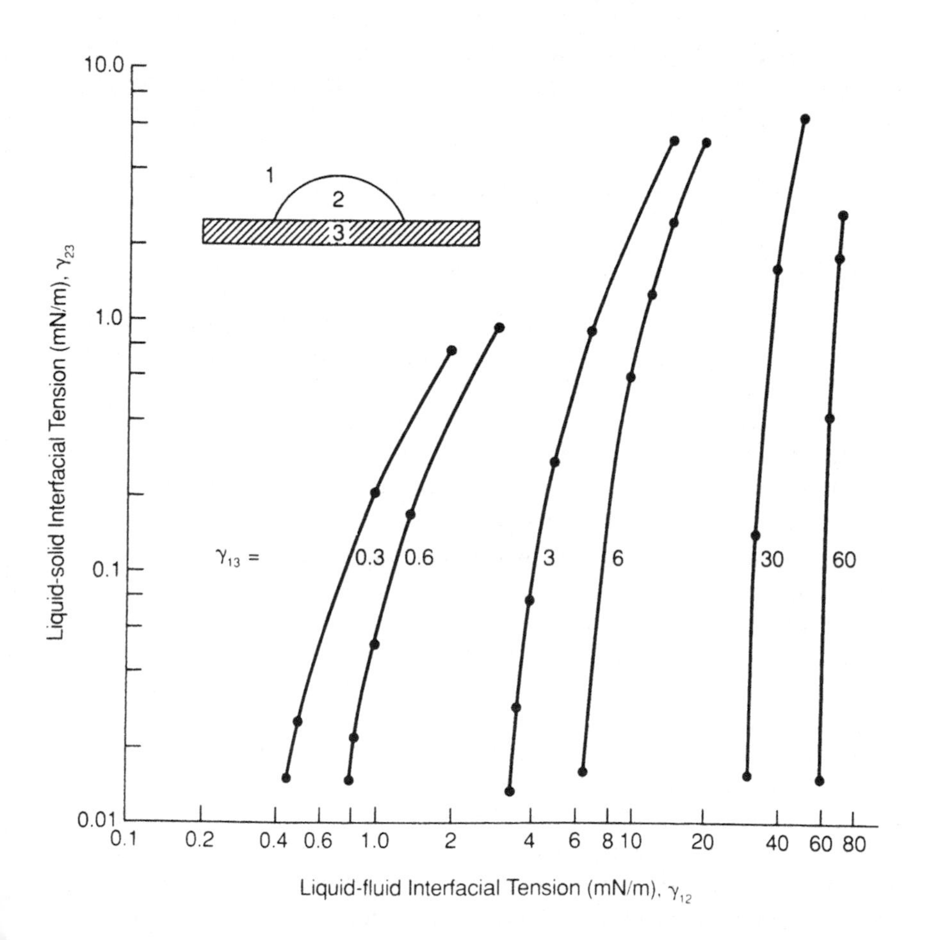

FIG. 2 Calculated values of liquid–solid interfacial tensions as a function of liquid–fluid interfacial tensions and fluid–solid interfacial tensions for three-phase systems identified by the inset. Each isopleth is for the indicated fluid–solid interfacial tension.

calculator or equivalent). Equations 3a and 3b differ only by the substitution of Young's equation.

Equation 3 has been calibrated with many fluids and solids, cells, other biological materials, and polyethylene glycol-dextran biphasic mixtures [3, 12, 18]. We have been able to determine that the results obtained with Eq. 3 are consistent with both internal controls and measurements made by other labs [10, 11]. Equation 3 also provides a good simulation of the results of a Zisman plot, in which the contact angles of a series of liquids are determined for a given solid sample and the data are extrapolated to the point of incipient spreading to estimate the solid surface energy of the sample [3].

Figure 2 gives the conventions used for the designation of phases 1, 2, and 3. Figure 2 also gives isopleths for given values of the solid surface tension, γ_{13}, showing the relation between γ_{12}, the liquid-fluid interfacial tension, and γ_{23}, the liquid-solid interfacial tension.

$$0 = [\gamma_{12}(\cos\theta + 1)]\,[0.5(\gamma_{12}\gamma_{13})^{-0.5}] - \exp[(\gamma_{13} - \gamma_{12}\cos\theta)\,(a\gamma_{13} + b)] \qquad (3a)$$

$$0 = (\gamma_{12} + \gamma_{13} - \gamma_{23})\,[0.5\,(\gamma_{12}\gamma_{13})^{-0.5}] - \exp\,[\gamma_{23}(a\gamma_{13} + \mathrm{b})] \qquad (3b)$$

where a = 0.000065, b = –0.01, γ is the interfacial free energy (mN/m), with subscripts denoting phases (see Fig. 2), and θ is the contact angle (degrees).

To determine cell surface energies, a lawn of cells is collected on a microporous filter, and contact angles are measured on the lawn with an appropriate liquid. For hydrocarbon fermentation, the three-phase contact angle among the cells, hydrocarbon, and growth medium is most important. To measure this, the lawn of cells is immersed upside down, and a hydrocarbon drop is floated against it from below. Each experimental situation will have a most suitable arrangement and choice of liquids.

To analyze the interaction of hydrocarbon droplets with the cell surface in hydrocarbon fermentations, Eq. 3 can be used to calculate the contact angle of a hydrocarbon droplet on the cell surface as a function of the interfacial tension between the hydrocarbon and aqueous phases. Throughout the time-course of a fermentation, the interfacial tensions will decrease as surfactant is produced. Complete spreading of the hydrocarbon on the cell surface occurs according to Eq. 2 when the sum of the cell–hydrocarbon and hydrocarbon–aqueous interfacial tensions is equal to or less than the aqueous–cell interfacial tension. Since the cell–hydrocarbon interfacial tension is low, this will happen when the hydrocarbon–aqueous interfacial tension is close to the aqueous–cell interfacial tension. Complete wetting of the cells by hydrocarbon will facilitate hydrocarbon uptake by allowing direct dissolution of the hydrocarbon into the cell membrane, providing a high concentration of substrate in direct contact with the cytoplasm. In addition, oxygen is very soluble in hydrocarbons, and this situation also acts to enhance oxygen transport to the cell.

The location of the cell with respect to the aqueous–hydrocarbon interface is also a consequence of the relative interfacial energies. Various possibilities are depicted in Fig. 1. A relatively simple analysis can be made with Eq. 4 below. If H, the relative height above the interface, is greater than or equal to +1, the cells will be located entirely in the hydrocarbon phase. If H is between +1 and −1, the cells will be located at the interface to varying degrees. If H is less than −1, the cells will be located entirely in the aqueous phase.

$$H = (\gamma_{\text{cell-aq}} - \gamma_{\text{cell-h}}) / \gamma_{\text{aq-h}} \quad (4)$$

where γ is the interfacial free energy at the indicated phase boundaries, aq is aqueous, and h is hydrocarbon.

Interfacial free energies can also be used to calculate equilibrium constants for the distribution of cells between liquid phases and the interface [14–18]. This approach has been used to generate the hypothetical case given in Fig. 3. In this example, a hypothetical cell with a surface tension in air of 50 mN/m is placed in

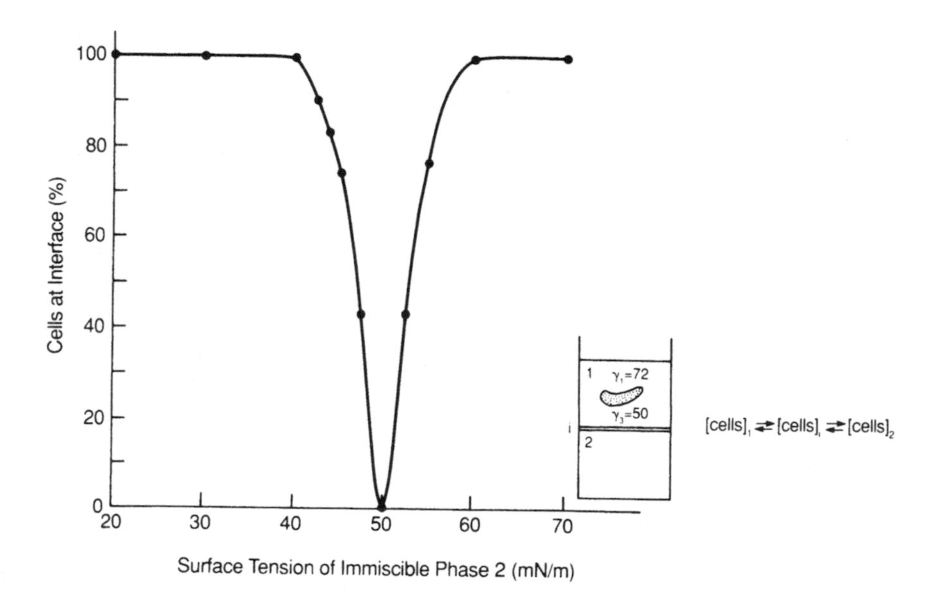

FIG. 3 Distribution of cells between two immiscible phases and the interface as a function of the surface tension of phase 2. For a cell with a surface energy against air of 50 mN/m, the fraction of cells at the interface is calculated as a function of the surface tension of phase 2. In this type of example, cells will be released from the interface when the surface tension of phase 2 equals the surface tension of the cell.

water with a surface tension of 72 mN/m. Immiscible liquids of various surface tensions are then mixed with separate samples of the cell suspension. Calculation of the interfacial tensions with Eq. 3 and the use of the interfacial tensions to estimate the distribution of cells between the phases and the interface generates Fig. 3. In this hypothetical example, liquids with surface tensions above 60 mN/m and below 40 mN/m cause concentration of the cells at the interface. In the region surrounding 50 mN/m, the cells detach from the interface and accumulate in the immiscible liquid phase. Different results would be obtained in other situations, and these tools for the analysis of interfacial interactions during the microbial utilization of insoluble substrates require adaptation to each situation.

The location of a cell with respect to the aqueous–air (cell–bubble) interface, as depicted in Fig. 1, is also described by Eq. 4. The extent of concentration of the cells at the air interface is dependent on the surface tension of the cell with respect to air and the liquid. Extremely hydrophobic cells such as *Mycobacterium* will concentrate at the air interface and form a pellicle.

Floc formation is common in hydrocarbon fermentations. Flocs are depicted in Fig. 1 but are often large, complex aggregates of cells, droplets, and bubbles. The adhesion of like or unlike particles depends on the free energy of adhesion as given by

$$\Delta G_a = -RT \ln K_{eq} = \gamma_{ij} - \gamma_i - \gamma_j \tag{5}$$

where ΔG_a is the free energy of adhesion between i and j, γ_i and γ_j are the interfacial free energy at surface i or j, γ_{ij} is the interfacial free energy between i and j, and K_{eq} is the equilibrium constant for adhesion.

These interfacial free energies can be calculated from the results of contact angle measurements with Eq. 3. Analyzing the free energy of all binary interactions will indicate which interaction involves the most negative free energy of adhesion. If the cells are the most hydrophobic particle, cell clumping will result. Of course, additional mechanisms, such as polymer bridging between particles in suspension and surface charge interactions, also affect floc formation.

VII. CELL DISTRIBUTION

Using Eqs. 3, 4, and 5, it is possible to calculate the distribution of cells between the phases of multiphasic systems. Knowledge of the interfacial tensions at all interfaces in the system allows calculation of the proportions of the total cell population in each phase and at each interface. The properties and concentrations of the biosurfactants being produced, the effect on surface and interfacial tensions, and the effect on the surface energy of the cell itself will affect cell distribution. These in turn will affect cell growth, biosurfactant production, and separation techniques used for product recovery.

These equations assume that there are randomizing effects leading to a steady-state distribution of cells. Thermal agitation is insufficient to cause the distribution of cells between phases according to the expectations of thermodynamics. For cells, this can only result from some sort of mixing or mechanical agitation. Since the degree of mixing is difficult to quantify, the results of calculations can only be expected to be proportional to the observed results, and different proportionality factors will be required in different situations. Nevertheless, very good correlations between results and expectations based on these calculations have been found in many studies of cell distribution between phases [14–16, 19].

The distribution of particles between two liquid phases and the interface between them can be described by the following equations:

$$\log \frac{C1}{C2} = 4(\gamma_{23} - \gamma_{13}) \tag{6}$$

$$\log \frac{C_i}{C_1} = \frac{(\gamma_{13} - \gamma_{23} - \gamma_{12})^2}{\gamma_{12}} \tag{7}$$

$$\log \frac{C_i}{C_2} = \frac{(\gamma_{23} - \gamma_{13} - \gamma_{12})^2}{\gamma_{12}} \tag{8}$$

$$t = V_1C_1 + V_2C_2 + V_iC_i \tag{9}$$

where t is the total number of particles, C is the concentration, V is the volume, and γ is the interfacial free energy. The subscripts 1 and 2 indicate liquid phases 1 (the top phase) and 2 (bottom phase), respectively, 3 indicates the cells, and i indicates the interface between phases 1 and 2.

Use of these equations requires assuming that the interface has a volume. Ideally, the interfacial volume is zero, but in practice there is some associated volume, for instance the volume of a packed monolayer of cells covering a unit of interfacial area. In the following, the volumes have been arbitrarily assumed to be as follows: $V_1 = V_2 = 1000V_i$.

Equations 3–9 allow calculation of the distribution of cells between two liquid phases and the interface on the basis of surface tensions. Figure 4 gives calculated values for the relative concentrations of cells at the air–water (A/W) and water–hydrocarbon (W/H) interfaces in a simulated fermentation as the surface tension decreases as biosurfactant is produced. Figure 5 serves to indicate the types of changes in cell location which could be observed in a similar experiment. The predominant feature of this three-phase system is that cell distribution is sharply dependent on surface and interfacial tensions. In many experimental systems, most cells tend to accumulate at the interface between immiscible fluids [20].

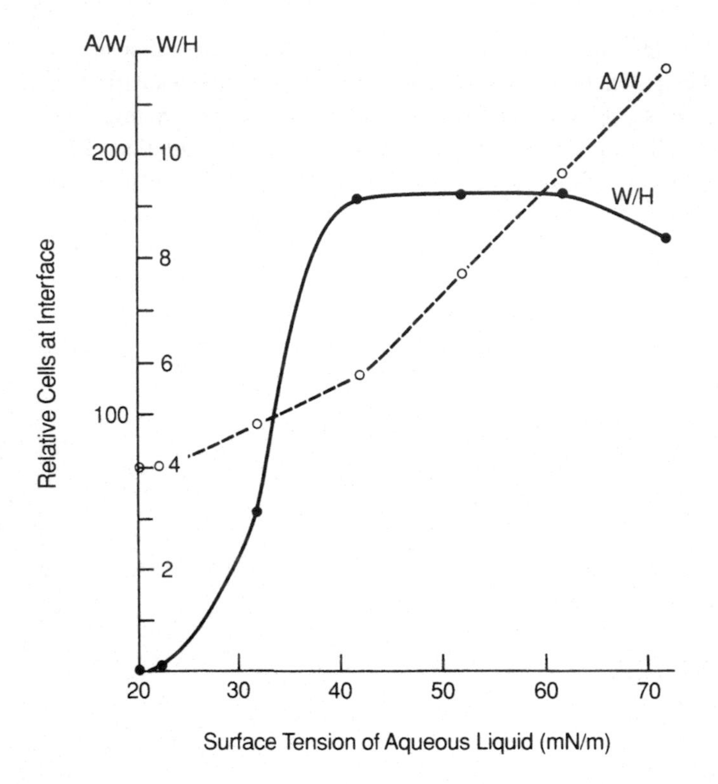

FIG. 4 Calculated values of the relative numbers of cells at the air–water (A/W) and water–hydrocarbon (W/H) interfaces for a typical hydrocarbon fermentation. Accumulation of microbes at the interfaces affects mass transport of both oxygen and hydrocarbon from the medium to the cells.

VIII. SOLID SUBSTRATES

Figure 6 diagrams microbial attachment to a solid substrate such as wax. Suspended cells adhere to the solid surface, probably by contact with the glycocalyx. If a steady state is achieved by mixing or flow, there is a distribution of the cells between the suspending medium and the solid surface in proportions depending on the interfacial free energies of the cell and the solid substrate [16, 19, 21]. The more negative the free energy of adhesion, the greater the proportion of cells attached to the surface (Fig. 6). The calculation of the free energy of adhesion is described by

$$\Delta G_{\text{adh}} = -\Delta G_{\text{sep}} = \gamma_{cs} - \gamma_{sl} - \gamma_{cl} = \alpha \log(C_{\text{attached}}/C_{\text{free}}) \tag{10}$$

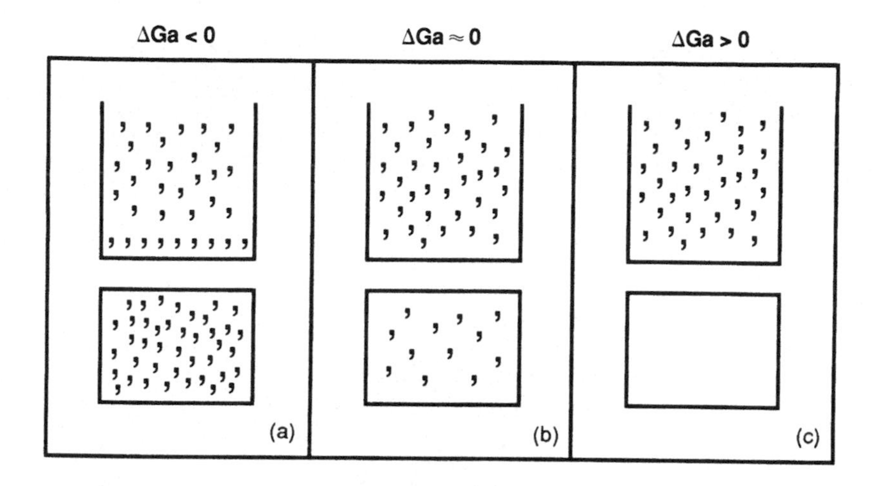

FIG. 5 Adhesion of microbes to insoluble substrates as a function of the free energy of adhesion. Side and bottom views of container with suspended microbes and various bottom surfaces. (a) Free energy of adhesion is negative and many microbes adhere. (b) Free energy of adhesion is close to zero and a few microbes adhere. (c) Free energy of adhesion is positive and no microbes adhere.

where ΔG_{adh} is the free energy of adhesion and ΔG_{sep} is the free energy of separation, c refers to cell, s to solid, and l to liquid, and α is the proportionality constant.

The production of surfactants by cells utilizing solid substrates results in the alteration of the interfacial free energies at both the cell and solid substrate surfaces. Depending on the nature of the surfactant and the solid surfaces, there can be a shift in the steady-state distribution of cells between free cell suspension and attachment to the solid. Some biosurfactants will coat the surfaces of the cells and the solids, as indicated in Fig. 6, either decreasing or increasing the degree of attachment to the surface of the solid substrate. Enhanced adhesion of the cells to the solid substrate would be expected to encourage biodegradation of the solid.

In many cases, it may be desirable to diminish microbial adhesion to solid surfaces. The prospects for the detachment of microbes from solid surfaces have many practical applications in both industrial and medical situations. For detachment to occur, the free energy of separation, ΔG_{sep}, must be negative. The free energy of separation can be influenced by alterations in the interfacial free energies of the system (Fig. 6).

There are several strategies that could enhance detachment:

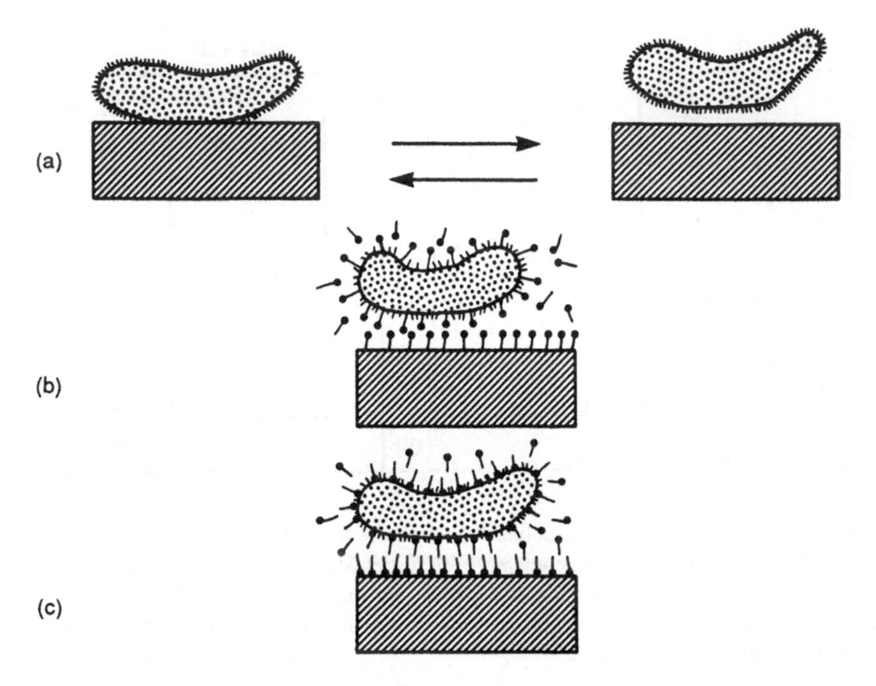

FIG. 6 The effect of surfactants on microbial adhesion to solid surfaces. Microbes attach or detach depending on the free energy of adhesion. Biosynthesis of biosurfactants can coat both the cell and the solid substratum, either enhancing or inhibiting attachment. The interaction of the surfactant with the cell surface and the solid substratum depends on the chemical nature of the surfactant and the surface energies of the cell and substratum.

1. A small solute could be added to the aqueous medium to reduce the surface tension to a value between that of the cell and the solid, which should result in separation if there are not major mechanical or surface charge components to the mechanism of adhesion [22]. If the solid and cell surface energies are known, the target values for the interfacial tensions can be calculated with the equations described below.

2. Amphiphilic molecules, such as water-soluble detergents, can also reduce the interfacial free energies or can coat both the bacteria and the substratum. If the resulting interfacial free energies cause the free energy of separation to become negative, separation will be enhanced. However, conventional surfactants are less effective than small solutes in penetrating into the interface of existing adhesive bonds.

3. In cases where the adhesion of the cells to the surface is enhanced by the production of biosurfactant, it may be possible to prevent attachment or cause

detachment by inhibiting the production of biosurfactant [23]. In medical situations, effective killing of certain adherent microbes by antibiotics may be difficult, and an alternative strategy for treatment may include consideration of metabolic inhibitors which do not kill the microbe but do inhibit the production of the relevant biosurfactants.

IX. CONCLUSIONS

Microbial utilization of insoluble substrates is often accompanied by the production of biosurfactants. The main effect of the biosurfactants is to reduce the interfacial tension at the surface of the insoluble substrate and to increase the availability of the substrate to the microbe. For liquid hydrocarbons, reduced interfacial tension facilitates emulsification, increasing the surface area available for dissolution, microbial attachment, and substrate absorption. For solid substrates, biosurfactants increase the wettability of the surface, enhancing microbial colonization and substrate utilization.

Thorough analysis of interfacial free energies by measurement of the liquid surface and interfacial tensions and determination of the surface energies of the cells and relevant solids provides an analytical procedure for developing an understanding of the physical processes involved in the microbial attack of insoluble substrates. By combining the techniques of microbiology and biochemistry with the methods of surface physics reviewed above, it is possible to increase our understanding of the physical and biological mechanisms involved in the microbial degradation and deterioration of insoluble liquids and solids.

REFERENCES

1. D. F. Gerson and J. E. Zajic, *Proc. Biochem. 14*: 20–29 (1979).
2. D. F. Gerson and J. E. Zajic, *ACS Symp. Ser. 106*: 29–57 (1979).
3. D. F. Gerson, in *Methods in Immunology*, Vol. II (I. Lefkovits and B. Pernis, eds.), Academic Press, New York, 1981, pp. 105–138.
4. D. F. Gerson and J. E. Zajic, *Dev. Ind. Microbiol. 19*: 577–799 (1977).
5. D. F. Gerson and J. E. Zajic, *Antonie van Leeuwenhoek 45*: 81–94 (1979).
6. L. A. Behie, J. E. Zajic, D. Berk, R. J. P. Brouze, and V. A. Naish, *Water Poll. Res. J. Can.* 27–49 (1977).
7. J. E. Zajic, H. Guignard, and D. F. Gerson, *Biotech. Bioeng. 19*: 1285–1301 (1977).
8. J. E. Zajic, H. Guignard, and D. F. Gerson, *Biotech. Bioeng. 19*: 1303–1320 (1977).
9. W. A. Zisman, *ACS Adv. Chem. Ser. 43*: 1–32 (1964).
10. A. W. Neumann, R. Good, C. Hope, and M. Sejpal, *J. Coll. Int. Sci. 49*: 291–304 (1974).
11. C. Van Oss, *Chem. Rev. 88*: 927 (1988).
12. D. F. Gerson, *Colloid Polymer Sci. 260*: 539–544 (1982).

13. C. Van Oss, C. Gillman, and A. W. Neumann, *Phagocytic Engulfment and Cell Adhesiveness as Cellular Surface Phenomena*, Dekker, New York, 1975.
14. D. F. Gerson, *Biochim. Biophys. Acta. 602*: 269–280 (1980).
15. D. F. Gerson and J. Akit, *Biochim. Biophys. Acta. 602*: 281–283 (1980).
16. D. F. Gerson and D. Scheer, *Biochim. Biophys. Acta. 602*: 506–570 (1980).
17. R. J. Neufeld, J. E. Zajic, and D. F. Gerson, *J. Ferment. Technol. 61*: 315–321 (1983).
18. D. F. Gerson, in *Physicochemical Aspects of Polymer Surfaces*, Vol. 1 (K. L. Mittal, ed.), Plenum, New York, 1983, pp. 229–340.
19. D. F. Gerson, C. Capo, A. M. Benoliel, and P. Bongrand, *Biochim. Biophys. Acta. 692*: 147–156 (1982).
20. H. Walter, D. E. Brooks, and D. Fisher, *Partitioning in Aqueous Two-Phase Systems*, Academic Press, Orlando (1985).
21. B. Mely-Goubert, D. Bellgrau, and D. F. Gerson, *Cell Biophysics. 13*: 65–73 (1988).
22. P. J. Facchini, F. DiCosmo, L. G. Radvanyi, and Y. Giguere, *Biotech. Bioeng. 32*: 935–938 (1988).
23. D. F. Gerson, D. Cooper, B. Ramsay, and J. E. Zajic, *Can. J. Microbiol. 26*: 1498–1500 (1980).

11

Surface Properties and Function of Alveolar and Airway Surfactant

SAMUEL SCHÜRCH Respiratory Research Group, The University of Calgary, Calgary, Alberta, Canada

MARIANNE GEISER and PETER GEHR Department of Anatomy, University of Berne, Berne, Switzerland

I. ALVEOLAR SURFACTANT

The pulmonary air–liquid interfacial film reduces the surface tension in the parenchyma to less than 1 mN/m on lung deflation. In addition to a low and stable surface tension, interdependence provided by the fibrous network enables the lung to maintain a large alveolar surface area (approximately 140 m^2 in the adult human lung) necessary for efficient gas exchange [1].

Pulmonary surfactant of mammalian lungs consists of a variety of macromolecular complexes comprised of lipids and specific proteins. Surfactant is synthesized by the alveolar type II cell, in which it is stored in lamellar bodies [2]. These lamellar bodies, when secreted into the alveoli, form tubular myelin [3], which appears to be the principal precursor of the surface film that lowers the surface tension. Lamellar bodies and tubular myelin, both contain lipid and protein components of surfactant. However, the compressed surface film is thought to consist primarily of dipalmitoylphosphatidylcholine (DPPC) [4]. Surfactant obtained through bronchiolar lavage contains approximately 90% lipid, 10% protein, and small amounts of carbohydrate. DPPC, which accounts for approximately half the lipid in surfactant, is primarily responsible for the surface tension-reducing property of the surfactant complex. The synthesis, secretion, and metabolism of DPPC and other surfactant lipids has been the subject of recent reviews [5]. The proteins associated with surfactant have been designated SP-A, SP-B, and SP-C by Possmayer and his associates [6]. The protein SP-A is variably glycosylated with a molecular mass of 28 to 36 kDa (reduced). It is relatively water-soluble, but SP-B and SP-C remain with the lipids extracted with organic solvents. Protein SP-B has a molecular mass of 15 kDa (nonreduced) and SP-C has a molecular mass of 3–5 kDa in the nonreduced or reduced states.

Surfactant proteins directly affect the interfacial properties of surfactant lipids. Recent studies imply that surfactant-associated protein A and the recently identified SP-D have other important roles in the lung, including immunological defence [7]. Rapid adsorption of surfactant phospholipids to the air–liquid interface is thought to be critical for maintaining the morphological integrity of the gas exchange region of the lung [8]. Purified SP-B [9], SP-C, or mixtures of the two proteins [10] substantially enhanced the rate of formation of a surface film at an air–liquid interface *in vitro*; this action was further enhanced by the addition of SP-A [11]. Preparations of phospholipids containing SP-B alone were more effective than similar preparations containing SP-C in reducing surface tension in a pulsating bubble surfactometer [10]. Preparations of surfactant lipids containing mixtures of SP-B and SP-C were shown to increase lung compliance and preserve the morphological integrity in the distal airways in prematurely delivered ventilated fetal rabbits [8]. Similar lipid extract surfactants have been widely tested in clinical trials and shown to improve oxygenation and decrease the need for respiratory support in infants suffering from respiratory distress syndrome [12]. However, the precise assignment of specific roles for SP-A, SP-B, and SP-C in the surface activity of surfactant lipids has not yet been clarified [7]. Ongoing research on the interactions between the purified surfactant proteins and individual surfactant lipid components will substantially enhance our knowledge about surfactant surface activity in the near future.

Lung surfactant films at 37°C reach lower surface tensions more readily and appear to be more stable in alveoli than in surface balances, in spite of considerable efforts to mimic natural film behavior in these test systems. This has been known since the pioneering studies of Pattle [13] and Clements [14]. Although films of lung surface-active material at 37°C can develop near-zero surface tensions upon dynamic compression in a Langmuir-Wilhelmy surface balance [15] or in a pulsating bubble surfactometer [16], a relatively high speed of film compression and a large film area reduction (50% or more) appear necessary. In addition, if the areas of compressed films with low surface tension are kept constant, the surface tension spontaneously and rapidly increases, presumably due to film collapse and/or leaks over and around surface barriers [4]. Within approximately 20 min, the surface tension reaches a high limiting value between 20 and 25 mN/m [14].

In contrast, data from intact lungs imply a far greater stability [17]. By inference from increases in transpulmonary pressure of lungs kept at 40% of total lung capacity, the spontaneous increase in alveolar surface tension appear to be very small, amounting to only 1–2 mN/m during a 20-min period. More direct evidence for high film stability in the lung *in situ* has been provided by measurements of alveolar surface tension, using the microdroplet method [18, 19] as shown in Fig. 1. In the intact lung, surface tensions below 1 mN/m can be achieved by very modest reductions of alveolar surface area by quasistatic deflation, regardless of whether the measurements are made at room or body temperature [20].

Detailed information about the microstructure of the lung [21] prompted Wilson to design a new lung model that explained the structural and functional relations observed in the lungs examined and allowed him to calculate surface tension from pressure–volume loops and morphometric data [22]. This model and the analysis

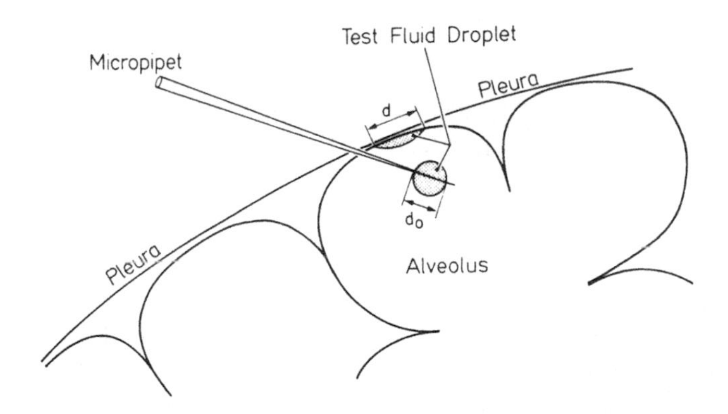

FIG. 1 Direct determination of alveolar surface tension by the microdroplet method. A test fluid droplet before (diameter d_0) and after (diameter d) placement onto an alveolar surface. From the relative diameter of the droplet, d/d_0, the surface tension can be determinated from a calibration curve.

of the relations among lung recoil, surface area, and surface tension takes into account the energy of distortion of the lung caused by surface tension. Not only does the surface tension depend on the surface area—this follows from the surface chemistry of insoluble monolayers—the structure of the lung is deformable; therefore, the surface area of the lung depends in turn on surface tension. This is called tissue–surface tension coupling [22]. Surface tension can act in two ways on lung mechanics: directly, by providing an additional contractile force, and indirectly, by modifying the alveolar geometry. Surface-tension forces tend to minimize the surface free energy by reducing the surface area; therefore, any local curvature will be reduced to the minimum compatible with counteracting forces provided by the septal fiber system. In normal air-filled lungs, at higher inflation levels, alveolar surfaces appear smooth because the relatively high surface tension flattens the capillaries. At low inflation levels, capillaries bulge toward the alveolar space, producing surface crumpling. This results in high local curvatures that are tolerated because the surface tension approaches near-zero values at low lung volumes [20].

Figure 2 illustrates the interrelationship of surface tension and lung structure on the alveolar level. The transpleural microscopic view of the same alveolus is shown on inflation (left) and on deflation (right) with corresponding scanning electron micrographs. The droplet demonstrates a higher surface tension on inflation (greater diameter) than on deflation for a given volume, here at 40% and 80% total lung capacity (TLC). Differences in alveolar surface texture are associated with differences in surface tension at the same lung volume at 40% TLC. The scanning electron micrographs of lungs fixed at 40% TLC on inflation (left) and deflation (right) demonstrate more pronounced surface irregularities due to bulging capillaries on deflation.

II. ESTIMATING THE TENSION AT THE PULMONARY AIR–WATER INTERFACE: *IN VITRO* METHODS

A. The Langmuir-Wilhelmy Method

Modification of the Langmuir surface balance, incorporating a tightly fitting barrier and a Wilhelmy dipping plate to measure surface tension, were among the first and have been the most popular [14]. In addition to measuring the lowest achievable surface tension, this apparatus can also record surface tension-area relations of surface films. Barrier systems require large samples and large subphase volumes and they tend to *leak*, that is to say, they allow surface contents to creep along constraining walls. The wall and barrier material is usually Teflon, which has a surface-free energy of approximately 18 mJ/m^2 at the Teflon–air interface, and an interfacial free energy of approximately 50 mJ/m^2 at the

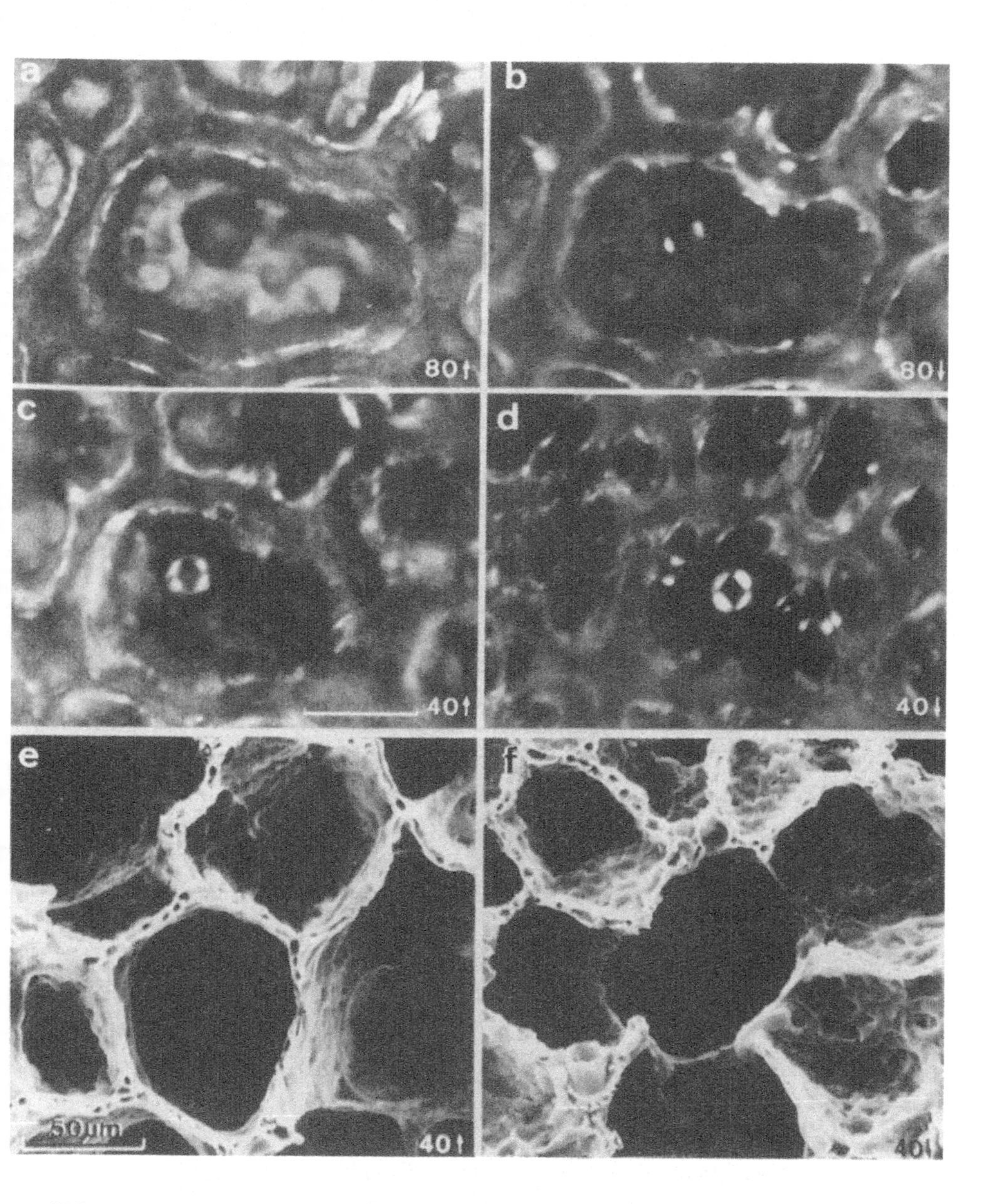

FIG. 2 A fluorocarbon (FC43, 3M Co) test fluid droplet placed onto a subpleural alveolar surface of a continuously perfused rabbit lung. The droplet was placed on inflation from zero pressure at 40% TLC, where the diameter indicates a surface tension of approximately 7 mN/m (c). At 80% TLC on inflation, the surface tension was about 23 mN/m (a). At 80% TLC on deflation the droplet diameter indicates a surface tension of 13 mN/m (b), and at 40% TLC on deflation, the surface tension was less than 2 mN/m (d). The difference in drop size is not perfectly visible because of light reflections by the almost hemispherical droplet, typical for close to zero surface tensions. Scanning electron micrographs of lungs fixed at 40% TLC on inflation (e) and deflation (f). Note more pronounced irregularities due to bulging capillaries in (f).

Teflon–water interface. At relatively low surface tension typically achieved by compressing pulmonary surfactant films, surface material tends to spread at the Teflon–water interface. This is the major pathway for film leaks (unpublished observation). This spreading of film material at the Teflon–water interface can be minimized by priming the barrier and trough by collapsing several layers of DPPC at minimum surface tension [15]. This promotes coating of the Teflon below the water level with the phospholipid and thus reduces the interfacial free energy of the Teflon–water interface from 50 to approximately 10 mJ/m^2. Films from DPPC or from surfactant material enriched with DPPC are remarkably stable when compressed to near-zero surface tensions in primed troughs. However, natural surfactant films are more fluid, and such films usually do not reach near-zero surface tension under quasistatic conditions.

B. The Pulsating Bubble Surfactometer

In an attempt to overcome difficulties with Langmuir balances, some investigators have been using small air bubbles formed on subsurface tubes in cuvettes containing as little as 20 μl of fluid [16]. Such bubbles can be pulsated, therefore allowing surface films to be expanded and compressed and lowest surface tension to be measured. It is also possible to record surfactant adsorption under dynamic conditions by following the surface tension at maximum bubble size at each pulsation. Surface tension-area relations could also be obtained; however, this would be quite difficult because of the small bubble size (approximately 0.5 mm diam). There are still problems with surface leaks up on the tube on which the bubbles are formed for the same reason as discussed above; the tubes are made of plastic materials. When surfactant films reach a surface tension below the free energy of the plastic, these films tend to creep along the plastic–air surface. This phenomenon was absent from Pattle's original bubble method [13], which made it more suitable for the measurement of film stability than the more modern apparatus just described.

C. The Captive Bubble Method

A refinement of Pattle's method, using bubbles 0.5–3 cm diam has now been developed to produce adsorption records, surface tension-area relations, and leak-free film stability measurements under quasistatic and dynamic conditions [23]. The apparatus contains a leak-proof bubble of controllable size to determine more reliable surface tension-area relationships (Figs. 3 and 4).

We reported recently [23] tests of the new captive bubble apparatus, using solvent-spread films of pure DPPC, and adsorbed films from lipid extract of bovine pulmonary surfactant and from purified rabbit surfactant. After only one to two compressions, the rabbit surfactant films exhibited the low surface tension,

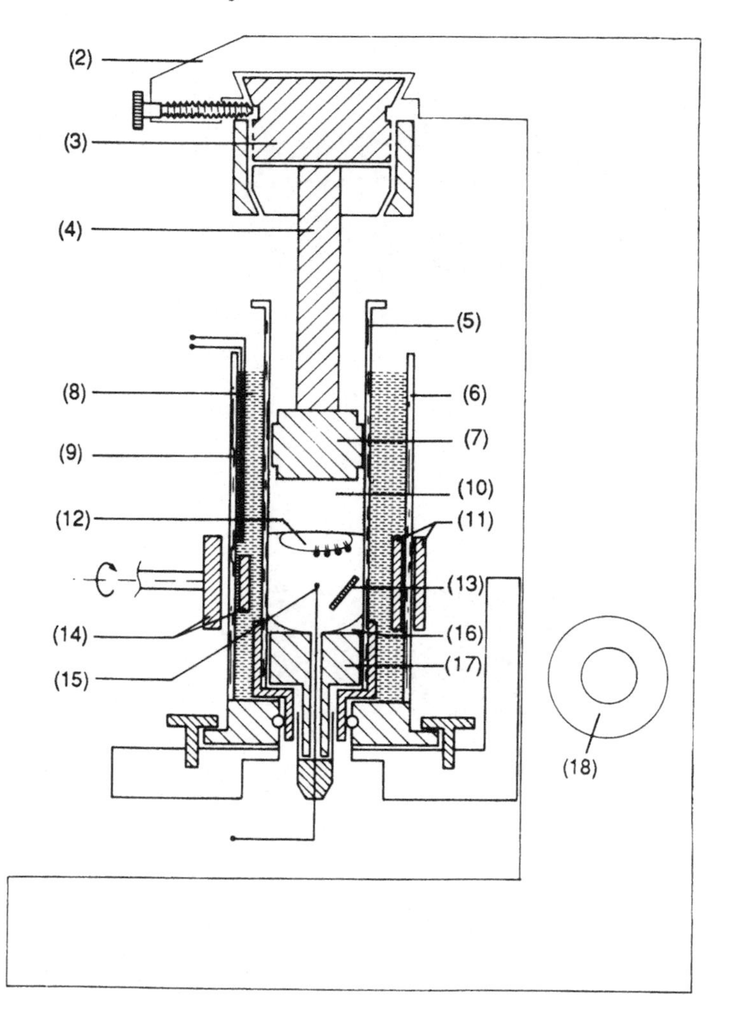

FIG. 3 The captive bubble surface tensiometer. (1), (2) Microscope stand. (3) Chuck for clamping syringe plunger onto microscope stand. (4) Plunger. (5) Glass cylinder (gastight syringe). (6) Plastic chamber with glass windows. (7) Teflon piston. (8) Water. (9) Heating element. (10) 1% agarose gel. (11) Magnets, position adjustable. (12) Air bubble. (13) Stir bar. (14) Magnetic stir bars. (15) Temperature probe. (16) 1% agarose meniscus or stainless steel funnel to prevent sticking of air bubbles. (17) Teflon plug. (18) Focusing knob. *Procedure*: A bubble of atmospheric air, 2 mm diam is formed below a slightly concave 1% agarose ceiling. The chamber, a 5-ml syringe (Unimetrics, gastight, Shorewood, IL) is closed pressure tight and mounted between the stage and nosepiece holder of a microscope stand. The focusing mechanism of the microscope stand is used to drive the syringe piston. The pressure in the chamber is reduced to expand the bubble to maximum size, 6–8 mm diam. The bubble area and therefore the surfactant film area decreases when pressure to the chamber is applied which decreases the bubble size.

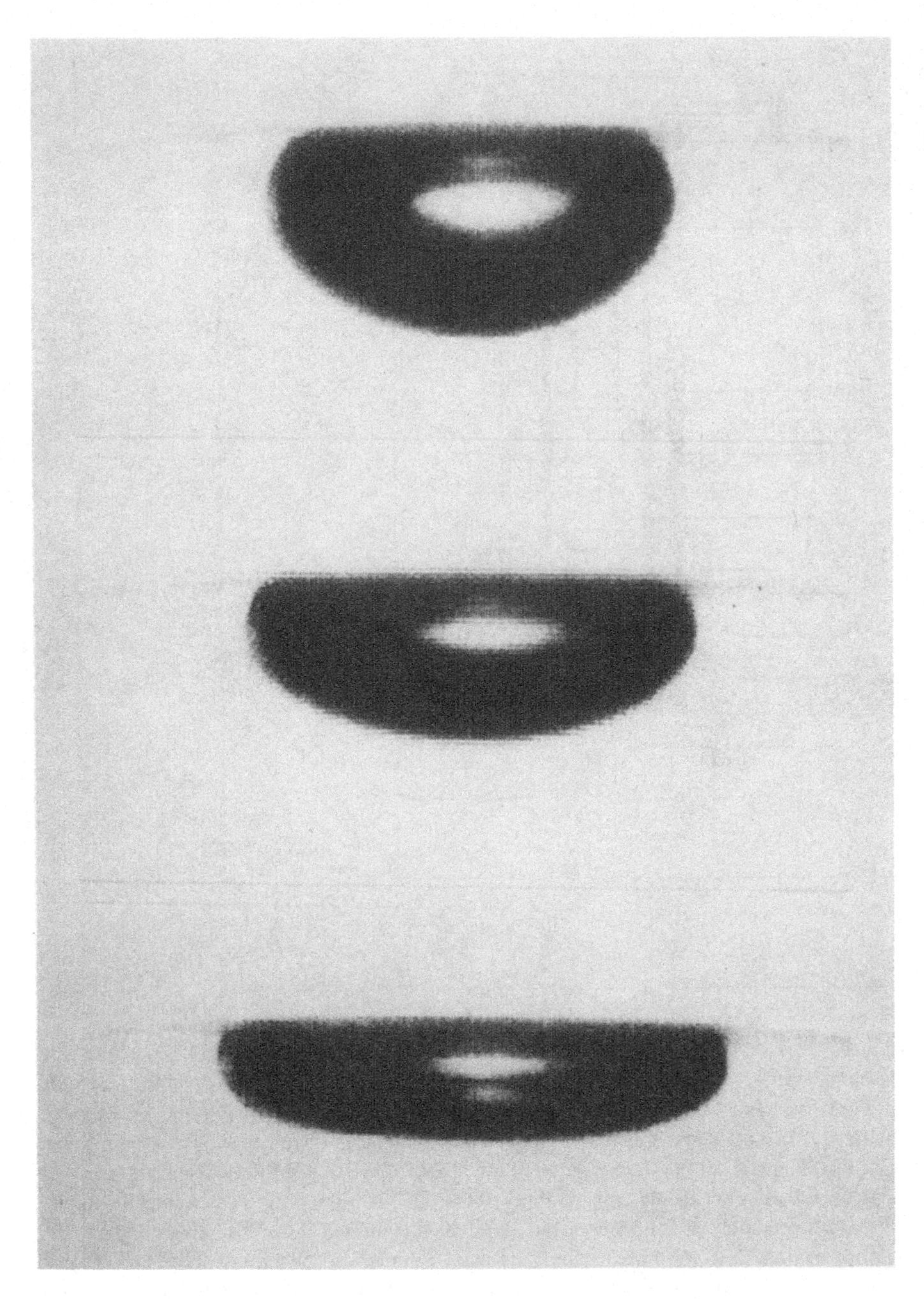

FIG. 4 Bubble shape at three different surface tensions, from top to bottom: 19.8, 11.8, 1.2 mN/m.

collapse rates, and compressibilities characteristic of the alveolar surface *in situ* and approaching the behavior of spread DPPC films.

III. SURFACE TENSION-AREA RELATIONS

Adsorption, hysteresis, and stability of films from purified rat pulmonary surfactant at phospholipid concentrations of 50, 200, and 400 μg/ml were studied in the captive bubble apparatus. At the highest concentration, adsorption was rapid, reaching surface tensions below 30 mN/m within 1 sec, whereas the lowest concentration required 3 min. Upon a first quasistatic (2–3 min) or dynamic (1 sec) compression, stable surface tensions below 1 mN/m could be obtained by a film area reduction of approximately 50%. After three or four cycles the surface tension-area relations became stationary, and the tension fell from 25–30 to 1 mN/m for a film area reduction of less than 20% (Fig. 5). Once the minimum tension was attained, hysteresis became negligible, provided the films were not collapsed at minimum surface tension. Under these conditions, the film could be cycled between 24 and 1 mN/m for more than 400 cycles without any noticeable loss of surface activity. After repeated cycling and collapse at minimum surface tension (collapse plateau), the results indicated that film material is displaced from the film, and the collapsed material does not participate in film formation upon bubble expansion. This observation might be relevant for artificial surfactant replacement therapies in cases of the respiratory distress syndrome in newborns. Maintaining a certain level of positive endexpiratory pressure might be beneficial for film stability, because with this procedure film collapse at minimum surface tension is avoided. It was also observed that, when the bubble clicks, the surface tension suddenly increased, and the bubble shape changed from flat to more spherical [23]. The associated isovolumetric decrease in surface area prevents the surface tension from rising as much as it would have in a constant area situation. This feedback mechanism could also have a favorable effect in stabilizing alveolar surface tension at low lung volumes, since Bachofen et al. [20] observed that alveolar septa tend to retract, decreasing alveolar surface area, when the surface tension is raised.

SP-A enhances adsorption and surface refinement of films from lipid extract surfactant. The effect of surfactant concentration and supplementation with SP-A on the surface activity of lipid extract surfactant (LES) was examined using the captive bubble technique. Adsorption of LES is strongly dependent on the phospholipid concentration. At 50 μg/ml, adsorption is slow, taking more than 30 min to reach values below 40 mN/m. At 100–200 μg/ml, adsorption to 25–26 mN/m was faster, taking between 20 and 30 sec, whereas at and above 800 μg/ml adsorption to 25–26 mN/m occurred within 1 sec. Addition of SP-A (1.0%–4.0%) to LES at low concentrations (200 μg/ml) dramatically increases the rate of

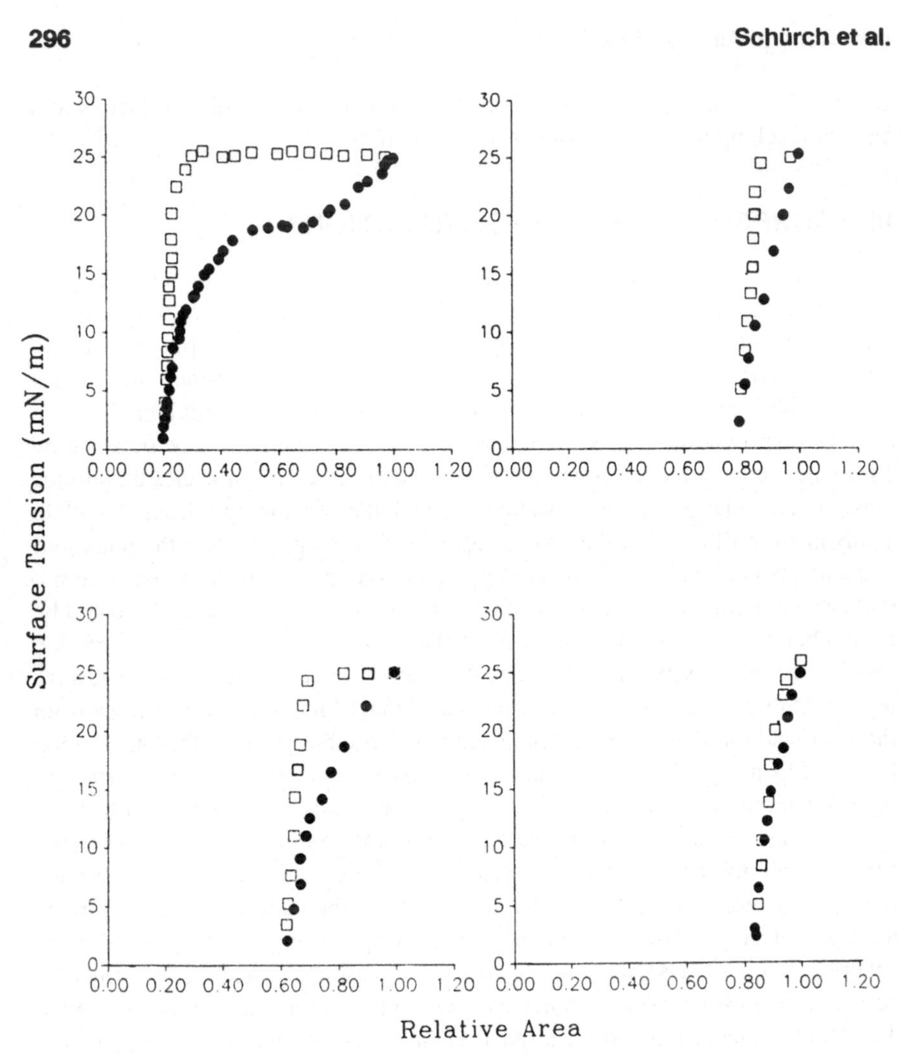

FIG. 5 Quasistatic surface tension-area relation obtained from rat pulmonary surfactant. Collapse at minimum surface tension was avoided by expanding the bubble after reaching a surface tension of 1–2 mN/m. Top left, first cycle; bottom left, second cycle; top right, third cycle; bottom right, fourth cycle. *Note*: decreasing area reduction necessary to reach near-zero tension upon consecutive cycling and reduced hysteresis. Filled circles, compression. Mean ± SEM; eight independent experiments.

adsorption (1 sec to 25–26 mN/m in the presence of calcium). In quasistatic cycling experiments, samples of relatively low phospholipid concentration (200 μg/ml) but with 1% SP-A added, require 20–30% less film area reduction than without SP-A to achieve minimum surface tensions of approximately 1 mN/m. The calculated film compressibilities at 15 mN/m imply that SP-A alters the surfactant film such that within four cycles, film compressibility is indistinguishable from that of a pure DPPC film.

In addition, SP-A reduces the incidence of clicking, indicating stabilization of the film at low surface tensions. In stationary surface tension-area hysteresis loops, produced by dynamic cycling, SP-A reduces the compression of the film area required to achieve low surface tension (1 mN/m) and eliminates the relatively flat compression part (squeeze-out plateau) at about 20 (mN/m) (Fig. 6). During dynamic hysteresis loops of LES films, the surface tension remains nearly constant as the film area starts to increase. This behavior is not compatible with that of a monolayer. The results imply that the surfactant film might change into a multilayer configuration, which can maintain very low surface tensions during initial area expansion.

IV. PARTICLE DISPLACEMENT AND AIRWAY SURFACTANT

Currently our knowledge of pulmonary surfactants is limited to the alveolar compartment and to large (accessible) airways. In both locations, the surface tension has been estimated by observing the spreading behavior of oil droplets. In alveoli, droplets were placed onto the surface film by micropipettes [19], whereas in large airways a bronchoscope was used for droplet placement and observation. By using this method, we have demonstrated the existence of a surfactant film lining the trachea and bronchi [24, 25]. This film has a surface tension of 30–32 mN/m.

In contrast to the extensive knowledge about alveolar surfactant, little is known with regard to surfactant in airways. Widdicombe [26] recently reviewed the possible sources and potential biological role of tracheal surfactants. He concluded that the trachea contains a complex mixture of lipids, including surface-active phospholipids. Since <1% of the alveolar phospholipid is carried up into the trachea and the composition of other lipids in the trachea is not characteristic of alveolar surfactant, Widdicombe [26] deducted that the lipids in the trachea must be derived from the tracheal epithelium and/or tracheobronchial glands. Gebhart et al. [27] suggested that cycling surface tension gradients provide a transport mechanism for clearance of inhaled particles. It seems likely that the peripheral airway zones are important in stabilizing the surfactant film and provide a

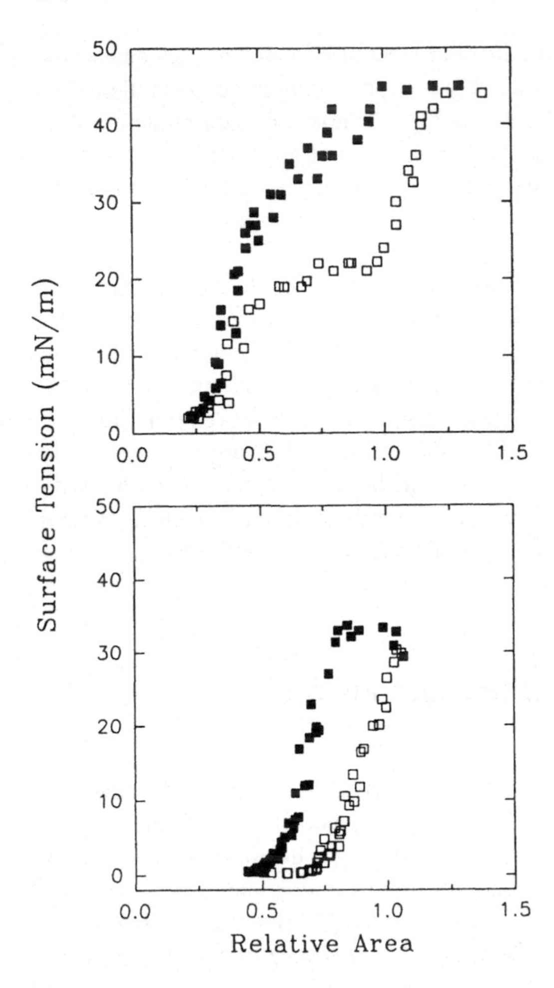

FIG. 6 The effect of SP-A on dynamic hysteresis loops of lipid extract surfactant (LES). Surface tension is plotted versus relative film area. The plots are the summaries of four consecutive dynamic cycles centering on cycle 20. Open symbols represent measurements during compression, filled symbols represent those made during expansion. Samples were suspended in 0.9% NaCl, 1.5 mM $CaCl_2$. Top: LES at 200 μg/ml. Bottom: LES at 200 μg/ml with 4% SP-A.

transport mechanism for particles from the peripheral airspaces to the mucociliary escalator.

V. PARTICLE SURFACTANT INTERACTIONS

Experiments conducted in the Langmuir-Wilhelmy balance and in *in vivo* experiments using aerosols of latex beads showed that the airway surfactant film promotes the displacement of particles from air into the aqueous phase and that the extent of particle immersion depends on the surface tension of the film [25]. Figure 7 shows particles placed onto a surfactant film at differing surface tensions.

An examination of electron micrographs with particles in peripheral airways or alveoli demonstrated that these particles are coated with an osmiophilic film (Fig. 8). This film extends over the whole surface of particles totally immersed in the aqueous phase as well as over the surface partially protruding into the airspace. It appears that our description of particle displacement has not been complete. We have dealt with the immersion process showing that the lower the film surface tension the greater is the extent of particle immersion because of more extensive wetting or film spreading over the particle surface. Regardless of the particle material, in the alveoli and likely in the peripheral airways, particles will be completely wetted by the film whose surface tension falls substantially (<1 mN/m) below that of the particle on expiration [19]. Remarkable, even in the central airways, we found the particles submerged in the aqueous phase, and the epithelium was frequently deformed, indicating that a force acting on the particle had been transmitted to the underlying cell layer.

To completely describe particle-surface film interactions, we have to refer to an additional force whose action is noticeable only for small particles of diameter less than 10μm. This force, called *line tension*, is a well-defined thermodynamic quantity whose action is characterized in the literature dealing with capillarity and nucleation in surface and colloid chemistry. In analogy to the surface tension defined for a two-dimensional surface, line tension is defined as the force operating in the one-dimensional three-phase line, or alternatively, as the free energy per unit length of the three-phase line [28]. An examination of the modified Young equation [28], which governs the equilibrium of a line element or edge when the three phases (particle, liquid, air) join, shows that the contact angle varies with the drop size because of the line tension. For a particle immersed more than 50% in the aqueous phase, the line tension promotes surface wetting. The excess free energy residing in the three-phase line is minimized by minimizing the length of this contact line. Therefore, the three-phase line is moved closer to the particle apex than it would have been without line tension.

In our work on line tension, we have used a system that can be applied to the polystyrene particles at the air–liquid interface modified by a surfactant film [29].

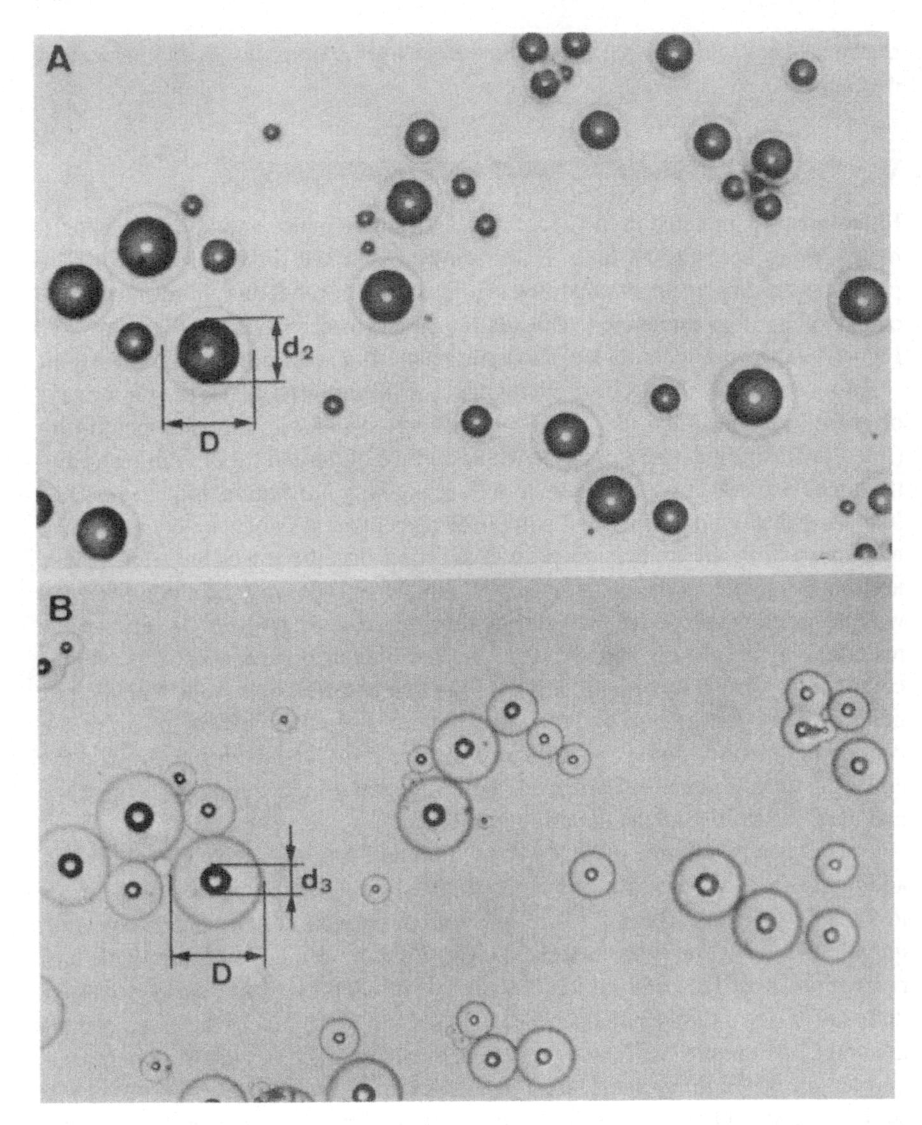

FIG. 7 Polymethylmethacrylate (PMMA) beads on a DPPC monolayer supported by an aqueous subphase (density 1.26 g/cm^3) at decreasing surface tensions. A: 40 mN/m, B: 30 mN/m. *Note*: The appearance of the labeled particle at decreasing film surface tensions, *D* indicates the total diameter (80 μm), *d* is the diameter of the segment exposed to air, this diameter is decreasing with the surface tensions, indicating increasing particle immersion.

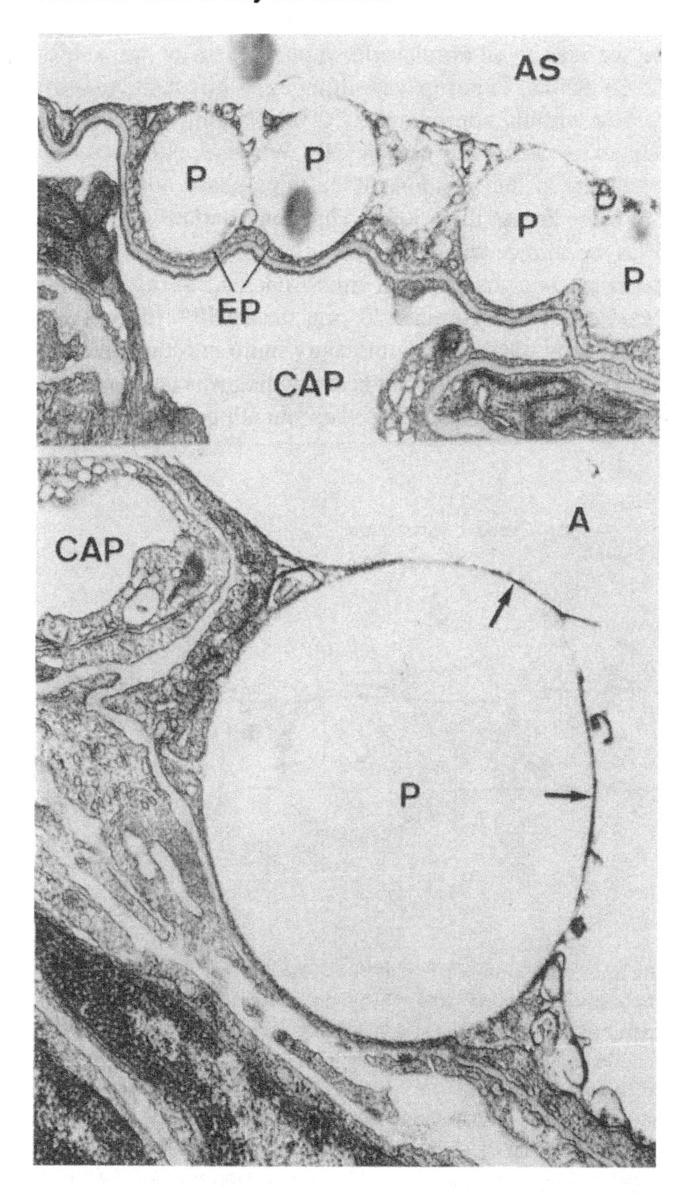

FIG. 8 TEM of alveolar wall showing 1-μm-diam polystyrene beads in an alveolar space (AS). The particles (P) have been displaced toward the epithelium (EP) and are completely covered by an osmiophilic layer (surfactant). Surface tension forces have caused deformation of the underlying capillary (CAP) by the particles (top). Higher magnification of same general area showing osmiophilic coating of a 3 μm particle, arrows (bottom).

Instead of a solid particle, we used an oil droplet whose surface tension was close to that of polystyrene, 32–35 mN/m. The drop was sitting on a surfactant layer of DPPC. Our value for the line tension, approximately 2×10^{-8} J/m, was in close agreement with that obtained by Torza and Mason [30]. We have estimated the contribution of the line tension to the position of the three-phase contact line (Fig. 9) for particles of a diameter less than 5 μm. This contribution was of the same order of magnitude as the surface tension forces.

In a recent pilot experiment, we placed small microspheres, average radius 4 μm, and larger microspheres, average radius 25 μm, on a DPPC film at 25 m/Nm. The smaller particles were displaced significantly more into the aqueous phase than the larger particles. In these experiments, everything was equal except for the particle size. Electron micrographs showed very smooth particle surfaces

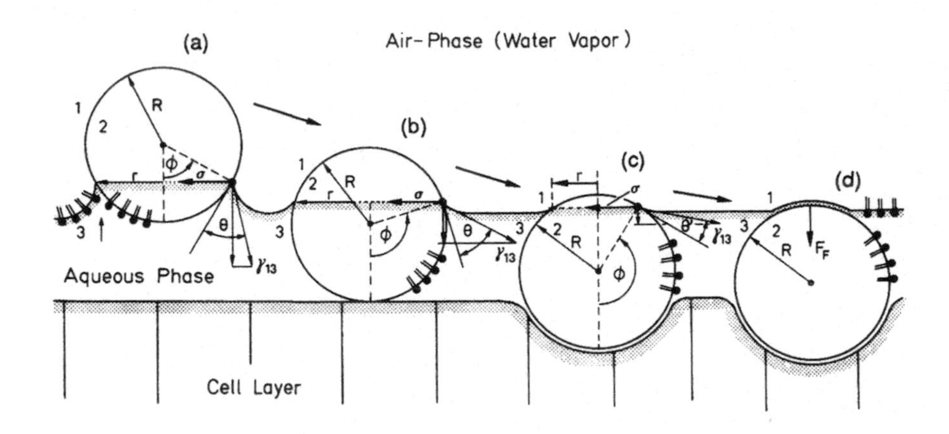

FIG. 9 (a) Situation during immersion immediately after deposition. The contact angle, θ, is characteristic for the particle surface tension (free energy, not shown) and the film surface tension, γ_{13}, at the air (1) particle (2) and water (3) contact line. The lower γ_{13} for a given particle surface material, the lower is θ. ϕ indicates the position at the three-phase line σ is the line tension, it is directed toward the center of the three-phase contact line (a circle here). In this situation, the particle equator is above the water level, σ tends to prevent wetting. However, the vertical component of γ_{13} is dominant here. (b) The particle is further displaced and contact with the cell layer is established. The line tension promotes wetting as the particle equator is below the water level. (c) The surface tension γ_{13} and the line tension σ promote further particle displacement, the cell layer is deformed by the particle. θ is substantially lower than the original θ because of the line tension contribution: σ promotes further particle wetting. (d) The particle is situated below the surfactant film, which may be considererd as an elastic film. The vertical force F_F proportional to the film surface tensions acts in vertical direction. Final deformation of the cell layer has occurred. There is no longer an air–particle–water three-phase line and, thus, no line tension.

with protrusions of less than 10 nm. Thus we attributed the above effect to the line tension, not to contact angle hysteresis.

In conclusion, line tension effects have to be considered for the interaction of micrometer-sized particles of different sizes, shapes, and surface characteristics with interfacial films and most likely for particle–membrane interactions. A summary of the forces involved in particle displacement at the airway surface are shown in Fig. 9.

VI. CONCLUSIONS

1. The studies on alveolar surfactant should enhance our understanding of the mechanics involved in film formation, in film structure, and film refinement. These should help to optimize the surfactant properties of exogenous surfactant to be used in the respiratory distress syndrome of the newborn and adult.

2. The studies on airway surfactant are focused on the ways in which particles interact with the site of initial deposition of inhaled particles, that is, on the mucous surface. This should provide new information on the structure and function of airway surfactant and particle kinetics. By comparing the interactions of particles with the mucous blanket under normal circumstances with their behavior under conditions of airway injury, new insights into the pathogenesis of airway disease can be obtained.

REFERENCES

1. P. Gehr, M. Bachofen, and E. R. Weibel, *Respir. Physiol. 32*: 121–140 (1978).
2. L. G. Dobbs, J. R. Wright, S. Hawgood, R. Gonzales, J. Venstrom, and J. Nellenbogen, *Proc. Natl. Acad. Sci. USA 84*: 1010–1014 (1987).
3. M. C. Williams, *J. Cell Biol. 72*: 260–277 (1987).
4. J. Goerke and J. A. Clements, *Handbook of Physiology*, Vol. III, 1986, pp. 247–261.
5. M. Post and L. M. G. Van Golde, *Biochim. Biophys. Acta 947*: 249–286 (1988).
6. F. Possmayer, *Am. Rev. Respir. Dis. 138*: 990–998 (1988).
7. T. E. Weaver and J. A. Whitsett, *Biochim. J. 273*: 249–264 (1991).
8. G. Grossman, R. Nilsson, and B. Robertson, *Eur. J. Pediatr. 145*: 361–367 (1986).
9. F. Possmayer and S.-H. Yu, *Biochim. Biophys. Acta 1046*: 233–241 (1990).
10. S.-H. Yu and F. Possmayer, *Biochim. Biophys. Acta 961*: 337–350 (1988).
11. A. M. Cockshutt, J. Weitz, and F. Possmayer, *Biochemistry 29*: 8424–8429 (1990).
12. A. Jobe and M. Ikegami, *Am. Rev. Respir. Dis. 136*: 1256–1275 (1987).
13. R. E. Pattle, *Nature 175*: 1125–1126 (1955).
14. J. A. Clements, *Physiologist 5*: 11–28 (1962).
15. J. N. Hildebran, J. Goerke, and J. A. Clements, *J. Appl. Physiol. 47*: 604–611 (1979).
16. G. Enhorning, *J. Appl. Physiol. 43*: 198–201 (1977).
17. T. Horie and J. Hildebrandt, *J. Appl. Physiol. 31*: 423–430 (1971).

18. S. Schürch, J. Goerke, and J. A. Clements, *Proc. Natl. Acad. Sci. USA 73*: 4698–4702 (1976).
19. S. Schürch, H. Bachofen, and E. R. Weibel, *Respir. Physiol. 62*: 31–45 (1985).
20. H. Bachofen, S. Schürch, M. Urbinelli, and E. R. Weibel, *J. Appl. Physiol. 62*: 1878–1887 (1987).
21. H. Bachofen, P. Gehr, and E. R. Weibel, *J. Appl. Physiol. 47*: 1002–1010 (1979).
22. T. A. Wilson, *J. Appl. Physiol. 50*: 921–926 (1981).
23. S. Schürch, H. Bachofen, J. Goerke, and F. Possmayer, *J. Appl. Physiol. 67*: 2389–2396 (1989).
24. P. Gehr, Y. Berthiaume, V. Im Hof, and M. Geiser, *J. Aerosol Med. 3*: 27–43 (1990).
25. S. Schürch, P. Gehr, V. Im Hof, M. Geiser, and F. Green, *Respir. Physiol. 80*: 17–32 (1990).
26. J. G. Widdicombe, *Eur. J. Respir. Dis. 67*: 1–5 (1985).
27. K. F. Gebhardt, H. Rensch, and H. Von Seefeld, *Progr. Respir. Res. 67*: 2234–2243 (1984).
28. D. Li and A. W. Neumann, *Colloids and Surfaces, 43*: 307–314 (1990).
29. J. A. Wallace and S. Schürch, *J. Colloids Interface Sci. 124*: 452–461 (1988).
30. S. Torza and S. G. Mason, *Kolloid-Z.u.A. Polymere 246*: 593–599 (1971).

12

Microbial Lipopolysaccharides

LINA CLOUTIER and NAIM KOSARIC Department of Chemical and Biochemical Engineering, University of Western Ontario, London, Ontario, Canada

I. INTRODUCTION

Because of the variety of unusual lipids found in bacteria, there is a great interest in studying these organisms as a potential source of microbial fats and oils. Bacterial lipids differ both quantitatively and qualitatively from the lipids of higher forms of life. Bacteria generally do not contain sterols because of their inability to form the steroid ring. They are, however, capable of synthesizing polyisoprenoids, which are found in quinone coenzymes, carotenoids, and undecaphenol phosphates (needed for peptidoglycan and lipopolysaccharide synthesis). Unlike plants and oleaginous species of yeasts and molds, bacteria do not accumulate triacylglycerols but instead produce specialized lipids like poly-8-hydroxybutyrate (PHB) or wax esters. Wax esters, like their animal and plant counterparts sperm whale oil and jojoba oil, are of interest in health care products and in the production of high-temperature and high-pressure lubricants [1]. Although high contents of triacylglycerol are found in some species of *Actinomycetales* (probably 80% of total lipids), bacteria still remain an unpopular source of edible lipids because of the presence of other complex lipids (some of which are potentially toxic) that make it difficult to extract and purify the desirable lipid component [2]. However, they are currently being studied as producers of biosurfactants [3]. As an example, the glycolipid, dimycolyltrehalose, isolated from these organisms has been studied for its potential application as a surfactant [4]. This lipid material contains complex C_{88} fatty acids (mycolic acids) attached to the disaccharide trehalose. Emulsan (an extracellular protein-associated lipopolysaccharide), produced by the bacterium *Acinetobacter calcoaceticus*, has been used for cleaning large oil tanks and tankers [2]. Another commercial application involving bacterial lipids is the intracellular biodegradable homopolymer of D-3-hydroxybutyrate, or PHB, produced principally by species of *Azotobacter, Pseudomonas*, and *Alcaligenes*. PHB offers an advantage over the chemical polymers, such as polyethylene and polypropylene, because it is biodegradable and has the ability to deemulsify both oil-in-water and water-in-oil emulsions.

Polyunsaturated fatty acids are generally absent in bacteria, with mono- and disaturated fatty acids being the predominant unsaturated fatty acids. The fatty acids of bacteria are generally 10–20 carbons in length with the 15–19 carbon chains predominating. Among the monosaturated fatty acids, the n-7 series (counting from the methyl group), is the predominant type with *cis*-11-vaccenic acid, $18{:}1\Delta^{11}$, being the most common. The C_{16} monounsaturated acid most frequently found is palmitoleic acid, $16{:}1\Delta^{9}$ [5]. Cyclopropane- and cyclopropene-containing fatty acids, branched fatty acids (iso and anteiso), and oxyfunctionalized fatty acids are also commonly found in bacteria [6]. Mycolic acids, for example found in the cell wall of actinomycetes are composed of β-hydroxy fatty acids substituted in the α position by relatively long hydrocarbon chains [7]. The simplest of them is corynomycolenic acid, which is found in *Corynebacterium diphtheriae*. Fatty acids have been shown to possess surface-active properties, and substituted fatty acids, like hydroxy acids produced during hydrocarbon fermentations, are much more effective in reducing surface tension than the normal types [3]. These fatty acids do not accumulate intracellularly but are found covalently linked to alcohols (e.g., glycerol and sugars) and amines (e.g., sphingolipids). With few exceptions (e.g., the extracellular rhamnolipid and storage polymers like poly-β-hydroxybutyrate), microbial lipids, both free and conjugated, are essentially associated with the cell's membrane systems [6].

Conjugated lipids, like glycolipids, are widely distributed in microorganisms. In bacteria, particularly in actinomycetes, mono- and oligosaccharides partially esterified by fatty acids (acylglycoses) are often encountered [6]. Acylated oligo- and polysaccharides, like lipopolysaccharides (or lipoglycans), are also found in Gram-negative and some Gram-positive bacteria. In Gram-positive bacteria (e.g., *Bacillus subtilis*), the most widely studied glycolipid, lipoteichoic acid, is located in the cytoplasmic membrane. The lipid moiety consists of a glycosyldiacylglycerol, which is attached to a hydrophilic glycerophosphate polymer [6]. In Gram-negative organisms, lipopolysaccharide in association with protein and phospholipids forms the external leaflet of the outer membrane bilayer. They account for approximately 1%–3% of the bacterial cell dry weight [8]. The lipid moiety, known as lipid A, anchors the polysaccharide portion to the membrane and is generally constructed from a disaccharide of D-glucosamine (2-amino-2-deoxy-D-glucose), to which are attached amide- and ester-linked fatty acids and of which many are D-3-hydroxyalkanoic and L-2-hydroxyalkanoic acids. The major hydroxy acids of lipid A are straight-chain, even-carbon compounds ranging from 10 to 21 carbons (the straight-chain, odd-carbon acids being the minor components). Branched-chain hydroxy acids have also been identified as part of lipid A in *Pseudomonas* spp., *Xanthomonas* spp., *Bacteroides* spp., *Desulfovibrio* spp. and some species of gliding bacteria. Another class of oxyfunctionalized fatty acids found in lipid A are the extremely rare 3-oxo-acids present in *Vibrio anguillarum* and two *Rhodopseudomonas* species. Table 1 lists a variety of fatty acids, found in the lipopolysaccharides of some Gram-negative bacteria [9]. Another type of lipopolysaccharide whose structure is poorly characterized is the extracellular emulsan synthesized by the bacterium *Acinetobacter calcoaceticus* RAG-1 grown on hydrocarbon. Its fatty acids (saturated and hydroxy saturated on C10 and C12) are joined mainly through ester linkages to polysaccharide that contains D-galactosamine and an unidentified aminouronic acid [10].

Glycosyldiacylglycerols are more widely distributed than acylglycoses and are found primarily in algae, photosynthetic bacteria, Gram-positive bacteria, and some fungi. In yeasts particularly, glycolipids are extracellular products. These lipids usually contain a mono- or disaccharide glycosidically linked to a hydroxy fatty acid.

Thus the naturally occurring microbial glycolipids appear to be a source of structurally unusual and sometimes extremely rare fatty acids. If compared to fatty acids of animal and plant origin, microbial fatty acids offer a potential for application as surface-active materials. The oxygenated fatty acids (in bacteria as well as in yeasts and molds) are very common and structurally very diverse. Table 2 lists

TABLE 1 Fatty Acid Composition of Lipopolysaccharides from Some Gram-Negative Organisms

	Fatty acid composition (% of total lipid A fatty acid)		
Organisms	3–OH Acids	2–OH Acids	Non-OH Acids
Escherichia coli	3–OH–14:0 (60%)	—	12:0 (16%) 14:0 (20%) 16:0 (4%)
Salmonella spp.	3–OH–14:0 (64%)	2–OH–14:0 (5%)	12:0 (13%) 14:0 (11%) 16:0 (7%)
Xanthomonas spp.	3–OH–i–13:0 (17%) 3–OH–12:0 (33%) 3–OH–i–11:0 (13%) 3–OH–10:0 (3%)	2–OH–i–11:0 (6%)	i– 11:0 (19%)
Pseudomonas maltophilia	3–OH–i–13:0 (32%) 3–OH–12:0 (19%) 3–OH–i–11:0 (16%) others (7%)	2–OH–i–11:0 (6%)	i– 11:0 (15%)
Bacteroides gingivalis	3–OH–i–17:0 (43%) 3–OH–16:0 (14%) 3–OH–i–15:0 (9%)	—	i– 15:0 (2%) 16:0 (25%) i– 17:0 (5%)
Vibrio paraheamolyticus	3–OH–14:0 (24%) 3:OH–12:0 (26%)	—	12:0 (16%) 14:0 (16%) 16:0 (11%)

Source: Ref. 9.

the naturally occurring oxygenated fatty acids and their respective sources. Some of these fatty acids are already in use commercially. For example wool grease (waxy coating on the surface of the fibers of sheep's wool), which is a source of α-hydroxy iso-acids and α-hydroxy normal acids, is used in soap manufacture [11].

In cosmetics and toiletries, fatty acids have a very broad range of application and essentially all categories of cosmetic products contain some type of fatty acid

TABLE 2 Oxygenated Fatty Acids

Total no. of carbon	Fatty acid: Systematic name	Fatty acid: Common name	Source
10	3-Hydroxydecanoic	—	*Pseudomonas pyocyanea*
12	12-Hydroxytetradecanoic	Sabinic	Savin juniper leaf wax
14	2-Hydroxytetradecanoic	—	Wool grease
	3,11-Dihydroxytetradecanoic	Ipurolic	*Ipomea purpurea*
16	2-Hydroxyhexadecanoic	—	Wool grease
	11-Hydroxyhexadecanoic	Jalapinolic	*Ipomea orizabensis,* Cruciferae seed oils
	16-Hydroxyhexadecanoic	Juniperic	Coniferous waxes
	16-Hydroxy-7-hexadecenoic	Ambrettolic	Musk seed oil
	15,16-Dihydroxyhexadecanoic	Ustilic A	*Ustilago zeae*
	9,10,16-Trihydroxyhexadecanoic	Aleuritic	Shellac
18	2-Hydroxyoctadecanoic	—	Wool grease
	12-Hydroxyoctadecanoic	—	Hydrogenation of ricinoleic acid
	18-Hydroxyoctadecanoic	—	Hydrogenation of kamlolenic acid
	9,10-Dihydroxyoctadecanoic	—	Castor oil
	9-Hydroxy-12-octadecenoic	—	Strophantus seed oil
	12-Hydroxy-9-octadecenoic	Ricinoleic	Castor oil
	18-Hydroxy-9,11,13-octadecatrienoic	Kamlolenic	Kamala oil
22	22-Hydroxydocosanoic	Phellonic	Cork
24	2-Hydroxytetracosanoic	Cerebronic	Brain cerobroside
	2-Hydroxy-15-tetracosenoic	—	Brain cerobroside

Source: Ref. 12.

or their derivatives. The three major categories of compounds derived from fatty acids and used in cosmetics are salt, esters, and amides. In Table 3 examples of fatty materials used in cosmetics and derived from fats and oils are given. Because of their wide range of solubility in water and oil, they function as emollients, emulsifiers, waxes, cleansers, and so on. Lanolin, which is a waxy substance obtained from wool grease that contains normal and branched alkanoic acids and hydroxy acids, finds wide application in creams. When combined with beeswax, which contains free cerotic acid, $C_{26}H_{52}O_2$ and can be neutralized with borax to form a water-in-oil emulsifier, an emollient cream results. Amine soaps of lanolin fatty acids are also used in shampoos and bubble baths because of their low level

TABLE 3 Cosmetic Materials Derived from Fats and Oils

Chemical type	Example
Triglyceride esters	Castor, sesame, peanut, safflower, soybean, coconut oils, and tallow
Fatty acids	Lauric, palmitic, myristic, stearic, oleic, linoleic, ricinoleic, hydroxystearic, arachidic, behenic, erucic, and arachiconic acids
Fatty alcohols	Lauryl, cetyl, stearyl, oleyl, ricinoleyl, linoleyl, lanolin, and tallow alcohols
Soaps	Sodium, potassium, ammonium, mono-, di-, and triethanolamine, mono-, di, and triisopropanolamine, and amino glycol salts of fatty acids
Detergents	Alkyl sulfates from coconut oil fatty acids, amide sulfonates, ester sulfonates, *N*-acyl sarcosinates, alkylolamides, amines, and alkyl β-amino propionates
Cationic antiseptics and softener rinses	Quaternary ammonium compounds, morpholinium compounds, and pyridinium compounds
Alkyl fatty acid esters	Isopropyl and butyl myristate, palmitate, stearate, oleate, and linoleate
Polyhydric alcohol esters	Propylene glycol, glyceryl, sorbitol, and sorbitan fatty acid esters
Ethoxylated fatty acids	Polyethylene glycol, mono- and difatty acid esters
Ethoxylated fatty alcohols	Polyethylene glycol ethers of cetyl, stearyl, oleyl, and lanolin alcohols
Ethoxylated sorbitan, esters	Tweens
Branched-chain high-molecular-weight alkyl esters	Hexadecyl myristate
Lanolin-derived fatty acids and fatty alcohols	Normal fatty acids (even-numbered C_{10} to C_{26}), iso fatty acids (even-numbered C_{10} to C_{28}), anteiso fatty acids (odd-numbered C_9 to C_{31}), and hydroxy fatty acids (even-numbered C_{12} to C_{20}) and fatty alcohols (aliphatic, sterol, and triterpenoid)
Lanolin derivatives	Lanolin fatty acids to form amine soaps, lanolin fatty alcohol mono- and polyesters of ricinoleic and linoleic acids, acetate esters, and ethoxylated ethers

of irritation, antimicrobial activity, and good foaming properties. Lanolin fatty acids esterified to monohydric alcohols (e.g., isopropanol) function as emollients and lubricants and also find utility in a wide range of cosmetic products [13].

Therefore, because of the wide range of applications of fatty acids in cosmetics, the use of microbial fatty acids as surface-active material becomes very attractive. These unusual acids or their derivatives could be used in all types of cosmetic products where they could function as emulsifiers, solubilizers, foaming agents, wetting agents, and detergents. Since microbes occur naturally, they are a renewable resource and a genetically manipulatable source of oleochemicals. With biotechnology, more economical or more effective processes for the production of this source of oleochemicals could be developed.

Lipopolysaccharides in particular are of interest as biosurfactants for application in cosmetics. They are found primarily in envelopes of Gram-negative bacteria.

II. ENVELOPE OF THE GRAM-NEGATIVE BACTERIA

As shown in Fig. 1 the envelope of the Gram-negative bacteria is composed of a cytoplasmic or inner membrane, a cell wall made up of a thin peptidoglycan layer and an outer membrane. Both membrane systems consist of a typical unit membrane having a thickness of about 75 Å. Gram-positive bacteria, on the other hand, are surrounded by a cytoplasmic membrane and a thick layer (about 200 Å) representing the cell wall or peptidoglycan [14].

Structurally the cell wall gives rigidity to the bacterial cell. This prevents cell lysis in an hypotonic environment, usually its natural habitat. Gram-positive bacteria have a very thick multilayered cell wall in comparison to the Gram-negative bacteria, which, however, possess an extra membrane system outside the peptidoglycan layer. In Gram-negative bacteria, only a monolayer of peptidoglycan surrounds the cytoplasmic membrane. The peptidoglycan enables the cells to withstand the osmotic pressure (approximately 3.5 atm) in the cytoplasm [15]. The outer membrane system, absent in Gram-positive bacteria, contains the lipopolysaccharide molecules, phospholipids, and a significant amount of proteins. The amount of phospholipid present was found to be large enough to cover one side of the lipid bilayer, and in *Salmonella typhimurium* it was calculated that there are approximately 2.5×10^6 molecules of lipopolysaccharide per cell. These molecules are located exclusively in the outer leaflet of the outer membrane and occupy approximately 45% of the surface of the outer membrane [14]. The segregation of lipopolysaccharide and phospholipid in the outer and inner membrane, respectively, has been demonstrated using ferritin-labeled antibodies [16] and confirmed by labeling with galactose oxidase [17]. The asymmetric arrangement of lipopolysaccharide across the outer membrane may not

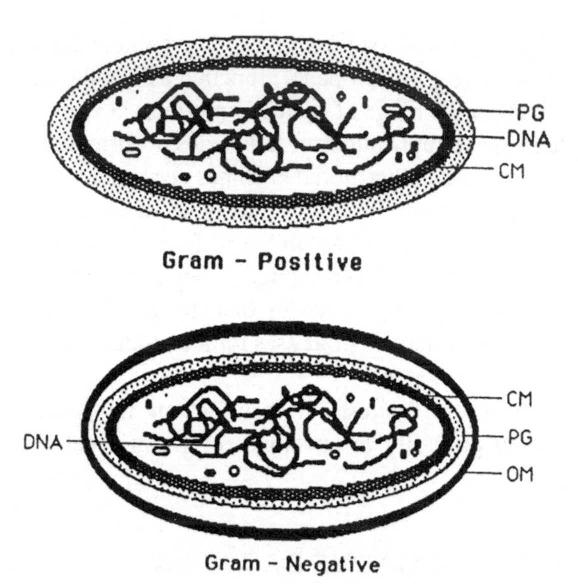

FIG. 1 Schematic representation of the membrane structure of Gram-positive bacteria.

have evolved in all Gram-negative bacteria as it has evolved in enteric bacteria as an adaptation to an environment containing hydrophobic compounds (e.g., bile salts) [18]. A schematic representation of a Gram-negative bacterial cell envelope is shown in Fig. 2 [9]. In this diagram the outer membrane is covalently attached to the peptidoglycan by lipoproteins.

One of the functions of the outer membrane of Gram-negative organisms is to confine hydrolytic enzymes (proteases, lipases, nucleases, glycosidases, phosphatases), binding proteins, and proteins involved in the degradation or modification of harmful components such as antibiotics and heavy metals. These proteins are contained outside the cytoplasmic membrane in the periplasmic space. Gram-positive bacteria excrete hydrolytic enzymes outside the cell. The outer membrane also provides passive channeling of nutrients and ions required for growth and serves as a selective barrier to the cell exterior (Gram-negative bacteria are more resistant to the action of certain dyes, chemicals, enzymes, and antibiotics) [14]. The low permeability to hydrophobic compounds is probably due to the lack of phospholipid in the outer membrane and the presence of long hydrophilic O-antigen polysaccharide chains. The permeability to hydrophilic compounds is regulated by the porin protein ($>10^5$ polypeptide per cell) [19].

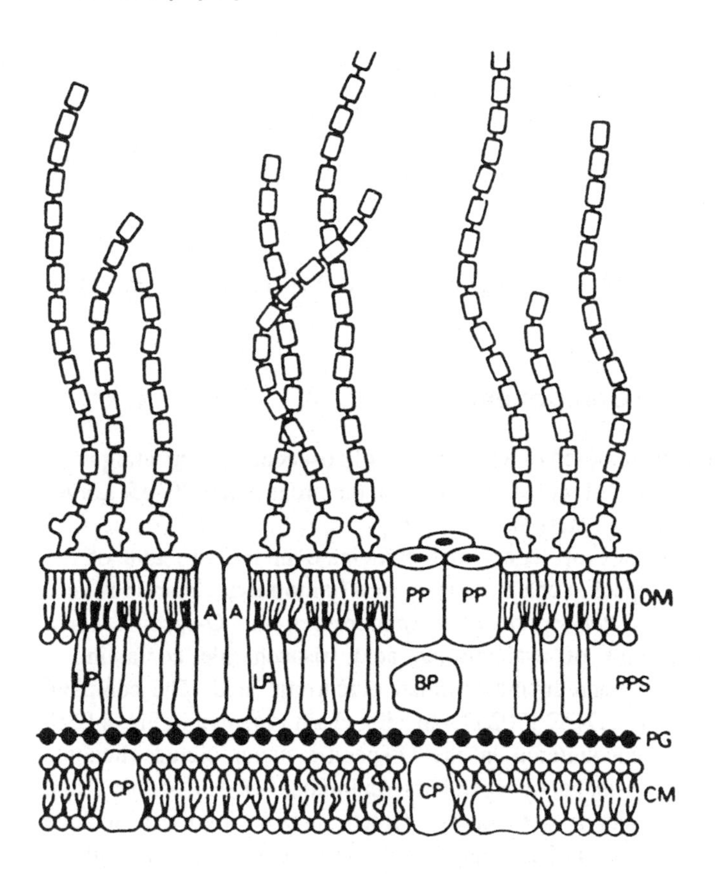

FIG. 2 Schematic representation of the molecular organization the cell envelope of Gram-negative bacteria. The components of the cell envelope are identified as follows; CM, cytoplasmic membrane; OM, outer membrane; CP carrier proteins; PP, pore forming trimeric proteins; A, transmembrane peptidoglycan-associated proteins; LP, lipoproteins; BP, nutrient-binding proteins; PG, peptidoglycan; PPS, periplasmic space. From Ref. 9.

III. CHEMICAL STRUCTURE OF LIPOPOLYSACCHARIDE

The general structure of lipopolysaccharide consists of a polysaccharide and a covalently bound lipid component termed lipid A. The polysaccharide portion is divided into two subregions, the O-antigen or O-specific chain and the core oligosaccharide. The core region is further divided into the inner and outer core. A schematic representation of the lipopolysaccharide structure is shown in Fig. 3. In the cell, the lipid A moiety is embedded in the outer leaflet of the outer bacterial

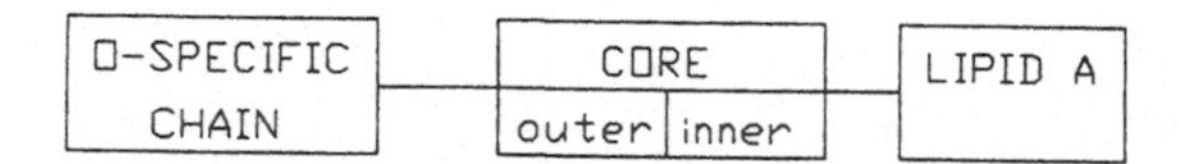

FIG. 3 General structure of lipopolysaccharide.

membrane through hydrophobic interactions with the O-specific chain projecting outside of the cell [20, 21].

A. Polysaccharide Component

The O-specific chain is composed of a polymer of oligosaccharide repeating units (n). On average, the number of repeating units is in the order of n = 20–35. Larger (up to n = 40) or smaller oligomers also exist [21].

There is a great diversity of sugars that participate in the structure of the O-specific chain. An O-specific chain made up of a homopolysaccharide is very uncommon. However, in *Rhodopseumonas sphaeroides*, the O-specific chain consists of repeating units of D-glucuronic acid trisaccharide containing a threonine residue on the nonreducing terminal glucuronic acid. The complete structure of *R. sphaeroides* ATCC 17023 O-specific chain is shown in Fig. 4 [22]. The sugars commonly encountered in the O-specific chain include hexoses (D-glucose, D-galactose, D-mannose), pentoses (arabinose, ribose, xylose), and their deoxy- and O-methyl derivatives. Other sugars involved include aminohexoses, aminopentoses, hexuronic acids, and hexosaminuronic acids, which may carry a variety of substituents like amino acids, phosphoryl, glyceryl, lactyl, and acetyl groups. Also the O-specific chain may differ in size due to a variation in the number of repeating units giving rise to a great heterogeneity in the lipopolysaccharide molecule [21].

For a given organism, the O-specific chain may or may not be present, giving rise to two types of lipopolysaccharide, the smooth and the rough lipopolysaccharide. In the smooth or S-form (wild-type lipopolysaccharide), the O-antigen is complete. In the rough or R-form lipopolysaccharide, the O-antigen is absent. The R-form lipopolysaccharide may also contain an incomplete core. The rough lipopolysaccharide arises from mutations in genes coding for enzymes involved in the biosynthesis of the O-antigenic chain and core oligosaccharide. Both the lipid A and proximal units in the inner core are necessary for the viability of the Gram-negative bacteria [23].

The core oligosaccharide is structurally less diverse than the O-specific chain. Although core types differ in their fine structure, they possess a general architecture. In *Salmonella*, the outer core region consists of a branched hexose

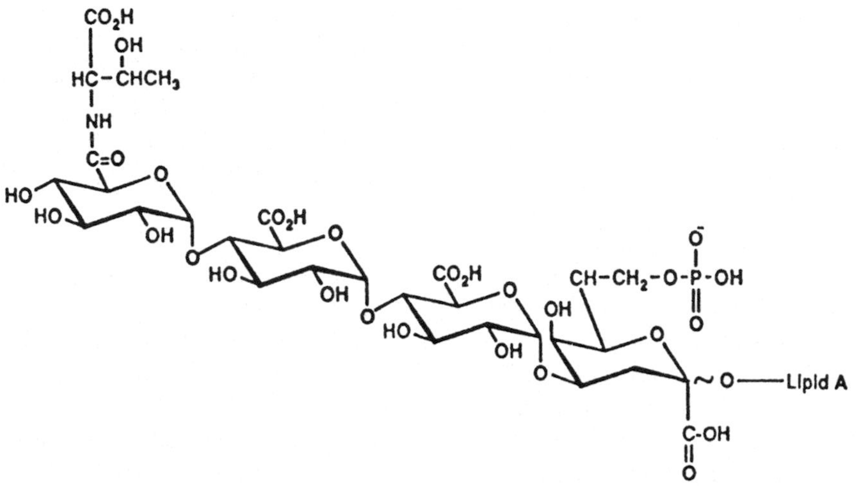

FIG. 4 O-specific chain and core from the lipopolysaccharide of *B. sphaeroides* ATCC 17023. From Ref. 22.

pentasaccharide of glucose and galactose, and the inner core region contains a trisaccharide of heptoses and 3-deoxy-D-mannooctulosonic acid (dOclA) residue that links the core to the lipid A. In some bacterial groups other than *Enterobacteriaceae* (e.g., *Pseudomonas, Bacteroides, Xanthomonas*), some core structures lack heptose or dOclA or both of these monosaccharides, suggesting a quite different core molecular structure [21]. As shown in Fig. 4, the core from *S. sphaeroides* ATCC 17023 consists of a single phosphorylated residue of dOclA.

B. Lipid Component

The lipid A moiety is structurally the most highly conserved part of the lipopolysaccharide molecule among bacterial species. It usually consists of phosphorylated β-1,6-linked D-glucosamine disaccharide backbone with ester- and amide-linked fatty acids. It is bound to the core oligosaccharide through a ketosidic linkage formed by the 3-deoxy-D-mannooctulosonate residue and the 6′-hydroxyl of the nonreducing glucosamine. In some organisms (e.g., *Rhodopseudomonas palustris, Rhodopseudomonas viridis*), the backbone is replaced by an *N*-acylated, 2,3-diamino-2,3-dideoxy-D-glucose [21].

In *Enterobacteriaceae*, lipid A consists of a B-1′-6-linked D-glucosamine disaccharide phosphorylated in position 1 and 4′. Four moles of (R)-3-hydroxytetradecanoic acids are linked to the two glucosamine residues of the backbone.

Each glucosamine (GlcNI and GlcII) is substituted by one ester and one amine-linked hydroxy fatty acid. The 3-hydroxyl group of the ester-linked 3-hydroxytetradecanoic acid of GlcNII is always esterified by tetradecanoic acid, whereas that of GlcNI is not. The amide-bound 3-hydroxytetradecanoic acid at GlcNII is usually 3-O-acylated by dodecanoic or tetradecanoic acid while that of GlcNI is only substituted in some cases (by hexadecanoic acid). The lipid A structure from *Escherichia coli* is shown in Fig. 5 [24].

Within nonenterobacterial lipid A, one group exists that possesses a very distinct lipid A structure. These unusual lipid As are for the most part from nontoxic lipopolysaccharide or from lipopolysaccharide of reduced toxicity. They have first been isolated in the purple phototrophic bacteria, the *Rhodospirillaceae* [24]. These lipid A structures may contain a backbone sugar other than D-glucosamine, the amide-linked fatty acid may not be the usual (R)-3-hydroxy type but the rare 3- (or 4-) oxo fatty acid type, and the glucosamine backbone may carry other sugar constituents glucosidically attached and may be missing the phosphate groups. In the following, only those lipid A structures from lipopolysaccharide studied so far that have exhibited almost no or low toxic effects will be discussed.

The lipid A from *R. viridis, R. palustris* and *R. sulfoviridis* contains a 2,3-diamino-2,3-dideoxy-D-glucose backbone [24]. One of the two amino groups of

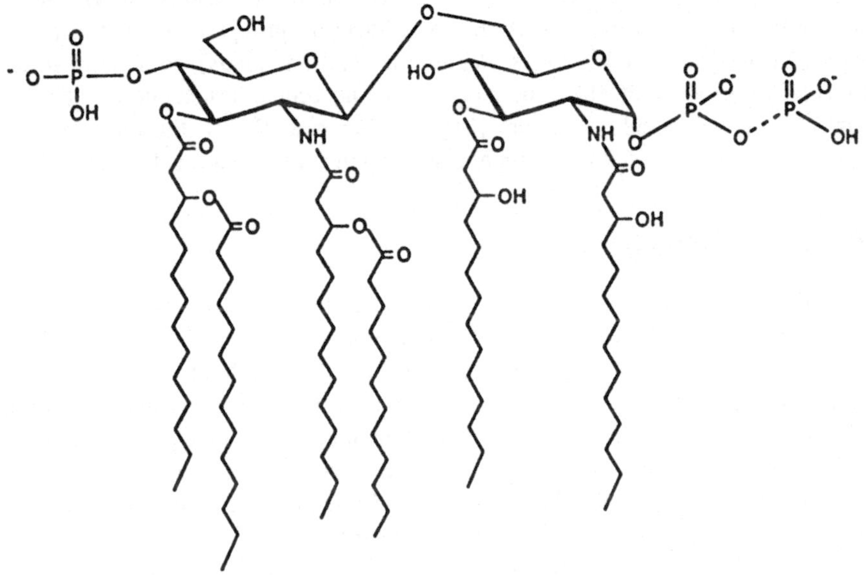

FIG. 5 Chemical structure of lipid A from *Escherichia coli*. From Ref. 24.

the backbone is substituted by a 3-hydroxy fatty acid (such as 3-hydroxy-tetradecanoic acid) and is essentially free of phosphorus. The chemical structure of free lipid A from *R. viridis* F is shown in Fig. 6. The same backbone disaccharide was found in lipid A from *Pseudomonas vesicularis, Nitrobacter hamburgensis, Nitrobacter winogradskyi*, and some soil bacteria. Lipid A from these nonphototrophic bacteria lack phosphate and contain ester- and amide-linked hydroxy- and oxo-fatty acids [24].

The lipid A from *Rhodopseudomonas capsulata* and *R. sphaeroides* was found to contain a 1,4′-biphosphorylated glucosamine disaccharide with an amide-bound 3-hydroxy-tetradecanoic acid and an ester-bound 3-hydroxy-decanoic acid. The chemical structure of *R. sphaeroides* ATCC 17023 is shown in Fig. 7. In this organism the amide-bound 3-hydroxy-tetradecanoic acid is O-acylated by an unsaturated fatty acid, $14{:}1\Delta^7$ [24].

A D-mannose containing and phosphate-free lipid A was found in *Rhodomicrobium vannielii, Rhodopseudomonas acidophila*, and in two species of the *Chromatiaceae* family, *Chromatium vinosum* and *Thiocapsa roseopersicina*. The rare 3-hydroxy-hexadecanoic acid is the dominating amide-bound fatty acid in *R. vannielii* and *R. acidophila*. The chemical structure of *R. vannielii* ATCC 17100 lipid A, whose backbone consist of a nonphosphorylated glucosamine disaccharide, is shown in Fig. 8 [24].

Lipid A structure from *Bacteroides* spp. was found to differ from *Salmonella*. The lipid A from *Bacteroides fragilis* and some closely related species (*B. thetaiotamicron, B. ovalis, B. distasonis*, and *B. vulgatus*), contain the following major fatty acids; isobranched pentadecanoic acid (exclusively ester-linked), 3-hydroxy-pentadecanoic acid, 3-hydroxy-hexadecanoic acid, 3-hydroxy-15-methyl-hexadecanoic acid, and 3-hydroxy-heptadecanoic acid. The

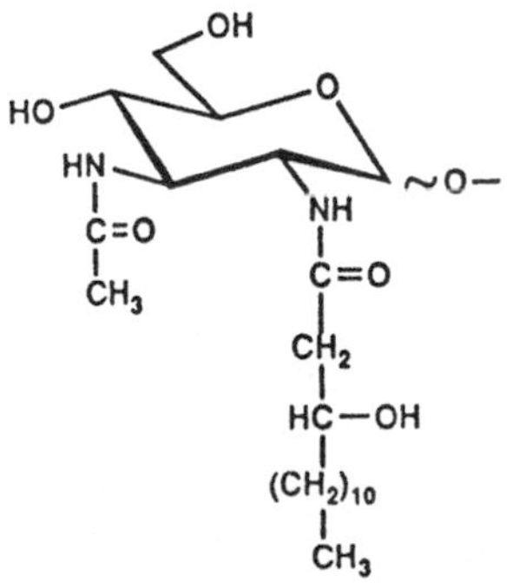

FIG. 6 Chemical structure of *Rhodopseudomonas viridis* F lipid A. From Ref. 25.

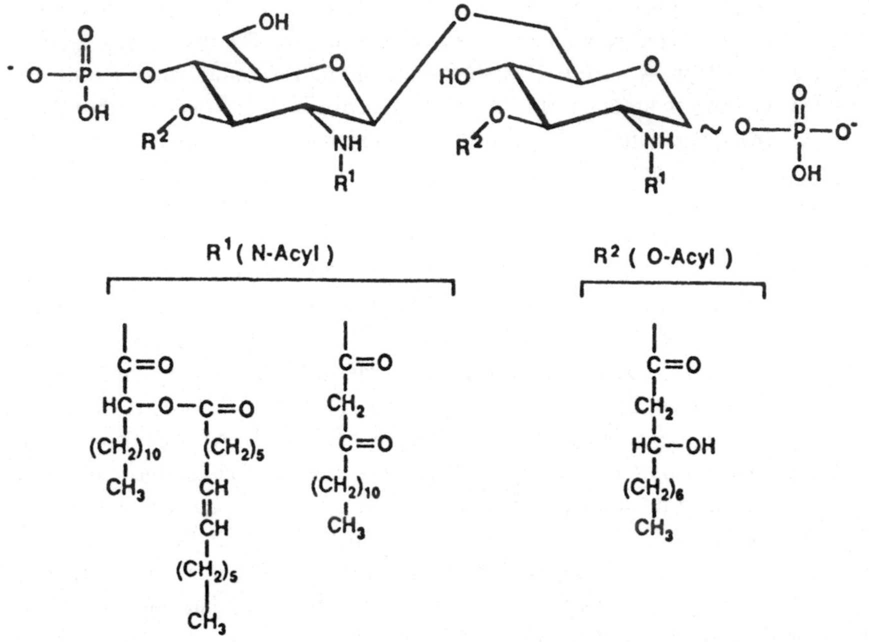

FIG. 7 Chemical structure of *Rhodopseudomonas sphaeroides* ATCC 17023 lipid A. From Ref. 25.

branched hydroxy fatty acids are exclusively amide-linked and the others are both ester- and amide-linked. The *Bacteroides* lipid A structure still remains unknown [24].

IV. LIPOPOLYSACCHARIDE BIOSYNTHESIS

The inner (cytoplasmic) membrane is the site of synthesis for lipopolysaccharide. Translocation to the outer leaflet of the outer membrane occurs at regions of contact (zones of adhesion) between the cytoplasmic membrane and outer membrane [14]. The O-specific chain, the core oligosaccharide, and the lipid A region of lipopolysaccharide are synthesized by different biosynthetic pathways. It is not known whether the pathway of lipopolysaccharide biosynthesis in *Rhodopseudomonas* follows the same pattern as in the *Enterobacteriaceae* which is the system most studied so far [20].

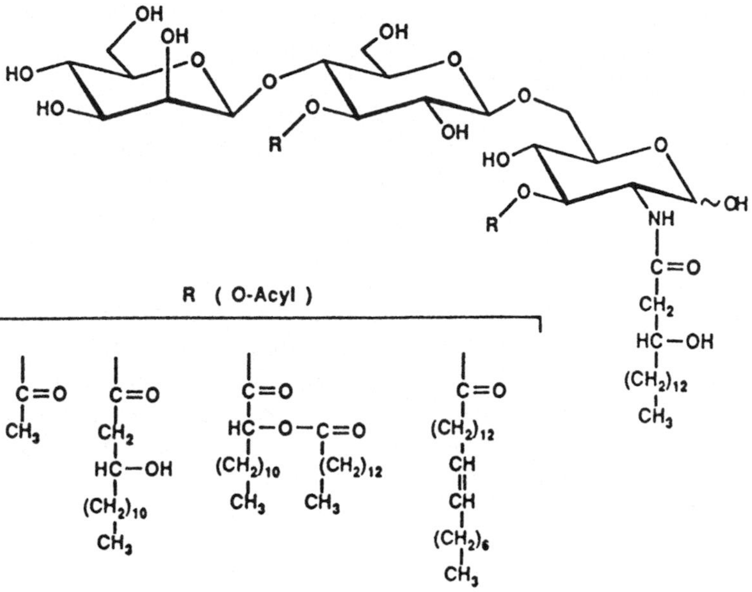

FIG. 8 Chemical structure of *Rhodomicrobium vannielii* ATCC 17100 lipid A. From Ref. 25.

A. Lipid A

Biosynthesis of lipopolysaccharide begins at the hydrophobic end, the lipid A portion. Failure to isolate mutants in the early steps of lipid A assembly has given researchers little information on this segment of lipopolysaccharide biosynthesis, suggesting that such mutations are lethal and that the lipid A structure and its proximal dOclA residues are indispensable for the growth and maintenance of the bacterial cells [20].

Lipid A-specific D-β-hydroxy fatty acids are derived from the normal pathway of fatty acid synthesis. However, it has not been established whether D-β-hydroxytetradecanoic acid arises directly from the major pathway of acyl chain elongation or is formed by a product-specific branch pathway. The nature of the acyl transferase reactions responsible for the addition of the ester-linked saturated fatty acids to lipid A and the enzymatic reactions involved in the formation of the glucosamine backbone also remain unknown. However, acylation of the lipid A backbone is complete prior to extension of the polysaccharide chain [14]. Changes in the O-acyl residues have been observed after cultivation at lower temperatures. At lower temperature (<15°C), the bacteria seem to incorporate larger amounts of unsaturated fatty acids in lipid A, however, the amount of

3-hydroxy-tetradecanoic acid was unaffected by growth temperatures. The incorporation of unsaturated fatty acids in lipid A indicates the need for the bacterial cell to maintain a certain fluidity in its outer membrane at low temperature [21].

The late steps of lipopolysaccharide biosynthesis in *Salmonella* are shown in Fig. 9. As shown, an incomplete lipid A backbone with two ester- and two amide-linked D-3-hydroxy-tetradecanoic acid serves as an acceptor to dOclA which is converted to its nucleotide sugar, CMP-dOclA, before its transfer. The transfer of dOclA is catalyzed by a membrane-bound enzyme system, the CMP-dOclA:lipid A dOclA transferase system. Transfer of the 4-aminoarabinose and phosphorylethanolamine residues may precede that of dOclA but is not necessary for dOclA incorporation into the backbone. The transfer of other fatty acid residues (dodecanoic acid, tetradecanoic acid) to the backbone takes place only after dOclA is covalently attached to the 6′ position of the nonreducing glucosamine residue. The next step is the transfer of the core sugars and O-antigen sugars, which is explained in the following sections [26].

B. Core Oligosaccharide

Biosynthesis of the core takes place by the sequential addition of monosaccharides to the incomplete lipopolysaccharide, which acts as the acceptor molecule for the glycosyl transferase enzyme. The glycosyl residues are converted to their nucleotide sugars (e.g., UDP-glucose and UDP-galactose) before participating in the glycosyl transferase reaction. It was shown that certain phospholipids are needed for this transferase reaction to occur. These lipids form a complex with the glycosyl acceptor (core lipid A or incomplete lipopolysaccharide) and the glycosyl transferase [27].

C. O-Specific Chain

The O-specific chain is synthesized on a carrier lipid called antigen carrier lipid (ACL). The sugars are converted to their nucleotide derivatives and added sequentially to the ACL, in the form of its monophosphate ester (ACL-P), by transferase enzymes. When the oligosaccharide-repeating unit ACL complex is synthesized, polymerization of the O-specific chain by a polymerase enzyme takes place on the ACL. The last step in the biosynthesis is the translocation of the finished O-specific chain from the ACL onto the core lipid A by a translocase enzyme. The ACL is recycled and reenters the biosynthetic pathway after it is reconverted to its ACL-P form by a specific membrane-bound phosphatase. The ACL was isolated from a *Salmonella* species and was found to be a C^{55}-polyisoprenoid alcohol. In other bacteria, an α-glycosyl diphosphoundecaprenol (α-Glc-P-P-ACL) was found to function as an acceptor for sugar incorporation. Here the polymerization process occurs in a single chain mechanism through the action of several

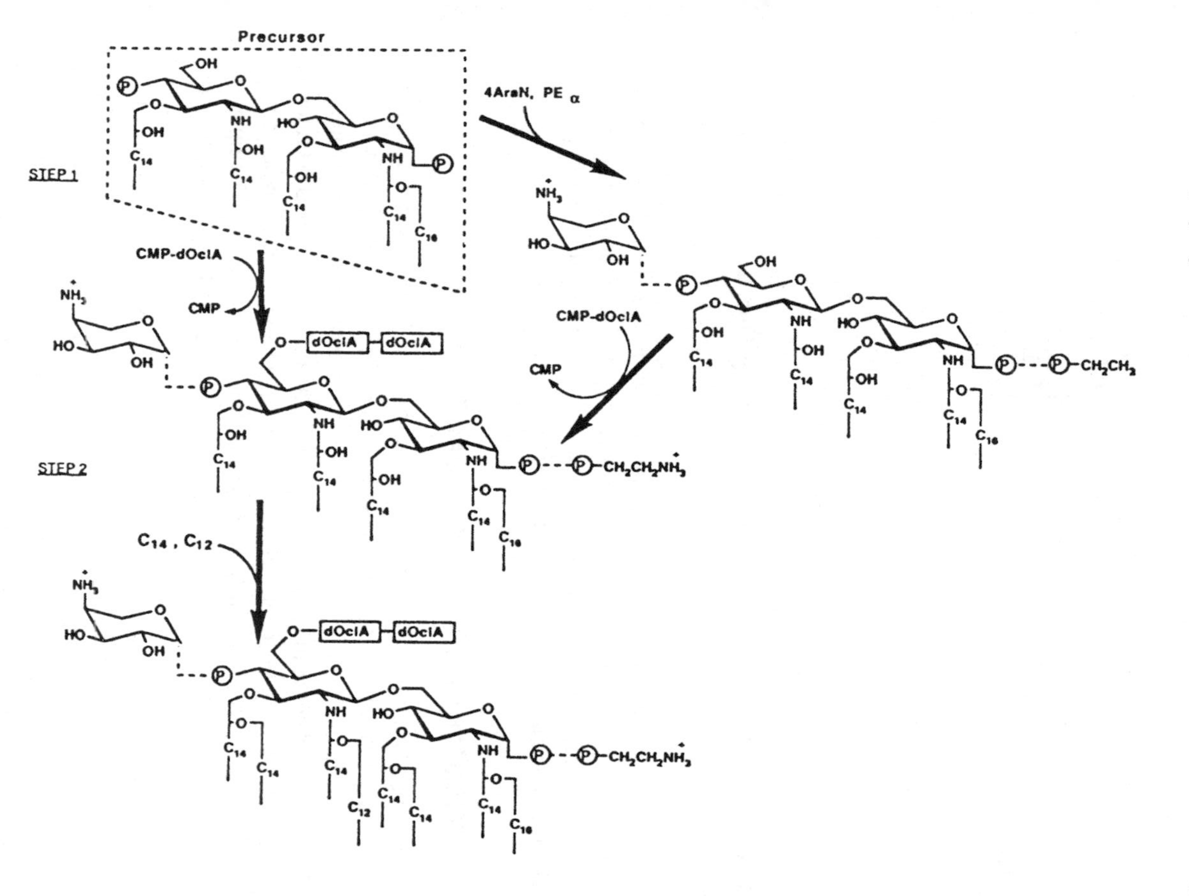

FIG. 9 Late steps in *Salmonella* lipid A biosynthesis. Step 1: Transfer of polar head groups to the precursor. Step 2: Transfer of Lauric and Myristic acid. From Ref. 26.

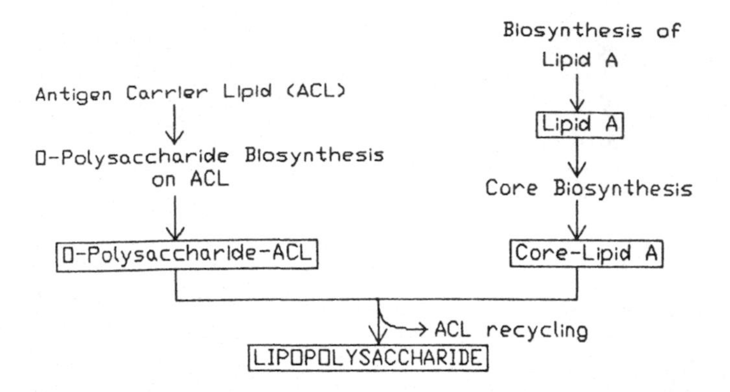

FIG. 10 General lipopolysaccharide biosynthetic pathway. From Ref. 27.

transferases acting in concert (no polymerase is involved). Translocation of the polysaccharide chain occurs along with transfer of an extra glucose residue from the ACL to a "precore" (missing one glucose). The general pathway of lipopolysaccharide biosynthesis is shown in Fig. 10 [27].

V. EXTRACTION OF LIPOPOLYSACCHARIDE

The most widely applied method for the extraction of lipopolysaccharide is the phenol–water (PW) procedure by Westphal et al. [28]. Until 1970, this method was used for the isolation of S-form and R-form lipopolysaccharide giving relatively pure extract containing small amounts of contaminants (1% to 3% proteins). This method involves treatment of dried bacterial cells with an aqueous mixture of 45% phenol at 68°C, which is monophasic at that temperature. The lipopolysaccharide extracted in the water phase can be purified by multiple ultracentrifugation [29].

R-form lipopolysaccharides, because of their long hydrophilic O-chains, have physicochemical properties quite distinct from the S-form lipopolysaccharides. Therefore the PW procedure was proven unsuitable for extracting this type of lipopolysaccharide. As a result, the phenol-chloroform-petroleum ether (PCP) extraction procedure was introduced in the late 1960s for extraction of R-form lipopolysaccharide [30]. In this method, the extraction mixture is monophasic and, because of their lipophilic nature, the R-form lipopolysaccharides are completely soluble in the mixture, whereas S-form lipopolysaccharide and contaminants such as proteins, nucleic acids, and glycans are insoluble and are thus excluded from the extracts. R-form lipopolysaccharide can be recovered by addition of water to the phenol mixture causing precipitation of lipopolysaccharide. Because this method is mild (can be carried out below 10°C), cleavages of covalent bonds are highly improbable and the final product is usually water soluble.

The PCP method can be applied further for the purification of S-form lipopolysaccharide preparations extracted by the PW method (many S-form lipopolysaccharide after extraction with the PW mixture were found to be soluble in the PCP mixture). Since the PCP mixture does not dissolve proteins, glycans, and nucleic acids, a higher yield of S-form lipopolysaccharide can be obtained (up to 5% of the bacterial dry weight). The combined PW/PCP method can also be applied for leaky mutants synthesizing both S- and R-form lipopolysaccharides. In this case, the PCP method is applied first to obtain R-form lipopolysaccharide, and the cell residue is then treated with the PW mixture, whereby the S-form can be obtained [29].

VI. PHYSICOCHEMICAL PROPERTIES OF LIPOPOLYSACCHARIDES

Because lipopolysaccharide is an amphiphathic molecule, its solubility in the water is expected to be dependent on the relative proportion of the hydrophilic polysaccharide and the hydrophobic lipid A. Thus S-form lipopolysaccharide, because of its higher polysaccharide content, was found to exhibit higher solubility than the R-form lipopolysaccharide. However, solubility of lipopolysaccharide in water cannot always be explained by the polysaccharide content alone. It also depends on the presence of negatively charged phosphate, pyrophosphate, and carboxyl groups like those found on dOclA residues. The types of cations neutralizing these negatively charged groups also play an important role in the solubility of lipopolysaccharide [29]. Recently, electrodialysis has been used to remove inorganic cations (Na^+, Mg^{2+}, K^+, Ca^{2+}) and low-molecular-weight basic amines (ethanolamine, putrescine, spermidine, spermine, cadaverin) which are present in lipopolysaccharide preparations regardless of the method of extraction used [29]. This leads to negatively charged or acidic lipopolysaccharide preparations of pH ranging from 2.7 to 3.6 that can be converted to a defined salt form by neutralization with a given base. The triethylamine salt form was shown to have the highest solubility. Solutions at concentration of 0.4 g/mL for the S-form and 0.2 g/mL for the R-form lipopolysaccharide had a low viscosity and opalescence. Sodium, potassium, pyridine, and ammonium salt forms exhibit a lower solubility with the divalent cations, Mg^{2+} and Ca^{2+}, being the lowest [29].

Free lipid A preparations are completely insoluble in water. Solubility can be achieved by addition of pyridine or triethylamine to a suspension of lipid A, but the solubility achieved is only temporary and lost on lyophilization. Presently, solubilization of free lipid A is achieved by converting lipid A to its uniform triethylamine salt. Solutions of lipid A up to a concentration of 50 mg/mL have been obtained [29]. Soluble lipid A preparations can also be prepared by complexing lipid A with bovine serum albumin [29].

VII. BIOLOGY OF LIPOPOLYSACCHARIDES

Lipopolysaccharides represent both the O-antigens and the endotoxins of Gram-negative bacteria. The O-specific chain plays a role in protecting the cell from phagocytosis and killing. However, its presence is not necessary for the survival of the bacteria *in vitro*. In *R. sphaeroides*, incomplete O-specific chains have been found, indicating that long O-specific chains are not needed in this host-independent bacterium. The lipopolysaccharide also carries immunodominant structures (O factors) against which the host immune system produces antibodies. The O-specific chain is, therefore, responsible for the O-antigenic properties of the lipopolysaccharide. The endotoxic properties of lipopolysaccharide are expressed by the lipid A moiety. Lipid A is responsible for many pathophysiological activities accompanying Gram-negative bacterial infections (e.g., lethal toxicity, pyrogenicity, and mitogenicity). In the intact bacteria, lipopolysaccharide is complexed with phospholipid and protein. Therefore, the immunological and pathological responses *in vivo* are probably caused by this lipopolysaccharide complex [31].

REFERENCES

1. J. L. Erwin, J. Geigert, S. L. Neidleman, and J. Wadsworth, in *Biotechnology for the Oils and Fats Industry*, (C. Ratledge, P. Dawson, and J. Rattray, eds.). American Oil Chemists' Society, Champaign, IL., 1984, Chap. 19.
2. C. Ratledge, in *Microbial Lipids*, Vol. 1 (C. Ratledge and S. G. Wilkinson, eds.), Academic Press, London, 1988, Chap. 22.
3. N. Kosaric, W. L. Cairns, and N. C. C. Gray, eds., *Biosurfactants and Biotechnology*. Marcel Dekker, New York.
4. C. Ratledge, in *Biotechnology for the Oils and Fats Industry* (C. Ratledge, P. Dawson, and J. Rattray, eds.), American Oil Chemists' Society, Champaign, IL., 1984, Chap. 12.
5. H. Goldfine, *Microbial Physiol. 8*: 1–58 (1972).
6. C. Ratledge and S. G. Wilkinson, in *Microbial Lipids*, Vol. 1 (C. Ratledge and S. G. Wilkinson, eds.), Academic Press, London, 1988, Chap. 2.
7. P. G. Brennan, in *Microbial Lipids*, Vol. 1 (C. Ratledge and S. G. Wilkinson, eds.), Academic Press, London, 1988, Chap. 6.
8. P. F. Smith, in *Microbial Lipids*, Vol. 1 (C. Ratledge and S. G. Wilkinson, eds.), Academic Press, London, 1988, Chap. 8.
9. S. G. Wilkinson, in *Microbial Lipids*, Vol. 1 (C. Ratledge and S. G. Wilkinson, eds.), Academic Press, London, 1988, Chap. 7.
10. C. A. Boulton, in *Microbial Lipids*, Vol. 2 (C. Ratledge and S. G. Wilkinson, eds.), Academic Press, London, 1988, Chap. 23.

11. A. M. Schwartz, J. W. Perry, and J. Berch, *Surface Active Agents and Detergents*, Vol. 2, Robert E. Krieger, New York, 1977, Chap. 2.
12. E. H. Pryde, in *Fatty Acids*, Monograph 7 (E. H. Pryde, ed.), The American Oil Chemists' Society, Champaign, IL., 1979, Chap. 1.
13. R. B. Hutchison and L. R. Mores, in *Fatty Acids*, Monograph 7 (E. H. Pryde, ed.), American Oil Chemists' Society, Champaign, IL., 1979, Chap. 29.
14. M. Inouye, *Bacterial Outer Membranes. Biogenesis and Functions*, John Wiley, New York, 1979.
15. Lugtenberg and Van Alphen, *Biochimica and Biophysica Acta, 737*: 51–115 (1983).
16. P. Mülradht and J. Golecki, *Eur. J. Biochem. 5*: 343–352 (1975).
17. U. Y. Funahara and H. Nikaido, *J. Bacteriol. 141*: 1463–1465 (1980).
18. H. Nikaido and M. Vaara, *Ann. Rev. Microbiol. 39*: 1–32 (1985).
19. N. J. Russell, in *Microbial Lipids*, Vol. 2 (C. Ratledge and S. G. Wilkinson, eds.), Academic Press, London, 1988, Chap. 17.
20. C. Galanos, O. Lüderitz, E. T. Rietschel, and O. Westphal, in *International Review of Biochemistry. Biochemistry of Lipids II*, Vol. 14 (T. W. Goodwin, ed.), University Park Press, Baltimore, 1977, Chap. 6.
21. E. T. Rietschel, C. Galanos, O. Lüderitz, and O. Westphal, in *Immunopharmacology and the Regulation of Leukocyte Function* (David R. Webb, ed.), Marcel Dekker, New York, 1982, Chap. 9.
22. P. V. Salimath, R. N. Tharanathan, J. Weckesser, and H. Mayer, *Eur. J. Biochem. 144*: 227–232 (1984).
23. P. H. Mäkelä and B. A. D. Stocker, in *Handbook of Endotoxin. Chemistry of Endotoxin*, Vol. 1 (E. T. Rietschel, ed.). Elsevier, New York, 1984, Chap. 3.
24. E. T. Rietschel, H. W. Wollenweber, H. Brade, U. Zähringer, B. Lindner, U. Seydel, H. Bradaczek, G. Barnickel, H. Labischinski, and P. Giesbrecht, in *Handbook of Endotoxin. Chemistry of Endotoxin*, Vol. 1 (E. T. Rietschel, ed.), Elsevier, New York, 1984, Chap. 5.
25. Mayer and Weckesser, in *Handbook of Endotoxin. Chemistry of Endotoxin 1*(7), (E. Th. Rietschel, ed.), Elsevier, New York, 1984.
26. V. Lehmann and T. Hansen-Hagge, in *Handbook of Endotoxin. Chemistry of Endotoxin*, Vol. 1 (E. T. Rietschel, ed.), Elsevier, New York, 1984, Chap. 8.
27. K. Jann and B. Jann, in *Handbook of Endotoxin. Chemistry of Endotoxin*, Vol. 1 (E. T. Rietschel, ed.), Elsevier, New York, 1984, Chap. 4.
28. O. Westphal, O. Lüderitz, and R. Blister, *Z. Naturforsch. Teil B7*: 148–155 (1952).
29. C. Galanos and O. Lüderitz, in *Handbook of Endotoxin. Chemistry of Endotoxin*, Vol. 1 (E. T. Rietschel, ed.), Elsevier, New York, 1984, Chap. 2.
30. C. Galanos, O. Lüderitz, and O. Westphal, *Eur. J. Biochem. 9*: 245–249, 1969.
31. R. J. Lynn, in *Microbial Lipids*, Vol. 2 (C. Ratledge and S. G. Wilkinson, eds.), Academic Press, London, 1988, Chap. 20.
32. Kalish, in *Fatty Acids and Their Industrial Applications*, (E. S. Pattison, ed.), Marcel Dekker, New York, 1968, Chap. 9.

III

Applications

13

Factors Influencing the Economics of Biosurfactants

CATHERINE N. MULLIGAN* Department of Biochemical Engineering, National Research Council of Canada, Montreal, Quebec, Canada

BERNARD F. GIBBS Department of Protein Engineering, National Research Council of Canada, Montreal, Quebec, Canada

**Current affiliation*: SNC Research Corporation, Montreal, Quebec, Canada.

I. INTRODUCTION

Surfactants are used as adhesives, flocculating, wetting, and foaming agents, deemulsifiers, and penetrants. They include both lower priced commodity and specialty chemicals. Sales of surfactants grow approximately 2%–3% per year [1]. Petroleum industries are the major users. Other uses include pulp and sludge-dewatering and slurry stabilization (Table 1). Desirable characteristics include solubility, surface tension reduction, low critical micelle concentrations, detergency power, wetting ability, and foaming capacity [2]. No surfactant can do all these functions equally well. Therefore, a wide variety of surfactants exists and the demand is high.

In 1982, the total use of these products in the United States was 2.5 million metric tons [3]. In 1989, the market reached $5.8 billion. Household detergents make up 41% of the market and petroleum additives 13%. They are classified as cationic, anionic, and nonionic. A breakdown of synthetic surfactant demands is shown in Fig. 1. Major types of surfactants in use are linear alkylbenzenesulfonates (LABS), alcohol sulfates (AS), alcohol ethers sulfates (AES), alcohol glyceryl ether sulfonates (AGES), α-olefin sulfonates (AOS), alcohol ethoxylates (AE), alkylphenol ethoxylates (APE), fatty alkanol amides

TABLE 1 Some Desirable Characteristics of Surfactants

Application	Characteristics
Detergency	Low CMC, good salt and pH stability, biodegradability, good foaming properties
Emulsification	Proper HLB, environmental safety
Lubrication	Chemical stability, adsorption at surfaces
Mineral flotation	Adsorption on specific ore(s), inexpensive
Petroleum recovery	Wetting of oil-bearing formations, microemulsion formation and solubilization, ease of breaking after recovery
Pharmaceuticals	Biocompatibility, low toxicity

Source: Ref. 2.

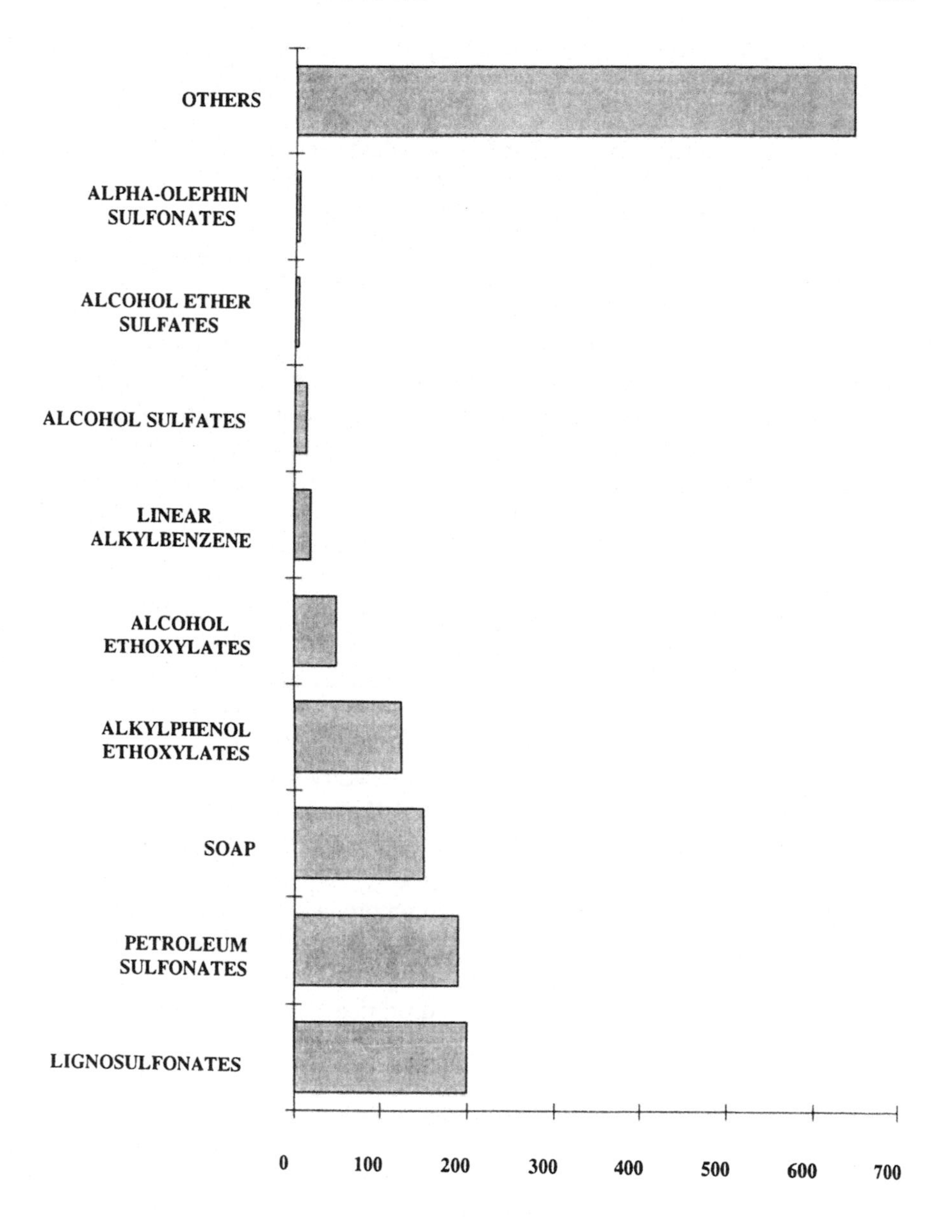

FIG. 1 U.S. production of synthetic surfactants by type in 1985 (metric tons × 10^3). (From Ref. 3.)

(FAA), and fatty amine oxides (FAO) [2]. They are derived mainly from petroleum products, although some are prepared from lignosulfonates and triglycerides.

Surfactants are potentially useful in every industry dealing with multiphase systems. These molecules contain both hydrophilic and lipophilic portions [4]. The effectiveness of a surfactant is determined by its ability to lower the surface tension. For example, a good surfactant can lower the surface tension of water (air–water interface) from 72 mN/m to 35 mN/m and the interfacial tension (oil–water interface) for water against *n*-hexadecane from 40 mN/m to 1 mN/m. These amphiphilic compounds concentrate at interfaces (solid–liquid, liquid–liquid, or liquid–vapor). The surface tension correlates with the concentration of the surface-active compound until the critical micelle concentration (CMC) is reached. The CMC is defined as the minimum concentration necessary to initiate micelle formation [5]. There is little further change in surface tension as any additional surfactant will be present in micellar form in the bulk phase. Efficient surfactants have very low critical micelle concentrations (i.e., less surfactant is necessary to decrease the surface tension).

An *emulsion* is defined as a "heterogeneous system, consisting of at least one immiscible liquid dispersed in another in the form of droplets, whose diameters, in general, exceed 0.1 mm. Such systems possess a minimal stability, which may be accented by such additives as surface active agents, finely defined solids, etc." [6]. It is necessary to specify water-in-oil (w/o) or oil-in-water (o/w) phases. The term hydrophilic–lipophilic balance (HLB) is used to classify which type of emulsion the emulsifier will favor. Emulsifiers and deemulsifiers stabilize or destabilize the emulsion. Many of these products are polymeric and decrease surface tension insignificantly.

The choice of surfactant is primarily based on product cost, unless the cost is insignificant in the process [2]. In general, surfactants are used to save energy and consequently energy costs (e.g., wetting or spreading energy). Charge-type, physicochemical behavior, solubility, and adsorption behavior are some of the selection criteria. In this competitive market, more economical and effective new products will replace the older ones. Although the market is well established, there will always be room for growth, particularly as new applications develop. One such application is the use of surfactants in the bioremediation of contaminated land sites [7, 8].

Biosurfactants are produced as metabolic by-products by bacteria, yeasts, and fungi. They can be produced extracellularly or as part of the cell wall. They are not only potentially as effective but offer some distinct advantages over the highly used synthetic surfactants [9]. Microbial surfactants exhibit high specificity and are consequently suited to new applications. Effective physicochemical properties (low interfacial tensions and critical micelle concentrations) and temperature stability are characteristic of these compounds. Other advantages include biodegradability, reduced toxicity, and a broad range of structures.

The chemical and physical properties of biosurfactants could be modified genetically, biologically or chemically for specific applications. A variety of simple potentially less expensive substrates can be used for their production (*n*-alkanes, carbohydrates, vegetable oil, and wastes) [10]. For example, *Pseudomonas aeruginosa* can produce rhamnolipids from C_{11} and C_{12} alkanes, succinate, pyruvate, citrate, fructose, glycerol, olive oil, glucose, and mannitol [11].

Biosurfactants have been tested in enhanced oil recovery and the transportation of crude oils [12]. They are effective in reducing the interfacial tension between oil and water *in situ*, reducing the viscosity of the oil removing water from emulsions prior to processing and releasing bitumen from tar sands. Emulsan has been commercialized for this purpose [13]. Its mean molecular weight is 9.9×10^5, consisting of a polysaccharide backbone to which fatty acids and proteins are attached. Multi-Biotech, a subsidiary of Geodyne Technologies, Calgary, is in the process of commercializing biosurfactants for enhanced oil recovery [14].

Applications for other biosurfactants, in the future, will be in pharmacology, biocosmetics, textiles, food, pulp and paper, coal beneficiation, and ore processing [15]. Since there are so many applications and different sets of conditions exist for each, new products will always have to be developed with unique characteristics [9]. Although most fermentations are aerobic, a few examples of anaerobic biosurfactant production exist. *Bacillus licheniformis* Strain JF-2 [16] is such an example. It would be well suited to *in situ* studies for enhanced oil recovery or soil decontamination.

These biological compounds are grouped as glycolipids, lipopeptides, phospholipids, fatty acids, and neutral lipids [17]. Most biosurfactants are either anionic or neutral. Only a few are cationic (those containing amine functions). The hydrophobic part of the molecule based on long-chain fatty acids, hydroxy fatty acids, or α-alkyl-β-hydroxy fatty acids. The hydrophilic portion can be a carbohydrate, amino acid, cyclic peptide, phosphate, carboxylic acid, alcohol, and so on. A wide variety of microorganisms can produce these compounds (Table 2). Most are either anionic or neutral. Only those containing amine groups are cationic.

Most biosurfactants are produced on hydrocarbons [18]. Production is most often growth associated. In this case, they can either use the emulsification of the substrate (extracellular) or facilitate the passage of the substrate through the membrane (cell wall associated). *Rhodococcus erythropolis* forms nonionic trehalose mono- and dimycolates (Fig. 2) during growth [20, 21]. Yields of 0.105 g/g substrate have been obtained for a biosurfactant concentration of 2.1 g/L.

TABLE 2 Classification and Microbial Origin of Biosurfactants

Surfactant classes	Microorganism
Trehalose lipids	*Arthrobacter paraffineus* *Corynebacterium* spp. *Mycobacterium* spp. *Rhodococcus erythropolis*
Rhamnolipids	*Pseudomonas aeruginosa*
Sophorose lipids	*Torulopsis bombicola*
Glucose-, fructose-, saccharose lipids	*Arthrobacter* spp. *Corynebacterium* spp. *R. erythropolis*
Cellobiose lipids	*Ustilago maydis*
Polyol lipids	*Rhodotorula glutinus* *Rhodotorula graminus*
Diglycosyl diglycerides	*Lactobacillus fermenti*
Lipopolysaccharides	*Acinetobacter calcoaceticus* (RAG-1) *Pseudomonas* spp. *Candida lipolytica*
Lipopeptides	*Bacillus subtilis* *Bacillus licheniformis*
Ornithine, lysine peptides	*Thiobacillus thiooxidans* *Streptomyces sioyaensis* *Gluconobacter cerinus*
Phospholipids	*T. thiooxidans* *Corynebacterium alkanolyticum*
Sulfonolipids	*Capnocytophaga* spp.
Fatty acids (corynomycolic acids, spiculisporic acid, etc.)	*Corynebacterium lepus* *Arthrobacter parrifineus* *Talaromyces trachyspermus*

Source: Ref. 17.

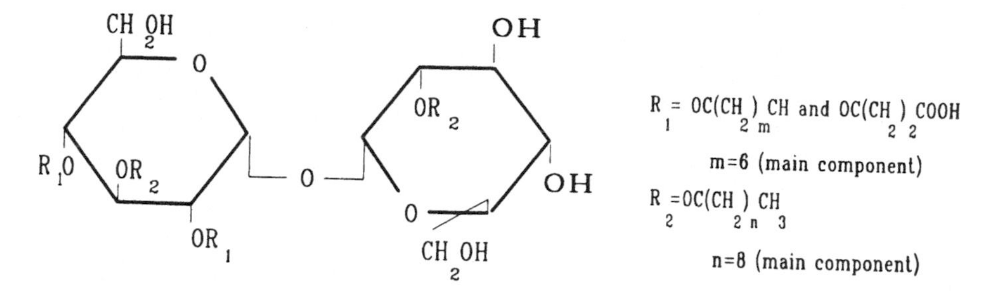

FIG. 2 Trehalose-2,3,4-2′-tetraester produced by *Rhodococcus erythropolis*. (From Ref. 19.)

However, this cannot explain the fact that biosurfactants are also produced from carbohydrates. They are usually secondary metabolites, being produced during late logarithmic and stationary growth phases. Several examples of these anionic compounds exist. The production of glycolipids by *P. aeruginosa* is well documented [22, 23]. These are the rhamnolipids R1 and R2 (Fig. 3), which can be produced from carbohydrates or hydrocarbons. The first contains two rhamnoses attached to β-hydroxydecanoic acid, whereas R2 consists of one rhamnose connected to the identical hydroxy fatty acid [22]. Pilot plant studies (using a 50-L bioreactor) have already been performed [24]. Final product concentrations have reached 2.0 g/L. Surface tensions of 29 mN/m are characteristic of these compounds. A surfactant produced by this microbe has been used in the enhanced removal of oil from gravel after the Exxon Valdez oil spill [25].

Torulopsis bombicola is one of the few yeasts known to produce biosurfactants [26]. Sophorose lipids (Fig. 4) can lower the surface tension to 33 mN/m. High yields have also been obtained when soybean oil is added with glucose (0.35 g/g substrate or 67 g/L). Since sophorolipids do not stabilize water-in-oil emulsions, they cannot be produced for this purpose. An application as a protective substance has been found in the cosmetics industry [27].

The case of surfactin [28], a lipopeptide produced by *Bacillus subtilis*, is also interesting (Fig. 5). It contains seven amino acids bonded to the carboxyl and hydroxyl groups of the 14-carbon acid [29]. Although a very effective surfactant is produced in glucose (27 mN/m at a concentration as low as 0.005%), hexadecane addition totally inhibits its synthesis. Yields in glucose of 0.02 g/g substrate have been obtained [30]. This compound has been shown to improve the mechanical dewatering of peat by greater than 50% [31] at a very low concentration (0.0013 g/g wet peat). The surfactant improves the dewatering by changing the flow characteristics of the trapped water within the peat waxes on the particle

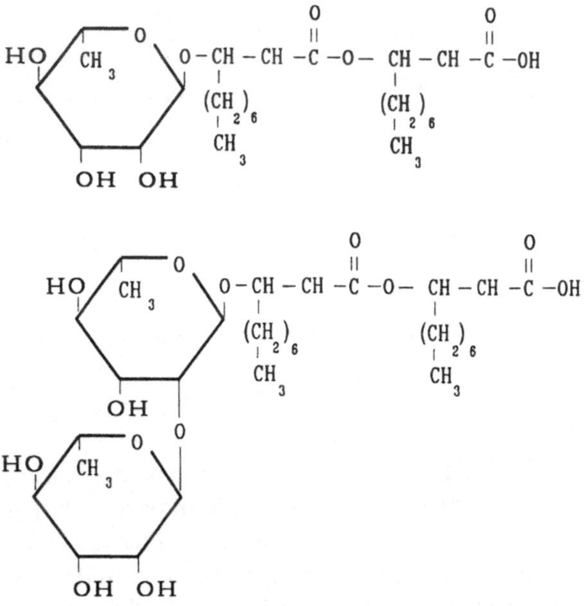

FIG. 3 Structures of rhamnolipids produced by *Pseudomonas aeruginosa*. (From Ref. 22.)

surfaces. It is sold by Calbiochem Corp. (San Diego, CA) for use in biochemical research for its ability to inhibit blood coagulation and protein denaturation and to accelerate fibrinolysis ($9.60/mg for 98% purity).

The main limiting factor to the commercialization of fermentation compounds is the economics of large-scale production [32]. This is also the case for biosurfactants. Although many biosurfactants and their production processes have been patented (Table 3), only emulsan, a lipopolysaccharide emulsifier produced by *Acinetobacter* spp., has been commercialized [13]. Economics is a major determinant in bringing a product to market since a monetary return has to be realized.

The surfactant market is extremely competitive. For example, for enhanced oil recovery, petroleum sulfonates cost $2/kg [33] and lignin derived surfactants are 40% less expensive. At this price, petroleum sulfonates are economical only if oil sells for $35 or more, whereas lignin-derived surfactants are economical at an oil price of $25 per barrel. Only in low-volume high-value applications (cosmetics, food, and pharmaceuticals) could high costs be tolerated. Biosurfactants must be produced less expensively or for the same price with distinct advantages.

The economics of biosurfactant production has not received considerable attention. Many reviews deal with the structures, production, and properties of

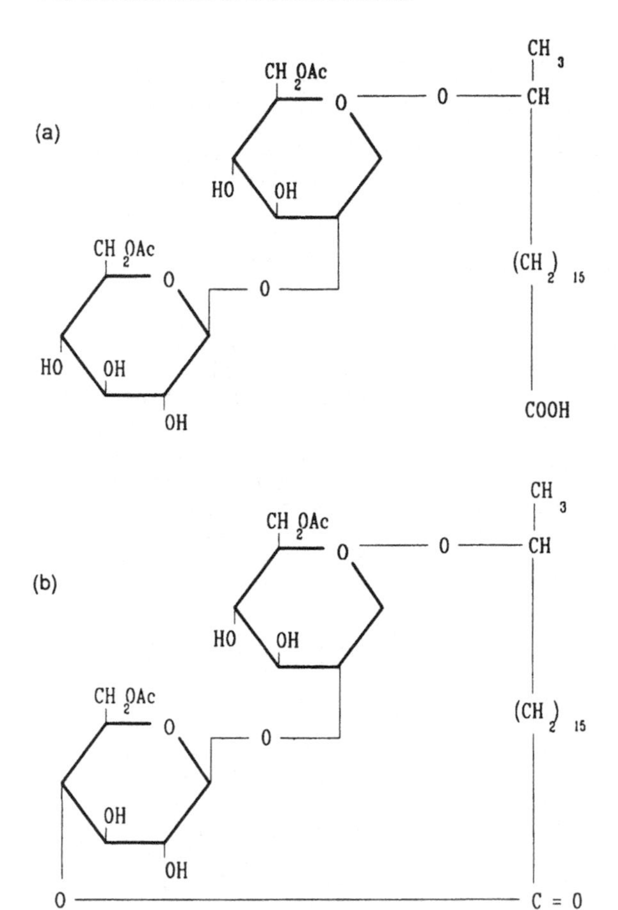

FIG. 4 Acidic (a) and lactonic (b) sophorose lipids of the yeast *Torulopsis bombicola*. (From Ref. 62.)

biosurfactants [34–36]. Economic strategies must be developed to enable competition with the synthetic surfactants. This chapter will detail the possibilities for the reduction of costs for all aspects of biosurfactant production. These would involve the following:

1. Choice of inexpensive raw materials
2. Increase of biosurfactant yield and production rate
3. Optimization of the fermentation process
4. Reduction of product recovery costs
5. Production of biosurfactants suitable for specified applications

TABLE 3 Patents Related to Biosurfactants

Product	Microorganism	Patent
Emulsan	*Arthrobacter* sp. ATCC 31012	Biotechnol. Aktienges., US 4,276,094 (1981)
Biosurfactant	*Corynebacterium hydroblastus* NRRL-B-5631	CPDL, US 3,997,398 (1976)
Biosurfactant	*Arthrobacter, Bacillus, Corynebacterium, Nocardia, Pseudomonas*	CPDL, CA 1,114,759 (1981)
Biosurfactant	*Arthrobacter* RAG 1	Gutnick, D., Rosenberg, E. DE 2,415,897 (1987)
Lipopeptide	*Methylomonas clara* ATCC 31226	Hoescht AG, DE 3,312,166 (1984)
Biosurfactant	*Penicillium spiculisporum*	Inoue-Japax Research Inc., Jpn Kokai 7837,189 (1987)
Sophorose lipid	*Torulopsis bombicola*	Kao Soap Ltd., DE 2,834,118 (1979), DE 2,938,383 (1980), Jpn. Kokai Tokkyo Koho 8192,786 (1981), EP 0005004 (1983)
Glycolipids (trehalose lipids)	*Arthrobacter paraffineus* ATCC 15591 *Corynebacterium hydroblastus* ATCC 15592	Kyowa Hakko Kogyo Co. Ltd. DE 1,905,472 (1970), US 3,637,461 (1972)
Fructose lipid	*Arthrobacter paraffineus* ATCC 15591	Kyowa Hakko Kogya Co. Ltd, DE 2,440,942 (1975)
Spiculisporic acid	*Penicillium spiculisporum* ATCC 16071	Kobayashi, T., Tabuchi, T., US 3,625,826 (1971)
Biosurfactant	*Thiobacilluus, Bacillus, Nocardia, Pseudomonas*	Phillips Petroleum Co. US 2,907,389 (1959) US 3,185216 (1965)
Emulsan	*Acinetobacter* sp. ATCC 31012	Petroleum Fermentation N.V. US 4,311,829 (1982), US 4,311,832 (1982)
Sophorose lipid	*Torulopsis magnoliae, Torulopsis apicola*	Spencer, J.F.T., Tullich, A.P., Gorin, P.A.J., US 3,205,150 (1965)
Surfactin	*Bacillus subtilis* ATTC 21331	Takeda Chemical Ind. Ltd, US 3,687,926 (1972)
Biosurfactant	*Candida*	VEB Petrol-chemisches Kombinant Schedt, DD 139,069 (1979)
Biosurfactant	*Candida, Pichia, Nocardia, Mycobacterium, Pseudomonas*	Wintershall AG, DE 2,401,267 (1975), DE 2,843,685 (1980), DE 2,911,016 (1980)
Trehalose lipid	*Rhodococcus erythropolis* DSM 43215	Wintershall AG, DE 3,248,167 (1984)
Biosurfactant	*Corynebacterium salvinum*	Zajic, J.E., Gerson, R. K. US 4,355,109 (1982)

Source: Ref. 17.

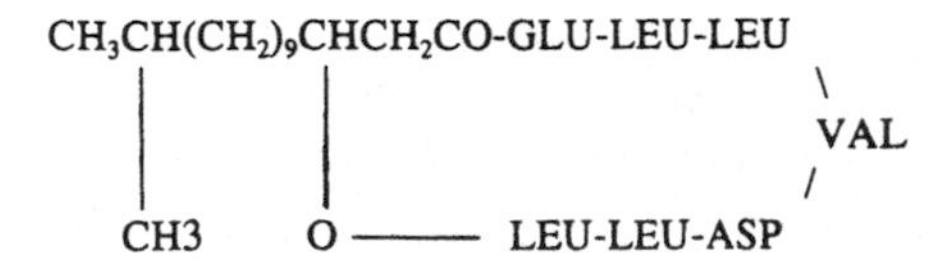

FIG. 5 Structure of the lipopeptide surfactin of *Bacillus subtilis*. (From Ref. 29.)

The research level is critical to the subsequent economic success of a product [32]. Strain improvement, medium development, and process optimization must be considered in both laboratory and pilot plant stages. For the process to be cost-effective, maximal production rates and yields must be obtained. In particular, downstream processing has traditionally received less attention in the literature than the fermentation process. However, these costs can dominate the process.

The aim in biosurfactant production studies is to develop an economic process by taking into consideration all aspects. In this article, emphasis will be placed on areas where research should be concentrated to accomplish this. The first section in this chapter deals with the choice of inexpensive raw materials including carbon, nitrogen, phosphate sources, and other components. Increasing biosurfactant yield by biosynthesis control, screening for overproducers, and genetic manipulation of microorganisms are discussed next. Methods to decrease costs in the fermentation process itself are examined (pH, temperature, agitation, and aeration). Various product recovery methods are compared (solvent extraction, *in situ* extraction, adsorption, flotation, ultrafiltration, etc.).

II. CHOICE OF INEXPENSIVE RAW MATERIALS USED IN BIOSURFACTANT PRODUCTION

A. Carbon Source

All microorganisms require for growth a source of carbon, hydrogen, nitrogen, oxygen, and, to a smaller degree, sulfur and phosphorus [32]. These materials are available in many forms. The choice of raw materials is very important to the overall economics of the process. Particularly in the bulk product market, production costs are influenced by the price of the feedstocks and other raw materials. Therefore, it is essential to utilize low-cost raw materials. Raw materials can make up 10%–50% of the final product cost [37]. Availability, stability, and variability of each component must also be considered. The amount to be utilized, form (solid or liquid), packaging, transportation, purity, and production as a by-product of a process are all factors influencing raw material costs. Lower purity materials are less expensive and can be tolerated in most cases.

In particular, the carbon source is very important in biosurfactant production. A wide variety has been used. They include hydrocarbon, carbohydrate, and vegetable oil sources (Table 4). Some organisms produce biosurfactants only in hydrocarbons, others only in carbohydrates, and still others utilize several substrates, in combination or separately. In general, optimal yields are obtained with hydrocarbon or carbohydrate and lipids. *Candida lipolytica* only produces biosurfactants on hydrocarbons [38]. Therefore, inexpensive crude hydrocarbon and carbohydrate sources must be considered when selecting a substrate. A range of commercially available substrates and their current market prices are shown in Table 4.

For carbohydrates, most biosurfactant production experiments have been performed using the more expensive pure forms of the sugar. For example, glucose has been used both in preliminary experiments for *B. subtilis* [28, 30] and for *P. aeruginosa* rhamnolipid pilot plant studies [24]. Although it has been postulated that less expensive sources can be used, they have not been evaluated [26]. Molasses is relatively inexpensive in the United States. In Europe, costs are elevated due to distribution costs. It is produced as a by-product from beet and cane sugar refining and can vary in quality. It contains 45%–60% sugar [40] in the form of sucrose and invert sugar. Vitamins, amino acids, and other components (i.e., metals) that may interfere with the fermentation are also present. The composition may vary as to total solids content, gums, waxes, ash, heavy metals, pH, and so on. Since its variation in price is not politically controlled, molasses should be investigated as a substrate.

In terms of hydrocarbon utilization, although synthetic surfactants require pure chemicals, biosurfactants do not. A wide variety of hydrocarbons have been used. Kerosene, consisting of naphthalenes and paraffins, has been evaluated as a substrate for *Nocardia erythropolis* [20]. *n*-Paraffins have been examined in rhamnolipid production by *P. aeruginosa* [41]. The chain length of the hydrocarbons did not effect the composition of the rhamnolipids formed in the latter. The price, however, is very dependent on the price of petroleum, which increased substantially at the end of 1990.

Product yields based on substrates must be considered when choosing a feedstock. For example, although biosurfactants from *Corynebacterium fascians* can be produced on both sucrose and hexadecane, yields were over 20 times higher on the latter, offsetting the cost of the hydrocarbon [42]. In the case of *T. bombicola*, the addition of soybean oil to the less expensive carbohydrate enhanced yields 35-fold [26]. Yields of 97 mg/g glucose were obtained, compared to 100 mg/g glycerol and 230 mg/g alkanes for rhamnolipid production by *P. aeruginosa* [29]. *Corynebacterium hydroblastus* prefers linear hydrocarbons C_{12}–C_{14}, and *R. erythropolis* produces biosurfactants best on C_{12}–C_{18} *n*-alkanes [35]. The production rate of surfactin by *B. subtilis* [30] increased with the

TABLE 4 Cost Comparison of Carbon Sources

Carbon source	Unit cost ($/kg)
Carbohydrates	
Corn syrup (95% glucose)	0.25
Glucose derived from corn	0.22[a]
Fructose derived from corn	0.48[a]
Corn starch	0.20[a]
Black strap molasses	0.08[b]
Ethanol	0.52
Methanol	0.12
Citric acid	1.64
Lactic acid (88%)	2.27
Succinic acid	9.57
Mannitol	7.31
Glycerol	1.61
Dextrose	0.52
Fructose	0.86
Sucrose	0.79
Lactose	0.48
Hydrocarbons	
Kerosene	0.49[c]
n-Paraffins	0.51
Hexane	0.30
Xylene	0.38
Hexadecane (87%)	4.50[d]
Vegetable oils	
Soybean oil	0.52
Corn oil	0.62
Safflower oil	1.17
Sunflower oil	0.62
Olive oil	2.33

[a]St. Lawrence Starch Co., Montreal, Canada
[b]Molasses Market News, New Orleans.
[c]Unocal, Clark, New Jersey.
[d]Humphrey Chemical, North Haven, CT.
Source: Ref. 39.

addition of citric acid (17 mg/h, 1% glucose and 1% citric acid compared to 12 mg/h for 2% glucose). Several substrates were compared for the production of a cellobiose containing lipid by *Ustilago maydis* [43]. Highest yields were obtained in glucose and coconut oil (0.79 g/g substrate). For *P. aeruginosa* 44T1, highest yields [11] were obtained in olive oil (7.56 g/L with a yield of 38.2%).

Attention must be paid, however, to the fact that different carbon sources can influence the composition of the biosurfactant formed and how it is produced. For example, *Arthrobacter* produces 75% extracellular biosurfactant when grown on ethanol or acetate but with hydrocarbons, it is totally extracellular [15]. The trehalose lipids produced by *Arthrobacter, Nocardia*, and *Corynebacterium* were replaced by sucrose and fructose lipids when grown on sucrose or fructose [35]. Also, *N. erythropolis* trehalose lipid formation was dependent on the chain length of the hydrocarbon [18].

The use of renewable carbon sources has not been extensively studied. Corn (60% starch), rice, wheat, and cassava are starch-bearing plants of plentiful nature (production increases every year). As in all natural products, the price fluctuates from year to year [44]. Conversion to mono- and oligosaccharides is required by acid or enzymatic treatment. In the United States, fermentation of starch to alcohol is performed on a large scale. Starch hydrolysates are a source of cheap inexpensive substrates for enzyme and antibiotic production. Starch is converted in large quantities for use as a sweetener (high fructose corn syrup). The cost of corn syrup (Table 5) is competitive with other carbon sources [45]. In 1981 in the United States, 3.6 million metric tons were produced [44]. Starch, as a source of dextrose, has not been thoroughly investigated for use as a substrate for biosurfactant production but could be profitable. However, Sandrin et al. [46] obtained a final surfactin concentration of 0.5 g/L by the direct use of starch by *B. subtilis*.

To obtain carbohydrates from wood, straw, and bagasse (plentiful in supply) hydrolysis must be performed. These sources are cheaper than corn but hydrolysis requirements offset any advantage. Fodder yeast from wood hydrolysis products is only performed on a small scale. The method of hydrolysis and conditions utilized can influence substantially the cost of the substrate (Table 5). Concentrated acid processes require recovery and recycling of the waste and extensive neutralization of the effluent before use as a substrate. Dilute acid processes with high temperature reduce acid loss but are not as efficient. New hydrolysis methods (Stake steam explosion process) could decrease costs further [45].

Jiménez and Chávez [47] made a comparison of the various biomass feedstocks to obtain glucose. Aquatic biomass, corn stover, sugar cane, wheat straw, wood, and molasses were compared. Various processing methods to obtain glucose were used. The list is shown in Table 5. Wood, corn stover, and aquatic biomass seem to be economical sources of glucose.

TABLE 5 Costs of Sugars from Biomass (¢/dry kg)

		1980	1985	1990
Corn Stover		3.3	4.6	6.8
Whole tree wood chips		2.9	3.3	5.7
Pretreated wood chips		3.1	5.0	7.2
Biosugar ex lignocellulosics				
Enzyme/acid pretreat		17.6	28.4	42.5
Concentrated acid/recycle		17.8	27.1	39.8
Dilute acid/extrusion		19.4	30.8	46.0
Concentrated acid/once-thru		27.7	41.1	59.0
Corn syrup (as glucose)		18.5	22.9	24.7
Glucose source				
Aquatic biomass	Enzymatic hydrolysis		12.3	
Corn stover	Hydrolysis (A)		36.1	
	Hydrolysis (B)		8.8	
Molasses	None		16.0	
Sugar cane	Mechanical		21.6	
Wheat straw	Enzymatic hydrolysis		41.8	
Wood	Hydrolysis (A)		24.9	
	Hydrolysis (B)		3.7	

Source: Refs. 45 and 47.

Peat has been examined as a source of carbohydrates [48] under various hydrolysis conditions. Concentrations of sulfuric acid between 0.5% and 2% for either 1 or 2 h at 124°C were investigated. Six monosaccharides were identified: ribose, xylose, arabinose, mannose, glucose, and galactose. At the maximum sugar concentrations (7.7 g/L), glucose accounted for 46% of the total.

It is more favorable to use agricultural-grade (higher cellulose content) than fuel-grade peats. Although the total sugar concentrations obtained by using concentrated acid hydrolysis were dilute (7.7 g/L maximum), *B. subtilis* was able to produce surfactin at high yields (0.14 g surfactin/g sugar). Other components such as amino acids and organic acids are also present in the peat hydrolysate [49]. By using peat of a lower decomposition level (H1 instead of H3 on the von Post scale), sugar concentrations of 19 g/L could be obtained (unpublished results). Surfactin yields were lower, however (0.04 g/g sugar). This is still higher than the results obtained in a glucose medium (0.02 g/g sugar) [30]. If the conditions for surfactin production could be optimized, this process would require serious consideration, particularly if production takes place near the site of peat harvesting.

Another potential source of substrates and the least expensive is waste streams (Table 6). In addition, waste treatment costs can be offset by the production of valuable products. These carbon sources can be obtained either at little (transportation cost only) or no cost. Waste streams containing fatty acids would be excellent for biosurfactant production. Rice hull hydrolysate, starch waste liquors, domestic waste, and potato-processing wastes could be used in a multiorganism conversion strategy (i.e., lipid-accumulating organisms are grown to produce triglycerides as a precursor for biosurfactant production by *T. bombicola*) [10].

Several examples of waste stream utilization have been demonstrated. Chicken fat can be used for the production of *P. aeruginosa* SB1 rhamnolipids [51]. Water expressed from fuel-grade peat (von Post scale H9) contains dissolved organics and thus could be used by *B. subtilis* as a substrate for surfactin production [52]. The surfactin produced from the pressate can then be added to wet peat to assist in the dewatering process, completing the production/utilization cycle [31]. The

TABLE 6 Estimated Supply of Potential Biomass Feedstocks

	Production (million dry metric tons/year)	
Biomass feedstock	(1980s)	(Potential 1990s)
Crops		
Wood	350	1420
Starch crops	94	1460
Sugar crops	12	—
Forage grasses	22	380
Subtotal	478	3260
Waste sources		
Agricultural	350	350
Animal/livestock	240	300
Industrial	160	250
Municipal (MSW)	160	240
Forestry/milling	85	130
Sewage	10	15
Miscellaneous	50	80
Subtotal	1055	1365
Total	1530	4625

Source: Ref. 50.

chemical oxygen demand (COD) is also reduced in the process from 690 to less than 50 mg/L, an effective waste treatment.

Considerable amounts of whey are obtained during cheese processing, causing considerable pollution (4%–5% lactose content). To utilize the lactose effectively, a chosen organism must be able to consume both the lactose and its breakdown products, glucose and galactose. Koch [53] has developed a strain of *P. aeruginosa* to utilize whey for the production of rhamnolipids and *B. subtilis* is also known to produce surfactin using this substrate (unpublished results).

A wide variety of substrates is available for evaluation as potential feedstocks for biosurfactant production. Very few have been examined other than the conventional pure sources. Inexpensive substrates could be very important in determining the economic feasibility of producing biosurfactants commercially.

B. Nitrogen Source

After carbon, nitrogen is usually the most plentiful substance in the medium for biosurfactant production. Amino acids, purines, pyriminides, proteins, DNA, and RNA are all forms of nitrogen. Organic nitrogen sources include corn steep liquor, fish meal, corn germ or gluten meal, urea, yeast, or yeast hydrolysates [44]. Inorganic nitrogen sources include gaseous ammonia, ammonium hydroxide, ammonium sulfate, ammonium nitrate, and so on. They may also contribute to pH control. Nitrogen content must be taken into account when evaluating the cost of a potential source.

The ratio of carbon to nitrogen and the source of nitrogen are important in biosurfactant production [18] during the growth phase of production and in stationary phase. In the former case, *Arthrobacter paraffineus* ATCC 19558 demonstrated increased yields with aspartic acid, asparagine, glycine, or glutamic acid, yeast extract, peptone, bactotryptone, and nutrient broth. Different strains of *Corynebacteria* showed positive effects with nitrate and ammonia. *Nocardia erythropolis* biosurfactant production required 0.02% yeast extract for optimum yields. Ustilagic acid production by *Ustilago zeae* was optimized with 0.06% corn steep liquor.

Pseudomonas aeruginosa produced the highest rhamnolipid yields if nitrate rather than ammonium was used at a specific carbon-to-nitrogen ratio [18 found by Guerra-Santos, Ref. 54, and 38 determined by Ramana, Ref. 55]. Limitation of ammonium (less than 5 mM) with small concentrations of glutamic acid (170 μM) present in proteose peptone were also shown to be beneficial for rhamnolipid production by *P. aeruginosa* [56]. The addition of the amino acid glutamine (6 mM) inhibited rhamnolipid production. However, the substitution of ammonia with nitrate (20 mM) improved surfactant production by 30% for strain ATCC 9027.

For surfactin production by *B. subtilis*, ammonium nitrate was a superior nitrogen source than ammonium chloride or sodium nitrate [30]. Although ammonium nitrate contains two types of nitrogen, nitrate depletion occurred first and seemed to be more important to surfactin production than ammonium. Also, doubling the ammonium nitrate from 0.4% to 0.8% increased the surfactin production rate by a factor of 1.6. Product yield was not favored by the addition of an organic nitrogen source such as nutrient broth [28].

Another organic nitrogen source, yeast extract, was required for glycolipid production by *T. bombicola* [26] but was very poor for *P. aeruginosa* [23]. In the former case, urea was very poor as a nitrogen source but has not been investigated extensively for other cases.

Ammonia, ammonium, and nitrate salts are used as fertilizers and are produced at very favorable prices (Table 7). The price of ammonia is based on natural gas. Annual worldwide production is 80 million tons [32]. Urea, another cheap source, is produced from ammonia for fertilizer but has not been utilized extensively in fermentation processes. Technical and agricultural grades, which are less expensive, contain enough of the desired component. Bulk purchasing in tank cars and truck loads is also less expensive.

TABLE 7 Cost of Nitrogen Sources

Nitrogen source	% N	Unit price ($/kg)	$/kg N
Ammonia	82.3	0.12	0.15
Ammonium hydroxide	24.2	0.26	1.08
Ammonium nitrate	33.5	0.14	0.42
Ammonium sulphate	14.2	0.14	0.99
Soybean meal	8.0	0.18	2.25
Cottonseed meal	5.2	0.18	3.46
Urea	47.7	0.15	0.31
Yeast extract[a]	8.0	3.30	41.25
Hy-Soy[b]	9.5	10.89	114.63
N-Z-Case[c]	13.1	10.89	83.13
Corn steep water[d]	6.7	0.14	2.09
Sodium glutamate	9.5	1.67	17.58
Brewer's yeast	8.0	2.40	30.00
Whey	1.5	0.60	40.00
Black strap molasses	1.5	0.08	5.32

[a]Amberex 1003, Universal Foods Corporation, Milwaukee, WI.
[b]Hydrolyzed soybean meal, Sheffield Products, Norwich, NY.
[c]Hydrolyzed casein, Sheffield Products, Norwich, NY.
[d]Corn Products, Englewood Cliffs, NJ.
Source: Ref. 39.

Amino acids can be obtained inexpensively from soybean meal, cottonseed meal, and peanut meal (0.18, 0.18, and 0.23 $/kg, respectively) [39]. They are added as solids to the fermentation and then are broken done enzymatically. Their initial compositions are compared in a review article on raw materials for the fermentation industry [44]. Their prices are dependent on the world market demand for animal feed. They are high in total amino acid content, but not as free amino acids. Various hydrolysate products are available for the industrial fermentation market (Hy-Soy from soybean meal). However, the hydrolysis process increases prices significantly.

Yeast extract, cornsteep liquor, and peptones (from casein or meat) are good sources of free amino acids (lysine, leucine, isoleucine, valine, threonine, and phenylalanine), nucleotides, and vitamins [44]. Corn steep liquor is the least expensive source, because it is a by-product of the production of starch from corn. It contains proteins, sugar, and acids (especially lactic acid) in addition to salts and vitamins and should contain a high solids content to reduce transportation costs. Its composition can vary depending on production conditions. Due to their expense, utilization of yeast extracts and peptones should be minimized as much as possible.

Complex carbon sources are also very good sources of nitrogen, particularly amino acids. The composition of whey, peat hydrolysate, and molasses are shown in Table 8 (unpublished data). In these cases, either little or no nitrogen

TABLE 8 Free Amino Acid Compositions of Various Complex Substrates

	Concentration (μg/mL)		
Amino acid	Whey permeate	Peat Hydrolysate[a]	Black strap molasses[b] (4% solution)
Aspartic acid	5.37	46.42	24.64
Alanine	2.49	8.71	39.92
Threonine	1.19	4.39	<0.01
Leucine	1.08	2.63	2.08
Phenylalanine	0.57	1.25	<0.01
Histidine	4.34	1.25	32.24
Arginine	3.91	3.88	21.04
Glycine	4.95	14.26	60.40
Glutamic acid	26.17	2.34	258.88

[a]Peat of H3 decomposition on the von Post scale was used.
[b]Obtained from Lantic Sugar, Montreal, Canada.
Source: From unpublished data.

supplementation is then required. For example, for surfactin production in peat hydrolysate, only diammonium phosphate (0.4%) was added.

Wastes (particularly agricultural) are potential nitrogen sources [32]. They include harvest residues, straw, coconut shells, banana skins, and animal wastes (manure and urine). Sources from the food industry include starch wastewater, seed hulls, and waste from fruit- and vegetable-processing plants. Stillage (distillation residue from the production of ethanol by fermentation of starch materials) and waste from the pulp and paper industry are other examples of nitrogen sources that could provide nitrogen for the inexpensive production of biosurfactants.

C. Phosphate Source

Phosphorus is another important element for bacterial growth and product development. For example, the ratio of carbon to inorganic phosphorous is very important in the production of rhamnolipids by *P. aeruginosa* [54]. Maximum production occurred at a ratio of 16 to 1. Inorganic phosphate is also used for its buffering capacity. Typically, various forms (sodium, potassium, or ammonium phosphates) of mono- or dibasic salts are used. Costs are compared in Table 9. Ammonium phosphates are the least expensive. Phosphoric acid has also been used to replace the phosphate salts in pilot studies for rhamnolipid production [24]. Wastes such as phosphate scum from sugar juice clarification can contain significant inexpensive amounts of phosphate. Various organic sources can also contain bound inorganic phosphates such as oils, proteose peptone (65 μM P_i) [57, 58] and whey (20 μM P_i) [59]. They are difficult to account for, as they are frequently complexed with calcium and magnesium and are released according to pH.

The influence of phosphate in a proteose peptone/glucose medium on rhamnolipid production by *P. aeruginosa* ATCC 9027 has been studied [58]. For surfactant production, this medium was superior to a phosphate buffered medium (37 mM P_i) and a phosphate sufficient nutrient broth (310 μM P_i). Little or no surfactant was produced in the latter media. The phosphate was completely utilized during growth in the proteose peptone medium and a concentration of surfactant of 30 times the CMC was obtained.

D. Metals

In general, the presence of metals does not have a significant effect on biosurfactant production. The addition of various metals did not effect product yields of *T. bombicola, Corynebacterium lepus, N. erythropolis*, and *Nocardia amarae* [60]. However, both *P. aeruginosa* [23] and *B. subtilis* [20, 30] were affected by

TABLE 9 Comparison of the Cost of Phosphates

Phosphate source	Unit cost ($/kg)	%P
Ammonium phosphate		
Monobasic	0.17	27
Dibasic	0.18	23
Sodium phosphate		
Monobasic	1.35	22
Dibasic	1.32	12
Potassium phosphate		
Monobasic	2.17	23
Phosphoric acid		
52–54%	0.29	17
70%	0.34	22

Source: Ref. 39.

the addition of $FeSO_4$. In the former case, the $FeSO_4$ concentration had to be minimized to 0.5 mg/L. For *B. subtilis*, surfactin yields could be increased with 0.2 g/L $FeSO_4$ [30] or 0.42 mg/L $MnSO_4$ [28] supplementation. The cost of $FeSO_4$ and $MnSO_4$ are $0.01/kg and $0.48/kg, respectively [39]. Many of the complex carbohydrates, in addition to water, contain these salts in the form of impurities. Since requirements are minimal, supplementation may not be necessary for these microorganisms. For example, peat hydrolysate contains phosphorus (56 mg/L) and manganese (6.2 mg/L) [49].

E. Other Components

Sulfur, potassium, calcium, sodium, and magnesium are also requirements for microbial growth. These seem less important to biosurfactant production. Magnesium, sulfur, sodium, and chloride is added in the form of magnesium sulfate ($0.35/kg) and sodium chloride ($0.06/kg salt) have been added [39] but are probably present in sufficient quantities in the water [32]. Vitamins can be found in nitrogen sources, such as yeast extract, and carbon sources, such as whey.

Acids and bases must also be added for pH control. Phosphoric acid, hydrochloric acid, and sulfuric acid are all used. Ammonium hydroxide, potassium hydroxide, and caustic soda can be chosen for base addition. Their costs are compared in Table 10.

TABLE 10 Comparison of Various Acids and Bases

Compound	Concentration (%)	Unit cost ($/ton)
Acids		
Hydrochloric	36.5	55
Sulfuric	98	75
Phosphoric	54	270
	75	350
Bases		
Potassium hydroxide	45	290
Ammonium hydroxide	29	260
Caustic soda	50	290
Potassium bicarbonate	98	630
Sodium bicarbonate	98	380

Source: Ref. 39.

III. INCREASE OF BIOSURFACTANT YIELD

A. Biosynthesis Control

An effective method of decreasing production costs is to increase product yields. Recovery, capital, and raw material costs can thus be decreased. Several methods can be used to achieve this objective. Insufficient information is available on the metabolic pathways involved in biosurfactant production. This knowledge is essential for biosynthesis control. Biosynthesis involves the synthesis of the lipid moiety, the nonlipid moiety (sugar, peptide, etc.), and the subsequent linkage of the two portions.

The pathways involved in biosynthesis are dependent on the carbon source and the type of biosurfactant produced. For example, a glycolipid surfactant synthesized from a carbohydrate will be regulated by both the lipogenic pathway and the formation of the sugar portion (glycolytic metabolism) [61]. It is for this reason that the addition of lipophilic compounds to the carbohydrate enhanced the production of *T. bombicola* glycolipid [26, 62]. The additional cost of precursors, such as soybean oil, can be offset by the increase in yield.

When a hydrocarbon substrate is used, the mechanism will be principally lipolytic and gluconeogenic [61]. Since this has been described previously, more attention will be paid in this chapter to the carbohydrate metabolism of *P. aeruginosa* and *B. subtilis*.

In the case of glycolipid formation by *P. aeruginosa*, it is probable that the formation of the lipid portion of the molecule is limiting [23]. Various factors regulating the synthesis of the lipid should be investigated. Once these are known, the synthesis of biosurfactant will no longer be limiting [61]. Nutrient limitation (i.e., nitrogen and phosphorus) may promote lipid and hence biosurfactant production. Some preliminary studies have been performed [56–58].

Nitrogen limitation has been shown to be a factor in *P. aeruginosa, U. maydis,* and *N. erythropolis* biosurfactant production [15]. In *P. aeruginosa*, growth proceeds until nitrogen limitation when carbon is in excess [56]. Glutamine synthetase (GS), an intracellular enzyme that catalyzes the conversion of ammonia to glutamine as nitrogen become limiting, was studied in relation to biosurfactant production. As growth slows, glutamate and ammonia are utilized while GS activity increases. Cell metabolism then switched from nitrogen (amino acid) to glucose metabolism. Rhamnolipid production was then induced. Glutamine synthetase activity, and subsequently biosurfactant production, are enhanced by the limitation of ammonia and utilization of nitrates or organic nitrogen such as glutamate. Although a complex nitrogen source (proteose peptone) was used in this study, small amounts of the less expensive glutamate should achieve the same effect.

Phosphate limitation is another way to influence the metabolism of *P. aeruginosa*. In an inorganic phosphate-limited medium (65 μM P_i) of proteose peptone/glucose/ammonium salts, biosurfactant production was induced upon phosphate depletion [58]. The change in activity of several intracellular enzymes (alkaline phosphatase, glucose-6-phosphate dehydrogenase, and transhydrogenase) that are also dependent on phosphate levels indicated a shift in metabolism as biosurfactant production occurred. Thus, regulating phosphate levels is a means of enhancing rhamnolipid production.

As the addition of the precursors, glutamic acid, leucine, aspartic acid, and valine did not enhance product yields, lipid accumulation also seems to be the major limiting factor in surfactin production by *B. subtilis* [28, 30]. Vater [63] showed that biosynthesis occurred during the logarithmic phase of growth and continued thereafter. However, optimal formation was observed at the end of logarithmic growth at low growth rates.

The control of the intermediary metabolism from carbohydrates in *B. subtilis* is thus very important to surfactin production (Fig. 6) [61]. The key to lipid formation is at the acetyl co-A/isocitrate dehydrogenase level. Surfactin production was initiated as isocitrate dehydrogenase (ICD) activity decreased (Fig. 7). De Roubin et al. [30] determined that decreasing isocitrate dehydrogenase activity by (1) decreasing the oxygen concentration, (2) increasing the growth rate, and (3) by adding citric acid, the metabolism of *B. subtilis* could be switched from the tricarboxylic acid (TCA) cycle (necessary for energy) to lipid accumulation and increased levels of surfactin. Since surfactin formation occurs throughout growth, the switching of the pathway from the TCA cycle is not

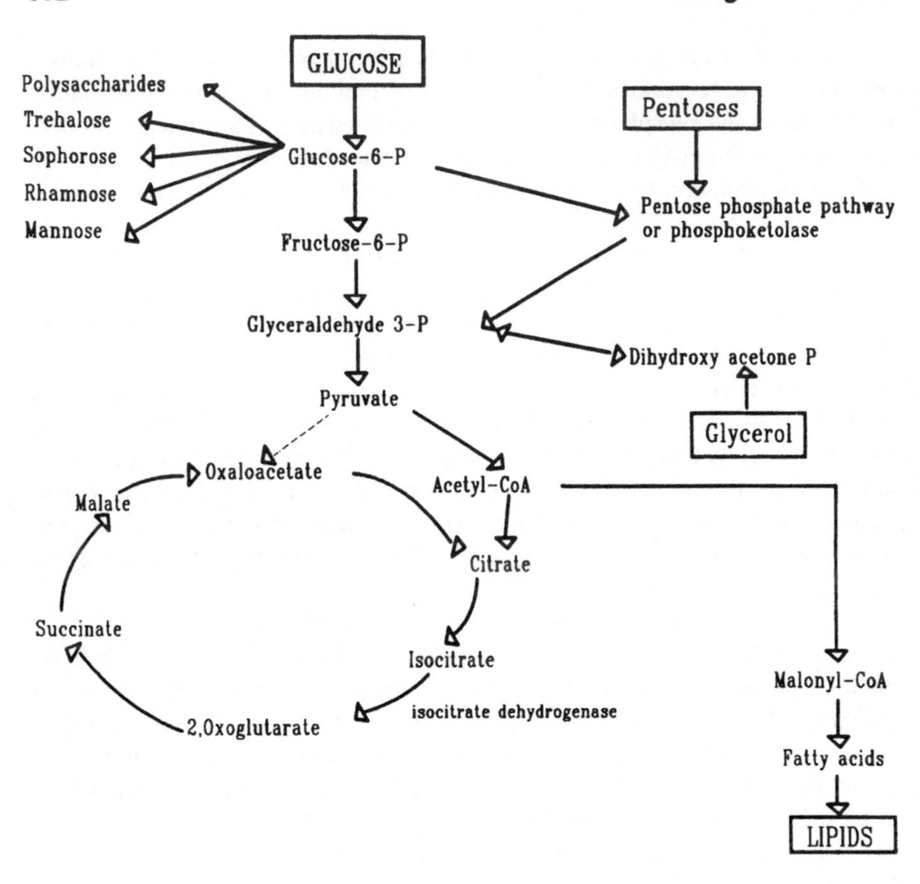

FIG. 6 Metabolic pathways of glucose utilization during biosurfactant production. (From Ref. 61.)

complete. However, production is favored as the pathway for lipid accumulation dominates.

To achieve maximal biosurfactant production, therefore, it is necessary to know the factors influencing the metabolism of the microorganism. A clearer understanding of this phenomenon must be realized so that one can manipulate culture conditions to overcome the rate-limiting synthesis steps.

B. Screening of Overproducers

Another method to increase biosurfactant yields is through the utilization of overproducing strains. For example, a chloramphenicol resistant strain, *P. aeruginosa*

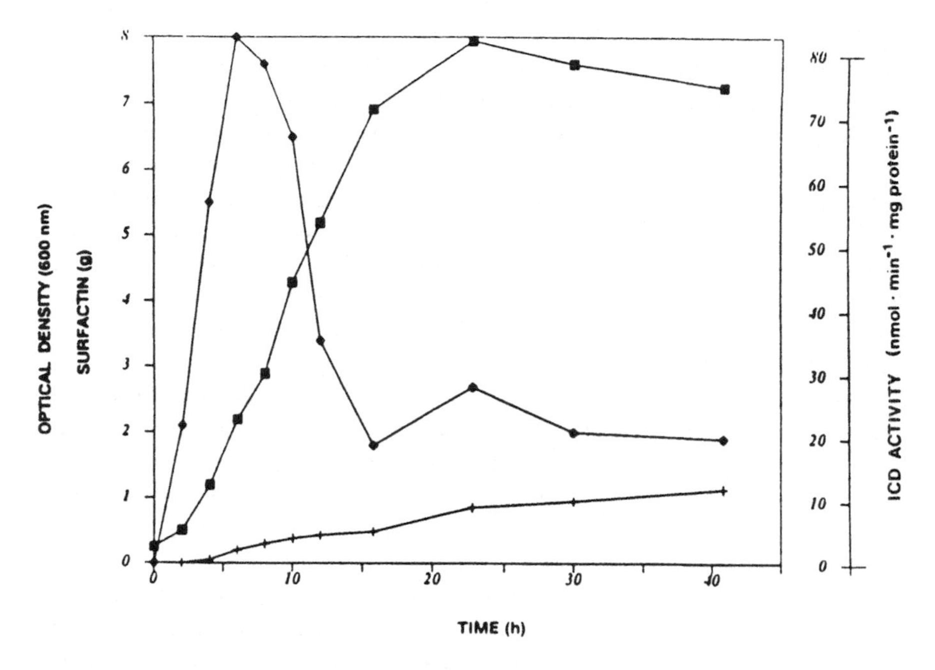

FIG. 7 Batch surfactin production by B. *subtilis* ATCC 51338 in 4% glucose, 0.4% ammonium nitrate, and 0.32 mM $FeSO_4$ medium. Optical density (■), ICD activity (◆) and surfactin (+) were followed during growth. (From Ref. 30.)

ATCC 9027 var. RCII, produces approximately twice the amount of rhamnolipid as does the *P. aeruginosa* ATCC 9027 strain for the same cultivation conditions [57, 58]. The activities of the enzymes involved in phosphate (alkaline phosphatase) and nitrogen (GS) metabolism were also heightened in the higher yield strain.

In general, the isolation process is very slow, because each strain must be purified, grown, and the production of surfactant monitored by surface or interfacial tension measurement or foam formation and/or an emulsion of a lipophilic substrate with the culture broth is observed [18]. Good surfactants are those that lower the surface tension below 35 mN/m.

It is particularly useful to isolate cultures from an oil or lipid medium where biosurfactant production is likely. An example is shown of the isolation of a *P. aeruginosa* strain from soil containing oil [64]. The isolates strains that indicated good characteristics would then be used in enhanced oil recovery. Mixed strains could also be useful to produce a blend of surfactants with interesting properties. This approach also leads to fewer problems with contamination, thus increasing productivity.

The isolation methods should be rapid and low cost to avoid dealing with thousands of nonproducing strains. An improved method of screening for microbes that produce biosurfactants from nonhydrocarbon medium involves the use of blood agar plates [65]. *Bacillus subtilis* produces surfactin on these plates and lyses the red blood cells in proportion to the amount of surfactin. Other surfactin producers were also found using this method. Cultures are streaked on the blood agar plates, incubated, and examined for clearing zones around the colonies. This could quickly indicate surfactant production. For confirmation, the microorganism can then be grown in liquid medium and the surface tension measured to monitor production. A large number of strains can be screened quickly by this method.

Another possible method involves the isolation of strains by indirect means. For example, *P. aeruginosa* strains with elevated GS activity levels produce higher yields of surfactants [56]. Strains that grow well in glutamic acid and limited ammonia medium agar have higher GS activity levels. These screening techniques are well known [66]. Once these strains are found, they can be tested for biosurfactant production in liquid medium.

Another simple screening method has been studied by Persson and Molin [67]. Cultures of *Pseudomonas* and *Vibrionaceae* were isolated from the environment and grown on agar plates. Each colony was suspended in water on a microscope slide. If the drop spread over the surface of the slide, the strain was positive and tested further. Over 2000 strains were tested in this manner.

C. Alteration of the Genetics of Producers

Genetically manipulating microorganisms to produce larger quantities of biosurfactant is another approach gaining interest. This includes methods to clone, amplify, delete, and transfer genes. Stable, high-expression biosurfactant production strains would be extremely useful. Very little is known on this subject, particularly since information is lacking regarding the pathways involved in biosynthesis. Genetic engineering to remove or alter biosurfactant regulatory enzymes is a possibility.

In the case of *B. subtilis*, enhanced biosurfactant production was achieved by ultraviolet radiation mutation [68]. A strain was isolated that produced approximately 3.5 times more surfactin than the parent strain. In addition, the production rate increased by the same factor. A mutation to enhance production occurred between *arg*C4 and *his*A1 on the genetic map. This information provides a target for future manipulation. Although protoplast fusion was used in this case to determine where the mutation had occurred, it could also be used to combine cultures of enhanced production with those of high substrate utilization.

Another approach is to tailor biosurfactant producing microorganisms to utilize inexpensive substrates they cannot normally use. For example, Koch et al. [53] have constructed strains *P. aeruginosa* PAO and PG201 to grow in lactose-based medium (whey) by inserting *E. coli lac*ZY genes and produce rhamnolipids. Chakrabarty [51] has isolated derepressed mutants of the hydrocarbon-degrading *P. aeruginosa* SB1 strain, which can produce rhamnolipids on hexadecane, glucose, or chicken fat. A mutated strain of *Pseudomonas* has been used to thrive on 2,4,5-trichlorophenoxyacetic acid and produce a glycoprotein to emulsify sludge trapped at the bottom of oil tankers [69].

Microorganisms could also be altered to improve a number of characteristics to improve the economics of fermentation. This could include a short start-up time, low nutrient requirements, high growth rates, and stability. Genetic enhancement of microbial strains to increase biosurfactant production using various substrates is a promising area in which extensive work is needed, provided that the cost is reasonable.

In addition, strains could be constructed to produce by-products, such as biomass, enzymes, and organics that could be recovered from the fermentation and sold to obtain a credit, thus reducing the product cost. For example, a strain of *B. subtilis* S499 has been found that coproduces an antifungal agent, iturin, and surfactin [46]. *Pseudomonas fluorescens* produces lipase and siderophores, in addition to biosurfactants [70]. These products could be sold together or separately.

IV. OPTIMIZATION OF THE FERMENTATION PROCESS

A. Reactor Design

Fermentation is the main process by which the product is synthesized. Both the type of reactor and the operating conditions must be optimized to achieve optimum production rates and yields economically. As in most fermentations, reactor development for biosurfactant production has been minimal. The vast majority of experiments have used the stainless steel stirred tank in batch mode.

An innovation to batch fermentation was made by Ramana and Karanth [64]. It was shown that, when *P. aeruginosa* CFTR-6 cells were transferred from a growth medium to a phosphate-limited medium, the glycolipid surfactant production increased by a factor of two. Fermentation time is also reduced from 96 to 24 h. Thus the productivity of this system (a resting cell system) compared to the conventional batch is significantly increased.

Continuous fermentations have several advantages over batch. Higher productivity is achieved by decreasing the time for start-up, harvesting, and cleaning. This translates into reduced labor costs [32]. Also, since biosurfactant production is optimal under nutrient limitation, this condition can be maintained. Pilot plant production (50 L) using *P. aeruginosa* was investigated in continuous mode using a novel compact loop (COLOR) bioreactor [24]. A dilution rate of 0.065/h was used to take advantage of optimum production by slowly growing cells. Continuous fermentations have been performed with *Pseudomonas fluorescens* 378 [70]. Dilution rates of 0.05–0.15/h were examined under nitrogen, phosphorous, iron (III), and oxygen limitation.

For a product inhibition reaction such as surfactant production, the use of a plug flow reactor [71] that can be approximated by several stirred tank reactors in series would be advantageous. The batch system of Ramana and Karanth [64], mentioned previously, is similar to this configuration. Different conditions can be maintained in each reactor. Growth would be emphasized in the first set of reactors and then the slower growing cells could then be optimized for surfactant

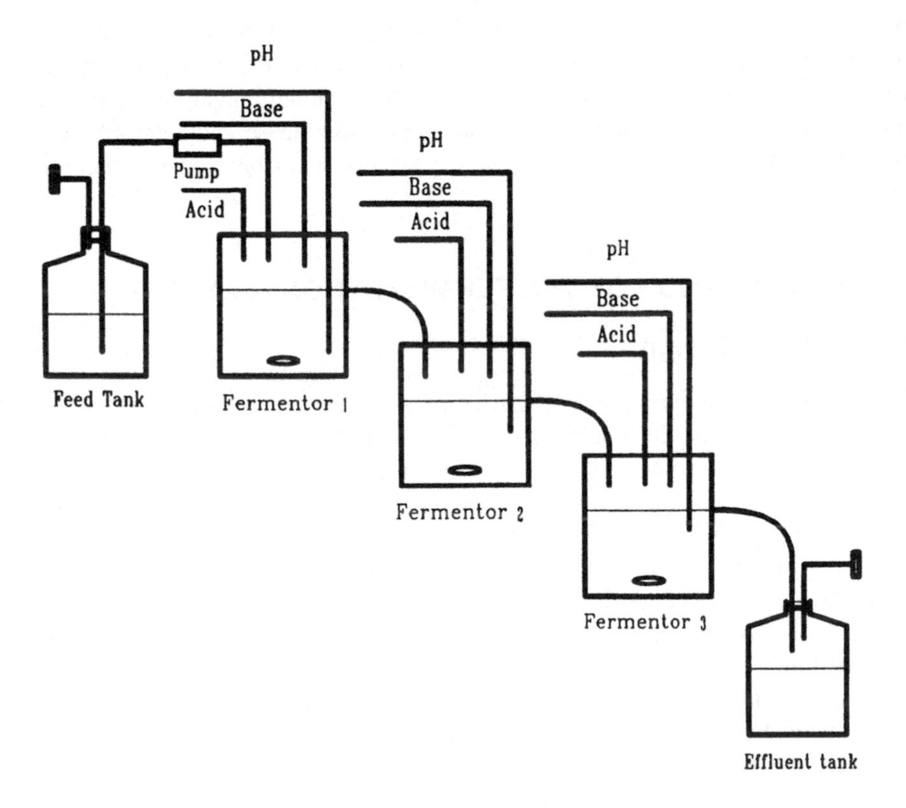

FIG. 8 Experimental set-up of a continuous three-stage fermentation. (From Ref. 72.)

production in the others. Growth, product yields, and productivity could be increased by such a three-stage system (Fig. 8) as shown recently for lactic acid production [72].

Optimum reactor configuration design can reduce aeration requirements. The height-to-diameter (H/D) ratio is an important factor. Tall narrow tanks are advantageous because of longer bubble residence time, higher dissolved oxygen from higher air pressure at the sparger, and the reduced volume of air [73]. Therefore, power consumption costs are less. Oxygen requirements are high as noted for *P. aeruginosa* [54].

Foaming is a major consideration during biosurfactant production. It decreases the working volume of the fermentor, leads to increased risk of contamination and losses of biomass and product. Glucose concentration had to be reduced during pilot plant studies because of excessive foaming [24]. Mechanical foam breakers have been used but are energy intensive (1000 rpm). Sonic devices have been used for other applications than biosurfactant production [32]. Other methods need to be investigated. Broth recycle can be used where the foam is collected, collapsed, and returned to the fermentor. Foam collection has been utilized during surfactin production [28] for batch fermentations. This is particularly effective since an inhibitory product is removed.

During continuous fermentation, cell concentration can be increased by recycling to the fermentor. This procedure in turn increases the productivity of the system. Mattei and Bertrand [74] have described a system with tangential ultrafiltration that removes the biosurfactants continually and recycles the cells to the reactor. Higher dilution rates (0.1–0.5/h) can then be utilized, decreasing the size of the reactor that is required. In general, methods to increase reactor productivity have not been studied intensively.

Heat removal is another aspect of fermentor design. Jacketed fermentors and cooling coils are used, but helical coils are cheaper and more efficient. Cooling towers are a more energy effective way (than mechanical compressors) of reducing the cooling water temperature so it can be returned to the fermentor. Hydrocarbon substrates have a much larger heat release than carbohydrates and thus should be considered when choosing a substrate [73].

A few other systems have been evaluated for biosurfactant production. Wagner et al. [75] investigated the use of immobilizing *Pseudomonas* spp. and *R. erythropolis* cells in calcium alginate, agar, carrageenan, chitosan, or polyacrylamide beads and placing them in a fluidized bed reactor. The best results were obtained with calcium, but they were no better than those using slow-growing cells. Therefore, the additional capital cost associated with bead-making equipment cannot be justified.

Another approach, continuous phasing in a cyclone fermentor, was investigated [76]. Oxygen transfer is superior in this type of reactor. Dosing by adding timed quantities of the limiting nutrient (nitrogen) was used to synchronize the growth of all the cells. This fermentation was computer controlled to ensure a constant relationship between the nutrient level and cell growth. Although this method is feasible (surfactin production by *B. subtilis* occurred), the true capability of this technique in increasing surfactin production is unknown.

Fermentors with air agitation such as the cyclone fermentor should be examined for biosurfactant production, because compared to mechanical agitation they are cheaper to fabricate, sterility is improved, heat of mechanical agitation is eliminated, no maintenance of bearings and seals is required, and large fermentors can be constructed since their size is not dependent on motor size and shaft length [73].

The use of an airlift fermentor was investigated for *N. erythropolis* and compared to a mechanically agitated fermentor [77]. Since the construction is simpler, five times less power for aeration and mixing is required for the airlift. However, product yield decreased by a factor of four compared to the mechanical agitation. An airlift fermentor with a different geometry could provide better results. Airlift percolators, as used in the petroleum industry, were successful for the growth of six bacterial strains to degrade furnace oil and produce a biosurfactant that could emulsify the oil in the water phase [78].

B. Fermentor Operating Conditions

Operating conditions of the fermentation must be optimized to increase product yield and decrease operating costs. Temperature and pH can significantly effect biosurfactant yields. The control of pH by the addition of acid and/or base control is a necessity in many cases. For *P. aeruginosa*, pH 6.25 ± 0.1 and 33 ± 2°C was optimal [23]. For *B. subtilis*, it was pH 6.7 ± 0.1 and 37 ± 1°C [30]. The surfactin production rate decreased by 50% from pH 6.7 to pH 7.2. For both *P. aeruginosa* and *B. subtilis*, production ceased at pH 7.5

Another factor is agitation–aeration. As can be seen in Fig. 9, to minimize operation costs a balance of the two must be achieved. Aeration requirements as a function of yield must be determined. For *B. subtilis*, by increasing the aeration rate from 0.6 vvm to 1.75 vvm, a 4.8-fold increase in the production rate was achieved [30]. In the case of *N. erythropolis* [18], the culture is shear sensitive. Biosurfactant yields decreased substantially as agitation increased from 250 rpm to 500 rpm. On the other hand, increased agitation increased the oxygen transfer rate, resulting in enhanced cellobioselipid production by *U. zeae*. Hydrocarbon substrates also require more agitation due to difficulty in dispersing them in the medium. Different agitator types that are more energy efficient have not been investigated.

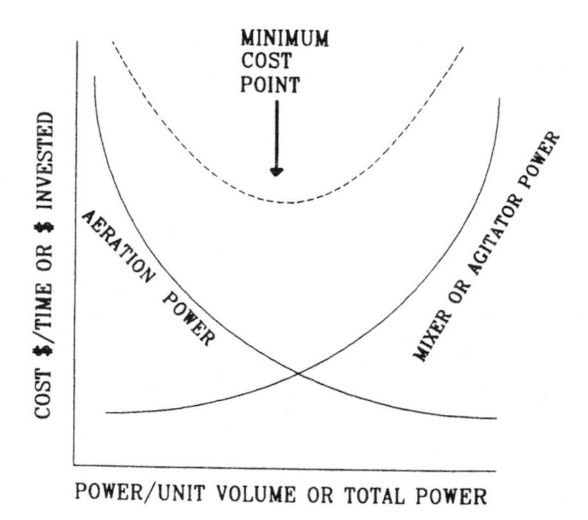

FIG. 9 Minimization of total aeration/agitation costs. (From Ref. 32.)

C. Recycling of Spent Medium

Very little attention has been paid to the utilization of the various components in the spent medium after biosurfactant production. For example, although the residual sugar concentration that was monitored (0.7 g/L of glucose) was found in pilot plant studies during rhamnolipid production [24], salts and other component concentrations in the spent medium are unknown. They could be recycled to the fermentor, thus significantly reducing the amount of fresh water and chemicals required in the process. Also, if recycling is used, dilution rates can be increased since higher residual concentrations can be tolerated. One disadvantage would be an accumulation of inhibitory substances in the fermentor. A purge could be used to offset this problem.

D. Production Scale

Current facilities for synthetic surfactant production from petroleum products are large-scale to exploit economies of scale (i.e., larger production plants are more cost effective). Biosurfactant production will have to follow this trend if it is to be profitable. Although operating costs are generally linear with production capacity, capital costs are exponential. To support a high production rate, the market will have to be large. For example, in 1984 in West Germany an estimation was made determining the influence of production scale on the price of a biosurfactant: 1000 t/a, $17.00/kg; 100 t/a, $18.00/kg, and 10 t/a, $50/kg for a concentration of 2.5 g/L [79]. These production levels are representative of the food industry.

V. PRODUCT RECOVERY

A. Factors Involved in Biosurfactant Recovery

Once the product has been synthesized, several factors influence the costs involved in product recovery. They include concentration of the product from fermentor, broth characteristics (pH, viscosity, complexity), product characteristics (molecular weight, solubility, charge, cell wall bound or extracellular), product yield by the recovery method, final product purity, and simplicity of the recovery method. Recovery costs can be reduced by improving product yield, low material costs, and combining steps.

The usual operations for chemical plants can be utilized for biosurfactant recovery [32]:

Broth conditioning (pH or heat change)
Centrifugation
Cell disruption
Filtration
Absorption or adsorption
Solvent extraction
Concentration (ultrafiltration, reverse osmosis, liquid membranes)
Distillation
Precipitation or crystallization
Drying

The effect of surfactant concentration in the fermentation broth on the final product cost has been shown by Kachholz et al. [79]. For example, the projected cost of a surfactant (2.5 g/L) would be $17.00/kg compared to $5.00/kg at a higher concentration (20 g/L) for the same plant size (1000 t/a). The complexity of recovery equipment increases as the initial product concentration decreases. More steps may be required. This demonstrates the importance of improved product yields during the fermentation and how it can influence downstream processing.

The glycolipids produced by *T. bombicola* can be recovered from the fermentation broth by gravity separation since they are denser than the broth. Contamination by medium components is minimal [26]. Recovery costs are reduced by not separating the individual surfactants. This approach is acceptable since many synthetic surfactants are sold as mixtures. This is a key example of a low cost and simple recovery system.

The choice of raw material or culture conditions can influence the ease of recovery of the product. The addition of either sodium citrate or hydroxide to increase the pH to 3 from 2, converted the water-soluble product to large amounts (0.46 g product/g substrate) of crystalline glycolipids from *Torulopsis apicola*, which can be removed by simple filtration [80]. Further purification, if desired, can be achieved by liquid chromatography on silica gal. Hydrocarbons are, in general, difficult to separate from the surfactants without leaving trace amounts in the product. This is particularly important if the product is to be used in the food industry. Also, complex media (e.g., molasses) may also require additional purification steps over high dextrose corn syrup. However, a change to a more expensive substrate that yields little waste or sellable by-products may be worthwhile. Streams could also be recycled to the fermentor for reuse.

In general, intracellular compounds are more expensive to recover, because their recovery is energy intensive. Cells must first be disrupted mechanically or enzymatically before purification, adding an additional step to the process. Some compounds, such as the yeast mannoprotein [81], a major component of the *Saccharomyces cerevisiae* cell wall, are recovered by heat or enzyme treatment.

The mode of production of several biosurfactants is dependent on the substrate utilized. For example, *Corynebacterium lepus* produces a cell-bound surfactant in glucose, which must be extracted by hexadecane or another solvent [82]. When grown in hexadecane it is extracellular. When grown on ethanol or acetate, 75% of the biosurfactant produced by RAG-1 is intracellular, whereas all is produced extracellularly when grown on hexadecane [15]. Since most biosurfactants are either extracellular or wall-bound, the possible recovery techniques for these components are discussed in the next three sections.

B. Classical Recovery Methods

Classical techniques are well suited to batch recovery and thus batch fermentations. They include solvent extraction, precipitation, and crystallization. The cells must first be separated and either the cell mass or the supernatant is extracted for biosurfactants. Settling, flotation, centrifugation, or rotary vacuum filtration are used for this step. Although settling and flotation are the least expensive, it is generally not feasible for bacterial cells. Centrifugation, although effective, requires a high capital investment and high maintenance costs; heat generation during centrifugation is significant, which may damage the product. For filter separation, filter aids (carbon, clays, diatomaceous earth, etc.) must be added for bacterial cells. Because it is difficult to separate cells from *n*-alkane substrates [19], this step is omitted.

A variety of solvents can be used for product recovery from the fermentation broth or cell slurries (Table 11). The choice is dependent on cost and effectiveness. Cellobiose lipids, sophorolipids, and rhamnolipids can be extracted by a

TABLE 11 Cost of Various Solvents Used for Biosurfactant Recovery

Solvent	Cost ($/kg)
Butanol	0.84
Ethyl acetate	0.90
Hexane	0.30
Chloroform	0.79
Methanol	0.12
Dichloromethane	0.64
Ethyl ether	1.13

Source: Ref. 39.

solvent, which is then evaporated and condensed for reuse. In general, the use of solvents is time consuming, expensive, and not very specific. Further purification must be done by column chromatography, thin-layer chromatography, and/or crystallization [18]. Rhamnolipids and cellobioselipids can also be obtained by acidification of the supernatants with acid and refrigerating.

An example is the isolation of surfactin produced by *B. subtilis* [28]. After clarification of the fermentation broth by centrifugation, the pH of the supernatant is reduced by the addition of concentrated hydrochloric acid. Dichloromethane is added in three steps to extract the surfactin from the aqueous phase. After evaporation of the organic solvent, the residue is redissolved in water (at a pH of 8.0) and filtered to remove undissolved impurities. Hydrochloric acid and dichloromethane are again added to further purify the surfactant. This is a key example of a solvent- and labor-intensive process.

Product recovery yields from the fermentation broth can be quite low. For example, the yield of *Corynebacterium lepus* lipids was 20%–25% by solvent extraction [42]. Product losses are an important factor to be considered in selecting an appropriate recovery process. Process economics can be significantly affected. Yields of 30%–50% from recovery steps, can effectively double or triple the cost of previous steps.

C. *In Situ* Methods

In situ methods in which the products are continuously removed from the culture broth can be used for continuous fermentations. They have several advantages: solvent use is reduced, foam problems are reduced, product degradation is minimized, and fermentor yields can be increased due to avoidance of end product inhibition [18].

Surfactin produced by *B. subtilis* can be removed in the air exhaust line in the foam, which is then collapsed before precipitation and solvent extraction [28]. A negligible amount of product remains in the fermentor. Cell-wall bound compounds of *C. lepus* can also be removed continuously by contact with dodecene or hexadecene for 2–4 h [83]. *Rhodococcus erythropolis* DSM 43215 bound compounds were similarly extracted by kerosene [18].

Ion-exchange resins have also been used in pilot plant production studies using *P. aeruginosa* [24]. After the cells were removed by centrifugation, rhamnolipids were enriched by absorption chromatography on a Amberlite XAD-2 resin and purification with ion-exchange chromatography, followed by evaporation, and freeze-drying. The product was 90% pure with a 60% recovery. Buffer, resin replacement, and solvent costs must be taken into consideration.

D. Newly Developed Recovery Methods

Ultrafiltration (UF) membranes have been found to be useful to concentrate and purify biosurfactants in one step [84]. Rhamnolipids and surfactin could be concentrated by this method, while eliminating any residual glucose, phosphates, amino acids, and small molecular weight compounds. Because of the ability of surfactant molecules to form aggregates, these aggregates can be retained by relatively high molecular weight cutoff membranes. The molecular weight of the surfactin has recently been confirmed by fast atom bombardment (FAB) mass spectrometry (Fig. 10). Since aggregates of 50 to 100 molecules of surfactin are formed, a 50,000 molecular weight cutoff Amicon XM 50 membrane was used successfully. A concentration factor of 160-fold can be rapidly achieved (98% of the surfactin is retained by the membrane). The purity of the surfactin was approximately 97%.

Rhamnolipids produced by *P. aeruginosa* can also be retained by ultrafiltration [84]. The nature of two rhamnolipids has also recently been verified by ion-spray mass spectrometry (unpublished results). The YM 10 membrane (10,000 molecular weight cutoff) was the most appropriate. In this case, 92% of the surfactant could be recovered. Increasing the molecular weight cutoff of the membrane, increased surfactant loss significantly. For a YM 30 membrane (30,000 molecular weight cutoff) the retention was 80% and for an XM 50 membrane (50,000 molecular weight cutoff), it was 58.9%. In summary, it is a technique useful for all biosurfactants whose concentration is higher than the critical micelle concentration.

The XM 50 membrane has also been shown to be able to retain glycolipids produced by *Rhodococcus* sp. strain H13A [85]. A coproduced protein is removed into the filtrate. Isopropanol precipitation of the retentate removed the remaining

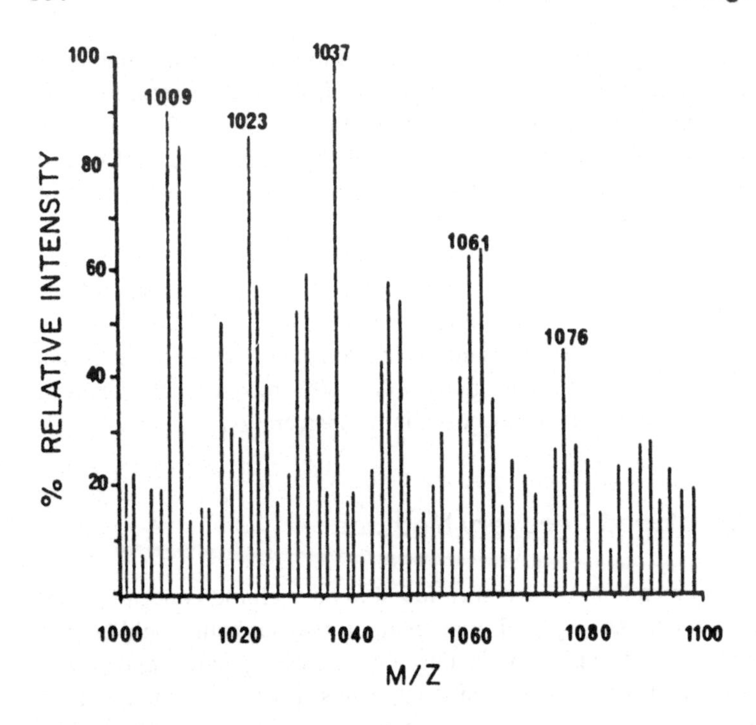

FIG. 10 FAB mass spectrum of purified *B. subtilis* surfactin. The protonated molecular ion is seen at 1037. (From Ref. 84.)

carbohydrate and protein. The use of other more appropriate membranes may eliminate the need for the isopropanol step.

The use of tangential flow filtration devices have been investigated by Mattei and Bertrand [86]. Mixed culture cells and residual hydrocarbons are recovered and recycled to the fermentor. Biosurfactants are accumulated in the filtrates and lyophilized. Large amounts of surfactants are produced continuously (3 g/L). A 60% product recovery efficiency was obtained. Since the product is collected from the filtrates of the membranes, an extra step is required to concentrate the product (lyophilization).

An integrated process for the continuous production of biosurfactant is proposed in Fig. 11. Cross-flow filtration is increasingly being used for cell recovery. The economics is superior to centrifugation [32] and the cells can be recycled without loss in viability. This technique has been studied for lactic acid production [87]. It is simple and can be used for many types of biosurfactants by matching any appropriate membrane (molecular cutoff and type of membrane). Adsorption of the biosurfactant on the membrane must be minimal. Energy

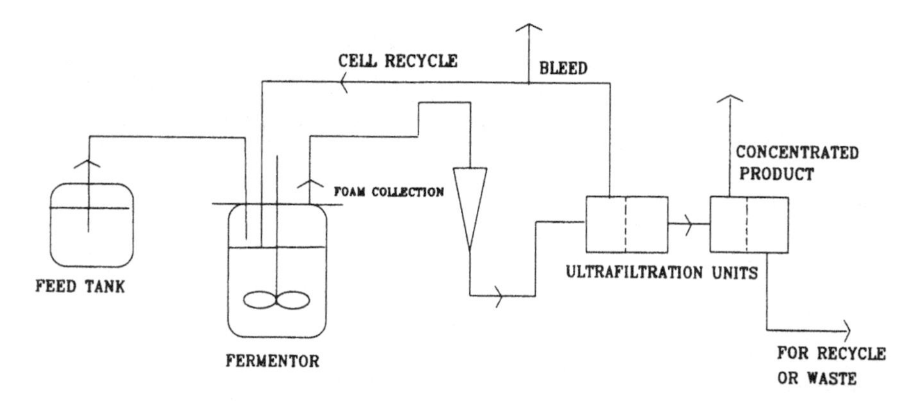

FIG. 11 Proposed process flowsheet for continuous biosurfactant production and recovery.

requirements, operating (cleaning) costs of the membrane system would be minimal if the correct membrane is used. Using mid-1985 costs, an operating cost for UF membranes was estimated to be $6 to $10/100 gal permeate [32].

Research in product harvesting and concentration is particularly required as little effort has been put into optimization to reduce product losses and decrease system complexity.

VI. BIOSURFACTANT PROPERTIES

A. Surfactant Efficiency and Effectiveness

If the quality of the biosurfactants is the same as synthetics, then high-capacity and lost-cost production are absolutely necessary. If the activity is higher than the synthetics, this is another matter. The future of biosurfactant commercialization will depend not only on the unit cost of the product but also on its effectiveness.

Critical micelle concentrations of biosurfactants (a measure of its efficiency) range from 1 to 2000 mg/L. Surface and interfacial tensions (measure of effectiveness) of good biosurfactants are less than 30 and 1 mM, respectively. For example, a commercial surfactant has a CMC of 8.0 mM and costs $1/kg. However, if a biosurfactant has a CMC such as surfactin of 0.01 mM and costs $20/kg, it would be less expensive to use since less of the surfactant is required.

A comparison of biosurfactants and synthetic surfactants is shown in Table 12. It is difficult to compare the properties of the biological with the synthetic, because the biological are poorly characterized due to mixtures of surfactants and interferences from broth components. For example, the CMC of surfactin

TABLE 12 Comparison of Biosurfactants with Synthetic Surfactants

	Surface tension (mN/m)	Critical micelle concentration (mg/L)	1990 Cost ($/kg)[a]
Producing organism			
R. erythropolis	37	15	12.20
P. aeruginosa	29	15	5.90
T. bombicola	37	82	2.80
B. subtilis	27	11 (0.01 mM)	20.32
Anionic synthetics			
Detergent alkylate dodecylbenzene (LABS)	47	590 (1.2 mM)	1.03
Sodium lauryl sulfate (SLS or SDS)	37	2023–2890 (7–10 mM)	0.95 (30%) 26.00 (98%)

[a]Biosurfactant costs are calculated on the basis of raw material accounting for 35% of the total cost, assuming kerosene, molasses and soybean meal as substrates.
[b]Accurate Chemical and Scientific Corp., Westbury, NY.
Source: Ref. 26.

produced by *B. subtilis* was determined initially by Cooper et al. [28] to be 25 mg/L. However, by measuring the concentration of the surfactin more accurately by acid hydrolysis of the compound and subsequent amino acid analysis (Fig. 12), it was determined to be 11 mg/L (0.011 mM) [84].

Rhamnolipids produced by *P. aeruginosa* have been measured by rhamnose content. This does not indicate the actual amount present, because two rhamnolipids are produced and a ratio of the two is assumed [24]. It makes it difficult to determine actual yields and properties. Once more accurate methods of measuring biosurfactant concentration are determined, it may be possible through genetic engineering, chemical modification, or culture conditions to alter the structures of the compounds so that they are more effective. The development cost, however, cannot be so high as to outweigh any gain in effectiveness.

B. Other Biosurfactant Properties

Several properties of the surfactant can also influence the economics of their use. They include biodegradability, specificity, temperature stability, toxicity, solubility, and pH range. If the process stream in which the surfactant is to be used must be adjusted for pH or temperature for the surfactant to be effective, this adds to the cost. Glycolipids produced by *A. paraffineus, T. bombicola*, and *P. fluorescens* are stable between 20°C and 90°C [88]. Salt concentration, impurities, and

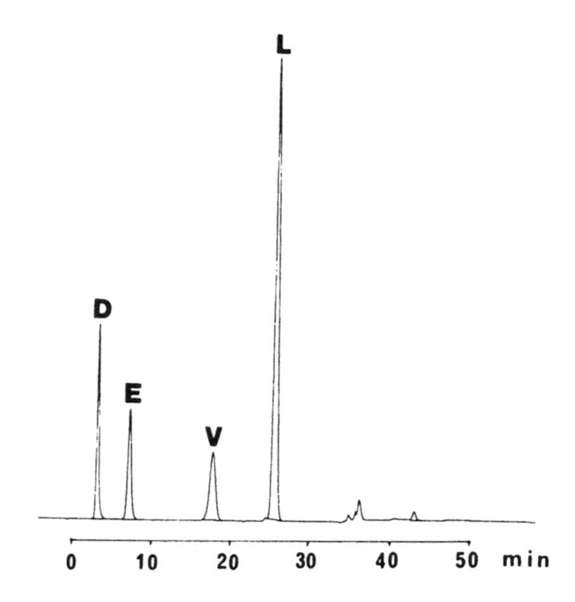

FIG. 12 Amino acid profile of purified surfactin after HCl hydrolysis. Amino acids represented are: D, aspartic acid; E, glutamic acid; V, valine; L, leucine. The surfactin concentration was then determined according to Mulligan and Gibbs [84].

pH can influence the CMC of a surfactant. For example, the rhamnolipid produced by *Pseudomonas* sp. MUB has a CMC of 5 mg/L at a pH of 3 and 30 mg/L at a pH of 9. It is more advantageous to have a surfactant that is not influenced significantly by process variations.

For enhanced oil recovery, it is desirable that the surfactant does not adsorb onto the reservoir rock [89]. A savings can be made if the surfactant can be recovered after flooding for reuse. In addition, less surfactant would be necessary, because the effectiveness is not reduced due to losses. Since synthetic compounds that adsorb onto the rocks are unstable at high salt concentrations and temperature, there is room for biosurfactant if they have these properties and their production costs are kept low enough to compete.

Biodegradability is also a requirement. If the surfactant is not toxic and is easily biodegradable, effluent treatment steps are not necessary before product release into the environment or after treatment with the surfactants. The latter case was shown in a study by Oberbremer et al. [8]. After the addition of sophorose lipids for the degradation of hydrocarbons, the biosurfactants were degraded. Also, avoidance of any toxic by-product formation during the fermentation process would also reduce waste treatment costs.

VII. CONCLUSIONS

Detailed economic analyses have not been performed to determine the cost of surfactant production. However, it can be concluded that there is much work to be done in the various aspects of biosurfactant production to make them economically competitive with synthetic surfactants. The choice of raw material can significantly influence the final product cost. In general, it is more economical to use bulk, readily available, technical grade materials.

Because of the rising cost of hydrocarbons and their subsequent depletion, alternative substrates will have to be used. The development of improved methods for the hydrolysis of lignocellulosics is especially necessary. Waste streams should also be investigated more thoroughly as no-cost substrates and as a way to reduce treatment costs. An optimization of product yield based on material cost must be achieved.

Product yields must also be enhanced to decrease product recovery costs resulting from dilute fermentation streams. New genetic techniques, biosynthetic control of the microorganisms involved in biosurfactant production through medium manipulation, and screening for overproducers must be exploited. By-product (protein, enzymes, antibiotics, etc.) sales would also enhance product economics.

Fermentation design has been limited with a few exceptions to the traditional stirred tank for biosurfactant production. Overall conversion of the substrate and increased productivity of the process by continuous fermentation should be the main concern of research efforts. *In situ* product removal is particularly desirable to increase inhibition, reduce solvent costs, and ease product recovery. Recycling of cells and spent medium should also be encouraged to reduce raw material and capital costs. Constant monitoring and control of nutrient levels through computerized systems could significantly increase product yields and consistency.

Microbial surfactants have a wide range of properties and structures. However, biosurfactants should also be well characterized to facilitate comparison with known synthetic surfactants. Its production cost, properties, and effectiveness must be determined. Scale-up information is lacking. Products could be used in applications either with or without modification. Only then can a market be developed for large-scale production of microbially produced surfactants. Multipurpose biosurfactants would also be desirable. High capacity, high activity, and low cost production and recovery are all essential.

REFERENCES

1. S. Wilkinson, C. Brady, L. Tantillo, N. Alperowicz, E. Chynoweth, D. Jackson, and L. Tattum, *Chem. Week*, Vol. 146, Jan. 31, 20 (1990).
2. D. Myers, in *Surfactant Science and Technology*, VCH Publishers, New York, 1988.
3. P. L. Layman, *Chem. Eng. News 63*: 23 (1985).

4. M. J. Rosen, in *Surfactants and Interfacial Phenomena*, Wiley Interscience, New York, 1975.
5. D. G. Cooper and J. E. Zajic, *Adv. Appl. Microbiol. 26*: 229 (1980).
6. P. Becher, in *Emulsions: Theory and Practice*, 2nd ed., Reinhold Publishing, New York, 1965.
7. R. Samson, T. Cseh, J. Hawari, C. W. Greer, and R. Zaloum, *Sci. Tech. Eau 23*: 15 (1990).
8. A. Oberbremer, R. Müller-Hurtig, and F. Wagner, *Appl. Microbiol. Biotechnol. 32*: 485 (1990).
9. D. G. Cooper, *Microbiol. Sci. 3*: 145 (1986).
10. N. Kosaric, W. L. Cairns, N. C. C. Gray, D. Stechey, and J. Wood, *J. Am. Chem. Soc. 51*: 1735 (1984).
11. M. Robert, M. E. Mercadé, M. P. Bosch, J. L. Parra, M. J. Espiny, M. A. Manresa, and J. Guinea, *Biotechnol. Lett. 11*: 871 (1989).
12. M. E. Hayes, E. Nestau, and K. R. Hrebenar, *Chemtech. 16*: 239 (1986).
13. Anonymous, *Chem. Week*, Vol. 135, Jan., 58 (1984).
14. Anonymous, *Res. Money 4*: 8 (1990).
15. J. Reiser, A. K. Koch, K. Jenny, and O. Käppeli, in *Biotechnology for Aerospace Applications*, Symposium on Biotechnology for Aerospace Applications March 1-2, 1989, US Air Force Academy, Colorado Springs, CO, Advances in Biotechnology Series, Vol. 3, 1989, pp. 85–97.
16. M. Javaheri, G. E. Jenneman, M. J. McInerrey, and R. J. Knapp, *Appl. Environ. Microbiol. 50*: 698 (1985).
17. M. Biermann, F. Lange, R. Piorr, U. Ploog, H. Rutzen, J. Schindler, and R. Schmid, in *Surfactants in Consumer Products, Theory, Technology and Application* (J. Falbe, ed.), Springer-Verlag, Heidelberg, 1987.
18. C. Syldatk and F. Wagner, in *Biosurfactants and Biotechnology*, Vol. 25, Surfactant Science Series (N. Kosaric, W. L. Cairns, and N. C. C. Gray, ed.), Marcel Dekker, New York, 1987, pp. 89–120.
19. S. Lang and F. Wagner, in *Biosurfactants and Biotechnology*, Vol. 25, Surfactant Science Series (N. Kosaric, W. L. Cairns, and N. C. C. Gray, eds.), Marcel Dekker, New York, 1987, pp. 21–45.
20. C. R. Macdonald, D. G. Cooper, and J. E. Zajic, *Appl. Environ. Microbiol. 41*: 117 (1981).
21. A. Kretschmer, H. Bock, V. Wray, and F. Wagner, *Appl. Environ. Microbiol. 44*: 864 (1982).
22. S. Itoh, H. Honda, F. Tomita, and T. Suzuki, *J. Antibiot. 24*: 855 (1971).
23. L. Guerra-Santos, O. Käppeli, and A. Fiechter, *Appl. Environ. Microbiol. 48*: 301 (1984).
24. H. E. Reiling, U. Thanei-Wyss, L. H. Guerra-Santos, R. Hirt, O. Käppeli, and A. Fiechter, *Appl. Environ. Microbiol. 51*: 985 (1986).
25. S. Harvey, I. Elashvili, J. J. Valdes, D. Kamley, and A. M. Chakrabarty, *Bio/Technology 8*: 228 (1990).
26. D. G. Cooper and D. A. Paddock, *Appl. Environ. Microbiol. 47*: 173 (1984).
27. Kao Soap Ltd., West German patent no. DE 2939519 (1980).

28. D. G. Cooper, C. R. Macdonald, S. J. B. Duff, and N. Kosaric, *Appl. Environ. Microbiol. 42*: 408 (1981).
29. A. Kakinuma, A. Oachida, T. Shima, H. Sugino, M. Isono, G. Tamura, and K. Arima, *Agric. Biol. Chem. 33*: 1669 (1969).
30. M. R. de Roubin, C. N. Mulligan, and B. F. Gibbs, *Can. J. Microbiol. 35*: 854 (1989).
31. D. G. Cooper, D. W. Pillon, C. N. Mulligan, and J. D. Sheppard, *Fuel 65*: 255 (1986).
32. H. Reisman, in *Economic Analysis of Fermentation Processes*, CRC Press, Boca Raton, 1988.
33. Anonymous, *Bioprocessing Technol. 12*(10): 1 (1990).
34. D. G. Cooper, J. E. Zajic, and D. F. Gerson, *Appl. Environ. Microbiol. 1*: 4 (1979).
35. M. Parkinson, *Biotech. Adv. 3*: 65 (1985).
36. E. Rosenberg, *Crit. Rev. Biotechnol. 3*: 109 (1986).
37. M. S. Peters and K. D. Timmerhaus, in *Plant Design and Economics for Chemical Engineers*, McGraw Hill, New York, 1980.
38. A. Pareilleux, *Eur. J. Appl. Microbiol. 8*: 91 (1979).
39. Chemical Marketing Reporter, October 1, 1990.
40. O. Brown, *Chem. Ind. (Lond.)* Feb. 7, pp. 95–97 (1983).
41. C. Syldatk, S. Lang, V. Matulovic, and F. Wagner, *Z. Naturfosch 40c*: 51 (1985).
42. D. G. Cooper, J. Akit, and N. Kosaric, *J. Ferment. Technol. 60*: 19 (1982).
43. B. Frautz, S. Lang, and F. Wagner, *Biotechnol. Lett. 8*: 757 (1986).
44. W. Dimmling and G. Nesemann, *Crit. Rev. Biotechnol. 2*: 233 (1985).
45. R. M. Busche, *Biotech. Prog. 1*: 165 (1985).
46. C. Sandrin, F. Peypoux, and G. Michel, *Biotech. Appl. Biochem. 12*: 370 (1990).
47. A. Jiménez and D. Chävez, *Chem. Eng. J. 37*: B1 (1988).
48. J. D. Sheppard and C. N. Mulligan, *Appl. Microbiol. Biotechnol. 27*: 110 (1987).
49. A. M. Martin and W. Manu-Tawiah, *J. Chem. Technol. Biotechnol. 45*: 171 (1989).
50. S. A. Leeper and G. F. Andrews, *A Critical Review and Evaluation of Bioproduction of Organic Chemicals*, DOE Report no. EGG-M-89498, 1989.
51. A. M. Chakrabarty, *Trends Biotechnol. 3*: 32 (1985).
52. C. N. Mulligan and D. G. Cooper, *Appl. Environ. Microbiol. 50*: 160 (1985).
53. A. K. Koch, J. Reiser, O. Käppeli, and A. Fiechter, *Bio/Technology 6*: 1335 (1988).
54. L. H. Guerra-Santos, O. Käppeli, and A. Fiechter, *Appl. Microbiol. Biotechnol. 24*: 443 (1986).
55. K. V. Ramana and N. G. Karanth, *J. Chem. Technol. Biotechnol. 45*: 249 (1989).
56. C. N. Mulligan and B. F. Gibbs, *Appl. Environ. Microbiol. 55*: 3016 (1989).
57. C. N. Mulligan, G. Mahmourides, and B. F. Gibbs, *J. Biotechnol. 12*: 37 (1989).
58. C. N. Mulligan, G. Mahmourides, and B. F. Gibbs, *J. Biotechnol. 12*: 199 (1989).
59. D. M. Irvine and A. R. Hill, in *Comprehensive Biotechnology*, Vol. 3 (M. Moo-Young, ed.), 1985, pp. 524–564.
60. D. G. Cooper, in *Biotechnology for the Oils and Fats Industry* (C. Ratledge, P. Dawson, and J. Rattrey, eds.), Monograph Series 11, American Oil Chemist's Society, Champaign, IL., 1984, pp. 281–287.
61. C. A. Boulton and C. Ratledge, in *Biosurfactants and Biotechnology*, Vol. 25, Surfactant Science Series (N. Kosaric, W. L. Cairns, and N. C. C. Gray, ed.), Marcel Dekker, New York, 1970, pp. 47–87.

62. A. P. Tulloch, J. F. T. Spencer, and A. J. Gorin, *Can. J. Chem. 40*: 1326 (1962).
63. J. Vater, *Progr. Coll. Polymer Sci. 72*: 12 (1986).
64. K. V. Ramana and N. G. Karanth, *Biotechnol. Lett. 11*: 437 (1989).
65. C. N. Mulligan, D. G. Cooper, and R. J. Neufeld, *J. Ferment. Technol. 62*: 311 (1984).
66. D. R. Dean, J. A. Hoch, and A. I. Aronson, *J. Bacteriol. 131*: 981 (1977).
67. A. Persson and G. Molin, *Appl. Microbiol. Biotechnol. 26*: 439 (1987).
68. C. N. Mulligan, T. Y.-K. Chow, and B. F. Gibbs, *Appl. Microbiol. Biotechnol. 31*: 486 (1989).
69. Anonymous, *Chem. Eng. News. 66*: 18 (1988).
70. A. Perrson, G. Molin, and C. Weibull, *Appl. Environ. Microbiol. 56*: 686 (1990).
71. O. Levenspiel, in *Chemical Reaction Engineering*, Wiley, New York, 1962.
72. C. Mulligan, B. Safi, and D. Groleau, *Biotech. Bioeng. 38*: 1173 (1991).
73. A. C. Soderberg, in *Fermentation and Biochemical Engineering Handbook* (H. C. Vogel, ed.), Noyes Publications, Park Ridge, NJ, 1983, pp. 77–119.
74. G. Mattei and J.-C. Bertrand, French Patent no. 2,578,552 (1985).
75. F. Wagner, J.-S. Kim, S. Lang, Z.-Y. Li, G. Marwede, U. Matulovic, E. Ristau, and C. Syldatk, Third Eur. Congr. Biotechnol. *I*: 3 (1984).
76. J. D. Sheppard and D. G. Cooper, *Biotechnol. Bioeng. 36*: 539 (1990).
77. A. Margaritis, K. Kennedy, and J. E. Zajic, *Diagn. Labour 23*: 285 (1980).
78. Z. M. Khalid and K. A. Malik, *Pak. J. Sci. Ind. Res. 31*: 714 (1988).
79. T. Kachholz and M. Schlingmann, in *Biosurfactants and Biotechnology*, Vol. 25, Surfactants Science Series (N. Kosaric, W. L. Cairns, and N. C. C. Gray, eds.), Marcel Dekker, New York, 1987, pp. 184–210.
80. O. Stuewer, R. Hommel, D. Haferburg, and H.-P. Kleber, *J. Biotech. 6*: 259 (1987).
81. D. R. Cameron, D. G. Cooper, and R. J. Neufeld, *Appl. Environ. Microbiol. 54*: 1425 (1988).
82. Z. Duvnjak and N. Kosaric, *Biotechnol. Lett. 7*: 793 (1985).
83. Z. Duvnjak and N. Kosaric, *Biotechnol. Lett. 3*: 583 (1981).
84. C. N. Mulligan and B. F. Gibbs, *J. Chem. Tech. Biotechnol. 47*: 23 (1990).
85. F. O. Bryant, *Appl. Environ. Microbiol. 56*: 1494 (1990).
86. G. Mattei and J.-C. Bertrand, *Biotechnol. Lett. 7*: 217 (1985).
87. M. Taniguchi, N. Kotani, and T. Kobayashi, *J. Ferment. Technol. 65*: 179 (1987).
88. S. Lang, A. Gibbon, C. Syldatk, and F. Wagner, in *Surfactants in Solution*, Vol. 2 (K. L. Mittal and B. Lindmann, eds.), Plenum, New York, 1984, pp. 1365–1376.
89. D. Gutnick, in *BIOTECH US '84*, Online Publications, Pinner, 1984, pp. 645–653.

14

Biosurfactants for Cosmetics

VÁCLAV KLEKNER Institute of Microbiology, Czechoslovak Academy of Sciences, Prague, Czechoslovakia

NAIM KOSARIC Department of Chemical and Biochemical Engineering, University of Western Ontario, London, Ontario, Canada

I. INTRODUCTION

Surfactants are of great importance for today's cosmetic industry, and as many as 1708 are listed in *CTFA Cosmetic Ingredients Handbook* (1988). Since that time, many new have been added to the list.

The early cosmetic industry used soap and bee wax to fulfill the role of the surface-active compounds. In the 1920s, glycerol monostearate (GMS) was synthesized and started to be used. Glycerol monostearate as well as soaps can still be found in many formulas and subsequently in many cosmetic products; however, in the last 30 years, a large number of new surfactants have been commercially applied. Table 1 shows the major industrial suppliers of surfactants. A summary of

TABLE 1 Worldwide Suppliers of Selected Surfactants

Supplier	Address (U.S. if available)
Akzo Chemicals BV	Akzo Chemie America 300 South Riverside Plaza Chicago, IL 60606 312-498-6700
Brooks Industries, Inc.	70 Tyler Place South Plainfield, NJ 07080 201-316-3938
Capital City Products Co.	525 West First Avenue P.O. Box 569 Columbus, OH 43216-0569 614-299-3131
Croda Inc.	183 Madison Avenue New York, NY 10016 212-683-3089
Durkee Chemical Prod.	925 Euclid Avenue Cleveland, OH 44115 216-344-8199
Eastman Chemical Prod., Inc.	P.O. Box 431 Kingsport, TN 37662 615-229-2000
GAF Chemicals Co.	1361 Alps Road Wayne, NJ 07470 201-628-3922
Goldschmidt Chemical Co.	914 East Randolph P.O. Box 1299 Hopewell, VA 23860 804-541-8058
Harcros Chemicals Inc.	P.O. Box 2383 Kansas City, KS 66110 913-321-3131
Henkel KGaA	300 Brookside Avenue Ambler, PA 19002 215-628-1481

TABLE 1 (Continued)

Supplier	Address (U.S. if available)
Hoechst AG	Postfach D-6230 Frankfurt/Main 80 Germany 063-305-5131
Huls America Inc.	Chemical Division 80 Centennial P.O. Box 456 Piscataway, NJ 08854-0456 201-980-6946
Inolex Chemical Co.	Jackson & Swanson Streets Philadelphia, PA 19148-3490 215-289-4717
ICI Specialty Chemicals	ICI Americas, Inc. Specialty Products Wilmington, DE 19897 302-575-3101
Lanaetex Prod. Inc.	151 Third Avenue Elizabeth, NJ 07206 201-351-9700
Lauricidin, Inc.	414 Green Street P.O. Box 339 Galena, IL 61036 815-777-1887
Lipo Chemicals, Inc.	207-19th Avenue Paterson, NJ 07504 201-345-8600
Lonza, Inc.	17-17 Route 208 Fair Lawn, NJ 07410 201-794-2400
Mazer Chemicals Co.	Div. of PPG Chemicals Groups 3938 Porett Drive Gurnee, IL 60031 312-244-3410

TABLE 1 (Continued)

Supplier	Address (U.S. if available)
McIntyre, Inc.	4851 St. Louis Avenue Chicago, IL 60632 312-927-2701
Nikko Chemicals Co.	14-6, Nihonbashi-Kodenmacho Chuo-Ku Tokyo, 103 Japan 03-663-1677
Quantum Chemical Co.	Emery Division 11501 Northlake Drive P.O. Box 429557 Cincinnati, OH 45249 513-530-7300
REWO Chemische Werke GnbH	Postfach 1160 Industriegebiet West D-6497 Steinau a.d. Str. Germany 6663/54-0
RITA Co.	P.O. Box 556 332 Virginia Street Crystal Lake, IL 60014 815-455-0530
Stepan	22 Frontage Road Northfield, IL 60093 312-446-7500
Van Dyk & Company	Main & Williams Streets Belleville, NJ 07109 201-759-3225
Vevy Europe S.P.A.	Via Padre Semeria 18 P.O. Box 716 16131 Genova, Italy 39-10-314193
Witco Co.	520 Madison Avenue New York, NY 10022 212-605-3941

Japanese research and development biosurfactants is shown in Table 2 and Japanese production of biosurfactants in Table 3.

It is expected that biotechnology will substantially contribute to the cosmetic industry as it has during recent years to the drug and pharmaceutical industry. New health-related properties are also being introduced into cosmetics, so that some cosmetic products can already be classified as pharmaceuticals. This shift should open new horizons for applying biotechnology to cosmetics in the near future.

Moreover, at present new ways to prepare emulsions as liquid crystals [1], liposomes [2], or multiple emulsions [3], are being employed that require the use of specific pure compounds.

Many of the new surfactants are natural products and could be classified as biosurfactants. However, for the purpose of this chapter, the biosurfactants are only those compounds produced by microorganisms or microbial enzymes.

Because there is a huge number of chemically synthesized surfactants, only a limited number of microbially produced biosurfactants has had the chance to penetrate the cosmetic market. On the other hand, the biosurfactants have the advantages of extremely low irritancy or antiirritating effects and compatibility with skin, both properties demanded by today's market. Moreover, they are produced from renewable sources and/or wastes. The disadvantage is their higher cost and lack of information on their application. As most chemically synthesized surfactants are derived from oil, the price relations may change in the future.

Also some traditional surfactants now being prepared from natural sources by chemical treatment could be obtained by enzyme or microbial conversion. The replacement of chemical procedures can improve specificity of conversion, and thus improve the yield and economy of a desired product while avoiding impurities of by-products, and can also improve treatability of wastewater, yielding more environmentally conscious production plants.

Several requirements are imposed on surfactants applicable in cosmetics. A 3-year shelf-life of surfactants as well as emulsions or suspensions is usually required. This implies that the preference is generally given to saturated acyl groups as opposed to unsaturated groups, which may require the use of additional antioxidants. The product has to have standard properties, a task which is sometimes difficult for fermentation products, especially, when the product is a mixture of several compounds. If biosurfactant is to replace conventional chemically synthesized surfactants, the possibility must exist for a broad range of HLB values to be prepared by mixing the biosurfactant with other surfactants. This information on microbially produced biosurfactants is nearly completely missing from the literature, which makes the introduction of these products to industry more difficult.

Cosmetic products must fulfill health and hygienic requirements according to regulations of both countries of production and targeted applications. These

TABLE 2 Some Japanese Research and Development on Biosurfactants (December, 1988)

Structure type	Product	Workers	Organization of R&D	Productivity	Estimated cost	Remarks
Glycolipid	Sophorolipid	Innoue	Kao Co. Ltd.	120 g/L	5,000 yen/kg	Humectant for cosmetics make-up agent named "SOFINA"
	Mannosyl erythritol lipid	Tabushi	Tsukuba Univ.	35.4 g/L (as lipid) 25.4 g/L (as *n*-alkane)	7,000–8,000 yen/kg	
	Rhamnolipid	Hisazuka	Tokyo Univ.	2.5 g/L		
	Trehalose	Suzuki	Kyowa Hakko	1.3 g/L		Difficult to modify because of ester bond
Lipo-amino acid	Ceriliin	Tahara	Shizouka Univ.			
	Emulsifying factor	Iguchi	Asahi Kasei Co. Ltd.			Suitable for emulsifying hydrocarbon

Fatty acid	Spiculisporic	Tabuchi	Tsukuba Univ.	110 g/L	2,000	Low foaming activity antielectro-static agent
Sugar ester		Suzuki	Kyowa Hakko Kogyo Co. Ltd.	1.5 g/30 g cell		Bioconversion from fructose and lauric acid
Sugar ester		Kiyono	Kitasato Univ.			Bioconversion using lipase recovery 88%
N-acyl amino acid		Tanaka	Ajinomoto Co. Ltd.	1.7 mg/0.47 g cell		Bioconversion from glutamic acid and fatty acid
Phosphatidyl glycerol		Kudo	Yekult Honsha Co.			From soy bean lecithin using phospholipase D
Lysolecithin			Kyowa Hakko Kogyo Co. Ltd.			
Oligosaccharide fatty acid ester		Watanabe	Nihon Surfactant Co. Ltd.	22.5 g/L		

TABLE 3 Production of Biosurfactants in Japan (December, 1988)

Biosurfactant	Price	Manufacturer
Spiculisporic acid	2,500 yen/kg	Iwata Chemical Co. Ltd.
Rhamnolipid	5,000 yen/kg	Iwata Chemical Co. Ltd.
Surfactin	100,000 yen/100 mg	Wako Pure Chemical Industries Ltd.
Sophorolipid	cosmetic use	Kao Co. Ltd.
N-acyl amino acid	2,500 yen/kg	Ajinomoto Co. Ltd.
Sherac	1,500 yen/kg	Koyo Chemical Co. Ltd. Nihon Sherac Co. Ltd. Gifu Sherac Co. Ltd.
Sodium casein	1,000 yen/kg	Nissei Kyoeki Co. Ltd.
Synthetic lecithin	10,000 yen/kg	several manufacturers
Chirayasaponin	8,000 yen/kg	Maruzen Kasei Co. ltd.
Glycyrrhizin	100,00 yen/kg	Maruzen Pharmaceutical Company Ltd.

regulations change often and, as cosmetics are becoming closer to pharmaceutical products, test demands are rising. A potential user would need to spend some time and money to completely evaluate the feasibility of the new product. This is, however, only possible in large companies as elaborate testing will probably be needed, as is the case with new drugs.

This is closely related to the development of suitable cosmetic formula, which is now not available for most biosurfactants. Formula development, which is still a trial-and-error process, is usually carried out by specialized laboratories.

The application of surfactants in cosmetics is very broad and the demands on surfactant properties differ substantially according to the application. It is out of scope of this chapter to discussed what property is needed for a given application. More on this subject can be found in Rieger's book *Surfactants in Cosmetics* [4]. Table 4 summarizes possible cosmetic applications of surfactants and groups of synthetic surfactants that may serve as an illustration of this wide diversity.

On the other hand, the variety of biosurfactants includes naturally occurring compounds as fatty acids, glycolipids, acylpeptides, phospholipids, proteins, and lipopolysaccharides.

II. COMPOUNDS PRODUCED BY MICROBES

Compounds produced by microbes have not been well documented in the literature for their application in cosmetics, although their properties may make their use more widespread. Broader acceptability of these rather untraditional surfactants depends on their widespread availability and on the cosmetic formulas and

TABLE 4 Applications of Surfactants in Cosmetics

Surfactants function as:
emulsifier, foaming agent, solubilizer, wetting agent, cleanser, antimicrobial agent, mediator of enzyme action

Forms of cosmetics:
cream, lotion, liquid, paste, powder, stick, gel, film, spray

Cosmetic products using surfactants:
insect repellent, antacid, bath products, acne pads, antidandruff products, contact lens solution, hair color and care products, deodorant, nail care, body massage accessories, lipstick, lipmaker, eye shadow, mascara, soap, tooth paste and polish, denture cleaner, adhesives, antiperspirant, lubricated condoms, baby products, foot care, mousse, antiseptics, shampoo, conditioner, shave and depilatory products, moisturizer, health and beauty products

Surfactant groups:
I. Amphoteric surfactants: acyl-amino acids, *N*-acyl amino acids
II. Anionic surfactants: acyl-amino acids and derivatives, carboxylic acids, ester and ether carboxylic acids, phosphoric acid esters, sulfonic acids (acyl isethionates, alkylaryl sulfonates, alkyl sulfonates, sulfosuccinates), sulfuric acid esters (alkyl ether sulfates, alkyl sulfates)
III. Cationic surfactants: alkylamines, alkyl imidazolines, ethoxylated amines, quaternaries (alkylbenzyldimethylammonium salts, alkyl betaines, heterocyclic ammonium salts, tetraalkylammonium salts)
IV. Nonionic surfactants: alcohols, alkanolamides (alkanolamine-derived amides, ethoxylated amides), amine oxides, esters (ethoxylated carboxylic acid esters, ethoxylated glycerides, glycol esters, monoglycerides, polyglyceryl esters, polyhydric alcohol esters, sorbitol esters, triesters of phosphoric acid), ethers (ethoxylated alcohols, ethoxylated lanolin, ethoxylated polysiloxanes, propoxylated POE ethers)

specific applications for a variety of products as shown in Table 4. A general scheme for the production of biosurfactants is shown in Figure 1.

A. Glycolipids

1. Sophorose Lipids

Torulopsis bombicola produces sophorose lipids (SLs) in large quantities [5]. There have been several studies on synthesis of the SLs, and the product has been considered for the use in many industries [6]. In fact, the mixture of

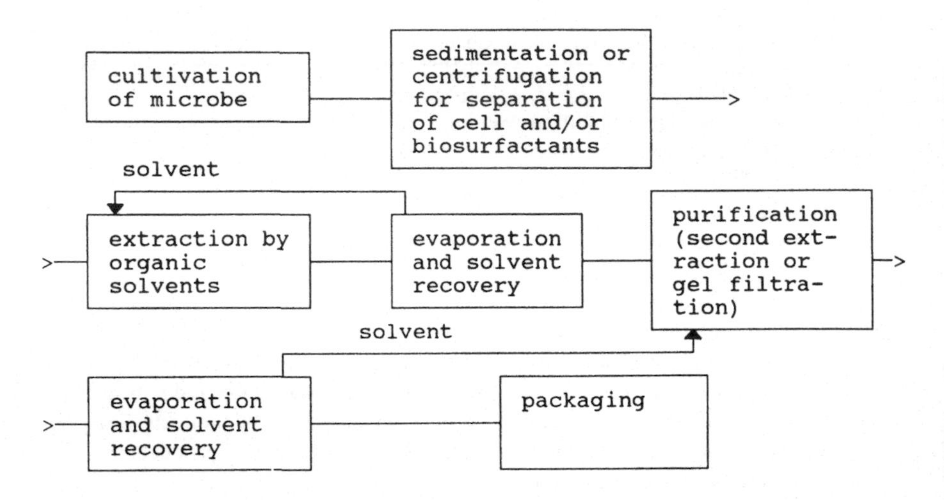

FIG. 1 General scheme for biosurfactant production by microbes.

different compounds having none, one or two acetyl groups attached at C6 of the sugar moieties and with different lactone linkage (usually C_{16} acid is also linked by glycosidic bound to ω-1 hydroxyl) is formed during fermentation. Composition of the mixture is dependent on cultivation conditions [7].

Kao Co. (Japan) developed a process for fermentation and isolation of SLs [9], and the modification of SLs by esterification. Esterification was done in two steps, first a methyl ester was prepared [10] that could be easily transesterified to other esters in the presence of MeONa [11]. Esterification of SLs is feasible because esterification of the mixture of SLs that is usually the result of fermentation and may not always be of standard properties, yields one compound. Alkyl-SLs cover HLB values from approximately 7 (oleyl-SL) to 25+ (methyl-SL) [12]. Several applications of esters and hydroxylakyl ethers of SLs in cosmetics were suggested, such as stick makeups (e.g., propolylated methyl ester [13]) and especially moisturizers (e.g., octyl, lauryl, and oleyl esters or propolylated derivatives [14, 15]) for skin and hair care products. Oleyl-SL and SLs modified by propylene oxide (polymerized product of 1 mole of SL with approximately 12 moles of propylene oxide) were reported to complement the natural skin moisturizing factor and to have found commercial application [12]. Nevertheless, what is the extent of applicability of SLs or modified SLs in cosmetics remains unclear as no widespread discussion of these compounds or the formulation using them is available in the literature.

2. Trehalose Lipids

Rhodococcus erythropolis is known to produce lipid derivatives of trehalose. Depending on the particular strain, the mixture of trehalose mono- and dicorynomycolates [16] or the mixture of mono- and disuccinoyl dialkanoyl trehalose [17] is produced. The latter seems to possess exceptional properties as surfactants [18]. Moreover, a yield of approximately 40 g/L has been achieved when *Rhodococcus erythropolis* SD-74 was grown on *n*-hexadecane in a high-osmotic-equivalent culture broth [19]. The process for preparation of succinoyl alkanoyl trehalose has been patented [20] and applications can be expected in the near future. No information is yet available on cosmetic use, but considering the process's valuable properties and reasonable production cost one may expect that work is under way.

3. Other

There are many other sugar derivatives, such as rhamnolipids produced by *Pseudomonas aeruginosa* [21], cellobiose lipids (*Ustilago* [22]), sucrose lipids (*Arthrobacter paraffineus* [23]), or acylglucoses (*Corynebacterium diphteriae* [24]) that could be considered for cosmetic applications, but low yields and high cost are for the time being serious obstacles for their broad use. Also, no information is currently available on their cosmetic applications.

B. Lipopolysaccharides

Emulsans are a family of high-molecular-weight lipoheteropolysaccharides (~1 million Mol. Wt.) composed of repeating units that contain trisaccharides esterified with fatty acids of the hydroxydodecanoic type [25]. The production of this biopolymer by *Acinetobacter calcoaceticus*, its properties, and proposed application for cleaning of oil-contaminated containers have been thoroughly reviewed [26] and patented [27]. The polymer forms intermolecular micelles and micellar coils. The lipophilic groups form the micellar core, and the hydrophilic groups are oriented outside. These biopolymers have an interesting potential as emulsifying and coemulsifying agents able to fluidize excessively heavy and difficult to spread emulsions.

Recently, α-emulsan, apo-α-emulsan, and other emulsan classes have been claimed as bioemulsifiers for cosmetics [28]. Typically, a cleansing cream comprising of 0.2% of α-emulsan together with other surfactants and an Aloe extract was developed. Nevertheless, even if the applicability of these lipopolysaccharides is directed to cosmetics, the basic data on cost and stability are not available.

III. COMPOUNDS PREPARED BY ENZYME CONVERSION

The use of hydrolytic enzymes for synthesis and/or trans conversions of hydrophobic molecules started with findings that enzymes need only very small amounts of water to keep their function in an otherwise nonaqueous environment [29]. In this respect, lipases especially have found many applications in the food and other industries [30]. Subsequently, the preparation of compounds applicable in cosmetics is being studied. This field without any doubt is very promising and is gaining recent interest as more fundamental knowledge, together with a broader spectrum of commercial lipases, are available. Table 5 shows surfactants that can be produced by enzymes and possible suppliers thereof.

A. Monoacylglycerides

Glycerol monostearate is still one of the most used surfactants in cosmetics. Monoglycerides (MG) are usually produced by glycerolysis of fat in the presence of basic inorganic catalyst at temperatures greater than 220°C. At high temperatures, dark-color by-products are formed requiring molecular distillation for separation of the product monoglycerides. Recently, glycerolysis at low temperatures using lipase was accomplished [31, 32]. A yield of 70% MG was obtained at 42°C when lipase from *Pseudomonas fluorescence* and glycerol/tallow (molar ratio 1.5:2.5) were used [31]. With temperature programming, a yield of approximately 90% MG was obtained. In this case, the initial temperature was 42°C for 8–16 h followed by incubation at 5°C for up to 4 days [32]. Even though several problems need to be overcome, such as the separation of enzyme from the solid MG phase, the procedure seems to be economically promising.

B. *N*-Acylaminoacids

Ajimonoto patented a preparation of *N*-long-chain acylamino acids, especially glutamates, by conversion of Na salts using whole microbes (*Pseudomonas, Xanthomonas*, or *Gluconobacter* [33]). N-long-chain acyl glutamates can be used for cosmetic formulations of stable oil-in-water emulsions [34].

Recently, commercially available immobilized lipase of *Mucor miehei* (Lipozyme, Novo Industry) has been used for synthesis of N-acyl bonds [35–36]. *N*-ε-oleyllysine, a new generation of surfactants, can be synthesized by transacetylation from triglycerides to lysine. A yield of up to 60% could be reached without any solvent at 90°C [38].

In this connection, it is interesting to note that *N*-acyl of basic amino acids as *N*-ε-cocoylornithine has been claimed to possess skin- and hair-moisturizing and protecting properties [39].

TABLE 5 Worldwide Suppliers of Surfactants Produced with Enzymes

Supplier	Product
Akzo Chemicals BV	Monoglycerides sugar esters and ethers sorbitan/sorbitol esters
Amerchol Co.	Acylpeptides sugar esters and ethers
Brooks Industries, Inc.	Acylpeptides
Capital City Products	Monoglycerides
Croda Inc.	Sugar esters and ethers
Durkee chemical Prod.	Monoglycerides Sugar esters and ethers sorbitan/sorbitol esters
Eastman Chemical Prod.	Monoglycerides sugar esters and ethers
GAF Chemicals Co.	Sorbitan/sorbitol esters
Goldschmidt Chemicals Co.	Monoglycerides
Harcross Chemicals Inc.	Sorbitan/sorbitol esters
Henkel KGaA	Acylpeptides monoglycerides sorbitan/sorbitol esters
Hoechst AG	Acylglutamates sorbitan/sorbitol esters
Hormel	Acylpeptides
Huls America Inc.	Monoglycerides
Inolex Chemical Co.	Acylpeptides
ICI Specialty Chemicals	Sugar esters and ethers sorbitan/sorbitol esters
Lanaetex Prod. Inc.	Monoglycerides sugar esters and ethers sorbitan/sorbitol esters
Lauricidin, Inc.	Monoglycerides
Lipo Chemicals, Inc.	Monoglycerides sugar esters and ethers sorbitan/sorbitol esters
Lonza, Inc.	Monoglycerides sorbitan/sorbitol esters
Mazer Chemicals Co.	Acylglutamates monoglycerides sugar esters and ethers sorbitan/sorbitol esters
McIntyre, Inc.	Monoglycerides
Nikko Chemicals Co.	Acylpeptides monoglycerides sugar esters ands ethers sorbitan/sorbitol esters

TABLE 5 (Continued)

Supplier	Product
Quantum Chemical Co.	Monoglycerides sugar esters and ethers sorbitan/sorbitol esters
REWO Chemische Werke GmbH	Monoglycerides
RITA Co.	sugar esters and ethers sorbitan/sorbitol esters
Stepan	Sugar esters and ethers
Van Dyk & Company	Monoglycerides
Vevy Europe S.P.A.	Sorbitan/sorbitol esters
Witco Co.	Monoglycerides sorbitan/sorbitol esters

C. Phospholipids

Soy phospholipids, obtained as a by-product in the production of soybean oil, can be modified by enzymes yielding lysophospholipids. These compounds alone or in mixture with monoglycerides and fatty acids have shown significant surface activity [40]. The enzyme employed can be of pancreatic or microbial origin. When the pancreatic enzyme was used, mainly hydrolysis proceeded [41]. Phospholipase D (M or Y1) derived from actinomycetes functions rather like transphophatidylase when a suitable acceptor is present. As an acceptor, it may serve primary, secondary, or phenolic hydroxyls, and so a broad spectrum of compounds can be synthesized. When glycerol is added, phosphatidylglycerol can be formed [42].

Apart from their direct use as surfactants, phospholipids can serve for construction of liposomes. Entrapment inside of liposomes of an active compound for delivery to the skin has been verified to be beneficial in cosmetics as well as in pharmaceuticals [43, 44]. The mildness of phospholipids and their compatibility with the skin make them primary sources for liposome creation. Liposomes can penetrate through skin layers and can be targeted to a specific layer by modifying their composition. As such, cosmetic applications may be considered dermatological, but care must be taken to use pure compounds. This, together with demand for specific compounds that enable construction of specific liposomes, gives a new opportunity for cosmetic application of enzyme-modified phospholipids.

D. Sugar Esters

Sucrose is a well-defined chemical substance and is abundant in nature. Glucose can be readily prepared from starch sources as well. These two sugars are inexpensive renewable resources having distinct advantages over oils.

Sugars esterified with long-chain fatty acids can serve as surfactants; for example blends of sucrose esters and glycerol can cover entirely all HLB values (2 to 16+) demanded by the cosmetic industry [45]. Moreover, they offer very low toxicity and irritancy.

Apart from that, sucrose cocoate has been used in wound-healing preparations [46] that increased epidermal layer thickness and content of DNA, glucosamine glycan, and lipids. Sucrose esters have been used also for formulation of multiple emulsions [47]. Also, esters of glucose have been studied as valuable cosmetic materials [48]. Sugar esters can be prepared by transesterification of methylesters with sucrose in the presence of a basic catalyst in dimethylformamid as a solvent, or more recently without a solvent [48]. Depending on the choice of the starting materials and reaction conditions, the products obtained are mono-, di-, tri-, and polyesters of sucrose. However, only mono- and diesters are of interest for cosmetic application. Chemically derived monoesters are usually C6 esters as this primary hydroxyl is readily attacked.

Recently, pancreatic lipase was used for preparation of C-6 esters by transesterification from trichloroethyl carboxylates to unprotected monosaccharides in anhydrous pyridine [49]. From these 6-O-acylmonosaccharides diesters can be prepared by next transesterification in tetrahydrofuran or methylchloride. Depending on the lipase used, C2 or C3 derivatives can be obtained. *Chromobacterium viscosum* and *Aspergillus niger* lipases attacked selectively C3 position of 6-O-butylglucose, in contrast to porcine pancreatic lipase that preferred C2 [49]. Sugar diesters can be selectively hydrolysed in the C6 position by *Candida cylindracea* lipase, and so pure C2 or C3 monoesters can be prepared [50].

To transfer these results to the industrial scale may take some time, as further experimentation is necessary, but lipases clearly offer a simple and inexpensive way for selective preparation of different sugar esters.

Apart from sugars, sugar alcohols can be esterified by lipases [51]. Thus, pancreatic lipase and lipase of *Chromobacterium viscosum* were used for transesterification between triglycerides as triolein or commercial oils and hexitols as well as pentitols dissolved in piridin. Primary hydroxylic groups were esterified (C1, C6, or C5) and monoesters counted for more than 92%. Surface activities of sorbitol monoesters were compared to commercially available sorbitan monoesters that are produced chemically by esterification of sorbitol (sorbitol undergoes dehydration prior to acylation as the result of high temperature and sulfuric acid) and were found to be superior to commercial products.

E. Other

One can expect that many other enzymatic conversions useful for cosmetics will emerge. Thus, cyclodextrin glucanotransferase (EC 2.4.1.19) produced by *Bacillus circulans* has been used for conversion of *Quillaja* saponin with maltodextrin to α-glycosyl *Quillaja* saponin [52], which had improved properties.

REFERENCES

1. G. Cioca and L. Calvo, *Cosmet. Toilet. 105*: 57 (1990).
2. K. Suzuki and K. Sakon, *Cosmet. Toilet. 105*: 65 (1990).
3. M. De Luca, J. L. Grossiord, J. M. Medard, and C. Vaution, *Cosmet. Toilet. 105*: 65 (1990).
4. M. M. Rieger, ed., *Surfactants in Cosmetics*, Marcel Dekker, New York, 1985.
5. P. A. J. Gorin, J. F. T. Spencer, and A. P. Tulloch, *Can. J. Chem. 39*: 846 (1961).
6. D. G. Cooper and D. A. Paddock, *Appl. Environ. Microbiol. 47*: 173 (1984).
7. H. J. Ashmer, S. Lang, F. Wagner, and V. Wray, *J. Am. Oil Chem. Soc. 65*: 1460 (1988).
8. V. Klekner, N. Kosaric, and Q. Zhou, *Biotechnol. Lett. 13*: 345 (1991).
9. S. Inoue, M. Kinta, and Y. Kimura, German Patent 2,834,118 to Kao Soap Co. (1979).
10. S. Inoue, Y. Kimura, and M. Kinta, German Patent 2,905,252 to Kao Soap Co. (1979).
11. S. Inoue, Y. Kimura, and M. Kinta, German Patent 2,905,295 to Kao Soap Co. (1979).
12. S. Inoue, *Biosurfactants in Cosmetic Applications*, Proceedings of the World Conference on Biotechnology for the Fat and Oils Industry, American Oil Chemists Society, 1988, pp. 206–209.
13. J. Kono, T. Suzuki, S. Inoue, and S. Hayashi, Japan Patent 80 43,042 to Kao Soap Co. (1980).
14. Y. Abe, S. Inoue, and A. Ishida, German Patent 2,938,383 to Kao Soap Co. (1980).
15. H. Tsutsumi, J. Kawano, S. Inoue, and S. Hayashi, German Patent 2,939,519 to Kao Soap Co. (1980).
16. A. Kretschmer, H. Bock, and F. Wagner, *Appl. Environ. Microbiol. 44*: 864 (1982).
17. Y. Uchida, R. Tsuchiya, M. Chino, J. Hirano, and T. Tabuchi, *Agric. Biol. Chem. 53*: 757 (1989).
18. Y. Ishigami, S. Suzuki, T. Funada, M. Chino, Y. Uchida, and T. Tabuchi, *Yukagaku 36*: 847 (1987).
19. Y. Uchida, S. Misawa, T. Nakahara, and T. Tabuchi, *Agric. Biol. Chem. 53*: 765 (1989).
20. T. Tabuchi and M. Kayano, Japan Patent 62 83,896 to Nippon Oils and Fats Co., Ltd. (1987).
21. H. E. Reiling, U. Thanei-Wyss, L. H. Guerra-Santos, R. Hirt, O. Kappeli, and A. Fiechter, *Appl. Environ. Microbiol. 51*: 985 (1986).
22. B. Frautz, S. Lang, and F. Wagner, *Biotechnol. Lett. 8*: 757 (1986).
23. Z. Duvnjak, D. G. Cooper, and N. Kosaric, *Biotechnol. Bioeng. 24*: 165 (1982).
24. P. J. Breunan, D. P. Lehane, and D. W. Thomas, *Eur. J. Biochem. 13*: 117 (1970).

25. E. H. Gans, *Cosmet. Toilet. 103*: 37 (1988).
26. D. L. Gutnick and Y. Shabtai, in *Biosurfactants and Biotechnology* (N. Kosaric, W. L. Cairns, and N. C. C. Gray, eds.), Marcel Dekker, New York, 1987, pp. 211–246.
27. Biotechnology A. G., Japan Patent 80 112,201 (1980).
28. M. E. Hayes, European Patent 242,296 to Petroleum Fermentation N.V. (1987).
29. A. Zaks and A. M. Klibanov, *Proc. Natl. Acad. Sci. USA 82*: 3192 (1985).
30. F. X. Malcata, H. R. Reyes, H. S. Garcia, and C. G. Hill, *J. Am. Oil Chem. Soc. 67*: 890 (1990).
31. G. P. McNeill, S. Shimizu, and T. Yamane, *J. Am. Oil Chem. Soc. 67*: 779 (1990).
32. G. P. McNeill and T. Yamane, *J. Am. Oil Chem. Soc. 68*: 6 (1991).
33. Ajimonoto Co., Inc., Japan Patent 57,129,696 (1982).
34. I. Hasegawa and K. Maeno; Japan Patent 61,271,029 to Kanebo, Ltd. (1986).
35. F. Servat, D. Montet, M. Pina, P. Galzy, A. Arnaud, H. Ledon, L. Marcou, and J. Graille, *J. Am. Oil Chem. Soc. 67*: 646 (1990).
36. D. Montet, M. Pina, J. Graille, G. Renard, and J. Grimaud, *Fett Wissenschaft Technol. 91*: 14 (1989).
37. J. Graille, D. Montet, F. Servat, J. Grimaud, G. Renard, P. Galzy, A. Arnaud, and L. Marcon, European Patent 298,796 (1988).
38. D. Monet, F. Servat, M. Pina, J. Graille, P. Galzy, A. Arnaud, H. Ledon, and L. Marcou, *J. Am. Oil Chem. Soc. 67*: 771 (1990).
39. K. Sagawa, H. Yokota, and M. Takehara, Japan Patent 61,137,808 to Ajimonoto Co., Inc. (1986).
40. S. Fujita and K. Suzuki, *J. Am. Oil Chem. Soc. 67*: 1008 (1990).
41. M. Egi, *Food Chem. 1*: 51 (1987).
42. S. Kudo, *Biosurfactants as Food Additives*, Proceedings of Word Conference on Biotechnology for Fat and Oils Industry, American Oil Chemists Society, 1988, pp. 195–201.
43. H. Lautenschlager, *Cosmet. Toilet. 105*: 89 (1990).
44. H. Lautenschlager, *Cosmet. Toilet. 105*: 63 (1990).
45. R. K. Gupta, K. Janus, and F. J. Smith, *J. Am. Oil Chem. Soc. 60*: 862 (1983).
46. S. T. Goode, R. R. Linton, and F. Baiocchi, World Patent 88 06,880 to R.I.T.A. Co. (1988).
47. M. De Luca, P. Rocha-Filho, J. L. Grossiord, A. Rabarou, C. Vantion, and M. Seiller, *Int. J. Cosm. Sci. 13*: 1 (1991).
48. N. B. Desai, *Cosmet. Toilet. 105*: 99 (1990).
49. M. Therisod and A. M. Klibanov, *J. Am. Chem. Soc. 108*: 5638 (1986).
50. M. Therisod and A. M. Klibanov, *J. Am. Chem. Soc. 109*: 3977 (1987).
51. J. Chopineau, F. D. McCafferty, M. Therisod, and A. M. Klibanov, *Biotechnol. Bioeng. 31*: 208 (1988).
52. T. Kuramoto, Japan Patent 63 42,729 to Mazuren Chemical Co. (1988).

15

Biosurfactants from Marine Microorganisms

SIEGMUND LANG and FRITZ WAGNER Institute of Biochemistry and Biotechnology, Technical University of Braunschweig, Braunschweig, Germany

I. INTRODUCTION

The most important sources of oil input into the world oceans are the following [1]:

Marine transportation
Tanker accidents
Coastal oil refineries
Offshore oil production
Industrial and municipal waste
River runoff
Urban runoff
Natural seep
Atmospheric rainout

In the last several years, special attention has focused on the tanker accidents of "Amoco Cadiz" (223,000 tons of crude oil were spilled at the Bretonic coast, 1983) and of "Exxon Valdez" (40,000 tons, Alaska, 1989) as well as on the premeditated input of about 1,500,000 tons of oil into the Persian Gulf (1991) by Iraq. This pollution exhibits a dramatic damaging effect upon plants, birds, and sea animals and also involves incalculable consequences all over the world.

To better understand the fate of oil after spreading on the sea surface the main components of crude oils should be mentioned [2]:

n- and *i*-alkanes, 15–35%
cycloalkanes, 30–50%
aromatics, 5–20%
oxygen-, sulfur-, and nitrogen-containing aromatics, 2–15%

Among these compounds, highly volatile substances (BP <200°C) are evaporated, and the low-boiling aromatics dissolve easily in water and may enter the food chain. To some extent aromatics and alkanes are degraded by microbial attack. Unfortunately, a large part of the residual oil forms water-in-oil emulsions, so called "chocolate mousse," later changing to tar balls with solid inclusions, which tend to sediment.

Although photooxidation plays a part in the degradation of oil, the degradation is mainly accomplished by mixed populations of microorganisms, bacteria being dominant in this function [3–5]. In comparison to laboratory conditions, their multiplication in the open sea is largely limited by low temperature and low concentrations of nitrogen and phosphorous salts. Because of missing the whole genetic information for all enzymes necessary for attacking both alkanes and aromatics, it is noteworthy to say that one single strain is not able to utilize the whole spectra of oil components.

Concerning the oil uptake mechanisms in the case of freshwater microorganisms, proofs are found in the literature [6–8] for:

1. An unmediated uptake with subsequent molecular diffusion of alkanes through pores of the microbial membrane.
2. A mediated uptake and facilitated diffusion after production of cell-associated or extracellular biosurfactants.

Little data exist on detailed characteristics of surfactants from marine bacteria, although the emulsification of hydrocarbons by surface-active agents is considered an essential step in hydrocarbon biodegradation in the marine environment [4, 9, 10].

If these biosurfactants could be overproduced by the aid of biochemical process engineering, they would be suitable in future to replace synthetic surfactants in combating oil spills. After an immediate mechanical collection (booms, skimmers) dispersants may be used to minimize oil contamination of birds and sea animals. The expected actions of such chemicals in the open sea are in prevention of coalescence of oil droplets after dispersion, the reduction of "chocolate mousse" formation, and the enhancement of the biodegradation processes. However, chemical dispersants can cause additional problems because of their toxicity and their persistence in the environment [3, 11, 12]. Biosurfactants, being of natural origin, seem to have advantages in this respect.

This chapter stresses the special response of marine bacteria—mixed populations or pure cultures—to hydrocarbons as illustrated by some specific laboratory experiments. Discussed are the production of biosurfactants, their molecular structures, their physicochemical characterization, and some tests on suitability as remedy for oil pollution in the marine environment.

II. SCREENING METHODS FOR BIOSURFACTANTS AMONG OIL-DEGRADING MARINE MICROORGANISMS

At the beginning of biosurfactant screening studies, various populations of mixed bacteria isolated from natural marine biotopes (sediments, sea water, foams) were tested for sufficient growth on crude oil. In view of specific osmotic phenomenon—many marine bacteria were found to lyse in dilute media [13]—they have to be cultivated in crude oil media based on natural sea water, synthetic sea water, or 3% NaCL. In this context, it is curious that many marine bacteria appear to grow better at 50–75% sea water concentration [14] and that the optimum concentration of Na^+ (70–300 mM) for the growth of a number of marine isolates is considerably lower than the Na^+ concentration in sea water (450–480 mM) [15].

Previous data [16–18] recommend adding to the sea water media nitrogen, phosphate, iron ions as well as protein hydrolysates. By using this enrichment culture technique [17] and serial transfers to fresh sterile media with crude or pure oils at different intervals (at least of 2 to 4 days), the oil became more dispersed and was degraded.

In the second stage of screening, these cultures have to be examined for the production of surface-active agents. Important methods for rapid determination of surfactant presence are the following:

1. Hemolysis of red blood cells [19].
2. Direct thin-layer chromatography of the biomass (1–10 mg) or of the culture supernatant [20].
3. Thin-layer chromatography of crude extracts after solvent extraction of cells/culture supernatant [21, 22].
4. Test of emulsifying activity [8, 23, 24].
5. Measurement of surface or/and interfacial tension [25–27].

In method 1, petri dishes with sheep blood agar are inoculated with small portions of microbial microorganisms. After incubation hemolysis of red blood cells around the colonies indicates potential biosurfactant production. Table 1 shows that this method is restricted to water-soluble agents. Among cultures/substances tested the rhamnolipid producer *Pseudomonas* sp. and also a synthetic detergent (used as a washing up liquid) gave positive results, whereas cultures

TABLE 1 Hemolytic Activity of Several Surfactants/Microorganisms after Incubation for 120 h at 27°C

Microorganism/surfactant	Hemolysis (+/–)
Arthrobacter sp. EK 1	—
Trehalose tetraester (TL-4)	+[a]
Marine strain MM 1	—
Glucose lipid	—
Arthrobacter sp. SI 1	—
Trehalose dicorynomycolate (TL-2)	—
Emulsifying agents of	
Arthrobacter sp. SI 1	—
Pseudomonas sp. DMS 2874	+
Rhamnolipids R-1 and R-3	+
Pril[b]	+

[a]Small colorless rings of hemolyzed red blood cells.
[b]Trademark of a detergent of Henkel (F.R.G.)
Source: Ref. 24.

producing surfactants with lower hydrophilic lipophilic balance (HLB) values failed in hemolysis.

The more common methods (methods 2 and 3) use silica gel plates and coloring reagents and are appropriate for the proof of low-molecular-weight substances, such as lipids. The emulsifying activity (method 4) of a solution of biosurfactants is measured by optical density as a function of the degree of stability of an oil-in-water emulsion obtained after a certain time of mechanical agitation or sonic oscillation. Only water-soluble extracellular substances can be detected.

With regard to the final method (5), the decrease of the surface and/or interfacial tension of the supernatant compared to the nutrient medium before cultivation has to be observed by means of a automatically working tensiomat. Zajic's group [25–27] used this type of rapid biosurfactant detection.

III. MICROBIAL SURFACTANTS FROM COASTAL AREAS OF THE ISRAELI MEDITERRANEAN SEA

A. *Acinetobacter calcoaceticus* RAG-1: Production of Emulsan

The first report on the best known marine biosurfactant, now exploited commercially as Emulsan, appeared in 1972 [17]. After incubating a sea water sample taken from a local beach (Tel Baruch, Israel) with 10 mg of KH_2PO_4, 1 g of $(NH_4)_2SO_4$, and varying quantities of Iranian crude oil per liter of sea water, a mixed population of microorganisms was obtained that catalyzed the dispersion of oil. From this enrichment culture, eight pure cultures were isolated and studied separately. Only one of the isolates (RAG-1) brought about a significant dispersion of crude oil. After previous tentative determination as an *Arthrobacter* sp., RAG-1 was characterized as *Acinetobacter calcoaceticus*, and is now commercially available with ATCC No. 31012. When sea water was supplemented with 0.029 mM K_2HPO_4 and 3.8 mM $(NH_4)_2SO_4$ after inoculation with the culture supernatant, oil dispersion occurred within 60 min.

In 1979, Rosenberg et al. [8] described the isolation and partial purification of the above emulsifying agent produced during growth at 30°C on a minimal medium containing 0.125% of urea, 0.125% of $MgSO_4 * 7H_2O$, 0.002% of $FeSO_4$, 0.001% of $CaCl_2$, 0.025% of K_2HPO_4, 0.2 M Tris-HCl buffer (pH 7.4), and either 0.2% (v/v) of hexadecane or 0.1% (v/v) of ethanol. As to the standard yield determination a quantitative functional assay for emulsan was developed on the basis of a linear relationship between the turbidity of the stable oil-in-water emulsion and the amount of emulsan present in the assay. The assay consists of emulsification of a mixture of hexadecane and 2-methylnaphthalene (1/1, v/v) in a buffered solution containing Mg^{2+} ions. One unit of activity is the amount of emulsan that gives rise to a turbidity of 100 Klett units (Klett Summerson colorimeter) [28]; this value corresponds to approximately 6 μg of the purified emulsifier [29].

Concerning isolation, the emulsifier was recoverable from the supernatant fluid of the culture by trapping at a heptane-water interface, by ammonium sulfate precipitation, or by precipitation with the positively charged cetyltrimethylammonium bromide [8, 30].

Studying the chemical and physical nature, the bioemulsifier was found to be a polyanionic heteropolysaccharide [29, 31–33]. In Table 2, average values for emulsans produced on different carbon sources are documented.

As for the fermentation process of *Acinetobacter calcoaceticus* RAG-1, some reports should be cited. First illustrations of growth and emulsan production showed maximum amounts of emulsifier activities of 14 U/mL of cell-free culture fluid in a hexadecane medium and 25 U/mL of cell-free culture fluid in an ethanol medium. All measurable activity was extracellular when RAG-1 was grown on hexadecane, whereas when grown on ethanol about 25% was cell-associated [8]. Goldman et al. [28] replaced the original sea water medium with K_2HPO_4 (22.2 g/L), KH_2PO_4 (7.26 g/L), $MgSO_4*7H_2O$ (0.2 g/L), $(NH_4)_2SO_4$ (4 g/L); ethanol (20 g/L) served as the carbon source. Additionally, developing a sensitive enzyme-linked immunosorbent assay (ELISA) with antibodies prepared against purified emulsan, the authors were able to detect emulsan associated with the cell surface and to monitor changes in the distribution of cell-free and cell-associated (capsule) emulsifier throughout the growth cycle. Kinetic measurements demonstrated that the increase of cell-free extracellular emulsan is accompanied by a corresponding decrease in cell-associated capsule. The yield of free emulsan after 72 h was about 200 μg/mL as measured by microplate ELISA. The content

TABLE 2 Molecular Composition of Emulsan

Parameter tested	Analysis data
Molecular weight	9.8×10^5 daltons
Polysaccharide backbone	D-Galactosamine D-Galactosamineuronic acid Unidentified 3rd amino sugar
Fatty acid content	5–15% (C_{12}; 2-,3-OH-C_{12}; C_{14}; C_{16}; C_{18}; others)
Protein association	5–15% (not necessary for emulsification)

Source: Refs. 29 and 31–33.

of cell-associated emulsan (measured as $A_{405nm}/10^7$ cells) dropped from 1.3 after 12 h to 0.2 after 72 h [28, 29].

The production of cell-free emulsan on ethanol was enhanced in the presence of chloramphenicol [34]. Developing a resting cell system (cells from the exponential growth phase; carbon and nitrogen sources) in the presence of this inhibitor of protein synthesis, cell growth was inhibited and an accelerated release of cell-free emulsan was observed. Radioactive-labeling studies indicated a *de novo* synthesis of the lipopolysaccharide after the stop of cell growth. In comparison, cells treated with antibiotic began to produce emulsan within 1 h and reached a maximum level at 6 h, whereas the parallel culture without the antibiotic grew normally but produced little emulsan during the initial 3 h. The specific emulsan productions defined as [emulsan (units/mL)/culture turbidity (Klett units)] × 100 were 40 and 3, respectively [34].

After the presence of the emulsan capsule on the cell surface was known to enhance the tolerance of *Acinetobacter calcoaceticus* RAG-1 cells to the cationic cetyltrimethylammonium bromide (CTAB) [30], mutants capable of growing in the presence of high concentrations of CTAB were isolated [35]. Such mutants showed maximum enhancement in both overall yield and increased the specific productivity two- to threefold over that of the wild type. Figure 1 demonstrates this behavior in the case of the mutant CTR-10-49. In addition, the authors showed that the mutation for CTAB resistance led to an increased capsule production.

Investigating the adherence characteristics of *A. calcoaceticus* RAG-1 to hydrophobic liquids and thereby examining the roles of cell-free and surface emulsan, Ng and Hu also tested continuous cultures (1-L working volume in a 2-L bioreactor) [36]. The increased production rate at higher dilution rate resulted in a higher content of specific surface-associated emulsan. At a dilution rate of 0.4/h, the specific surface emulsan concentration was almost twofold that at the low dilution rate of 0.025/h.

In further development of the fermentation process, the capability of *Acinetobacter* RAG-1 (ATCC 31012) to utilize triglycerides and fatty acids was investigated to improve yield and productivity. Kottutz [37] observed that using rapeseed oil (main components: C_{18}, 39%; C_{22}, 48%) or 12-hydroxy stearic acid, the yield of emulsan was similar to ethanol. After 27 h of incubation on comparable quantities of carbon sources, the following yields on emulsan were derived: 1.12 g/L (rapeseed oil), 0.78 g/L (12 OH-stearic acid), and 1.19 g/L (ethanol). Preliminary studies on structural and physicochemical properties showed that the emulsifier produced from ethanol and 12-hydroxy stearic acid seem to be the same, whereas the product from rapeseed oil could be different (larger changes in sugar, fatty acid, and protein content; more unfavorable values of surface and interfacial tensions).

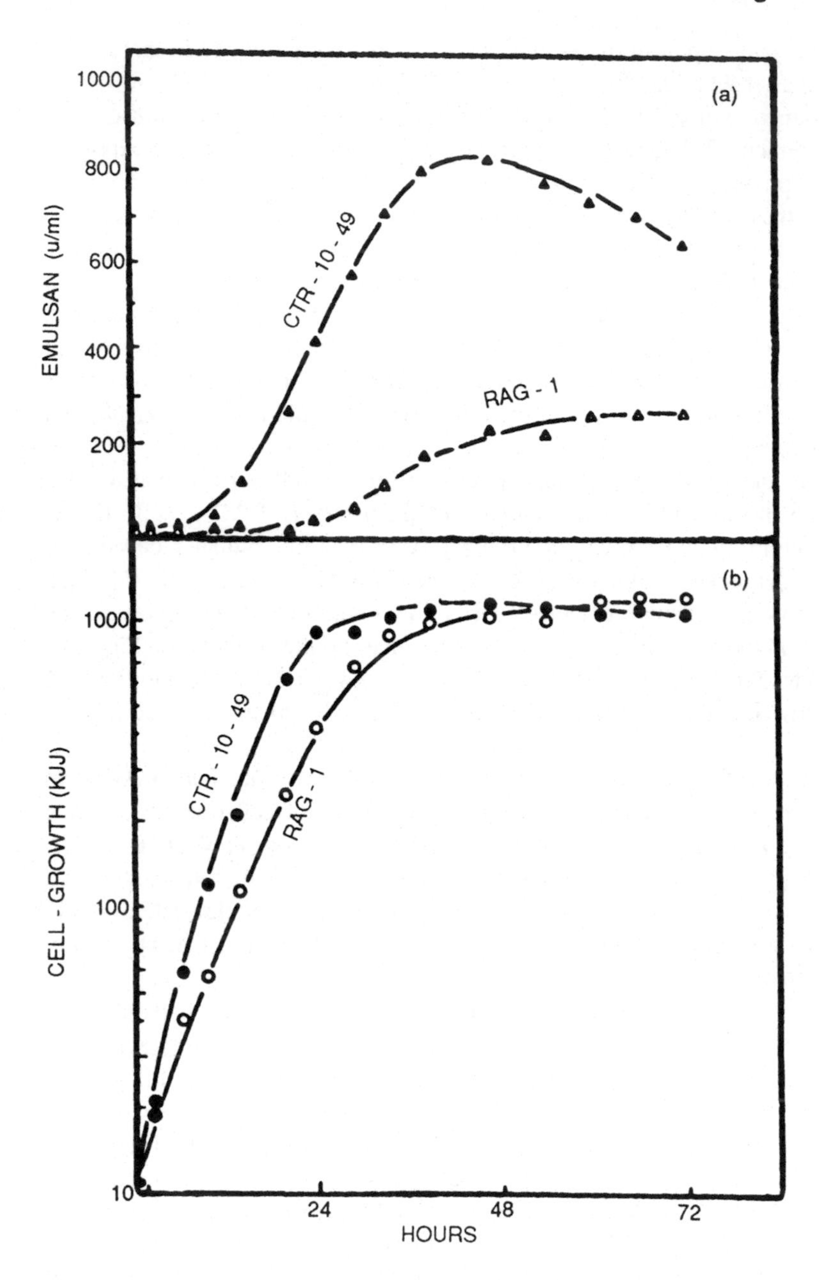

FIG. 1 Cell growth and emulsan production of *Acinetobacter calcoaceticus* RAG-1 and of the mutant CTR-10-49. K.U., Klett units. *Source*: Ref. 35 with permission.

Recently, batch and fed batch fermentation processes using a minimal medium (no sea water; only 15 g/L of salts) and soybean oil were established by Shabtai and Wang [38]. In initial experiments with a 14-L bioreactor, they observed that at an ammonium ion concentration of 60 mM and at a soybean oil concentration of 50 g/L (expressed as fatty acids, after alkaline hydrolysis and GC measurement) the specific growth rate at the onset of the fermentation was close to 0.4/h, resulting in a fast buildup of cell mass (estimated by measuring protein in the culture). Figure 2 shows that 20 g/L of emulsan were produced within 40 h giving a productivity of about 0.5 g/L h^{-1}.

As *Acinetobacter* only utilizes the fatty acid portions of the triglycerides and because of an observed lipolysis slowdown, the cultivation method was improved by switching from soybean oil to free fatty acids in the medium (oleic acid). The yield of 0.2 g emulsan/g fatty acid is about twofold higher than that obtained with ethanol as the carbon source. A final concentration of more than 20 g/L was reached. Besides some important advantages of triglycerides or fatty acids as carbon sources (cheap, nontoxic, nonvolatile), there are also disadvantages, such as complications during the fermentation process (foam) as well as in the downstream processing (separation of the wanted polymer from the cells and the lipophilic substrate).

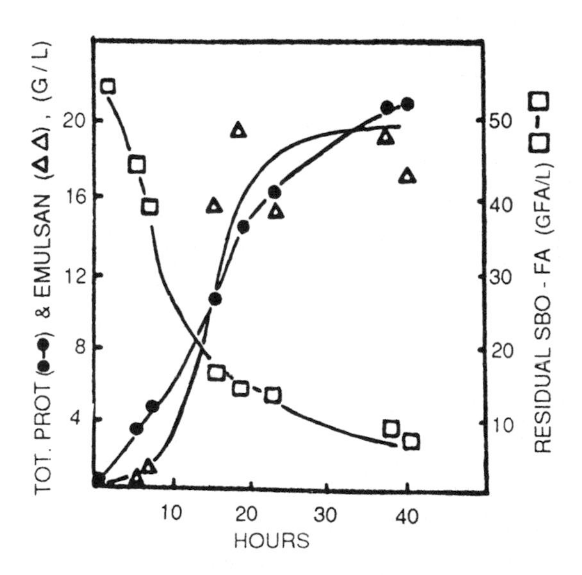

FIG. 2 Growth, emulsan production, and soybean oil utilization in an 8-L batch fermentation of *A. calcoaceticus* using high-ammonium-ion concentration. *Source*: Ref. 38 with permission.

B. Emulsan Characterization: Survey of the Literature

Gutnick and Shabtai gave an excellent overview [29] of the microbiology of emulsan production and the interaction of emulsan with hydrocarbons. In summary, the most important features were the following:

Emulsan as a RAG-1 capsule [28, 34, 39,40]
RAG-1 esterase [33]
Cell surface and emulsan mutants of RAG-1 [28, 30, 41]
Role of emulsan in growth on crude oil [42, 43]
Adherence of RAG-1 to hydrophobic substrates [36, 44, 45]
Binding of emulsan to hydrocarbon interfaces [8, 46]
Hydrocarbon substrate specificity [46]
Emulsanosols [47, 48]
Altered properties of emulsan at oil-to-water interfaces [48]

C. Emulsan: Binding to Hydrocarbon Interfaces

In view of our main subject—the possible use of biosurfactants in combating oil pollution—the studies on binding to hydrocarbon interfaces should be reviewed. Emulsan reduces the interfacial tension between water and *n*-hexadecane to only 15 mN/m and binds tightly to the hydrocarbon interfaces. It has a good ability for emulsification of hydrocarbons and for the stabilization of such emulsions [8, 46–48]. As to these last properties it possesses following advantages:

1. Low concentration of biopolymer required
2. Hydrocarbon substrate specificity (e.g., hexadecane/methylnaphthalene mixture)
3. Reversibility
4. Cation binding to emulsan-stabilized emulsions

It was observed that emulsan is only located at the hydrocarbon–water interface. Only small quantities of biopolymer were found in either the water or the oil phase. Oil-in-water emulsions (including 0.01 to 0.1 mL of hydrocarbons) were stabilized by 0.02–0.2 mg/mL of emulsan.

D. *A. calcoaceticus* RAG-1/Emulsan: Application Tests

Possible applications of emulsan-producing bacterium and emulsan itself were tested. For the protection of the marine environment, especially in the case of oil tanker cleaning procedures, an alternative approach with a growing microbial culture was tested leading to oil emulsification and degradation [9]. However, even under the best laboratory conditions (e.g., with nitrogen and phosphorus supplementation) microorganisms degraded only up to 60%–80% of the crude oil; the residual material still remained a major pollutant problem. For heavy-oil transportation, emulsan served as an emulsifier to form a 70% oil-in-water emulsion. In one experiment at a surfactant-to-oil ratio of 1:500 the emulsion was pumped successfully for 380 miles through a pipeline [29].

E. *Phormidium* J-1: Production of Emulcyan

Hydrophobic carbon sources are not always absolute presuppositions for the microbial overproduction of surface-active materials; examples are the foregoing formation of emulsan on ethanol, the rhamnolipid excretion from the freshwater bacterium *Pseudomonas* sp. (glycerol [49, 50]) or the cellobiose lipid production by *Ustilago maydis* (glucose [51]).

Sometimes, the necessity for the production of these chemicals in the absence of oily substrates seems not to be clear, but in the case of *Phormidium* sp. a hydrophobic, benthic cyanobacterium, Fattom and Shilo gave plausible arguments [52]. They suggest that the emulsifying polymer synthesized by this strain masks cell-surface hydrophobicity, thus causing detachment of the cells from hydrocarbon interfaces.

The benthic *Phormidium* sp. strain J-1 was isolated from a drainage channel in the Hulch swamp in Israel and is now commercially available as ATCC 39161. The ingredients of nutrient BG 11 described by Stanier et al. [53], dissolved in water or added to a special salt solution, served as the growth media [54]. The strain was cultivated at pH 7.5, temperature 26°C, and under continuous illumination. During the stationary phase of growth, *Phormidium* sp. excreted a high-molecular-weight polymer. This substance was found to be able to flocculate bentonite particles from suspensions [55].

Later on, in emulsifying activity tests done by Rosenberg et al. [8], this substance, called emulcyan, indicated a considerable stabilization of oil-in-water emulsions [52]. The chemical analysis of emulcyan resulted as the following:

Sugar moieties
Protein moieties
Fatty acids: 14:0 (16.5% of total content), 16:0 (29%), 18:0 (8.6%), 14:1 (11.6%), 16:1 (19%), 18:1 (15.1%)
Molecular weight: >100,000 (gel chromatography)

Emulcyan started to accumulate in the growth medium when the cells entered the stationary phase, reaching concentrations of 2.5 units/mL after 12 days and 8 units/mL after 28 days [for yield determination in detail, see Ref. 52]. Fattom and Shilo [52] proposed that the production of emulcyan by *Phormidium* cells serves as a dispersal strategy by this nonhormogonia-producing cyanobacterium.

IV. MICROBIAL SURFACTANTS FROM COASTAL AREAS OF THE FRENCH MEDITERRANEAN SEA

Most published results related to petroleum biodegradation in a marine biotope were obtained from samples originating either from sea water or from sea sediments. A French group [23] studied bacteria from the foam formed at the outlet of a Petroleum refinery (Gulf of Fos, France). Beginning with mixed populations, the aim was to obtain a mixed culture capable of attacking compounds such as aromatic hydrocarbons and asphaltenes commonly known as hard degradable compounds.

Bertrand et al. [23] cultivated a mixed population EM-4 composed of eight strains in a continuous culture (17 L working volume, 30°C). The medium contained sea water enriched with ammonium chloride (4 g/L), sodium phosphate (0.4 mM), and crude oil (1–12 g/L, Asthart type from Gabes Bay, Tunisia); the pH was adjusted to 8. After 8 days of batch culture, the continuous culture was started. Growth was realized under nonaseptic conditions. For the isolation of biosurfactants and determination of their emulsifying activity, the culture medium was centrifuged and filtered on a series of filters of decreasing porosity. The emulsifying activity of the supernatant was studied by measuring the intensity and stability of an emulsion (containing pure hydrocarbons or crude oil) prepared by mechanical stirring or by ultrasonic treatment. The results were the following:

The degradation percentage reached 83% with a 0.05/h dilution rate (D) and an initial 6 g/L crude oil concentration.
The different crude oil compounds (saturated, aromatics, polar hydrocarbons, and asphaltenes) were degraded at 97%, 81%, 52%, and 74%, respectively.
The culture supernatant generated stable emulsions of high optical density values (O.D. at 610 nm).
An analysis of the compounds produced during crude oil utilization revealed the presence of sugars (710 mg/L) and of lipids (96 mg/L).

Improving this continuous culture process (D = 0.11/h and S_0 = 12.2 g/L) by a tangential-flow filtration device for a better recovery of the biosurfactants, Mattei and Bertrand [10, 56] found these compounds to be highly concentrated in the residual filtrates. Then 2.7 g/L of total sugars and 0.32 g/L of lipids could be detected. Among the lipid fraction, monoglycerides, diglycerides, and fatty acids were major compounds; C_{14}, C_{16} and C_{18} chain lengths dominated in the case of fatty acids. The authors reported that after extrapolation of the laboratory tests, 1.6 g/L of total lipids were able to emulsify 1 kg of petroleum in the emulsion activity tests; for this purpose, 5 L of used medium are necessary.

The same French group reported the following concerning batch cultivations of the mixed population EM-4 [57]. After 12 days of incubation, 92% and 83% of saturated and aromatic crude oil ingredients were degraded, respectively, as well as 63% of polar products and 48.5% of asphaltenes. Maximum degradation occurred at a sodium chloride concentration of between 400 and 800 mM in the medium. Measuring the emulsifying activity of the culture supernatant during different stages of growth, it appeared that the activity was strongest and the stability was best at the end of the exponential phase, after 8 days (see Fig. 3). For comparison only the 5-day values are documented.

Investigating the nature of such emulsifying agents excreted into the culture medium, the authors detected sugar and lipid concentrations that were higher after growth on hydrocarbons than on a medium containing acetate or protein hydrolysates. A good correlation was found between emulsifying activity and concentrations of sugars and lipids. After separation, the bacterial strains of population EM-4 were identified as two *Acinetobacter*, two *Alcaligenes*, one

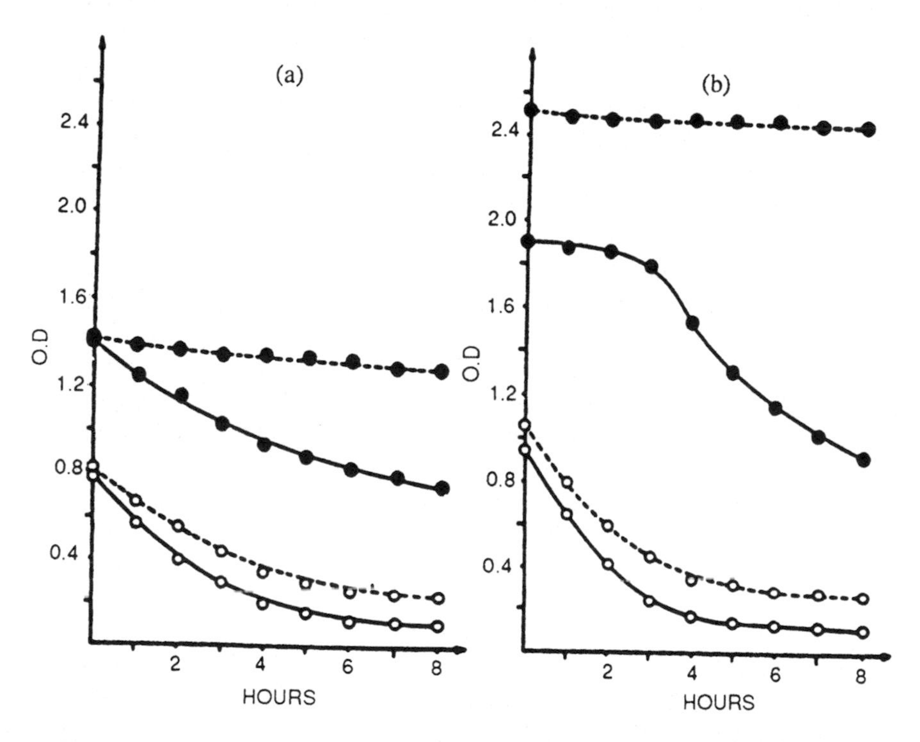

FIG. 3 Emulsifying activity of culture medium after 5 d (a) and 8 d (b) growth of mixed bacteria population EM-4 on crude oil. ●——● crude oil emulsified with mechanical stirring; ●- - - -● crude oil emulsified with sonic oscillation; ○——○ tetradecane emulsified with mechanical stirring; ○- - - ○ tetradecane emulsified with sonic oscillation. *Source*: Ref. 57 with permission.

Micrococcus, one *Flavobacterium*, one *Moraxella*, and one *Bacillus*. The last three did not develop growth on hydrocarbons tested. A precise study of the capacity of degradation of different hydrocarbons by the other five pure species showed that the percentage degradation is different from strain to strain. Especially, the weak halophile (NaCl concentration <5% favored) from *Alcaligenes* and *Acinetobacter* species were studied with respect to emulsifying activity and composition of culture medium after growth.

One of the above mentioned strains, *Alcaligenes* sp. PHY 9 L.86 exhibited a maximum efficiency for degrading hydrocarbon substrates and therefore was selected to study in greater detail the production of emulsifying agents [58]. Growth was not observed at NaCl concentrations between 0 and 100 mM. Maximum growth was achieved at 400 mM; *Alcaligenes* sp. PHY 9 is thus a marine bacteria according to the definition of Larsen [59].

The lipid composition of cell pellets and supernatants were examined throughout the bacterial growth (tetradecane was utilized within 48 h) using thin-layer chromatography coupled with flame ionization detection. Phospholipids were major components (94.6% of total) of cellular lipids, but in culture supernatant, phospholipids, and free fatty acids were dominating moieties. Maximum values for lipid content were 8 mg/L and additionally 13 mg/L for carbohydrate content at the early stationary phase (27 h). At the same time in examining emulsification ability, the optical density of the emulsion was 0.98 and remained stable with 80% of the initial O.D. after a 3-h incubation. There was a good relationship between lipid production and emulsifying activity.

Using electron microscopy with specific staining for acid polysaccharides, large exopolysaccharide fibers associated with vesicles were detected at the end of the growth phase, which also caused foam formation by the cells themselves. The production of an oil-emulsifying agent that was polysaccharide in nature has already been described by Floodgate [60] for a marine bacterium that degraded oil rapidly.

Recently Goutx et al. [61] tested three other marine bacteria, strains No. 8 and No. 17 (belonging to the genera *Pseudomonas* or *Alteromonas*) and *Pseudomonas nautica* for growth, emulsifying activity, carbohydrates, and lipids. The bacteria were grown at 32°C in a synthetic mineral salts medium composed of NaCl (600 mM), Tris-hydroxymethylamino methane (50 mM), KCl (10 mM), $CaCl_2$ (10 mM), NH_4Cl (70 mM), $MgSO_4*7H_2O$ (100 mM), Na_2HPO_4 (1 mM), and $FeSO_4$ (0.3 mM) at pH 7.5. Carbon sources were 1–3 g/L of acetate or lactate and 0.7–1 g/L of tetradecane and eicosane. Table 3 shows degradation rates and generation times of these bacteria together with those of *Alcaligenes* sp. PHY 9 L.86, all grown on soluble and insoluble substrates. Generation times were longer on insoluble substrates than on water-soluble carbon sources. The same table also indicates that the emulsifying activity (O.D. values) of the corresponding

TABLE 3 Characteristics of Four Marine Bacteria Grown on Water-Soluble and Water-Insoluble Substrates

Bacterial strain	Substrate (g/L)		Degradation rate (%)	Generation time (h)	Emulsifying capacity (3 h) (O.D. 610 nm)
Strain no. 8	Acetate	3	99	4.5	0.63
	Eicosane	1	56	22	1.15
Strain no. 17	Lactate	2	99	2	0.46
	Tetradecane	1	99	5	1.5
	Eicosane	1	65	8	1.4
Alcaligenes sp. PHY 9 L 86	Acetate	2	99	4	0.30
	Tetradecane	0.7	98	6	0.80
	Eicosane	1	88	6	0.42
Pseudomonas nautica 617	Acetate	2	99	5	0.56
	Tetradecane	1	97	6	1.8
	Eicosane	1	40	28	0.8

Source: Ref. 61 with permission.

supernatants were 1.3- to 3-fold higher in a culture medium of hydrocarbon-grown bacteria than in those of acetate- or lactate-grown strains. Nevertheless, this activity varied with the hydrocarbon type.

The protein values were low, averaging 3.65 mg/L, whereas carbohydrates and lipids were found in concentrations (μg/mg protein) ranging from 16 to 290 and 3 to 190, respectively. Compared to carbohydrates the lipid content increased 1.1- to 23-fold in supernatant after changing the carbon source in the medium (from water-soluble to water-insoluble carbon-source). The lipid composition of cellular material was different from that of the extracellular material. Besides phospholipids (dominant in cellular lipids), polar lipids (e.g., glycolipids, monoglycerides) and free fatty acids in special cases reached values equal to or higher than the first group. An overview is presented in Table 4, which shows, for example, that free fatty acids were major components in *Alcaligenes* sp. PHY 9 supernatants, whereas they were only slightly concentrated in supernatants of other strains.

Another important result of Bertrand's group is of interest [57]. Studying the chemical composition of the surface foam of the sea, out of which the potent bacteria were isolated, they detected carbohydrates and lipids in the range of 2 to 5 g/L and 0.2 to 1.2 g/L, respectively. Similar to culture supernatants, the major

TABLE 4 Composition of Extracellular Lipids (%) from Acetate, Lactate, or Alkane-Grown Cultures[a]

Bacterial strain	Substrate	PL	AMPL	DG	TG	FFA	ME+KE	WE+SE	UNK
Strain no. 8	Acetate	43.1	56.8	—	—	—	—	—	—
	Eicosane	48.7	51.2	—	—	—	—	—	—
Strain no. 17	Lactate	29.5	16.4	—	15.3	—	—	—	24.4
	Tetradecane	61.3	11.0	2.0	4.1	8.1	—	10.2	3.3
	Eicosane	43.0	53.4	—	3.53	—	—	—	—
Alcaligenes sp. PHY 9	Acetate	46.8	Tr.	1.9	0.9	51.7	0.9	—	—
	Tetradecane	11.4	3.8	—	8.0	73.2	—	3.5	—
	Eicosane	74.8	20.3	—	—	Tr.	—	4.8	—
Pseudomonas nautica 617	Acetate	82.7	17.2	—	—	—	—	—	—
	Tetradecane	70.2	6.4	—	1.5	5.2	—	15.1	1.3
	Eicosane	48.5	7.8	—	—	17.7	—	25.9	—

[a]Abbreviations: phospholipids (PL), acetone mobile polar lipids including glycolipids and monoglycerides (AMPL), diglycerides (DG), free fatty acids (FFA), triglycerides (TG), methyl ester and ketone (ME+KE), wax esters and sterol esters (WE+SE); unknown compounds with intermediate retention time (UNK)
Source: Ref. 61 with permission.

fatty acids were C_{16}, C_{17}, and C_{18}, saturated and unsaturated. After emulsion preparation, these emulsions were stable for many hours. As synthetic detergents have not been observed, the authors assume that in the natural environment microorganisms produce surface-active agents that enable the biodegradation of hydrocarbons.

V. MICROBIAL SURFACTANTS FROM THE EUROPEAN NORTH SEA

To combat possible oil spills in the German wadden sea (part of the European North sea), various methods were tested in field experiments [62], for example, the application of chemically synthesized surfactants as well as of surfactants from freshwater microorganisms. As the last ones seemed to be very effective and almost nontoxic in prevention of oil pollution in sensitive coastal biotopes, Wagner's and Gunkel's groups screened for surfactant producers among marine bacteria [63]. Their intention was to get biosurfactants of improved compatibility concerning the marine environment.

Using marine mixed cultures obtained from the sea around the Isle of Helgoland, they succeeded in enrichment of oil degrading bacteria. After nitrogen-/phosphorus-/yeast extract supplementation of a 100% or 75% seawater medium containing 1% of $C_{14,15}$-*n*-alkanes (technical grade), some different biosurfactants could be detected by careful screening using the methods described in part II of this Chapter. At last three potent bacteria were separated from the mixed community by direct plating of dilutions of enriched cultures on petri dishes containing either silica gel/agar/*n*-alkanes or agar/gas-phase *n*-alkanes. Two strains were identified as different *Arthrobacter* species. The first data on the nature of the biosurfactants were published in 1989 by Schmidt et al. [63]. In-detail information followed in 1991, Schulz et al. [24] and Passeri et al. [64] presented both the molecular structures of two glycolipids and the approximate composition of an extracellular emulsifier, as well as the cultivation conditions and some properties.

A. *Arthrobacter* sp. EK1: Production of Trehalose Lipids

After initial cultivation on the above-mentioned supplemented 75% sea water medium, the incubation time could be reduced from about 20 down to 3 days (early stationary phase) by using a mineral salts medium including only about 10 g/L of salts [64]. Figure 4 shows growth and biosurfactant production of *Arthrobacter* sp. EK1 during a cultivation of 30 g/L of *n*-alkanes in a 20-L bioreactor. The surfactant formation began in the late exponential phase and continued in the stationary phase. After 3 days, 4.8 g/L of mainly cell-associated glycolipids could be detected. The specific production was 0.52 g/g biomass. Major compounds of the glycolipid mixture (detected by TLC) were as shown in Fig. 5: (1) α,α-trehalose-dicorynomycolate (TL-2), about 5%, and (2) α,α-trehalose-tetraester (TL-4), about 90%.

Both compounds are known from the freshwater bacterium *Rhodococcus erythropolis* DSM 43215 [65–67]. With regard to TL-4, the exact position of the succinate group was missing until now. By chemical methods (hydroboration, acidic hydrolysis) and subsequent nuclear magnetic resonance (NMR) analysis (^{1}H, 1D and 2D COSY spectra) it was found that in the native compound the succinate moiety was attached to C2. High field NMR studies on the native glycolipid directly confirmed this result. The fatty acids attached to positions 2′, 3, and 4 of trehalose were octanoic (37.8%), nonanoic (5.7%), decanoic (53.3%), and undecanoic, lauric, and myristic acid (3% together).

B. Bacterial Strain MM1: Production of a Novel Glucose Lipid

Another low-molecular-weight cell-bound glycolipid was produced by a second marine isolate tentatively but probably wrong determined as *Alcaligenes* sp.; an

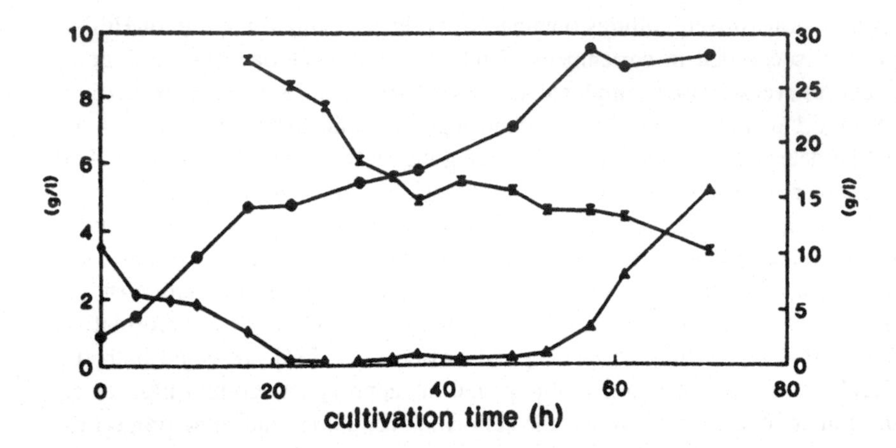

FIG. 4 Cultivation of the marine bacterium *Arthrobacter* sp. EK 1 in a 20-L bioreactor. Conditions: pH 7.0, T = 21°C, $C_{14,15}$-*n*-alkanes, nitrogen/phosphorus/iron/YE-supplementation. (▲) tetraester (g/L); (●) biomass (g/L); (◆) nitrate (g/L); (⧗) *n*-alkane (g/L). *Source*: Ref. 64 with permission

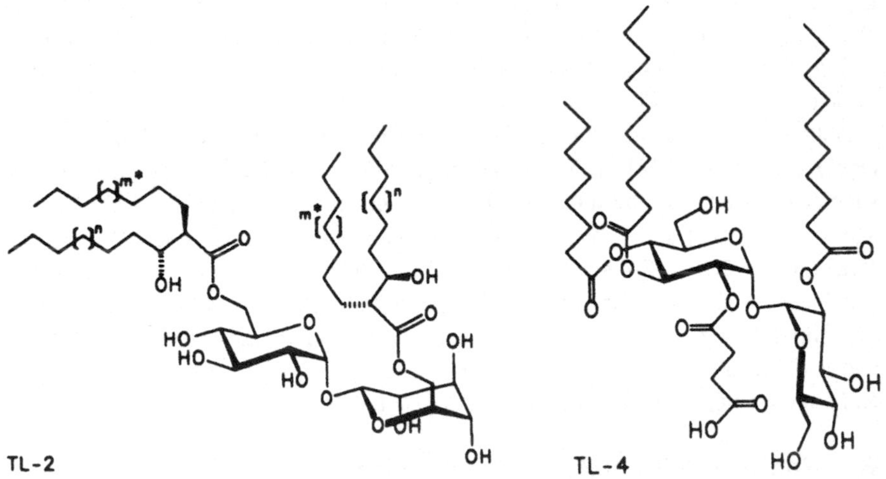

TL-2

TL-4

FIG. 5 Trehalose dicorynomycolate (TL-2) and trehalose tetraester (TL-4) from *Arthrobacter* sp. EK 1. (*) Partly unsaturated, m + n = 18-22. *Source*: Refs. 24 and 64 with permission.

exact confirmation of the genus until now is missing. Similar to the trehalose lipids, the novel glucose lipid was detected during TLC studies of crude products from extraction procedures. Figure 6 shows the molecular structure [24], whose elucidation based on HPLC, GC/MS, ^{1}H- and ^{13}C-NMR [68]. β-Hydroxydecanoic acid moieties but of different quantities are also known from rhamnose lipids of *Pseudomonas* sp. [22, 49] or from other glucose lipids of *Serratia rubidaea* [69].

After preliminary studies on a poorly supplemented sea water medium, a synthetic sea water medium, containing, for example, NaCl (23 g/L), nitrate and phosphate ions as well as $C_{14,15}$-*n*-alkanes (30 g/L), served as nutrients in fed-batch processes. Figure 7 indicates that the production of the cell-bound glucose lipid was growth-associated and that its yield reached more than 1.5 g/L during feeding of sodium nitrate; the specific production amounted to 0.065 g/g of biomass after 90 h. Cultivations on higher alkane concentrations without further nitrogen feeding (nitrogen-limiting conditions) did not alter the biosurfactant yield [68].

With respect to the function of this glycolipid for its own producer, experiments were achieved involving the cultivation of both precultured *n*-alkane as well as of precultured pyruvate cells. Pyruvate cells failing in glucose lipid production, indicated a long lag phase for *n*-alkane adaption, whereas a glucose lipid supplementation at the beginning effected a rapid growth (Fig. 8). From this observation, the authors assume that the hydrocarbon uptake into the cells could be facilitated by this glycolipid [68].

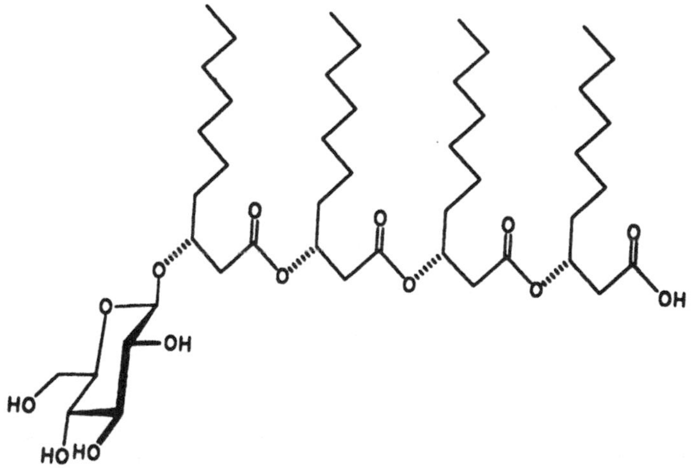

FIG. 6 Glucose lipid from the marine strain MM 1. *Source*: Ref. 24 with permission.

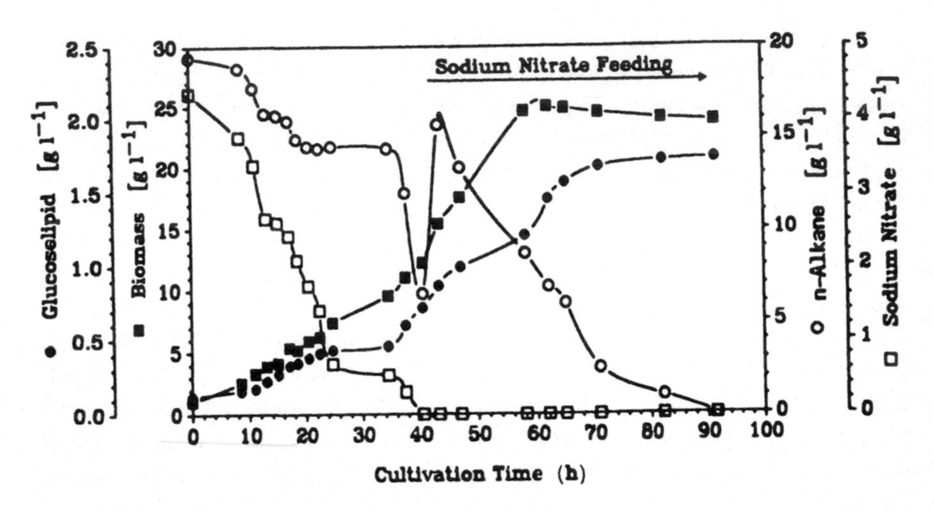

FIG. 7 Fed-bath cultivation of the marine strain MM 1 in a 10-L bioreactor. Conditions: Synthetic sea water medium, 3% $C_{14,15}$-*n*-alkanes; partly continuously feeding of $NaNO_3$; T = 27°C, pH 7.0, 500–1000 rpm, 0.6 v/v/m. *Source*: Ref. 68.

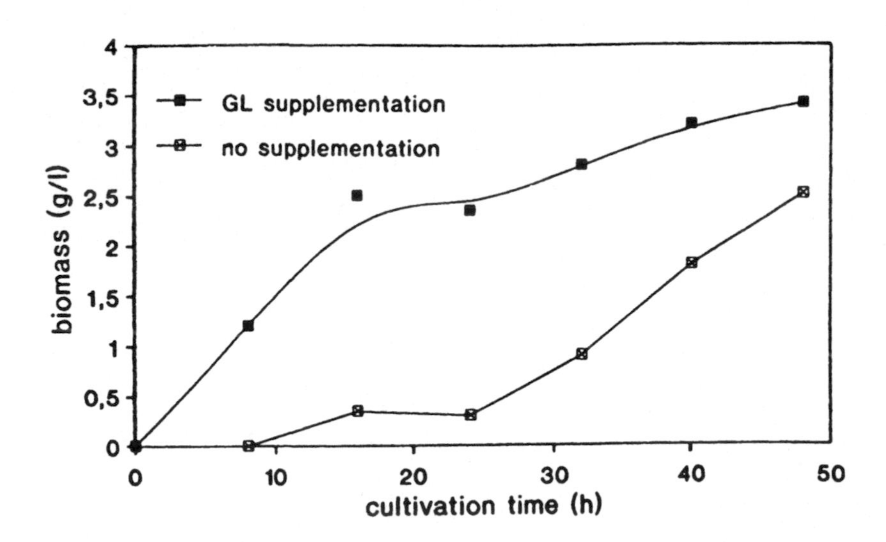

FIG. 8 Influence of the glucose lipid (0.1 g/L) on the biomass formation of the marine strain MM 1 from $C_{14,15}$-*n*-alkanes (cells precultured on pyruvate). Conditions: 100-mL cultures shake flasks, synthetic sea water, 20 mL/L of alkanes, 100 rpm, 27°C, pH 6.5–7.5. *Source*: Ref. 68.

C. *Arthrobacter* sp. SI1: Production of Trehalose Lipids and of Extracellular Emulsifying Agents

Using *n*-alkanes as the carbon source, trehalose lipids (TL-2, minor product; TL-4, major component; molecular structures, see *Arthrobacter* sp. EK1) and extracellular emulsifying agents (high molecular weight) were detected during the cultivation of *Arthrobacter* sp. SI1. When the carbon source was changed from $C_{14,15}$-*n*-alkanes to ethanol only TL-2 and the extracellular emulsifiers were found [24]; surprisingly, some quantities of the commonly cell-wall-bound trehalose dicorynomycolates (TL-2) excreted into the supernatant, probably because of the emulsifying agents and ethanol.

Figure 9 shows the time course of the formation of the emulsifying factors in a 50-L bioreactor using a synthetic sea water medium with nitrogen-/phosphorus-/yeast extract supplementation. The formation of the emulsifier appeared to be partly growth associated. After 80 h, the yield on crude emulsifier was highest expressed as 3 h-O.D. values (623 nm) of an oil-in-water emulsion. After removal of the cells and ultrafiltration of the supernatant (molecular weight exclusion: 100,000) 1.5 g/L of crude product were isolated.

Preliminary structure elucidation studies including enzymatic digestion and GC/MS, suggest the occurrence of a lipoprotein as emulsifying agent. Values on exact composition of the emulsifier should come in the near future.

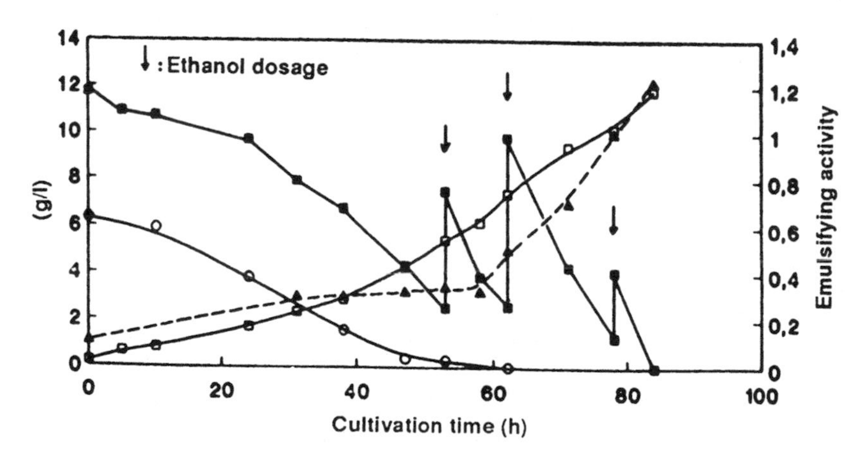

FIG. 9 Cultivation of *Arthrobacter* sp. SI1 on ethanol in a 50-L bioreactor. Conditions: Synthesis sea water with nitrogen/phosphorus/YE supplementation, pH 7.2–7.5, 20°C, intensor system, 800 rpm, 0.5 v/v/v/m. (□) biomass; (○) $NaNO_3$; (■) ethanol; (▲) OD 623 nm (60 min). *Source*: Ref. 24.

D. Properties of Glycolipids and Emulsifiers

The purified glycolipids from *Arthrobacter* sp. EK1 and strain MM1 were examined with respect to their behavior regarding surface and interfacial tension. The results are summarized in Table 5 [24]. Especially TL-4 and the novel glucose lipid are able to reach very low values indicating good properties for special surfactant purposes.

In stabilization of oil-in-water emulsions, the crude emulsifying agents of *Arthrobacter* sp. SI1 dominated over the above mentioned glycolipids. Testing several pure hydrocarbons and hydrocarbon mixtures in natural sea water, the authors found mixtures containing aliphatics and aromatics such as $C_{14,15}$-*n*-alkanes/1-methylnaphthalene (1/1; v/v) or kerosene, respectively, most efficiently emulsified (Fig. 10, also see Ref. 24). Cations of Mg^{2+} are necessary for emulsification.

E. Field Tests of Biosurfactants in Combat of Oil Pollution in the Wadden Sea

The German wadden sea of the European North sea is rich in mollusks and other slow moving species, spawning grounds, nursery grounds, and salt marshes. Possible oil accidents near to these highly sensitive areas could have dramatic consequences for survival of flora and fauna. Therefore, field tests were performed in the wadden sea [62].

Prior to these tests, the state of the art combating oil pollutions was to collect the major quantities of oil by mechanical methods (booms, skimmers) and after this to apply suitable surfactants. Comparing the influence of synthetic (which are stored for such purposes) and microbial surfactants on artificial oil pollution in limited small coastal areas (sediments), the following observations were made (see Table 6):

TABLE 5 Surface and Interfacial Properties of the Purified Glycolipids from the Marine Isolates

Biosurfactant	Minimum surface tension (mN/m)	Minimum interfacial tension[a] (mN/m)	CMC (mg/L)
TL-2	36	17	4
TL-4	26	<5	15
Glucose lipid	28	<5	25

[a]Against n-hexadecane
Source: Ref. 24.

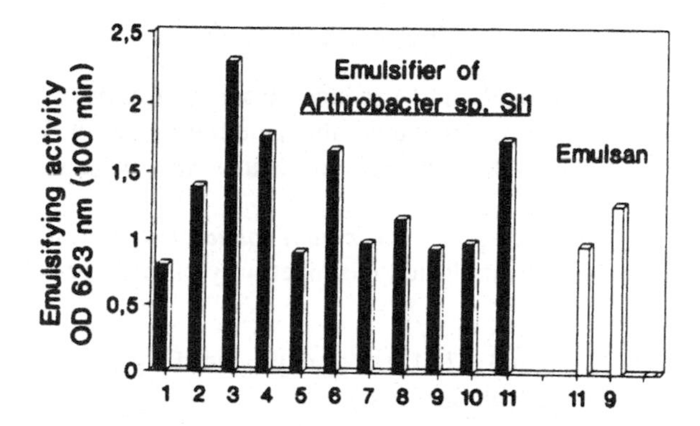

FIG. 10 Influence of several hydrocarbons on the emulsifying activity of the emulsifier from *Arthrobacter* sp. SI1 compared to emulsan. Conditions: 0.5 mg emulsifier/ml sea water; 1. mihagol-S; 2. 1-methylnaphthalene; 3. mihagol-S/1-methylnaphthalene (1:1); 4. model oil; 5. cyclohexane; 6. 1,2,4-trimethylcyclohexane; 7. 1-phenyldecane; 8. hexadecene; 9. hexadecane; 10. decane; 11. kerosene. *Source*: Ref. 24 with permission.

TABLE 6 Influence of Surfactants on the Artificial Oil Pollution of Sediments (Wadden Sea) and Their Effects on Special Marine Organisms

Crude oil + surfactant	Penetration (3d) (mg oil/kg dry mass)	Residual oil, 7 months (%)	Survival of *Corophium volutator* (%)	LC_{50}/48 h *Artemia*[b] (ppm)
TL-2	n.d.	0	n.d.	n.d.
TL-4	30	n.d.	75	7.500
Rhamnolipid	100	n.d.	72	3.000
Finasol[a]	200	88	0	64
Corexit[a]	160	n.d.	5	440
Without surfactant	100	67	35	n.d.

[a]Trade name of commercially available surfactants
[b]Without crude oil
Source: Refs. 11 and 70–72.

1. The dispersion by chemical surfactants increased the penetration of oil into the sediment [11, 62].
2. The microbial trehalose lipids TL-2 and TL-4 produced from the freshwater bacterium, *Rhodococcus erythropolis*, prevented the penetration; after 7 months a total elimination of all toxic polycyclic aromatics was detected [11, 70, 71].
3. Data on toxicity tests using the brine crayfish *Corophium volutator* (field test) or with *Artemia* larva (laboratory test) indicated the superiority of the biosurfactants [70–72].

It should be noted that both TL-2 and TL-4 are also produced by marine *Arthrobacter* species recently described (see Sec. V.A).

VI. CONCLUSION

The emulsification of hydrocarbons by surface- and interfacial-active agents is an essential step in hydrocarbon biodegradation, particularly in the marine environment. Comprehensive data on molecular composition, physicochemical properties, bioprocess engineering, and possible applications are reported from the following marine surfactants:

Lipopolysaccharide emulsan (protein content not necessary for emulsification)
Lipid and carbohydrate mixtures
Trehalose and glucose lipids
Lipoproteins

Acinetobacter and *Arthrobacter* species mainly belong to their microbial producers.

The application of biosurfactants as well as their microbial producers could make an important contribution in control and management of oil pollution.

REFERENCES

1. W. Gunkel and G. Gassmann, *Helgoländer Meeresuntersuchungen 33*: 164 (1980).
2. W. Gunkel, in *Angewandte Mikrobiologie der Kohlenwasserstoffe* (R. Schweibfurth, ed.), Expert Verlag, Sindelfingen, 1985.
3. R. M. Atlas, *Microbiol. Rev. 45*: 180 (1981).
4. G. D. Floodgate, in *Petroleum Microbiology* (R. M. Atlas, ed.), Macmillan, New York, 1984, pp. 355–397.
5. B. Austin, in *Marine Microbiology* (B. Austin, ed.), University Press, Cambridge, 1988, pp. 149–156.
6. O. Käppeli, A. Fiechter, and W. R. Finnerty, in *Advances in Biotechnology* (M. Moo-Young, C. W. Robinson, C. Vezina, eds.), Vol. I, Pergamon Press, New York, 1981, pp. 177–182.

7. T. Suzuki, K. Tanaka, I. Matsubara, and S. Kinoshita, *Agric. Biol. Chem. 33*: 1619 (1969).
8. E. Rosenberg, A. Zuckerberg, C. Rubinovitz, and D. L. Gutnick, *Appl. Environ. Microbiol. 37*: 402 (1979).
9. D. Gutnick and E. Rosenberg, *Ann. Rev. Microbiol. 31*: 379 (1977).
10. G. Mattei, E. Rambeloarisoa, G. Giusti, J. F. Rontani, and J. C. Bertrand, *Appl. Microbiol. Biotechnol. 23*: 302 (1986).
11. K. H. v. Bernem, *Senckenbergiana Marit. 16*: 13 (1984).
12. P. D. Abel, *J. Fish Biol. 6*: 279 (1974).
13. D. Pratt, in *Effect of the Ocean Environment on Microbial Activities* (R. R. Colwell and R. Y. Morita, eds.), Park Press, Baltimore, 1974, pp. 3–15.
14. K. Gundersen, in *Marine Ecology*, Vol. 3 (O. Kinne, ed.), John Wiley, London, 1976, pp. 301–356.
15. J. L. Reichelt and P. Baumann, *Arch. Microbiol. 97*: 329 (1974).
16. R. M. Atlas and R. Bartha, *Biotechnol. Bioeng. 14*: 309 (1972).
17. A. E. Reisfeld, E. Rosenberg, and D. Gutnick, *Appl. Microbiol. 24*: 363 (1972).
18. J. Le Petit and M. H. N'Guyen, *Can. J. Microbiol. 22*: 1364 (1976).
19. C. N. Mulligan, D. G. Cooper, and J. R. Neufeld, *J. Ferment. Technol. 62*: 311 (1984).
20. T. Matsuyama, M. Sogawa, and I. Yano, *Appl. Environ. Microbiol. 53*: 1186 (1987).
21. U. Göbbert, A. Schmeichel, S. Lang, and F. Wagner, *J. Am. Oil Chem. Soc. 65*: 1519 (1988).
22. C. Syldatk, S. Lang, F. Wagner, V. Wray, and L. Witte, *Z. Naturforsch. 40c*: 51 (1985).
23. J. C. Bertrand, E. Rambeloarisoa, J. F. Rontani, G. Giusti, and G. Mattei, *Biotechnol. Lett. 5*: 567 (1983).
24. D. Schulz, A. Passeri, M. Schmidt, S. Lang, F. Wagner, V. Wray, and W. Gunkel, *Z. Naturforsch, 46c*: 197 (1991).
25. J. E. Zajic, H. Guignard, and D. F. Gerson, *Biotechnol. Bioeng. 19*: 1303 (1977).
26. D. G. Cooper, J. E. Zajic, and D. F. Gerson, *Appl. Environ. Microbiol. 37*: 4 (1979).
27. D. F. Gerson and J. E. Zajic, *Antonie van Leeuwenhoek 45*: 81 (1979).
28. S. Goldman, Y. Shabtai, C. Rubinovitz, E. Rosenberg, and D. L. Gutnick, *Appl. Environ. Microbiol. 44*: 165 (1982).
29. D. L. Gutnick and Y. Shabtai, in *Biosurfactants and Biotechnology*, Surfactant Science Series, Vol. 25 (N. Kosaric, W. L. Cairns, and N. C. C. Gray, eds.), Marcel Dekker, New York, 1987, pp. 211–246.
30. Y. Shabtai and D. L. Gutnick, *Appl. Environ. Microbiol. 49*: 192 (1985).
31. A. Zuckerberg, A. Diver, Z. Peeri, D. L. Gutnick, and E. Rosenberg, *Appl. Environ. Microbiol. 37*: 414 (1979).
32. I. Belsky, D. L. Gutnick, and E. Rosenberg, *FEBS Lett. 101*: 175 (1979).
33. Y. Shabtai and D. L. Gutnick, *J. Bacteriol. 161*: 1176 (1985).
34. C. Rubinovitz, D. L. Gutnick, and E. Rosenberg, *J. Bacteriol. 152*: 126 (1982).
35. Y. Shabtai and D. L. Gutnick, *Appl. Environ. Microbiol. 52*: 146 (1986).
36. T. K. Ng and W. S. Hu, *Appl. Microbiol. Biotechnol. 31*: 480 (1989).
37. E. Kottutz, Diploma Thesis, Technical University of Braunschweig (1984).
38. Y. Shabtai and D. I. C. Wang, *Biotechnol. Bioeng. 35*: 753 (1990).

39. D. L. Gutnick, E. A. Bayer, C. Rubinovitz, O. Pines, Y. Shabtai, S. Goldman, and E. Rosenberg, *Adv. Biotechnol. 11*: 455 (1980).
40. O. Pines, E. A. Bayer, and D. L. Gutnick, *J. Bacteriol. 154*: 893 (1983).
41. O. Pines, Y. Shoham, E. Rosenberg, and D. L. Gutnick, *Appl. Microbiol. Biotechnol. 28*: 93 (1988).
42. O. Pines and D. L. Gutnick, *Appl. Environ. Microbiol. 51*: 661 (1986).
43. J. M. Foght, D. L. Gutnick, and D. W. S. Westlake, *Appl. Environ. Microbiol. 55*: 36 (1989).
44. E. Rosenberg, N. Kaplan, O. Pines, M. Rosenberg, and D. L. Gutnick, *FEMS Microbiol. Lett. 17*: 157 (1983).
45. M. Rosenberg, D. L. Gutnick, and E. Rosenberg, in *Microbial Enhanced Oil Recovery* (J. E. Zajic, D. G. Cooper, T. R. Jack, and N. Kosaric, eds.), Penn Well Books, Tulsa, OK, 1983, pp. 114–123.
46. E. Rosenberg, A. Perry, T. Gibson, and D. L. Gutnick, *Appl. Environ. Microbiol. 37*: 409 (1979).
47. Z. Zosim, D. L. Gutnick, and E. Rosenberg, *Biotechnol. Bioeng. 24*: 281 (1982).
48. Z. Zosim, D. L. Gutnick, and E. Rosenberg, *Biotechnol. Bioeng. 25*: 1725 (1983).
49. G. Hauser and M. L. Karnovsky, *J. Bacteriol. 68*: 645 (1954).
50. C. Syldatk, S. Lang, U. Matulovic, and F. Wagner, *Z. Naturforsch, 40c*: 61 (1985).
51. R. U. Lemieux, J. A. Thorn, and H. F. Bauer, *Can. J. Chem. 31*: 1054 (1953).
52. A. Fattom and M. Shilo, *FEMS Microbiol. Ecol. 31*: 3 (1985).
53. R. Y. Stanier, R. Kunisawa, M. Mandel, and G. Cohen-Bazire, *Bacteriol. Rev. 35*: 171 (1971).
54. A. Fattom and M. Shilo, *Appl. Environ Microbiol. 47*: 135 (1984).
55. A. Fattom and M. Shilo, *Arch. Microbiol. 139*: 421 (1984).
56. G. Mattei and J. C. Bertrand, *Biotechnol. Lett. 7*: 217 (1985).
57. E. Rambeloarisoa, J. F. Rontani, G. Giusti, Z. Duvnjak, and J. C. Bertrand, *Mar. Biol. 83*: 69 (1984).
58. M. Goutx, S. Mutaftshiev, and J. C. Bertrand, *Mar. Ecol. Prog. Ser. 40*: 259 (1987).
59. H. Larsen, *FEMS Microbiol. Rev. 39*: 3 (1986).
60. G. D. Floodgate, in *Microbial Ecology* (M. W. Loutit and J. A. R. Miles, eds.), Springer Verlag, Berlin, 1978, pp. 82–85.
61. M. Goutx, M. Acquaviva, and J. C. Bertrand, *Mar. Ecol. Prog. Ser. 61*: 291 (1990).
62. *Texte 6/87 Umweltbundesamt der Bundesrepublik Deutschland, Meereskundliche Untersuchung von Ölunfällen* (1987).
63. M. Schmidt, A. Passeri, D. Schulz, S. Lang, F. Wagner, K. Poremba, W. Gunkel, and V. Wray, in *Dechema Biotechnology Conferences* (D. Behrens, ed.), VCH Verlagsgesellschaft, 1989, pp. 811–814.
64. A. Passeri, S. Lang, F. Wagner, and V. Wray, *Z. Naturforsch. 46c*: 204 (1991).
65. P. Rapp, H. Bock, V. Wray, and F. Wagner, *J. Gen. Microbiol. 115*: 491 (1979).
66. E. Ristau and F. Wagner, *Biotechnol. Lett. 5*: 95 (1983).
67. J. S. Kim, M. Powalla, S. Lang, F. Wagner, H. Lünsdorf, and V. Wray, *J. Biotechnol. 13*: 257 (1990).
68. A. Passeri, M. Schmidt, T. Haffner, S. Lang, F. Wagner, and V. Wray, *Appl. Microbiol. Biotechnol. 37*: 281 (1992).

69. T. Matsuyama, K. Kaneda, I. Ishizuka, T. Toida, and T. Yano, *J. Bacteriol. 172*: 3015 (1990).
70. K. H. v. Bernem, in *Texte 6/87, Umweltbundesamt der Bundesrepublik Deutschland, Meereskundliche Untersuchung von Ölunfällen*, 1987, pp. 22–23.
71. K. H. v. Bernem, *Texte 6/87, Umweltbundesamt der Bundesrepublik Deutschland, Meereskundliche Untersuchung von Ölunfällen*, 1987, pp. 64–74.
72. S. Lang, B. Frautz, G. A. Henke, J. S. Kim, C. Syldatk, and F. Wagner, *Biol. Chem. Hoppe Seyler 367* (Supplement 17th FEBS Meeting, Berlin, 24.-29.8.1986): 212 (1986).

16

Biosurfactants in Food Applications

JORAN VELIKONJA and NAIM KOSARIC Department of Chemical and Biochemical Engineering, University of Western Ontario, London, Ontario, Canada

I. INTRODUCTION

Modern man's food is certainly different from the food that was prepared and consumed since time immemorial. There are several reasons for this change. One major cause is the tremendous shift in the ratio of rural to urban population that started to take place with the onset of industrialization in the early 19th century and culminated about a hundred years later. This process is still going on in the Third World. Industrialization also resulted in an increased employment of women, who no longer were able to spend much of their time preparing food for the entire family. A need for pre-prepared food products was artificially created, and an ever-growing and developing food industry was the direct consequence.

Another reason for the change was the concern about nutrition and hygienic values of food brought about by the average population's higher level of education, as well as a global, post–World War II orientation to a better quality of life in general ("health foods") and concerns about physical appearance ("low calorie foods"). Such distinct aspects of life through fashion and tourism certainly contributed to this trend. The popularity of foreign and exotic foods in the West made them domesticated and widely accepted as industrial products. A bizarre mixture of hedonism and asceticism, the philosophy of the modern consumer, is to a considerable extent directly influenced by contradictory advertizement messages.

The struggle for the new consumer led to the development of new, previously unknown foods (snack foods, convenience foods), as well as to the alteration of traditional foods which now have to be prepared on an industrial scale and with the greatest possible economic efficiency. Therefore, processes in the food industry are being constantly improved and are becoming more sophisticated. The inclusion of more than 2,500 various additives in the United States alone enabled the food industry to produce more than 15,000 different food items [1].

Additives are included in food to alter the existing functionalities of food constituents or to bring about new ones. The aim of this chapter is to give an overview of some of the physical properties of food that can be directly influenced by surface-active substances and to elucidate the potential of microbial biosurfactants as food additives through examples from patent literature and studies of their application in test recipes.

II. FUNCTIONAL PROPERTIES OF INTEREST IN FOOD

Food, like any other material, is characterized by a number of physical and chemical properties. Depending on the expectations of subjects who judge a food product, the score of multitude of properties determines the overall quality of that particular item.

The quality of food has been defined as the "composite of those characteristics that differentiate individual units of a product, and have significance in determining the degree of acceptability of that unit by the buyer" [2]. Qualities are classified as *quantitative* (e.g., proportion of ingredients and drained weight), *hidden* (e.g., nutritive value, adulterants, and toxic substances) and *sensory* (e.g., appearance, kinesthetics, and flavor).

Apart from the nutritional quality of foods and the raw materials for their production, all other properties affecting their utilization are known as functionalities or functional properties [3]. Foods and food ingredients have been more broadly termed *biological products*, and the sum of their properties could be subdivided into two major classes [4]:

1. *Nutritional properties.* Those properties of food that are related to its nutritive value or to the content of substances that can adversely influence the health affecting the animal [or human] body after passage into the alimentary tract, including chemurgic [catabolic], anabolic, metabolism-enhancing, and toxic functionalities.
2. *Functional properties.* Those properties influencing foods prior to the onset of digestion, including enzymatic and nonenzymatic functionalities (e.g., sense affecting and manipulative), and industrial properties.

The three major groups of nutritive food constituents (carbohydrates, proteins, and lipids) are bearers of the chemurgic and anabolic properties. Metabolic enhancers are present in food in smaller quantities. They are either vital for a wholesome diet (such as salt and other electrolytes and vitamins) or promote digestion, for example, the acrid (piperin, capsaicin, bitter glycosides, their aglycones, etc.) or aromatic principles (terpenes, terpenoids, and other volatiles) of spices and condiments, enzymes added to food as *in vivo* digestion aids, or dietary fiber.

The second class of properties comprises the functions of those enzymes (constitutive or added) that alter the properties of foods prior to consumption (e.g., meat tenderizers, clarifiers for beverages, milk-clotting enzymes, polysaccharide hydrolases). The other functionalities (nonenzymic) belonging into this class are the most important determiners of the physicochemical properties of foods. They can be subdivided into two categories:

1. *Sense affecting.* All those properties perceived by one or more of the five senses.
2. *Manipulative.* Those properties that facilitate or hinder various process steps during industrial or domestic food preparation.

Industrial properties pertain mainly to nonfood uses of biological products, although some enzymatic and manipulative features are of importance for successful industrial processing of foods.

In many instances, characteristics of a particular food constituent influence more than one functional property. As an example, the emulsifying action of egg yolk lecithin and proteins affects sense perception (smooth mouthfeel of mayonnaise), and allows homogeneous distribution of oil droplets in the final product. Physicochemical properties of the various food ingredients can be further classified according to the following properties:

Hydrophilicity. Affinity to water and other polar solvents, including solubility, water uptake, and wettability.

Interphasic properties. Ability to accumulate at the phase boundary of two immiscible media, including emulsification, fat uptake, foaming, adsorption, and coacervation (micelle formation).

Intermolecular properties. Ability to enter noncovalent or (to a lesser degree) covalent interactions among like or different molecules, including viscosity, thickening, gelation, film formation, foaming, adhesion, cohesion, stickiness, hardness, complex formation, spreading, elasticity, and plasticity.

Organoleptic properties. Properties perceived by the senses: vision (surface appearance and form, color), olfaction (odor, aroma), gustation (sweetness, saltiness, sourness, bitterness, and umami, the "fifth taste" produced by the glutamate receptors in the buccal mucosa [5]), palpation, and mastication (texture, mouthfeel, hardness, etc.) and audition (brittleness).

Other properties. For example, incompatibilities and other special chemical and physical properties.

Color is certainly one of the most important physical properties of food and the primary sensory quality to be evaluated by the consumer. If a certain food fails to match the traditional expectations regarding color, it can easily be rejected. The psychological basis of food color perception is documented [6].

Mechanical properties of food, that is, "those pertaining to the behavior of the material under applied forces" [7], are of no less importance than the chromatic ones, because they affect the behavior of the raw materials and products during storage, transport, handling, and processing, as well as the organoleptic property known as texture (the complex of mechanical sensations registered during mastication).

The mechanical characteristics of food can be systematized as shown in Table 1 [8]. The parameters in Table 1 are defined as follows:

Hardness. The force necessary to attain a given deformation.

Cohesiveness. The strength of the internal bonds making up the body of the product.

Viscosity. The rate of flow per unit force.

Springiness. The rate at which a deformed material goes back to its undeformed condition, after the deforming force is removed.

Adhesiveness. The work necessary to overcome the attractive forces between the surface of the food and the surface of other materials with which the food comes in contact (e.g., tongue, teeth, palate).

Fracturability. The force with which the material fractures. It is related to the primary parameters of hardness and cohesiveness. In fracturable materials, cohesiveness is low and hardness can be low to high.

Chewiness. The energy required to masticate a solid food product to a state ready for swallowing. It is related to the primary parameters of hardness, cohesiveness, and springiness.

TABLE 1 Mechanical Characteristics of Food

Parameters		
Primary	Secondary	Popular terms
Hardness		Soft → firm → hard
Cohesiveness	Fracturability	Crumby → crunchy → brittle
	Chewiness	Tender → chewy → tough
	Gumminess	Short → mealy → pasty → gummy
Viscosity		Thin → viscous
Springiness		Plastic → elastic
Adhesiveness		Sticky → tacky → gooey

Source: Ref. 8.

Gumminess. The energy required to disintegrate a semisolid food product to a state ready for swallowing. It is related to the primary parameters of hardness and cohesiveness.

The functional properties of various types of foods (apart from the organoleptic ones) are listed in Table 2 [9].

This chapter deals mainly with emulsion properties as the principal food functionalities to be influenced through the addition of surfactants.

III. EMULSIFICATION

"An emulsion is a heterogeneous system, consisting of at least one immiscible liquid intimately dispersed in another in the form of droplets, whose diameter, in general, exceeds 0.1 μm. Such systems possess a minimal stability, which may be accentuated by such additives as surface-active agents, finely divided solids, etc." [10].

This rather comprehensive definition of emulsions emphasizes the transient nature of this type of colloid and stresses the importance of the emulsifying agent, which improves the thermodynamically unfavorable situation arising from the contact of fat and water. Thus stable emulsions can be produced, with a lifespan of months or even years.

Many biological compounds are inherently insoluble or only sparingly soluble in water because of large hydrophobic parts in their molecules. These compounds are called lipids. Nonetheless, they are found intimately mixed together in a number of homogeneous food products, both natural and prepared. They can be dispersed into one another in the form of emulsions. Examples of natural emulsions are milk, egg yolk, and coconut milk. Examples of processed food emulsions

TABLE 2 Functional Properties of Various Foods

Type	Functionality
Beverages	Solubility, grittiness, color
Baked goods	Emulsification, complex formation, foaming, viscoelastic properties, matrix and film formation, gelation, hardness, absorption
Dairy substitutes	Gelation, coagulation, foaming, fat holding capacity
Egg substitutes	Foaming, gelation
Meat emulsion products	Emulsification, gelation, liquid holding capacity, adhesion, cohesion, absorption
Meat extenders	Liquid holding capacity, hardness, chewiness, cohesion, adhesion
Soups and gravies	Viscosity, emulsification, water absorption
Topping	Foaming, emulsification
Whipped desserts	Foaming, gelation, emulsification

Source: Ref. 9.

are cream, butter, margarine, mayonnaise, salad dressing, sausages, frankfurter, ice cream, cakes, chocolate, coffee whitener, fluid shortenings, spreads, and so on.

The phase present in the form of droplets is called the dispersed, discontinuous, or internal phase. These droplets are surrounded by the continuous or external phase. There are two basic possibilities: oil-in-water (o/w) and water-in-oil emulsions (w/o). Depending on the desired emulsion type, a surfactant soluble in what will be the continuous phase has to be chosen. Although not generally true, this is a good rule of thumb (Bancroft's rule, see below). If the dispersed phase contains additional discrete phases, the resultant systems are called multiple emulsions, which are either (o/w/o) or (w/o/w).

Emulsions are only one general type of the colloidal systems encountered with foods and summarized in Table 3. Food emulsions are usually much more complex systems than the definitions in Table 3 indicate, containing a host of compounds with a marked influence on colloidal systems (proteins, oligo- and polysaccharides, polar lipids, etc.), as well as solid and semisolid particles and gas inclusions.

TABLE 3 Types of Colloidal Dispersions and Examples in Food

Dispersed phase	Continuous phase	Common names	Examples
Solid	Liquid	Sols	Proteins in aqueous solution
Liquid	Liquid	Emulsions	Milk, mayonnaise
Gas	Liquid	Foams	Whipped toppings
Liquid	Solid	Gels, solid emulsions	Jellies, meat products
Gas	Solid	Open structures	Baked goods, textured products

Source: Ref. 3.

Another possible classification of emulsions is according to droplet size:

Macroemulsions. 0.2–50 μm, thermodynamically unstable.
Miniemulsions. 0.1–0.4 μm, thermodynamically unstable, but more than macroemulsions.
Microemulsions. 0.01–0.1 μm, thermodynamically stable, transparent.

Almost all food emulsions fall into the first category, whereas microemulsions only recently became interesting as systems for introducing water-soluble compounds (nutrients and flavors) into foods [11].

A. Formation of Emulsions

The most common way of preparing an emulsion is by agitation. The application of mechanical energy deforms and disrupts the interface between two immiscible or mutually slightly soluble liquids, creating droplets of the dispersed phase surrounded by the continuous phase. Only about 0.1% of the applied mechanical work is used up in the surface enlargement of the disperse-phase droplets. Deformation of these droplets is needed to achieve their further fragmentation. Forces that bring about the deformation of droplets act against the forces that tend to give the droplet a perfectly spherical shape and that make up the Laplace pressure. It is well known that the pressure at the concave side of an interface is higher than the one at the convex side (Laplace equation):

$$\Delta p = \gamma \left(\frac{1}{R_1} + \frac{1}{R_2} \right) = \frac{2\gamma}{R} \qquad (1)$$

Here γ is the interfacial tension; R_1 and R_2 are the principal radii of curvature for a droplet surface that can be defined by two such radii. For a spherical surface ($R_1 = R_2$) the second equality of the preceding equation is valid.

Such particles of liquid can be disintegrated into smaller ones if a pressure greater than Δp is applied over a distance r. The necessary pressure gradient, $2\gamma/R^2$, is decreased if the interfacial tension has been lowered through the addition of a surfactant. The surface free energy (E_s) of the dispersed phase is

$$E_s = \frac{\gamma \cdot \Delta A}{V_a} \tag{2}$$

where ΔA is the increase in interfacial area and V_d the volume of the dispersed phase. An emulsion of oil droplets with a radius of 1 μm in water has an E_s of approximately 3 kJ/m^3. The energy necessary to produce such an emulsion would be at least 3 MJ/m^3 (1000 times more) [12]. The high input of energy is needed to achieve the required Laplace pressure gradient (~2×10^{10} Pa/m in this case) or velocity gradient (about 10^7 s^{-1}, if η = 1 mPas), and is eventually dissipated as heat.

A high net power input per unit volume is needed for an efficient emulsification. With a mechanical energy density of approximately 10^{10} W/m^3 required in the above example, the water-based emulsion would be boiling in 0.03 s. Therefore an efficient dissipation of heat is needed, although peak energy densities persist only locally and for a very short time. Surfactants can reduce energy requirements 10-fold or more, but this decrease is not always directly dependent on concentration, particularly if the energy density is sufficiently high.

The Kelvin equation relates the chemical potential μ_B of the dispersed substance B and its activity a_B in the continuous phase to the droplet radius R:

$$\mu_B^r - \mu_B^\infty = kT \ln \frac{a_B^r}{a_B^\infty} = 2 V_B^0 \frac{\gamma}{R} \tag{3}$$

Here again, γ is the interfacial tension, V_B^0 is the molar volume of liquid B in the pure state, and the superscripts r and ∞ denote values at a spherical interface (radius = R) and at a plane interface (radius is infinite). k is the Boltzmann constant.

The shear stress, $\dot{\gamma}\eta_c$, resulting from viscous flow of the continuous phase around the dispersed droplets, is facilitating their disruption. Here $\dot{\gamma}$ is the velocity gradient, and η_c is the viscosity of the continuous phase. If η_c is not very high, the flow pattern that contributes to fragmentation is turbulent (high Reynolds number).

The aforementioned addition of surfactants is of crucial importance in emulsification. Surfactant molecules are transported to the interface where they accumulate in a surface layer. The nature of the surfactant determines which of the phases will be dispersed and which will be continuous. Bancroft's rule tells that the phase in which the surfactant is soluble will most likely be the continuous one.

Coalescence of droplets may spontaneously take place after a time. Whether and when this will happen is largely determined by the nature and concentration of the surfactant. It is a complex and incompletely understood process.

Emulsification usually starts with the formation of pockets (films) of the continuous phase within cut-off particles from the disperse phase. Surfactants prolong the life of such films, enabling interfacial tension gradients to build up along the interface of the two phases flowing past each other. At equilibrium, the interfacial tension gradient (in the z direction) equals the shear stress in the opposite direction:

$$-\frac{d\gamma}{dz} = \tau \tag{4}$$

Here $-d\gamma/dz$ is the interfacial tension gradient and τ is the shear stress, represented by Newton's law:

$$\tau = \eta \left(\frac{du_z}{dy}\right)_{y=o} \tag{5}$$

where η is the viscosity and du_z/dz is the gradient of the z components of velocity perpendicular to the interface measured at the distance y from the interface. As a consequence, the interfacial tension gradient counterbalances the viscous flow of a film and can support a film of liquid with a thickness h and a density ρ_c if

$$|2d\gamma/dz| \geq \rho_c gh \tag{6}$$

where g is acceleration due to gravity. It follows that the lower the interfacial tension, the thinner the film it can support. But the conclusion that with no surfactant added much thicker films could be supported is wrong. Only the addition of surfactants enables the formation of the necessary interfacial tension gradient in a fluid. When an emulsion is formed, no actual equilibrium between shear and tension forces is established. Streaming leads to locally compressed and expanded areas, and the surfactant that is transported to the newly formed interfaces reestablishes the low interfacial tension. The surface dilational modulus $\mathcal{E}$, which is dependent on the type of surfactant, the rate of transport to the interface, and the rate of interface dilation/ compression, gives the extent to which γ differs from the equilibrium:

$$\mathcal{E} = \frac{d\gamma}{d(\ln A)} \tag{7}$$

With thin films, the transport of surfactant molecules to the surface is very fast, and the Gibbs' elasticity of the film is of interest:

$$\mathcal{E}_f = 2\mathcal{E} = -\frac{2d\gamma/d(\ln \Gamma)}{1 + 0.5h\frac{d(f_s \cdot c_s)}{d\Gamma}} \tag{8}$$

Here Γ is the surface concentration or surface excess of the surfactant, expressed in mol/m^2, and the activity of the surfactant (multiplied by the appropriate unit of concentration, in this case by mol/m^3) is concealed as $(f_s \cdot c_s)$, since

$$a_s = f_s \cdot \frac{c_s}{c^\ominus} \tag{8a}$$

where a_s and f_s are activity and activity coefficient of the surfactant (both dimensionless), c_s is surfactant molar concentration, and $c^\ominus$ is the unit molar concentration (in this case $c^\ominus = 1$ mol/m^3). When surfactant concentration is sufficiently low ($f_s \rightarrow 1$), activity can be approximated with the numerical value of c_s.

If $\mathcal{E}_f = 0$, no interfacial tension gradient can build up, and, if it is high, the film drainage is inhibited. As can be seen in the formula above, a decrease in film thickness h increases the film elasticity and in turn this stabilizes the dilated film. But if the surfactant concentration is very low, film dilation will cause a drop in Γ and, consequently, $\mathcal{E}_f$ can reach a minimum at the thinnest part of the film. Thus, surfactant concentration must not be kept too low.

If the surfactant is dissolved in the disperse instead in the continuous phase, the film that is formed from the latter will be unstable (Bancroft's rule). Various causes make the plant interface unstable, so that the process of droplet formation can take place. High shear stress velocities u_e of turbulent eddies can cause pressure differences on the interface, which must be greater than the Laplace pressure:

$$\Delta p = (\rho_1 - \rho_2)u_e^2 > 2\gamma/R \tag{9}$$

where ρ_1 and ρ_2 are densities of the two phases ($\rho_1 \geq \rho_2$). The size of the eddy must be approximately $2R$.

The interface can be disrupted also by the formation of surface waves with a wavelength λ, for example, by sonication. The diameters of the formed droplets are approximately $\lambda/2$ or smaller.

B. Influence of Surfactants on Emulsion Formation

Surfactants influence emulsion formation in many ways simultaneously, leading to a complex pattern of effects difficult to study, partly because of the short duration of most of them ($<10^{-4}$ s).

1. Surfactants Diminish Interfacial Tension

This value directly influences the radius (R) of droplets formed, depending on which type of flow, that is, which forces prevail during emulsification:

laminary flow (viscous shear): $R \propto \gamma$
turbulent flow (inertial forces): $R \propto \gamma^{3/5}$

The distribution of the surfactant between the continuous phase and the interface tends to reach an equilibrium during protracted emulsification. However, from the equation for the surface dilation modulus $\mathcal{E}$ (Eq. 7) it follows that the interfacial tension γ changes as the interfacial area A is increased.

For the simple case of a plane interface periodically compressed and dilated by longitudinal waves of frequency ν (angular frequency $\omega = 2\pi\nu$) there exists an expression relating $\mathcal{E}$ to several parameters (γ, a_s, Γ, $\mathcal{D}_s$, ω):

$$\mathcal{E} = -\frac{d\gamma/d(\ln \Gamma)}{\sqrt{1 + 2\zeta + 2\zeta^2}} \tag{10}$$

where

$$\zeta = \frac{d(f_s \cdot c_s)}{d\Gamma} \sqrt{\frac{\mathcal{D}_s}{2\omega}} \tag{10a}$$

The diffusion coefficient $\mathcal{D}_s$ of the surfactant is usually approximately 10^{-9} m^2/s. Activity can be approximated with the numerical value of c_s as in Eq. 8 if c_s is less than the critical micelle concentration.

For actual emulsification, where curved interfaces are formed in a stream of liquid and where the transport of surfactant proceeds by convection rather than by diffusion, ultimately resulting in depletion from the continuous phase when enough active surface has been created, no similar theoretical relations exist. However, a decrease in $\mathcal{E}$ can be expected, as a consequence of decreased surfactant concentration in the bulk (smaller a), and a rising rate of surface formation [larger $d(\ln a)/dt$], because the smaller the droplets are, the larger their surface is when formed in unit time.

By the end of emulsification the interface activity γ will be

$$\gamma_0 < \gamma < \gamma_{equil.} \tag{10b}$$

2. Surfactants Reduce the Necessary Surface Free Energy

The work required to increase the interface area A of an emulsion can be represented as

$$W = \gamma\,\Delta A + A\,\Delta\gamma \tag{11}$$

The second term in the preceding equation is due to the extra energy needed because of the nonuniform elasticity of the film (i.e., variable interface tension). This extra energy as well as an extra energy needed to overcome the viscous resistance to the surface expansion, are supposed to be insignificant.

3. Surfactants Increase the Interfacial Tension Gradient

The interfacial tension gradient (ITG, or τ) has been defined in Eqs. 4 and 5 as the tangential stress at the boundary of two fluids moving past each other. As surfactant molecules are adsorbed onto the interface, a relaxation of the built-up stress occurs. However, if the time for such relaxation is limited, the stress can amount to the following, for example

$$\frac{\gamma_0 - \gamma}{d} \approx \frac{10\text{mN/m}}{1\mu\text{m}} \approx 10 \text{ kPa} \tag{12}$$

Surfactants increase the difference in the numerator, creating a difference in stress and velocity across the interface. Therefore, a free circulation within the droplet is no longer possible, which results in an easier deformation of the spherical shape of the droplet and thus emulsification is promoted.

4. Surfactants Retard Droplet Coalescence

This phenomenon takes place during droplet formation. It is due to the uneven distribution of the surfactant molecules in the interface of films separating adjacent droplets. It was pointed out above (Eq. 8) that the thinnest portion of a film (h min) becomes its most elastic part ($\mathcal{E}_{fmax}$) with the highest surface tension. There the surface concentration (Γ) of surfactant is minimal, since its transport back into the bulk of the continuous phase is more rapid from a thin spot. The resulting tangential gradient of Γ causes surfactant molecules to migrate from thicker portions of the film to the thinnest spots, for example, between two approaching droplets. These molecules, in turn, drag an atmosphere of solvent molecules with them, causing a flow of the continuous phase to take place (the Marangoni effect), intercalating more liquid between the approaching particles and preventing them from coalescence. This stabilizing effect manifests only with the surfactant in the continuous phase (Bancroft's rule, see Sec. III). The low Gibbs' elasticity on the concave side of the interface (in the disperse phase) does not promote the buildup of an interfacial-tension gradient and a Marangoni effect.

5. Surfactants May Cause Interfacial Instability

If a gradient of the chemical potential (μ) of a solute across an interface exists, the interface is thermodynamically unstable. A number of events can cause the unstable surface to disrupt and droplets to form or to fragment already existing ones.

A μ-gradient driven transport of solutes to and from the interface results in convection currents near the interface. This convection can disturb the interface facilitating emulsification, or it can be even of such a small wavelength as to break the interface into sufficiently small droplets and cause spontaneous emulsification.

Concave curving of the surfactant-supplying phase (lowest surface concentration of surfactant, thus highest interfacial tension γ) presents an unstable state in which the ensuing Marangoni effect increases the likelihood of droplet shredding.

Viscous shear can lead to accumulation of the surfactant at certain points on the droplet (especially in the case of proteins as surface-active agents), decreasing γ sharply and promoting further dispersion of the droplet.

As regards the influence of surfactants on the droplet size, it can be generally stated that increasing the surfactant concentration decreases the droplet size, a phenomenon strongly dependent on the surfactant type. Several simple models were studied with anionic, nonionic, and polymeric surfactants, but conclusions derived therefrom cannot easily be applied to more complex systems, such as food emulsions.

C. Emulsion Stability

As stated earlier, emulsions (with the exception of microemulsion) are more or less unstable systems, the phases having a tendency to separate. Instability arises from the excess energy introduced into the system to enlarge the surface area of the phases. The total free energy (G) of the oil/water system prior to emulsification is given by:

$$G^i = G^i_\sigma + G^i_w + G^i_{\sigma/w} + G^i_{l/s} \tag{13a}$$

whereas after emulsification it is

$$G^f = G^f_\sigma + G^f_w + G^f_{\sigma/w} + G^f_{l/s} - (TS_{\text{config}})^f \tag{13b}$$

Here G is the Gibbs free energy, the superscripts i and f stand for the initial and final state, respectively, and the subscripts o, w, o/w and l/s denote oil, water, the oil/water interface, and the liquid/solid interface at the container walls. The last term in Eq. 13b comprises the configurational entropy of suspended droplets (S_{config}), which approximates zero when no emulsion exists.

The term $G^i_{\sigma/w}$ in Eq. 13a can be represented by (see Eq. 11):

$$G^i_{o/w} = \gamma_{o/w} \cdot A^i \tag{14}$$

It is much smaller than the corresponding term in Eq. 13b and can be neglected in their difference. So can be the terms $G^i_{l/s}$ and $G^f_{l/s}$ in both equations because of the relatively small liquid/container interfacial area. The first two terms in both equations are equal:

$$G^i_o = G^f_o \qquad G^i_w = G^f_w \tag{15}$$

and the free energy of emulsion formation becomes

$$\Delta G_{\text{form}} = G^f - G^i = (G^f_i - G^i_i) - T\,\Delta S_{\text{config}} \tag{16}$$

or (cf. Eq. 11)

$$\Delta G_{\text{form}} = \gamma\, \Delta A - T\, \Delta S_{\text{config}} \tag{17}$$

The free energy of emulsion formation is positive, because of the large increase in interfacial area (typically 10^5 times that of an unemulsified system [13], which makes ($\gamma\ \Delta A$) usually much larger than ($T\ \Delta S_{\text{config}}$) and the system is thermodynamically unstable. The breakup of an emulsion reduces the interfacial area and ΔG_{form}.

The various effects influencing the overall thermodynamic and kinetic stability of an emulsion come from the following:

van der Waals/London forces
Electrostatic forces
Steric interactions
Hydration forces
The free polysaccharide effect

The first two types of forces, known for a long time and well characterized on a molecular level, served as a basis for the development of an interaction theory of colloid stability by Deryagin and Landau [14], and independently by Verway and Overbeek [15] (DLVO theory).

1. Van der Waals/London Forces

Attractive dipole-dipole forces between polar molecules are known as van der Waals interactions. There are also weak interactions between dipoles induced on otherwise nonpolar molecules by other neighboring dipoles and the London dispersion interactions from transient dipoles created by changes in the electron cloud density. Attractive forces generally tend to decrease emulsion stability. The attraction energy (negative sign) of permanent and induced dipoles per molecule is

$$E_D = -\frac{\mu_D^2}{l^6}\left(\alpha_0 + \frac{\mu_D^2}{3kT}\right) \tag{18}$$

Here μ_D is the dipole moment of the molecule, l is the distance between the molecules, and α_o is the distortion polarizability of the molecule. The first term represents an interaction between a permanent and an induced dipole; the second term represents the attraction between favorably oriented permanent dipoles.

The London dispersion interaction arises from transient dipoles of nonpolar molecules brought close to one another. The interaction energy is

$$E_L = -\frac{3}{4}h\nu_0\left(\frac{\alpha_0^2}{l^6}\right) = -\frac{L}{l^6} \tag{19}$$

where $\boldsymbol{h}$ is the Planck's constant and ν_o is the principal specific frequency. The characteristic energy ($\boldsymbol{h}\nu_o$) is equal to the ionization energy of many simple molecules.

The combined attraction energy E_A of these two types of molecular interaction (the sum of Eqs. 18 and 19) is then

$$E_A = -\frac{1}{l^6}\left(2\mu_D^2\alpha_0 + \frac{2\mu_D^4}{3kT} + \frac{3\alpha_0^2 \boldsymbol{h}\nu_0}{4}\right) = -\frac{\mathcal{A}}{l^6} \tag{20}$$

London forces generally are the most important source of attraction between molecules of nonpolar or slightly polar substances. Because of their additivity, they become long-range forces between macroscopic particles, such as droplets of an emulsion. A similar expression is obtained for dispersion force attractions between two equal spherical particles of radius R at a distance $D = H + 2R$ between their centers (or H between their surfaces) [17]:

$$E_L = -\frac{\mathcal{A}_H}{6}\left(\frac{2}{\Phi^2 - 4} + \frac{2}{\Phi^2} + \ln\frac{\Phi^2 - 4}{\Phi^2}\right) \tag{21}$$

In the above equation, $\Phi = D/R = (2 + H)/R$, whereas $\mathcal{A}_H$ is the Hamaker constant (see Eqs. 22 and 23). If $H << R$, a simplified equation is obtained:

$$E_L = -\frac{\mathcal{A}_H R}{12H} = \frac{\mathcal{A}_H}{12(\Phi - 2)} \tag{21a}$$

For two unequal spheres with radii R_1 and R_2 ($R_1 < R_2$):

$$E_L = -\frac{\mathcal{A}_H}{12}\left(\frac{\psi}{\chi^2 + \chi\psi + \chi} + \frac{\psi}{\chi^2 + \chi\psi + \chi + \psi} + \ln\frac{\chi^2 + \chi\psi + \chi}{\chi^2 + \chi\psi + \chi + \psi}\right) \tag{22}$$

where $\chi = H/(2R_1)$ and $\psi = R_2/R_1$, respectively. For $H << R_1$ the simplified equation is

$$E_L = -\frac{A_H R_1 R_2}{6H(R_1 + R_2)} \tag{22a}$$

If the spheres are in a vacuum, the Hamaker constant, which is of the order of magnitude of 10^{-19}–10^{-20} J, is

$$\mathcal{A}_H = \pi^2 C^2 L \tag{23}$$

C is the number of atoms contained in unit volume of the substance, and L is the London interaction constant thereof (see Eq. 19). However, in the case of an emulsion, droplets of the disperse phase are separated by the continuous phase, and the effective Hamaker constant becomes

$$\mathcal{A}_H \approx \pi^2\left(C_d^2 L_d + C_c^2 L_c - 2C_d C_c \sqrt{L_d L_c}\right) \tag{24}$$

where the subscripts c and d stand for the continuous and disperse phases, respectively.

For large separation distances (>20 nm) interaction energies calculated with the above equations begin to differ from actual values. This "retardation effect" is a consequence of the finite velocity of electromagnetic waves, becoming increasingly important at such distances and not allowed for in the original Hamaker equations. Therefore modifications have been suggested; for example, for a pair of spherical droplets with radius R and interparticle distance H [17]:

$$E_L = -\mathcal{A}_H R\left(\frac{2.45}{120\pi H^2} + \frac{2.17}{720\pi^2 H^3} + \frac{0.59}{3360\pi^3 H^4}\right) \tag{25}$$

A clustering of droplets known as flocculation (see below) is the result of the aforementioned attraction forces. The stabilization of an emulsion requires repulsive interactions (electrostatic, steric, and hydration interactions) to counterbalance van der Waals–London forces.

2. Electrostatic Forces

Ions on the surface of food emulsion droplets make them electrically charged. The charge can come from simple ions (e.g., H_3O^+ and OH^-), charged precursors of macromolecules (e.g., amino acids), simple surfactants (e.g., lecithins) and proteins with their ionic amino acid side chains and terminal amino acids. Around the charged droplet an atmosphere of counterions from the bulk of the continuous phase forms an electrical double layer. The concentration of ions in the double layer decreases exponentially with distance from the droplet surface. Both the surface charge of the droplet and the electrolyte concentration in the continuous phase affect the interaction force. When a charged particle undergoing Brownian motion collides with another particle, this interaction can occur under conditions of constant surface potential (if there is always enough time for the counterions to maintain an electrochemical equilibrium of the distorted double layer), or under constant surface charge. It was shown that the time of a Brownian collision ($\approx 10^{-7}$ s) is much less than the relaxation time of an electrical discharge restoring the double layer (≈ 1 s) [18]. Therefore, the potential energy of interaction at constant surface charge for two spherical particles with radii R_1 and R_2 at a distance H is [19]

$$E_\sigma = \frac{\varepsilon R_1 R_2}{4(R_1 + R_2)}(\psi_1^2 + \psi_2^2)\left\{\frac{2\psi_1\psi_2}{\psi_1^2 + \psi_2^2} \ln \frac{1 + \exp(-\kappa H)}{1 - \exp(-\kappa H)} - \ln\left[1 - \exp(-2\kappa H)\right]\right\} \tag{26}$$

Here $\varepsilon = 4\pi\varepsilon_0\varepsilon_r$ is the dielectric constant of the continuous phase, ψ_1 and ψ_2 are surface potentials of the two spheres, and κ is the so-called Debye reciprocal length (thickness of the double layer), defined as

$$\kappa = \left(\frac{8\pi e^2 C_0 \nu^2}{\varepsilon kT}\right)^{1/2} \qquad (27)$$

Here e is the unit electric charge, C_0 is the number concentration of the electrolyte, and ν is the charge number of either the cation or the anion in a symmetrical (1:1, 2:2, etc) electrolyte.

A similar equation applies to the interaction of two spheres of radii R_1 and R_2 in the constant surface potential (ψ) model [20]:

$$E_\psi = \frac{\varepsilon R_1 R_2}{4(R_1 + R_2)} (\psi_1^2 + \psi_2^2) \left\{ \frac{2\psi_1\psi_2}{\psi_1^2 + \psi_2^2} \ln\frac{1 + \exp(-\kappa H)}{1 - \exp(-\kappa H)} + \ln\left[1 - \exp(-2\kappa H)\right] \right\} \qquad (28)$$

Double layer thicknesses calculated for various concentrations of a 1:1 electrolyte from Eq. 26 (at 25°C) range from 30.5 nm (10^{-4} mol/L) to 0.305 nm (1 mol/L).

In a 0.1 M solution of a 1:1 electrolyte the double layer thickness is approximately 1 nm thick and the droplets are not efficiently protected from aggregation by repulsive electrostatic forces. The higher the charge number of either of the ions, the more drastically decreases the double layer thickness. This explains why food emulsions are coagulated if a sufficient amount of salt is added, as well as why charged proteins as part of the droplet surface stabilize emulsions.

3. Steric Interactions

Adsorbed protein or polysaccharide (gum) molecules stabilize emulsions against flocculation by creating a sufficiently thick layer of chains and loops, and thereby minimizing van der Waals attraction. The orientation of segments of the macromolecule at the oil–water interface (surface-aligned trains, protruding loops, and tails) depends, in the case of proteins, on the hydrophilic and hydrophobic amino acid side chains and their distribution within the molecule. The layers on two approaching or colliding droplets can interpenetrate and/or compress each other. There are expressions for the steric repulsion energy for simplified models such as the one for a pair of spherical droplets with radii R_1 and R_2, a minimal interparticle distance H and a uniform segment distribution in an adsorbed layer of uniform thickness L_s and given that $L_s < H < 2L_s$ [21]:

$$E_s = 8\pi \boldsymbol{R}T \frac{R_1 R_2}{R_1 + R_2} \Gamma^2 \frac{\upsilon_p^2}{\upsilon_c} (0.5 - X) \left(1 - \frac{H}{2L_s}\right) \qquad (29)$$

Here $\boldsymbol{R}$ is the gas constant, Γ is the surface concentration of the adsorbed protein, and X is called the Flory-Huggins interaction parameter. The partial specific volumes of the protein and continuous phase are υ_p and υ_c, respectively. If the adsorbed layer has a thickness greater than the interparticle distance ($L_s < H$), then

$$E_s = 4\pi RT \frac{R_1 R_2}{(R_1 + R_2)^2} \Gamma^2 \frac{\upsilon_p^2}{\upsilon_c} (0.5 - X) \left(3 \ln \frac{L_s}{H} + \frac{2H}{L_s} - 1.5\right) \tag{29a}$$

The surface concentration of proteins is largest at their isoelectric point, when their stabilizing influence on emulsions is due to the steric forces.

4. Hydration Forces

These short-range (<5 nm) repulsion forces between hydrophobic droplets in aqueous media are thought to originate either in the energy needed to rearrange molecules of water at the surface of and around approaching particles, or in the repulsion of permanent and induced dipoles. At very small interparticle distances (<1 nm) a thin film of water molecules is kept in place by hydration forces, preventing the droplets from coalescence. From direct measurements [22] an exponential dependence of the hydration forces to the distance has been established:

$$E_H = \mathcal{A} \exp(-bl) \tag{30}$$

5. Free Polymer Effect

Polysaccharides from the continuous phase can act as both stabilizers and destabilizers. The effect is explained by the fact that the approaching of two particles decreases the local concentration of the adsorbed polymers, and thereby an osmotic repulsive force is created. But the displacement of water from the region between the spheres produces the opposite effect.

6. Summary Effect of the Interaction Forces

A sum of the described partial interactions between droplets gives the total energy of interaction:

$$E_T = E_{LD} + E_E + E_S + E_H \tag{31}$$

When the London–van der Waals attraction is superposed over the electrostatic repulsion, a graph as the one in Fig. 1 is obtained [23]. It can be seen that an energy barrier at a distance of about 5 nm prevents the droplets from coalescence, and at 15–20 nm there is an energy minimum, allowing aggregation of the droplets. If uncharged polymers protect the droplets ($E_E = 0$), other repulsive interactions would produce a similar behavior. The role of polysaccharides as stabilizers of protein-emulsified systems can be explained in this way [19].

D. Emulsion Breakup

As was shown above, the free energy of emulsions is generally higher than that of its separated constituents, because of the tremendous, energy-consuming increase

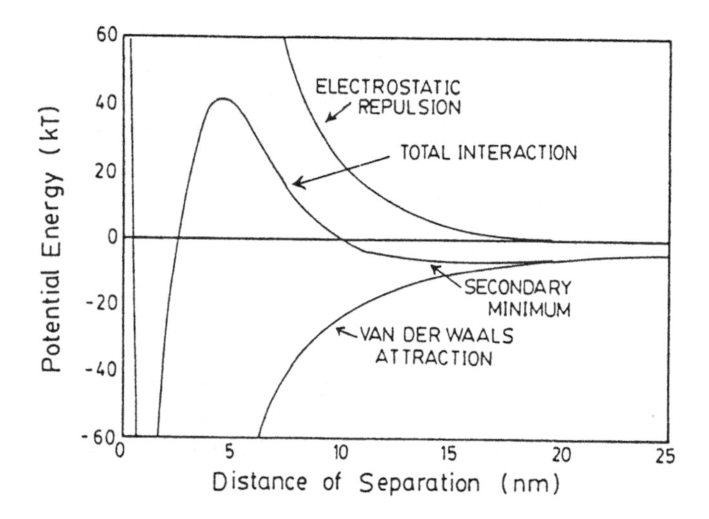

FIG. 1 Dependence of the potential energy of a pair of electrostatically stabilized droplets on their distance. Energy is in multiples of $kT = 1{,}381 \times 10^{-23}$ (T/K) J. From Ref. 23.

in interfacial area. Therefore, emulsions are not thermodynamically stable and tend to break up in a finite amount of time. The breakup of emulsions can proceed in different ways: flocculation, coalescence, sedimentation or creaming, Ostwald ripening, and phase inversion [24].

1. Sedimentation and Creaming

External force fields (gravitational, centrifugal, or electrostatic) can build up a droplet concentration gradient within the emulsion, which can ultimately result in the accumulation (random or ordered) of the disperse phase either on top of the continuous phase (creaming) or on the bottom (sedimentation). Thereby, the volume of individual droplets and their size distribution do not change. The velocity of falling or rising spherical droplets under the influence of the forces of gravity, buoyancy, and viscous drag (rate of sedimentation or creaming) is given by the Stokes' equation:

$$u_{sc} = \frac{4(\rho_c - \rho_d)gR^2}{18\eta_c} \tag{32}$$

where ρ_c and ρ_d are densities of the continuous and disperse phase, respectively, g is acceleration due to gravity, R is the droplet radius, and η_c is the viscosity of the continuous phase. It is assumed that there is no interaction between the droplets, that the system is monodisperse, and that the continuous phase is a

Newtonian liquid, the viscosity of which is not affected by the presence of particles in it. In real food emulsions, none of these criteria is met. Prolonged creaming can lead to an irreversible droplet aggregation and, ultimately, to phase separation. Although the actual creaming rate cannot be calculated by the preceding equation, it can be seen that increasing the medium viscosity (e.g., by a polysaccharide or gum) or reducing the droplet size by high-pressure homogenization can prolong the life of an emulsion.

2. Flocculation

When attractive forces (such as London and van der Waals forces, as well as Coulomb interactions) exceed the forces of repulsion between the droplets, droplet aggregates can form. No disruption of the interface occurs that would lead to droplet fusion. Flocculation can proceed to such an extent that a quasi separation of phases takes place.

3. Coalescence

The disruption of an unstable interface film makes the droplets join into larger ones, reducing their concentration and changing the particle size distribution. The limiting case would be a true phase separation. Coalescence occurs if an improper emulsifier has been chosen or if its concentration is too low.

4. Ostwald Ripening

This phenomenon taking place in polydisperse systems is due to the fact that all so-called immiscible liquids are to some extent soluble in each other. Since substances in droplets of smaller sizes have higher chemical potentials (Eq. 3), there is a tendency to attain a uniform distribution of chemical potentials and droplet sizes through growth of the smaller ones. The eventual outcome of Ostwald ripening would be one single, large drop of the disperse phase suspended in the continuous phase, that is, phase separation.

5. Phase Inversion

The interface layer of the continuous phase is concavely curved, whereas that of the disperse phase is convex. This can suddenly change due to a temperature elevation, an increase in concentration of one of the phases, or the addition of a new component. Thus the former continuous phase becomes dispersed in the former disperse phase, sometimes even in a higher state of dispersion than in the original emulsion. Strictly speaking, this is not an emulsion breakdown, and it is brought about by other processes mentioned above, especially coalescence and Ostwald ripening.

IV. FOOD-RELATED USES OF SURFACTANTS

Some food constituents, such as free fatty acids and their salts, lecithin, and soluble proteins exert surface-active properties and act as surfactants. Making mayonnaise from pure egg yolk and vegetable oil without any extraneous agent may serve as an example.

Apart from their obvious role as agents that decrease interfacial tension, thus promoting the formation and stabilization of emulsions, food emulsifiers can have several other functions in food. For example, they can change the functional properties of other food components (crumb softeners and dough conditioners in bakery products, etc.) and they can modify fat crystallization (e.g., efflorescence reduction in confectionery products).

Some other substances play an equally important role in the stabilization of emulsions, but they do so by creating a stabilizing network around the disperse phase. Proteins and polysaccharides, as well as particulate solids, belong in this group.

The choice of a food emulsifier depends on its hydrophilic/lipophilic balance [25, 26]. This semiempirical value represents the proportion of hydrophilic to hydrophobic groups in the surfactant molecule. It is consistent with the aforementioned Bancroft's rule. Values of 1 to 20 are common, but can be as high as 40. Higher HLB values indicate water solubility and better suitability to prepare oil-in-water emulsions. Another criterion for choice is the phase inversion temperature (PIT) for nonionic emulsifiers [27], which in a narrow temperature interval change from being preferentially water-soluble to being more oil-soluble.

Food emulsifiers as additives can be naturally derived, synthetic, or chemically modified biological compounds. Emulsifier classes currently in use are summarized in Table 4 [28]. Table 5 summarizes the most important functionalities of surfactants used in food production [28]. Among the various naturally derived surfactants, those produced by microorganisms deserve special attention.

TABLE 4 Food Emulsifier Categories

Lecithin and lecithin derivatives
Glycerol fatty acid esters
Hydroxycarboxylic acid and fatty acid esters
Lactylate fatty acid esters
Polyglycerol fatty acid esters
Ethylene or propylene glycol fatty acid esters
Ethoxylated derivatives of monoglycerides
Sorbitan fatty acid esters
Miscellaneous derivatives

Source: Ref. 28.

TABLE 5 Functional Properties of Food Emulsifiers

Functions	Product examples
Emulsification (water-in-oil)	Margarine
Emulsification (oil-in-water)	Mayonnaise
Aeration	Whipped toppings
Improvement of whippability	Whipped toppings
Inhibition of fat crystallization	Candy
Softening	Candy
Antistaling	Bread
Dough conditioning	Bread dough
Improvement of loaf volume	Bread
Reduction of shortening requirements	Bread
Pan release agent	Yeast-leavened and other dough and batter products
Fat stabilizer	Food oils
Antispattering agent	Margarine and frying oils
Antisticking agent	Caramel candy
Protective coating	Fresh fruits and vegetables
Surfactant	Molasses
Viscosity control	Molten chocolate
Improvement of solubility	Instant drinks
Starch complexation	Instant potatoes
Humectant	Cake icings
Plasticizer	Cake icings
Defoaming agent	Sugar production
Stabilization of flavor oils	Flavor emulsification
Promotion of "dryness"	Ice cream
Freeze–thaw stability	Whipped toppings
Improved wetting ability	Instant soups
Inhibition of sugar crystallization	Panned coatings

Source: Ref. 28.

Although all natural surfactants and their chemical derivatives have recently been designated as biosurfactants [29], the term generally applies only to microbial products.

The most important microbially derived surfactants are listed in Table 6. In principle, biosurfactants vary from industrial surfactants only by their origin. Representatives of both groups fulfill the same basic task: they partition on interfaces of immiscible phases and lower the interfacial tension, thereby creating all the conceivable possibilities of their application. Their principal role in the microbial cell is to enable the degradation and uptake of hydrophobic nutrients. Possibilities for the use of biosurfactants in foods were discussed earlier [31].

TABLE 6 Important Microbial Surfactants and Their Main Producers

Surfactant	Producer
Glycolipid	*Arthrobacter* sp.
Glycolipid and/or protein	*Torulopsis petrophilum*
Sophorose lipids	*Torulopsis bombicola*
	Torulopsis apicola
Trehalose dicorynomycolates	*Rhodococcus erythropolis*
Rhamnose lipids	*Pseudomonas* spp.
Sucrose lipids	*Arthrobacter paraffineus*
Fructose lipids	*Arthrobacter paraffineus*
Corynomycolic acids	*Corynebacterium lepus*
Spiculisporic acid	*Penicillium spiculisporum*
Fatty acids, mono- and diglycerides	*Acinetobacter* sp.
Polysaccharide–fatty acid complex	*Candida tropicalis*
"Liposan" (mostly carbohydrate)	*Candida lipolytica*
Lipoheteropolysaccharide (Emulsan)	*Acinetobacter calcoaceticus*
Neutral lipids	*Nocardia erythropolis*
	Corynebacterium salvonicum
Lipoprotein (surfactin)	*Bacillus subtilis*
	Bacillus licheniformis
Peptidolipid	*Candida petrophylum*
Polysaccharide–protein complex	*Corynebacterium hydrocarboclastus*
Whole cell (lipopeptide)	*Acinetobacter calcoaceticus*

Source: Modified from Ref. 30.

Specialty products made in bioreactors are relatively common in diverse industrial, medical, and household applications. Table 7 lists only the important biotechnology products used in the food industry. Some of them are purified end products of fermentation; others are subsequently chemically modified.

A. Some Food-Related Applications of Food Surfactants

The food industry does not yet use biosurfactants as food additives on a large scale as many biosurfactant properties and regulations regarding the approval of new food ingredients have to be resolved. Elaborate testing and evaluation of any new food ingredient is required according to the U.S. Food and Drug Administration regulations, and this process can be quite long. Issues that have to be addressed are related to nutritional, functional, sensory, biological, and toxicological properties of the new ingredient, production economics as compared to the synthetic surfactants for the same use, consumer acceptability of the new ingredient, legal

TABLE 7 Biotechnology Products Associated with Food Production and Preparation

Product	Uses
Organic acids, their salts and derivatives	pH control agents, acidulants, preservatives, flavoring agents, flavor enhancers, adjuvants, color stabilizers, gelling enhancers, melt modifiers, turbidity reducers, etc.
Mono/oligosaccharides	Sweeteners for diet and health food
Polysaccharides	Thickeners, water-binding agents, gellants, foaming agents, rheology modifiers, nutritive supplements
Amino acids, peptides	Constituents of protein hydrolyzates for flavoring, antimicrobial agents (nisin, bacteriocins), monosodium glutamate as taste enhancer
Proteins	Single-cell proteins (SCP) as food and feed additives
Enzymes	Microbial rennets, meat tenderizers, flour modifying proteases, beer stabilizers/clarifiers, amylases, glycoamylases, and pullulanases for starch hydrolysis, glycose isomerase for fructose and high-fructose syrup production, pectin-degrading enzymes (fruit juice prodn.), lipases as interesterification catalysts (e.g., in food surfactant production), glucose oxidase as oxygen scavenger, invertase for confectionery products
Lipids and derivatives	Specialty fats and oils, emulsifying and deemulsifying agents, lubricants, die-releasing aids, wetting agents, fat-blooming preventers, etc.
Other substances of interest in food production	B-group vitamins, L-ascorbic acid, special flavors (vanilla, fruit, mushroom, mint, onion, etc.), coloring agents, taste enhancers (5′-nucleotides)

regulation, and general eating habits and customs. However, it has to be pointed out that glycolipid biosurfactants are analogous to synthetic fatty acid esters of mono- and oligosaccharides, which are applied worldwide on an industrial scale. The successful application of sophorolipid biosurfactants and their acylated and alkoxylated derivatives in cosmetic preparations [32] could make them competitive alternatives to conventional synthetic food surfactants, at least from the functional, hygienic, and legal points of view.

Some of the results of acute [33] and subacute [34] toxicity tests of sophorolipid surfactants were as follows: oral LD_{50} (rat, mouse) of 10–16 g/kg, no eye irritation (rabbit) or abraded skin (guinea pig) at 50% concentration, no mutagenicity in tests with *S. typhimurium* TA 98 and TA 100, a no-effect outcome of a 1 month study on rats fed 53 mg/kg d^{-1} (0.06% in their diet). A prerequisite of the use of new substances as additives in food is to make sure that no toxic intermediates or products of their metabolism appear in the organism during degradation or excretion. Similar products synthetically produced from sucrose and fatty acids have been known and used for a long time [38]. Besides, sophorose [or cyclic (1→2)-β-D-glucan] is already being used as a sweetening agent [36].

In Japan, where legal restrictions concerning the use of novel naturally derived (or biotechnologically produced) components in the food industry are not as strict as elsewhere, sophorolipids are patented as flour additives for quality improvement and better shelf life of bakery goods (e.g., 0.1 parts of sophorolipids and/or their lactones to 70 parts of flour) [37]. Surface-active properties of hydrolyzed and freeze-dried yeast cell walls (*Saccharomyces uvarum*) have been patented as surfactants in margarine production [38].

Other glycolipids were also patented as additives for cosmetics and pharmaceuticals. Thus, rhamnolipids produced by *Pseudomonas* BOP 100 were used to obtain liposomes [39]. Emulsions suitable for cosmetic, food and pharmaceutical applications were prepared with a 2-*O*-(2-decenoyl)rhamnolipid [40]. Rhamnolipids of *Pseudomonas aeruginosa* UI 29791 were produced in high yield (46 g/L in batch culture, 6.4–10 g/L d^{-1} in semicontinuous culture on corn oil (40 g/L) as the sole carbon source and recommended, among others, for use in foods [41, 42].

Useful surface active properties and antitumor activity in mice [43] were shown by 2,2′,3,3′-tetra-*O*-alkyl-α,α′-trehalose prepared by alkylation of 4,6:4′,6′-di-*O*-benzylidene-α,α′-trehalose with alkylbromides (C_2–C_{25}). Promising as excellent surfactants, dispersants, and stabilizers are also succinoyl trehalose lipids and their sodium salts [44]. An antiviral succinylated trehalose glycolipid efficient against *Herpes simplex* Type I virus was produced by *Rhodococcus erythropolis* on a glycerol-containing medium [45].

Some novel mannosylerythrytol lipids might be considered as potential food surfactants. One example is 4-*O*-(di-*O*-acetyl-di-*O*-alkanoyl-β-D-mannopyranosyl)erythrytol and the monoacetyl derivative as the two major constituents (~80%) of the total lipids produced with a yield of approximately 40 g/L by *Candida antarctica* T 34 grown on soybean oil as the sole carbon source.

High-molecular-weight emulsifiers from procaryotes have also been proposed as food emulsifiers. For example, emulcyan, the extracellular complex of >200 kD, consists of carbohydrates, fatty acids, and a protein moiety and is produced by the benthic cyanobacterium *Phormidium* J-1 (ATCC 39161) [46].

The carbohydrate-emulsifying activity of lipopolyheterosaccharides of the genus *Acinetobacter* is well documented [47]. In addition to the proposed use of emulsans and apoemulsans from *A. calcoaceticus* in personal care products, the microbial surfactant combined with synthetic emulsifiers (e.g., a polyoxyethylene oleyl ether) was patented also for use in food emulsions [48].

V. CONCLUSION

The preparation and processing of various foods, especially on an industrial scale, is often greatly facilitated by the inclusion of small amounts of selected additives. Surface-active agents play an important role, among them producing significant effects although present in minute amounts. Their importance will only grow with time, enabling the manufacture of new food products and the development of new technologies for the existing ones. Despite the highly efficient commercial synthetic food surfactants, a search for still newer and better products will continue. The list of attempts to use equally efficient microbially derived surfactants presented in this chapter should by no means discourage such a search through the methods of biotechnology, especially under the prevailing public opinion that everything chemical is bad [49]. It is a favorable time for the promotion and general acceptance of these useful substances. More investigation is needed to establish their usefulness as food additives and to evaluate their safety as regards metabolism, biodegradation, nutritional aspects, and human health. With the application of highly productive mutant or recombinant strains and the use of inexpensive and abundant secondary raw materials as culture media (e.g., whey) [50] biosurfactants may successfully compete with their synthetic analogs.

REFERENCES

1. A. L. Branen, P. M. Davidson, and S. Salminen, in *Food Additives*. Marcel Dekker, New York, 1989, pp. iii–v.
2. A. Kramer and B. A. Twigg, *Quality Control for the Food Industry*, Vol. 1, AVI Publishing Co., Westport, CT, 1970.
3. A. Pour-El, in *Functionality and Protein Structure* (A. Pour-El, ed.), Symposium Series 92, 1979, pp. ix–xi.
4. A. Pour-El, in *Protein Functionality in Foods* (J. P. Cherry, ed.), American Chemical Society Symposium Series 147, 1981, pp. 1–19.
5. Y. Kawamura and M. R. Kare (eds.), *Umami: A Basic Taste*, Marcel Dekker, New York, 1987.
6. M. A. Amerine, R. M. Pangborn, and E. B. Roessler, *Principles of Sensory Evaluation of Food*, Academic Press, New York, 1965.

7. N. N. Mohsenin, *Physical Properties of Plant and Animal Materials, Vol. I: Structure, Physical Characteristics and Mechanical Properties*, Gordon and Breach, New York, 1970.
8. A. Surmacka Szczesniak, in *Physical Properties of Foods* (M. Peleg and E. B. Bagley, eds.), AVI Publishing, Westport, CT, 1983, pp. 1–41.
9. J. E. Kinsella, *CRC Crit. Rev. Food Sci. Nutrit. 7*: 219–280 (1976).
10. P. Becher, *Emulsions: Theory and Practice*, American Chemical Society Monograph No. 135, Reinhold, New York, 1957.
11. M. El-Nokaly, G. Hiler Sr., and J. McGrady, in *Microemulsions and Emulsions in Foods* (M. El-Nokaly and D. Cornell, eds.), American Chemical Society Symposium Series 448, pp. 26–43 (1991).
12. P. Walstra, in *Encyclopedia of Emulsion Technology*, Vol. 1 (P. Becher, ed.), Marcel Dekker, New York, 1983, pp. 57–127.
13. K. P. Das and J. E. Kinsella, *Adv. Food Nutrition Res. 34*: 81–201 (1990).
14. B. V. Deryagin and L. Landau, *Acta Physicochim. URSS 14*: 633–662 (1941).
15. E. J. W. Verway and J. T. G. Overbeek, *Theory of the Stability of Lipophobic Colloids*, Elsevier, Amsterdam, 1948.
16. H. C. Hamaker, *Physica (Amsterdam) 4*: 1058–1072 (1937).
17. J. H. Schenkel and J. A. Kitchener, *Trans. Faraday Soc. 56*: 161–173 (1960).
18. G. Frens, D. J. C. Engel, and J. T. G. Overbeek, *Trans. Faraday Soc. 63*: 418–423 (1967).
19. G. R. Wiese and T. W. Healy, *Trans Faraday Soc. 66*: 490–499 (1970).
20. B. V. Deryagin, *Kolloid-Z. 69*: 155–164 (1934).
21. D. H. Napper, *Polymeric Stabilization of Colloidal Dispersions*, Academic Press, New York, 1983.
22. V. A. Parsegian, N. L. Fuller, and R. P. Rand, *Proc. Natl. Acad. Sci. USA 76*: 2750–2754 (1978).
23. N. S. Parker, *CRC Crit. Rev. Food Sci. Nutr. 25*: 285–315 (1987).
24. T. F. Tadros and B. Vincent, in *Encyclopedia of Emulsion Technology*, Vol. 1 (P. Becher, ed.), Marcel Dekker, New York, 1983, pp. 129–285.
25. W. C. Griffin, *J. Soc. Cosmet. Chem. 1*: 311–326 (1949).
26. W. C. Griffin, *J. Soc. Cosmet. Chem. 5*: 249–256 (1954).
27. K. Shinoda and H. Arai, *J. Phys. Chem. 68*: 3485–3490 (1964).
28. W. E. Artz, in *Food Additives* (A. L. Branen, P. M. Davidson, and S. Salminen, eds.), Marcel Dekker, New York, 1989, pp. 347–393.
29. S. Kudo, *Proceedings of the World Conference on Biotechnology in the Fats and Oils Industry 1987* (T. H. Applewhite, ed.), American Oil Chemists Society, Champaign, IL, 1988, pp. 195–201.
30. N. Kosaric, N. C. C. Gray, and W. L. Cairns, in *Biosurfactants and Biotechnology* (N. Kosaric, W. L. Cairns, and N. C. C. Gray, eds.). Marcel Dekker, New York, 1987, pp. 1–19.
31. T. Kachholz and M. Schlingmann, in *Biosurfactants and Biotechnology* (N. Kosaric, W. L. Cairns, and N. C. C. Gray, eds.), Marcel Dekker, New York, 1987, pp. 183–210.
32. S. Inoue, *Proceedings of the World Conference on Biotechnology in the Fats and Oils Industry 1987* (T. H. Applewhite, ed.), American Oil Chemists Society, Champaign, IL, 1988, pp. 206–209.

33. Y. Ikeda, T. Sunakawa, S. Tsuchiya, M. Kondo, and K. Okamoto, *J. Toxicol. Sci. 11*: 197–211 (1986).
34. Y. Ikeda, T. Sunakawa, K. Okamoto, and A. Hirayama, *J. Toxicol. Sci. 11*: 213–224 (1986).
35. L. Osipow, F. D. Snell, W. C. York, and A. Finchler, *Ind. Eng. Chem. 48*: 1459–1462 (1956).
36. S. Kitahata, S. Okada, and S. Edakawa, Japanese Patent Kokai 63-116,668 (1988).
37. A. Shigeta and A. Yamashita, Japanese Patent Kokai 61-205,449 (1986).
38. K. Ohata and K. Kamata, Japanese Patent Kokai 61-227,827 (1986).
39. Y. Ishigami, Y. Gama, H. Nagahara, T. Motomiya, and M. Yamaguchi, Japanese Patent Kokai 63-182,029 (1988).
40. Y. Ishigami, Y. Gama, Y. Uji, K. Masui, and Y. Shibayama, Japanese Patent Kokai 63-77,535 (1988).
41. L. Daniels, R. J. Linhardt, B. A. Bryan, F. Mayerl, and W. Pickenhagen, European Patent Appl. 282,942 (1988).
42. R. J. Linhardt, R. Bakhit, L. Daniels, F. Mayerl, and W. Pickenhagen, *Biotechnol. Bioeng. 33*: 365–368 (1989).
43. E. Nagashima, H. Matsuda, T. Kuraishi, T. Katori, and K. Kukita, Japanese Patent Kokai 62-19,598 (1987).
44. Y. Ishigami, S. Suzuki, T. Funada, M. Chino, Y. Uchida, and T. Takeshi, *Yukagaku 36*: 847–851 (1987).
45. A. Kawai, M. Kayano, T. Funada, and J. Hirano, Japanese Patent Kokai 63-126,493 (1988).
46. M. Shilo and A. Fattom, U.S. Patent 4,826,624 (1989).
47. D. L. Gutnick and Y. Shabtai, in *Biosurfactants and Biotechnology* (N. Kosaric, W. L. Cairns, and N. C. C. Gray, eds.), Marcel Dekker, New York, 1987, pp. 211–246.
48. K. Miyata, T. Tsuchida, and K. Tawara, Japanese Patent Kokai 62-155,931 (1987).
49. A. Turner, *Food Manufact. 61*: 40–41, 45 (1986).
50. A. K. Koch, J. Reiser, O. Käppeli, and A. Fiechter, *Bio/Technology 6*: 1335–1339 (1988).

17

Biosurfactants for Environmental Control

REINHARD MÜLLER-HURTIG and FRITZ WAGNER Institute of Biochemistry and Biotechnology, Technical University of Braunschweig, Braunschweig, Germany

ROMAN BLASZCZYK and NAIM KOSARIC Department of Chemical and Biochemical Engineering, University of Western Ontario, London, Ontario, Canada

I. INTRODUCTION

The ability of microorganisms to degrade hydrocarbons was first described in 1895 by Miyoshi who reported on the microbial utilization of paraffins [1]. Shortly thereafter, the microbial consumption of methane has been described. However, for many years hydrocarbons were considered more or less biologically inert, and hydrocarbon-utilizing microorganisms were mainly of academic interest with little applicability other than in prospecting for oil and gas deposits, use in the fermentation industry, and for clean up of refinery effluents [2, 3]. It was the Torrey Canyon accident in 1967 that aroused public concern about the ecological effects of oil pollution and focused scientific interest on the role of microorganisms in removing oil from the environment. In 1969, ZoBell [4] reported that more than 100 species of bacteria, yeasts, and fungi were capable of oxidizing hydrocarbons; however, most microbial species are limited to the kinds of hydrocarbons they can attack [5, 6]. Genetically engineered microorganisms can degrade a much wider spectrum of hydrocarbons [7]; however, the stability of such species in the natural environment is still unknown. This suggests that a group of microorganisms, rather than a single species, would be required to degrade crude oil.

There are two large-scale processes in which the hydrocarbon-degrading microorganisms can be employed: (1) degradation of spilt oil in the seas and (2) biodegradation of spilt hydrocarbons in soil.

II. MECHANISMS

A. Transport of Hydrocarbons

Three modes of hydrocarbon transport to microbial cells [8] are generally considered: (1) interaction of cells with hydrocarbons dissolved in the aqueous phase, (2) direct contact of cells with large hydrocarbon drops, and (3) interaction of cells with solubilized/pseudosolubilized or accommodated hydrocarbon droplets that are much smaller than the cells. The first mechanism is considered to be valid for more water-soluble aromatic [9] and gaseous [10] hydrocarbons. It is generally agreed that the rate of solubilization of long-chain alkanes in the aqueous medium is too low to support the observed rate of growth of microorganisms.

According to the second mechanism, microbial cells attach to the surface of hydrocarbon drops that are much larger than the cells, and substrate uptake presumably takes place through diffusion or active transport at the point of contact [11–13]. The availability of substrate surface area for cell attachment is a limiting factor for microbial growth. Extracellular production of biosurfactants–bioemulsifiers by hydrocarbon utilizing microorganisms is well documented [14–16].

Emulsification promoted by these factors would result in a dispersion of hydrocarbon droplets in the aqueous medium and thereby increase the surface area between the two phases. The propensity of hydrocarbon-grown cells to adhere to hydrocarbon phase was widely reported [13, 17–19]. Addition of biosurfactants to the hydrocarbon medium stimulated growth of microorganisms [20,21].

In the third mechanism, microbial cells may interact with minute particles of solubilized/pseudosolubilized/microemulsified hydrocarbons, which are much smaller than the cells. Hydrocarbon particles attach to the surface of the cells rather than cells attaching to the surface of hydrocarbon drops as in the second mechanism. With the decrease in substrate particle size, the interfacial area between hydrocarbon and water is expected to increase significantly [22]. A hydrocarbon-solubilizing factor was isolated [23,24], and it was found that the rate of hydrocarbon solubilization was high enough to fully account for the rate of hydrocarbon uptake during growth of microorganisms [25].

Depending on the organism, hydrocarbon uptake may take place through either one or combination of the above mechanisms. It was reported that two strains of the same species (*Pseudomonas*), grown on *n*-hexadecane, demonstrated different mechanisms of hydrocarbon uptake [26].

B. Biochemical Pathways

The actual biochemical pathway for bacterial degradation of hydrocarbon contamination depends on the particular substrate metabolized and the type of microorganisms involved. However, the preferred pathway is via the electron transport chain using molecular oxygen as the terminal acceptor.

Degradation of aliphatic and aromatic compounds is realized in a stepwise fashion. Aliphatic terminal carbon oxidation is the first stage in the conversion, followed by a dehydrogenation reaction to the corresponding aldehyde. Oxidation continues in the third stage conversion to the corresponding fatty acids, which then undergo bacterial oxidation to yield the fatty acid plus acetic acid. The acetic acid is then degraded further to yield carbon and energy for assimilatory purposes (Fig. 1).

The first phase of aromatic metabolism is often the modification or removal of substituents on the benzene ring followed by a stepwise conversion to catechol. Catechol is of primary importance as it represents the hydroxylated forms of benzene and phenol. Further on, aromatic rings such as toluene are assimilated via the extradial or meta cleavage pathway, following identical R-group oxidation, as with the aliphatics (Fig. 2).

Figure 3 illustrates the degradation of catechol, which completes the bioconversion process generating organic acids used in the TCA cycle. This in itself acts as an efficient receptor for the input of biochemical intermediates from catabolic

$CH_3(CH_2)_nCH_2CH_2CH_3$

↓

$CH_3(CH_2)_nCH_2CH_2CH_2OH$

↓

$CH_3(CH_2)_nCH_2CH_2CHO$

↓

$CH_3(CH_2)_nCH_2CH_2COOH$

↓

$CH_3(CH_2)_nCOCH_2COOH$

↓

$CH_3(CH_2)_nCOOH + CH_3COOH$

FIG. 1 Steps of *n*-alkanes biodegradation. Adapted from Ref. 30.

O_2

H

OH

OH

H

OH

OH

Benzene

cis-Benzene dihydrodiol

Catechol

O_2

meta

ortho

CHO

COOH

OH

COOH

COOH

2-Hydroxy-*cis,cis*-Muconic semialdehyde

cis,cis-Muconic acid

O_2

O

COOH

COOH

β-Ketoadipic acid

CH_3–CHO + CH_3–CO–COOH

COOH–CH_2–CH_2–COOH + COOH–CH_3

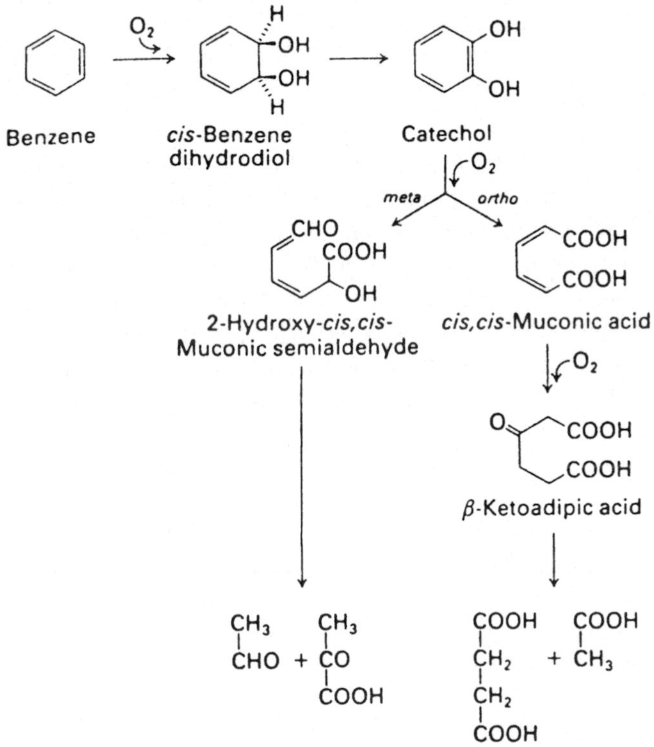

FIG. 2 Steps of benzene biodegradation. Adapted from Ref. 30.

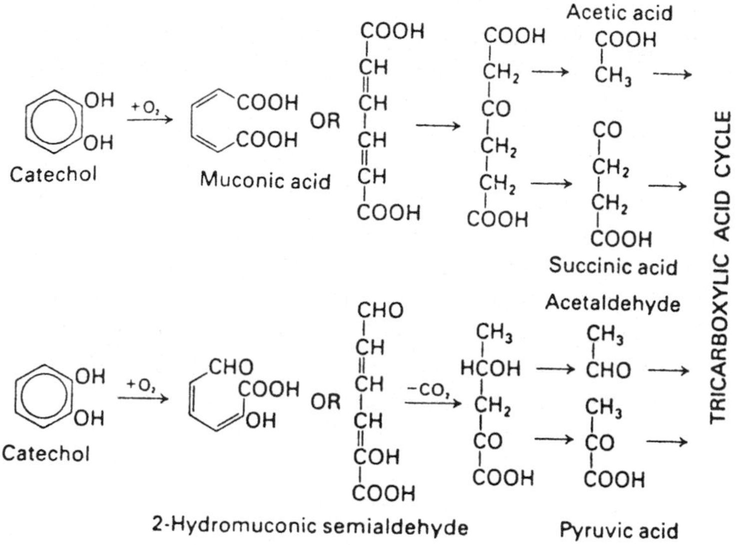

FIG. 3 Steps of catechol biodegradation. Adapted from 30.

pathways and is a principal source of metabolic energy in the form of adenosine triphosphate (ATP) [27].

III. BIODEGRADATION OF CRUDE OIL IN THE SEAS

When crude oil is spilled into the aquatic environment, the lighter components volatilize and the more soluble components begin to dissolve. Heavier components of oil agglomerate and have a tendency to settle to the bottom. On the bottom, they are utilized by anaerobic microorganisms with the production of biogas (CH_4 + CO_2).

Both laboratory and field data confirm the chemical changes in crude oil during microbial degradation. The short-chain alkanes ($<C_{10}$) are lost by evaporation or by dissolution in water; they are also the first compounds attacked by microorganisms. Alkanes with chains of C_{10}–C_{24} are rapidly degraded by microorganisms according to the mechanism shown in Fig. 1. The greater the chain length and the amount of branching, the more resistant the compound is to microbial attack.

According to Pilpel [28], oil in the sea is destroyed at the rate of hundreds of grams per cubic meter of contaminated ocean per year. Gibbs [29] calculated, on the basis of laboratory experiments, that the degradation rate of Kuwait crude oil in Irish seawater was 30 and 11 mg/L yr^{-1} in summer and winter, respectively. The rate of degradation of oil in Gulf of Mexico sediments was estimated as 56 μg hydrocarbon carbon/mL of sediment per day [3]. Under favorable laboratory conditions, it has been estimated that 100–960 mg of oil/m^3 of seawater was degraded per day, however, in situ measurements showed that only 1–50 mg of hydrocarbon/m^3 day^{-1} was degraded [30]. Nitrate and phosphate are known to greatly accelerate oil degradation and mineralization [3]. Most of commercial products seeded in the seawater to speed up the oil degradation contain nitrogen and phosphorus. An oil-degrading mixture known as Pentrodez contained 20 hydrocarbonoclastic microorganisms and chemical nutrients for microbial growth, particularly compounds of nitrogen and phosphorus [31]. A list of some commercial products containing biological additives is given in Table 1.

TABLE 1 Commercial Products Containing Biological Additives

Product Name	Manufacturer
HYDROBAC	Polybac Corporation 954 Marcon Blv., Allentown, PA 18103
NO-SCUM	Natural Hydrocarbon Elimination Company 206 Paul Revere Street, Houston, TX 77024
PETROBAC	Polybac Corporation 954 Marcon Blv., Allentown, PA 18103
PHENOBAC	Polybac Corporation 954 Marcon Blv., Allentown, PA 18103
Petrodeg-100	Bioteknika International Inc. 7835 Greeley Blv., Springfield, VA 22152
TYPE L, DBC PLUS	Flow Laboratories, Inc., Enviroflow 7655 Old Springhouse Road, McLean, VA 22102
ROLFZYME	International Enzymes, Inc. 1706 Industrial Road, Las Vegas, NV 89102
INPOL EAP 22	CECA, S.A. 11, Avenue Morane Saulnier, 78141 Velizy-Villacoublay, France

Adapted from Ref. 3.

IV. BIODEGRADATION OF SPILT HYDROCARBONS IN SOIL (SOIL BIOTREATMENT)

Soil biotreatment is a viable alternative to physicochemical methods of soil treatment. It has been proven effective not only in the laboratory but in pilot plant studies [32] and in full-scale commercial applications [27, 31, 33].

Biodegradation naturally occurs in all soils. A large number of bacteria, molds, yeasts, and certain cyanobacteria is involved in aerobic oxidation of petroleum hydrocarbons. Nutrients are required to optimize this natural process. The type and amount of nutrients depends on the natural composition of soil and the concentration of contaminants. When the limiting factor for the biodegradation process is oxygen, the addition of hydrogen peroxide can increase the degradation rate significantly [34, 35]. It was found that, if the values for the maximum oxygen demand and the site-specific degree of contamination are known, the time required for degradation can be reduced up to 70% [36]. The addition of oleophilic products to the soil can also enhance the biodegradation process. It was found that treatment of oily wastes by addition of an adsorbent originally used for spilled oil (Oclansorb produced by High Point Industries), increased the biodegradation by 15% [37]. Enhanced degradation was also achieved when selected autochthonous soil bacteria were used for inoculation [38, 39].

In situ biodegradation involves the stimulation of natural microbial activity. Performance is generally appropriate in areas where the water table is relatively close to the surface. Design of the system includes an examination of the site hydrogeology, compounds to be degraded, site conditions, process operation, and site microbiology. It may be necessary to install wells to inject the biologically treated water as an alternative to an infiltration gallery. Oxygen is often provided via hydrogen peroxide addition to the groundwater prior to reinjection or the use of air spargers located in the wells. Typical clean-up periods can range from 3–18 months and can cost from $20,000–$200,000. This type of biodegradation is most effective for large areas of soil contamination [39–41].

Approximately 30,500 m^3 of soil polluted with coal tar and phenols was treated at the former Greenbank Gas Works (east of Blackburn, Lancashire, UK) [33]. Coal tar was quantified as a sum of 16 polycyclic aromatic hydrocarbons (PAHs), which were defined as priority pollutants by the U.S. Environmental Protection Agency [42], and phenols as the sum of nine individual components including phenols, cresols, xylenols, naphthols, and 2,4,6-trimethylphenol. Active organisms capable of growing on specific PAHs or phenols were isolated in batch culture by incorporation of the corresponding chemicals as sole carbon sources into mineral salt media, and inoculating with a suspension of soil from the Greenbank site. Although many of the PAHs were potentially biodegradable, one of the major impediments to their effective removal was their inherently low bioavailability. A series of surfactants were therefore screened to select a suitable product to increase the availability of the hydrocarbon to the microbial inoculum. It was found that turbidity of coal tar after 1 h of standing was 3.5–5 times higher when surfactants were added to the inoculum compared to the turbidity after adding only inoculum.

During full-scale work, contaminated material was carefully processed prior to placement in treatment beds to maximize homogeneity and achieve a suitable reduction in the particle size for optimizing biodegradation. These beds were inoculated layer by layer with a mixture of microorganisms specific to the contaminants in question (phenols or PAHs), nutrients, and surfactants by using an agricultural boom sprayer mounted on the back of a tractor. Subsequent rotovation of the bed ensured adequate mixing of the supplements with the contaminated soil. This procedure was continued throughout the treatment period, together with periodic application of water or booster levels of nutrients and inocula when necessary. The data for the biodegradable contaminants across the site, prior to reclamation and following independent validation are shown in Table 2.

When in situ bioremediation is not applicable or a small amount of contaminated soil is to be treated, a concept of enhanced land farming is used. This concept involves intimate mixing of the soil contaminant, microbial inoculum, and the correct nutrient blend in order to achieve maximum degradation of the offending pollutant. To obtain total optimization of the enhanced land farming process, a specially developed analytical testing program is implemented to establish the following: (1) identification and quantification of contaminants, (2) determination of the required carbon-phosphorus-nitrogen ratio with respect to nutrient

TABLE 2 Concentration of Polyaromatic Hydrocarbons and Total Phenols Across Gas Works Site Before and After Site Remediation

	Polyaromatic hydrocarbons (mg/kg)		Phenols (mg/kg)	
Volume of soil (m^3)	Before reclamation, estimated mean	After validation, mean	Before reclamation, mean	After validation, mean
5,060	11,500	171	22	<4
5,060	19,600	249	27	<4
750	3,700	243	29	<3
3,940	11,100	144	37	<2
10,100	36,900	110	470	<3
560	9,900	76	52	<4

Adapted from Ref. 33.

selection and application rates, and (3) biofeasibility study including determination of the substrate/microbe oxygen consumption rate and substrate biodegradation assay. The major disadvantage of a traditional land farming system is that only a maximum soil depth of 40–50 cm can be treated; therefore a large surface treatment area is required to treat a relatively limited volume of soil. Sufficient aeration and maintenance of constant moisture levels are other restrictions in this process.

A successful modification of the land farming concept can be a soil banking system, schematically presented in Fig. 4. Of major importance is the use of the high-density polyethylene impermeable liner to capture run-off or leachate from the soil bank. Full scale, in situ, soil bioremediation is an alternative method, especially for large tracts of land, which formerly housed gas works, oil refineries, chemical factories, and waste tips. The capital cost for land farming can generally run from \$100,000–\$200,000 plus the cost of the land and the cost of excavation. Operating costs are generally confined to the cost of tilling operations and nutrient addition [40].

A pilot scale bioremediation of petroleum contaminated soil was described by Barnhart and Myers [32]. From the Carlow Road, Port Stanley site (Ontario, Canada), 4800 m^3 of oil tar contaminated soil was collected in the biotreatment facility composed of three 15-cm lifts of clay compacted after each lift, resulting in a permeability of 10^{-7} cm/second. Clay-faced berms surrounded the biotreatment pad completing the containment facility. The bacterial suspension was prepared including nutrients sufficient for the rapid growth. Bacteria were supplied 4 days/week (1200 gal of bacterial suspension, consisted of approximately one-third cell mass yield) through the treatment period. The suspension was supplied through a high-pressure distribution system in the soil. Biodegradation of various petroleum hydrocarbons occurred during a 4-month period. Intensive

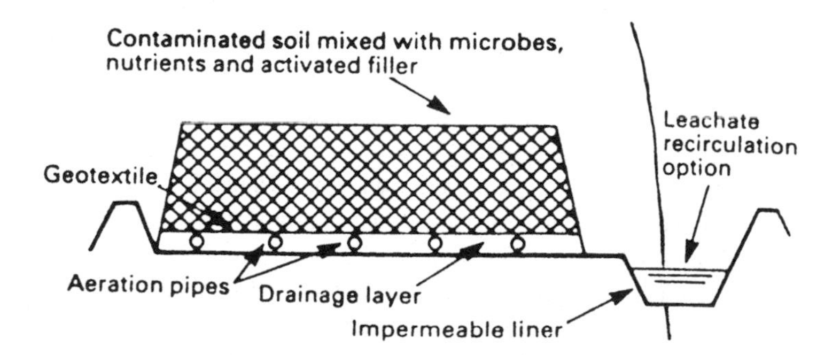

FIG. 4 Engineered soil bank cross section. Additional features include: Irrigation system, soil bank cover, microbial addition, and nutrient addition. Adapted from Ref. 30.

biological and physical operations resulted in a decrease of all monitored contaminants. Benzene, toluene, and xylene (BTX) decreased by 73%, oil and grease by 36%, total polycyclic aromatic hydrocarbons (PAH) by 86%, and five-ring polycyclic aromatic hydrocarbon benzo(a)pyrene (BAP) by 60%. Bioremediation of the soil did not impact air quality within the work area.

V. ROLE OF BIOSURFACTANTS IN HYDROCARBON DEGRADATION

A. Effect of Surfactants and Biosurfactants Producing Microorganisms in Submerged Culture

Studies with chemical surfactants show that the degradation of phenanthrene by a not identified isolate could be increased by a higher substrate surface per volume ratio after sonification or by a nonionic surfactant based on ethyleneglycol. The optimum mean degradation rate of phenanthrene was achieved with 0.05% surfactant and an additional organic phase of heptanonylnonane [43]. Altogether in submerged culture chemical dispersants enhance hydrocarbon oxidation in some cases while others inhibit oxidation. Mulkins-Phillips and Stewart reported [44] that only the dispersant Sugee 2 (Handy Chemical Ltd., LaPrairie, Que.) with the poorest emulsifying capacity of four tested (Corexit 8666, Gamlen Sea Clean, G. H. Woods Degreaser-Formula 11470, Sugee 2) promoted *n*-alkane degradation. Crude oil degradation at 8°C was also temporarily retarded by the dispersant Corexit 9527 [45]. After 2 weeks of incubation the utilization of the *n*-alkane and isoprenoid compounds was lowered, but after 12 weeks this effect was no longer evident.

Adding rhamnolipids as biosurfactants, Hisatsuka et al. [46] could only find a specific growth stimulation for 6 tested *Pseudomonas* strains producing also rhamnose lipids, but not for 11 strains of *Corynebacterium, Achromobacter, Micrococcus,* or *Bacillus* growing on *n*-hexadecane. The added concentration of 75 mg/L of rhamnolipids may be toxic as the growth yield after 2 days was more than 30% lower for 3 of these 11 strains. Also the stimulation of hydrocarbon degradation by added extracellular sophorose lipids—produced by *Torulopsis bombicola* growing on water-insoluble alkanes—was only found for *Torulopsis* yeasts but not for other typical alkane-utilizing yeasts. Forty-three synthetic nonionic surfactants with hydrophilic–lipophilic balance values ranging from 3.6 to 19.5 (from Kao Soap Co. Ltd, Tokyo) were unable to replace the glycolipid [47]. Furthermore, the biosurfactants of *Candida lipolytica* formed during the growth on hexadecane seemed to be specific for the degradation of hexadecane (Fig. 5) [48]; this means that specific biosurfactants should be produced by microorganisms growing on the contaminating crude oil.

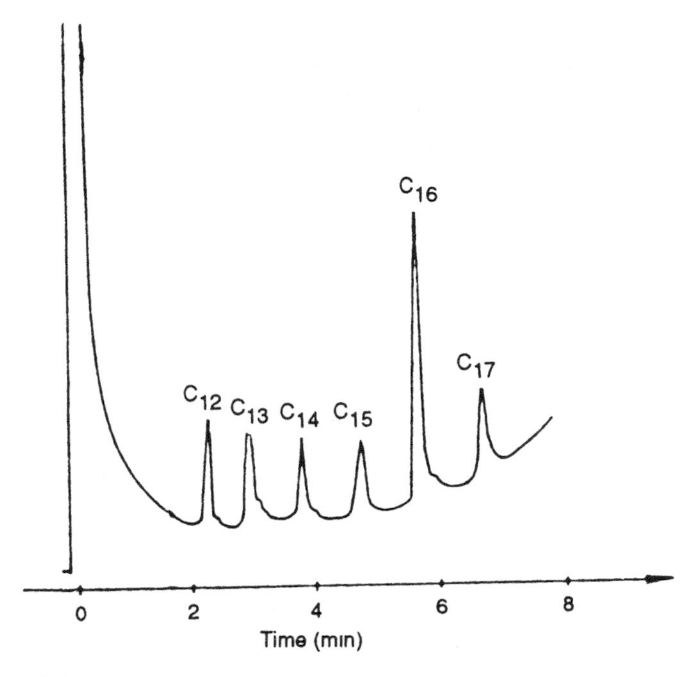

FIG. 5 Gas chromatogram of the *n*-alkane C_{11}–C_{17} mixture solubilized by the cultivation medium after the growth on hexadecane. From Ref. 48.

The effect of various glycolipids on oil degradation was tested in soil model systems. In submerged culture with 10% soil content, the interfacial tension had to be lowered by the formation of glycolipids to enable the degradation of all added model oil components of low water solubility (tetradecane, pentadecane, hexadecene, pristane, phenyldecane, and trimethylcyclohexane). These hydrocarbons were degraded by an original soil population in a second degradation phase after a naphthalene degradation phase (Fig. 6) [49].

All tested glycolipid biosurfactants enhanced the degradation of the model oil when the initial interfacial tension was reduced from 21 mN to a range of 2 to 16 mN. Probably because the cells did not have to produce biosurfactants for a better uptake of these hydrocarbons of lower water solubility, all glycolipids shortened the timespan of the second adaptation phase from 21 to 0–8 hours. Furthermore hydrocarbon elimination was increased from 81% to 93–99% (Table 3) [49] and hydrocarbon mineralization by adding sophorose lipids from 17% CO_2 to 49% CO_2 [50]. In contrast the extent of mineralization was not raised by adding chemical surfactants [51]. Altogether the degradation capacity was increased by the addition of biosurfactants from 16.3 to the range of 23.8–39.0 g

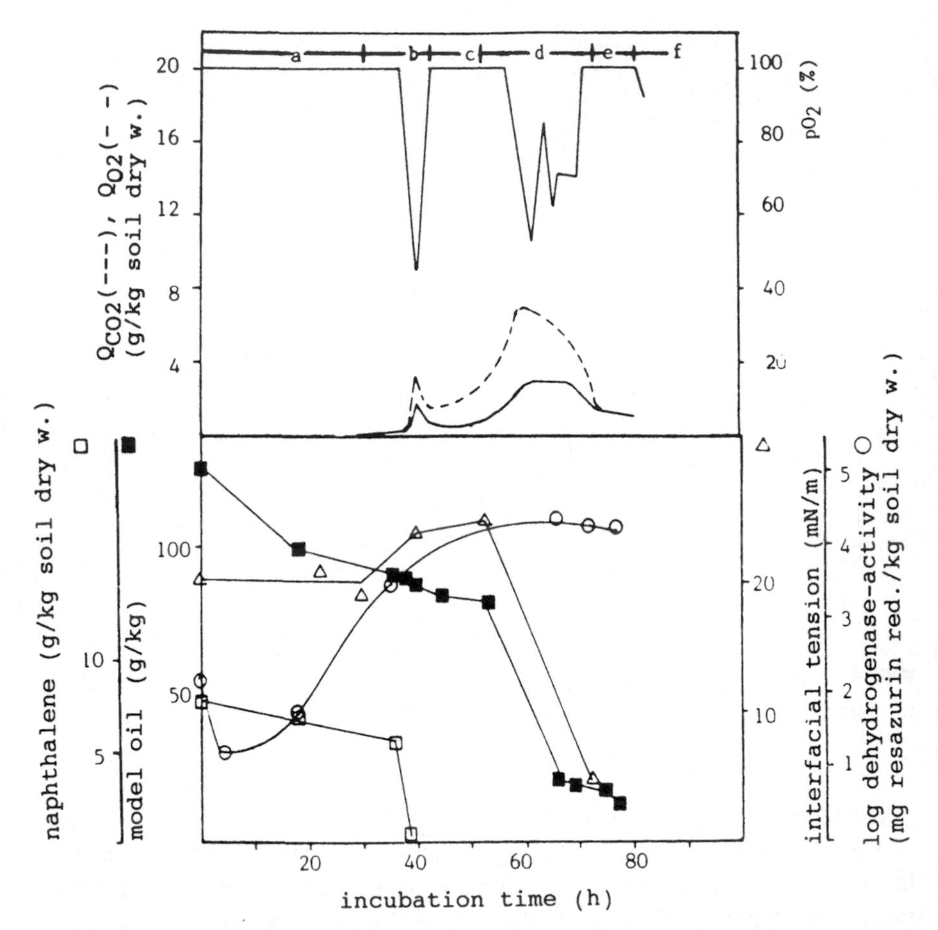

FIG. 6 Degradation of a model oil with a natural soil population in submerged culture. Conditions: mineral salts medium with 1.35 g/L model oil; 48% tetradecane and pentadecane, 20% 1,2,4-trimethylcyclohexane, 10% 1-phenyldecane, 6.5% naphthalene, and 5.5% pristane, inoculated with 99 g/L soil dry weight, 25°C, pH 7.4–7.6, aeration 27 Nl/h. (a-1) Adaptation phase, (b-1) degradation phase (naphthalene), (c-2) adaptation phase, (d-2) degradation phase (all other model oil components), (e-3) adaptation phase, (f-3) degradation phase (metabolites).

hydrocarbon/kg soil d.w. day^{-1}. More effective were the more hydrophobic surfactants sophorose lipids, cellobiose lipids, and trehalose-6,6′-dicorynomycolates having HLB values from 4.05 to 8.0. Surprisingly trehalose-6.6′-dicorynomycolates and –2,3,4,2′-tetraesters were not most effective, although approximately 50% of the selected soil population could produce similar trehalose lipids [52].

TABLE 3 Influence of Glycolipids on the Degradation Efficiency

Cultivation	Hydrocarbon elimination Durance h	Hydrocarbon elimination Degree %	HLB	Degradation capacity (g hydrocarbon kg $sdw^{-1}d^{-1}$)
	Under oxygen-limited conditions			
Without surfactant	114	81	—	16.3
Trehalose-6,6′ corynomycolate	71	93	4.05	37.2
Sophorose lipids	75	97	6.87	39.0
Cellubioselipids	79	99	8.0	32.3
Rhamnose lipids	77	94	9.5	28.6
Trehalose-2,3,4,2′-tetraester			10.02	23.8
	Without any oxygen limitation			
Without surfactant	79	89	—	25.7
Sophorolipids	57	95	6.87	46.5

Efficiently all added biosurfactants except cellobiose lipids were degraded within the degradation phase of the hydrocarbons of lower solubility in water.

Under conditions without any oxygen limitation the degradation capacity of the soil was 25.7 g hydrocarbon/kg soil d.w. (sdw) day^{-1}, that means 1–3 orders of magnitude higher than measured in soils with mineralization rates of 0.02–1.38 g hydrocarbon/kg sdw day^{-1} [53]. The addition of biosurfactants could increase this rate up to 46.5 g hydrocarbon/kg sdw day^{-1}, when sophorose lipids were added. The added surfactant concentration of 200 mg/L was most effective. The stability of an emulsion of 1.35% model oil in water was highest when the sophorose lipid concentration was above 150 mg/L [54].

The biosurfactant emulsan, a polyanionic heteropolysaccharide from *Arthrobacter calcoaceticus* RAG-1 brings about a significant dispersion of crude oil [55]. However, biodegradation of saturated and aromatic hydrocarbons from crude oil was reduced 50%–90%, investigating a mixed population respectively of eight isolated strains capable of degrading saturates, aromatics, or both. For some pure cultures, aromatic biodegradation was either unaffected or slightly stimulated by emulsification of the oil [56]. Mutants of *Arthrobacter calcoaceticus* RAG-1 defective in emulsan production showed that the cell-bound form of emulsan is required for growth on crude oil because neither added emulsan nor emulsan-producing wild-type cells stimulated the growth of the mutant on these hydrocarbons [57]. The authors argued that emulsan coated oil may prevent access of other biosurfactants that lower the interfacial tension for the formation of a macroemulsion i.e., for hydrocarbon uptake.

To reduce the cost of producing biosurfactants, the effect of glycolipid overproducing strains was investigated. The trehalosedicorynomycolates are the only tested biosurfactants that are overproduced under growing conditions [58]. However, the formed cell-bound trehalose lipid could not reduce the adaptation period of the soil population like the isolated form. Trehalosedicorynomycolates are tightly bound as described by Koronelli [59]. Bound to autoclaved cells they could not reduce the adaptation period for the model oil degradation by the soil population. The degradation rate, the extent of elimination, and mineralization of carbon atoms were enhanced by this hydrocarbon-utilizing strain. However, pristane and naphthalene were metabolized to a lesser extent by *Rhodococcus erythropolis* alone (Table 4) [60].

Pseudomonas sp. DSM 2874 cells produce fewer rhamnolipids under growth conditions than under limited conditions, but this production was enough for emulsification of *n*-alkanes, so their utilization by hydrocarbon degrading soil organisms was also facilitated. Another reason was that a low concentration of rhamnolipids of 20 mg/L accelerated hydrocarbon degradation almost as well as 200 mg/L sophorose lipids. When 7×10^7 cells of this strain was added, the first adaptation phase before the soil microorganisms begin to degrade naphthalene was reduced from 45 to 36 hours. *Pseudomonas* sp. DSM2874 was not able to

TABLE 4 Characteristics of Degradation After Adding Trehalosedicorynomycolates or the Glycolipid Producing the Strain *Rhodococcus erythropolis*

Cultivation	Degradation rate[f] (g oil/g sdw/d)	Elimination of hydrocarbons (%)	Mineralized C atoms (%)
Soil[a]	30.3	93.0	29
Soil, Di[b]	49.7	98.5	28
Soil, aut, Rh. er.[c]	29.5	96.0	26
Rh. er.[d]	43.5	90.0	30
Soil, Rh. er.[e]	45.2	>99.0	39

[a]Soil as inoculum.
[b]Soil and purified trehalose dicorynomycolates (Di).
[c]Soil and autoclaved (aut) biomass of *Rhodococcus erythropolis* (Rh. er.) with trehalose dicorynomycolates bound to the cells.
[d]Autoclaved soil and *Rhodococcus erythropolis*.
[e]Soil and *Rhodococcus erythropolis*.
[f]The degradation capacity was determined up to an elimination of 90% of the hydrocarbon mixture.

degrade naphthalene. The second adaptation phase, up to the utilization of other hydrocarbons was shortened from 49 to 6 hours. This effect, however, need not result from the supply of rhamnolipids, because the strain of *Pseudomonas* employed is able to degrade these hydrocarbons in the presence of autoclaved soil after the same period of time has passed (Table 5) [61]. A mixture of a *Rhodococcus* and a *Pseudomonas*—both glycolipid-producing strains—was also used as a starter culture, but neither the extent of degradation nor the effect of other starter cultures were reported [62].

B. Effect of Biosurfactants on Oil Degradation in Soil

The microbial degradation of lubricating oil (15% paraffins, 54% cycloparaffins, and 31% aromatics) was only stimulated when a chemical dispersant Corexit 7664 (0.11–0.22% v/v) was added into a slurry of the contaminated soil, but not when it was added to the soil alone [63]. At low concentrations (10 μg/g soil) nonionic ethoxylate surfactants enhanced the rate and the extent of phenanthrene biodegradation in both a mineral and an organic soil at 70% of water field capacity, but a surfactant-induced desorption was not appreciable [64].

In a model system of percolated contaminated soil, the added biosurfactants enhanced degradation of a model oil or a fuel oil. However, in the case of oxygen limitation, there was no effect of the added biosurfactants. Furthermore, the nitrite formed was toxic in concentrations above 100–250 mg/L, stopping further degradation of the model oil [65]. Therefore, oxygen was supplied by hydrogen peroxide for laboratory studies of oil degradation in soils [36] and in field experiments [66]. The peroxide was added whenever the oxygen concentration

TABLE 5 Effect of the Rhamnolipid Concentration on the Adaptation Phases of the Oil-Degrading Soil Population

Rhamnolipid concentration (mg/L)	1. Adaptation phase (h)	Degradation phase (h)	2. Adaptation phase (h)
0	29	9	29
20	15	10	16
50	18	7	7
100	13	12	6
200	18	10	5
7×10^9 *Pseudomonas* cells	23	11	4

percolating mineral salts medium was below 2 mg/L after passage of the soil fixed bed. This limit was chosen because the growth of some microorganisms is retarded at a lower oxygen concentration. The inhibition of the soil population by hydrogen peroxide was low and exceeded 5% only at concentrations above 100 mg/L [67].

The elimination of the model oil was increased by biosurfactants added to the percolated medium with the same dependence on the hydrophilic lipophylic balance values as in submerged culture (see Table 3). Added sophorose lipids caused 50% elimination within 6 days in comparison to 38% in the control experiment (Fig. 7). Furthermore the extent of mineralization calculated from H_2O_2 use was 46% instead of 28.3% of the eliminated hydrocarbons.

This acceleration of biodegradation when adding sophoroselipids was confirmed in three of four different soil types (a light loamy sand, a loamy sand, and a silty loam with an annealing loss between 1.1% and 4.5). Only in a soil of high organic content (annealing loss 8.4%) was there no positive effect [68].

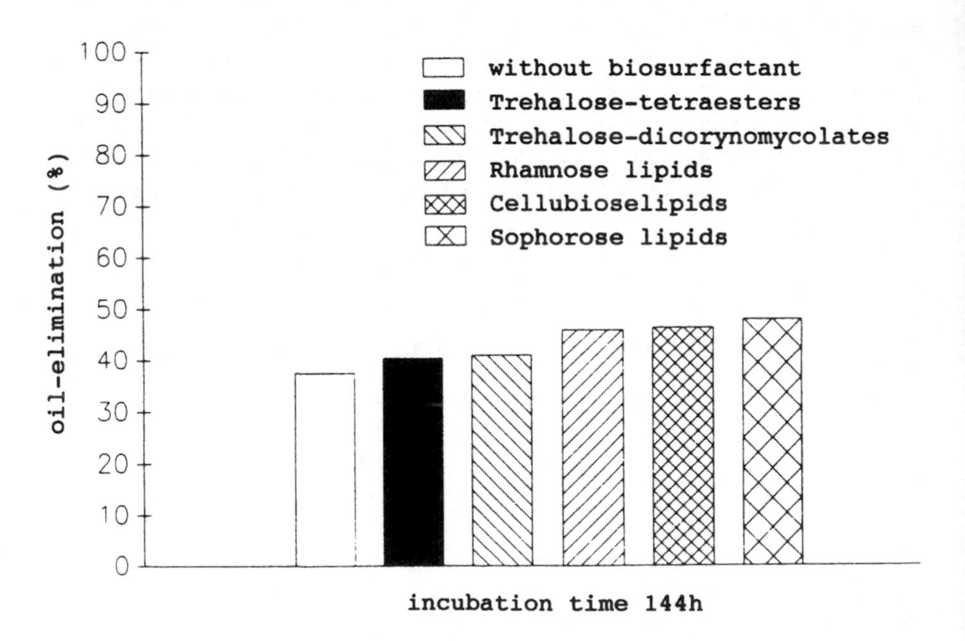

FIG. 7 Effect of biosurfactants added to the mineral salts medium on the elimination of a model oil during 144 h (the lower half of the soil fixed bed was contaminated with 15 mg oil/g soil wet weight).

VI. OIL DEGRADATION WITH ADDED BIOSURFACTANTS IN CONTINUOUS CULTURE

In dual-phase continuous culture systems with an oil phase, oil degradation was studied but not enhanced. In a one-stage [69] and a three-stage continuous culture system [70] at low dilution rates of up to 0.05 hour, the degradation of a layer of 8 ml physically not dispersed topped diesel oil by microorganisms of Ontario lake water lasted a very long time (2–3 months). Furthermore, open flow-through systems were used as a better model for oil elimination in the Arctic Ocean than batch cultivation [71], where the oil was probably not lost by emulsification.

For better oil degradation rates, the oil must be preemulsified [72]. At a dilution rate of $D = 0.05$ hour^{-1}, 8 g crude oil/L hour^{-1} could be degraded. The profile of this continuous culture is shown in Fig. 8. For preemulsification, the large amount of biosurfactants (up to 3 g/L) continuously produced could be recuperated using a tangential flow filtration device. The biosurfactants contained no proteins or phospholipids but did contain sugars and total lipids (65.4% C_{14}–C_{16} fatty acids) in a ratio of 8.4:1.0 [73]. With cell recycling, the dilution rate could be raised to 0.11 hour^{-1} so that the oil degradation rate was further increased from 0.24 g crude oil to 1.06 g/L hour^{-1}. For 82% degradation, a rate of 20.3 g crude oil/L day^{-1} is possible at a substrate concentration of 9 g/L [74]. Such optimization could be carried out faster with the Doehlert experimental design [75]. In batch culture with the selected yeast *Rhodotorula* sp., the degradation rate was five times lower [76] as in the case of natural soil population degrading a model oil [52].

VII. EFFECT OF ADDED SURFACTANTS ON OIL ELIMINATION

Percolating chemical surfactants can reduce up to 93% of the hydrocarbon concentration by mobilization in artificial soils of higher water permeability [77]. In a native soil, chemical surfactants can mobilize more than 80% of oil [78]. An initial fuel oil concentration of 6 g/L in a sandy soil was mobilized to the same extent by a 3% solution of polyglycolester of a fatty acid, but this solution was toxic to microorganisms and to fishes. By a suitable not explained mixture of surfactants with an overall surfactant concentration below 1%, the toxicity of these one or two nonionic and one cationic surfactants was low and their degradability in the standard test was above 80%, but their mineralization and their residual concentration in the soil was not described [79].

In oil-contaminated mud flats, the elimination of polycyclic aromatics from the crude oil Arabian light was due to wave action or to microbial degradation. The chemical surfactant Finasol OSR-5 doubled the initial content of aromatics and decreased the amount of aromatics removed after 6 month, whereas adding the

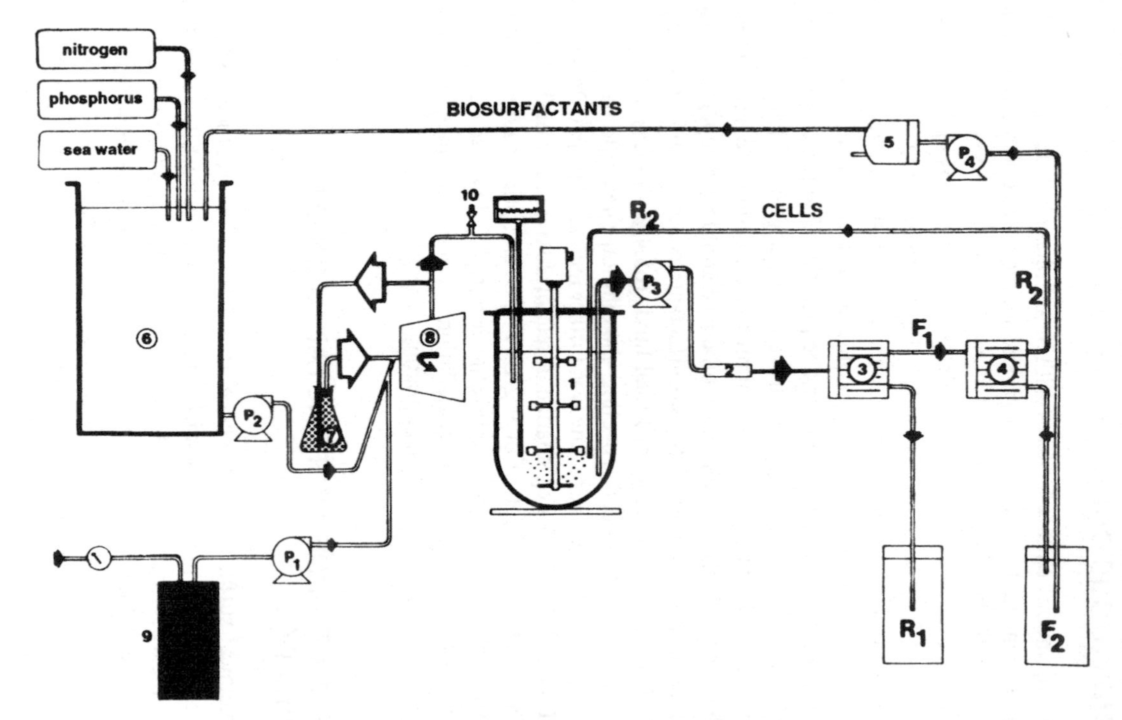

FIG. 8 Fermentation procedure of a crude oil in continuous culture on seawater. (1) Reactor, (2) prefilter (porosity <10 μm), (3–4) tangential-flow system (1.2 μm and 0.2 μm, respectively), (5) Fluorex cartridge (porosity = 0.22 μm), (6) tank with the growth medium, (7) preemulsification reservoir, (8) pump, (9) tank with crude oil (10) sampling for emulsifying activity measurements, P^1, P^2, P^3, P^4: pumps. From Ref. 72.

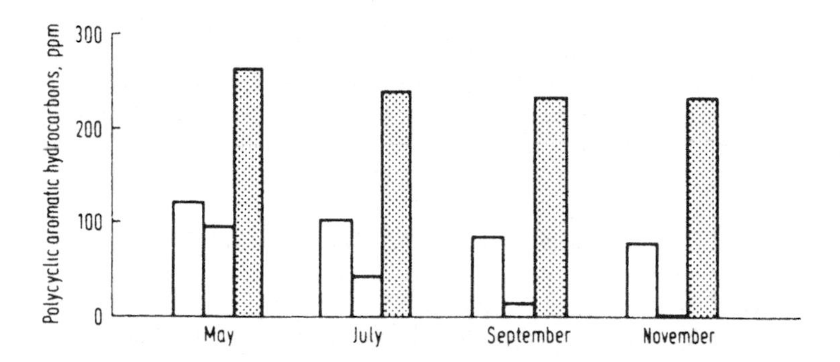

FIG. 9 Elimination of crude oil from flat tidal environments. 2 m^2 Field contaminated: 10× with 1 L Aramco crude oil; 10× with Aramco oil supplemented with 1 g/L trehalose dicorynomycolates: 10× with 1 L Aramco oil and afterwards 10× with 100 mL Finasol OSR5 in 1 L seawater.

biosurfactant trehalose-6,6′-dicorynomycolates caused complete elimination within the period (see Fig. 9) [80]. The reason for that was not the difference in hydrophobicity, as is shown by the fact that biosurfactants of various hydrophobicities had a similar positive effect in contrast to the above chemical surfactant.

VIII. CONCLUSIONS

The mechanisms of hydrocarbon degradation are well known for *n*-alkanes and smaller aromatics, although some variation was found. Crude oil degradation was shown both in the seas and in soil.

Biosurfactants are necessary for the degradation of hydrophobic hydrocarbons. Added biosurfactants can enhance hydrocarbon degradation in a specific or unspecific manner. Chemical surfactants suitable for mobilization cannot replace them. Starter cultures of biosurfactant-producing strains may be used to enhance biodegradation in the environment.

REFERENCES

1. E. Beersteacher, *Petroleum Microbiology,* Elsevier, Houston, 1954.
2. J. B. Davis, *Ind. Eng. Chem. 48*: 1444 (1956).
3. L. R. Brown, *Chem. Eng. Prog. 83*(10): 35 (1987).
4. C. E. ZoBell, Microbial Modification of Crude Oil in the Sea, Join Conference on Prevention and Control of Oil Spills, New York, 1969, p. 317.
5. L. R. Brown, R. J. Strawinski, and C. S. McCleskey, *Can. J. Microbiol. 10*: 792 (1964).

6. G. L. MacMichael, *Metabolism of Gaseous Alcanes by Nocardia Paraffinicum,* Ph.D. Dissertation, Mississippi State University, 1984.
7. D. A. Frello, J. B. Mylroie, and A. M. Chakrabarty, *Proceedings of the International Biodegradation Symposium,* (J. M. Sharpley and A. M. Kaplan, eds.), Applied Science Pub., Essex, England, 1976, p. 79.
8. M. E. Singer and W. R. Finnerty, in *Petroleum Microbiology* (R. M. Atlas, ed.), Macmillan, New York, 1984, pp. 1–59.
9. R. S. Wodzinski and D. Larocca, *Appl. Environ. Microbiol. 33*: 660 (1977).
10. N. Uemura, J. Takahashi, and K. Ueda, *J. Ferment. Technol. 47*: 220 (1977).
11. H. W. D. Katinger, *Biotechnol. Bioeng. Symp. 4*: 485 (1973).
12. Y. Miura, M. Okazaki, S.-I. Mura-kawa, S.-I. Hamada, and K. Ohno, *Biotechnol. Bioeng. 19*: 715 (1977).
13. M. Rosenberg and E. Rosenberg, *J. Bacteriol. 148*: 51 (1981).
14. J. E. Zajic and W. Seffens, *CRC Crit. Rev. Biotechnol. 1*: 87 (1984).
15. D. Haferberg, R. Hommel, R. Claus, and H.-P. Kleber, *Adv. Biochem. Eng./Biotechnol. 33*: 53 (1986).
16. U. Göbbert, S. Lang, and F. Wagner, *Biochem. Lett. 6*: 225 (1984).
17. R. S. Kennedy, W. R. Finnerty, K. Sudarsanan, and R. A. Young, *Arch. Microbiol. 33*: 53 (1975).
18. A. Mimura, S. Watanabe, and I. Takeda, *J. Ferment. Technol. 49*: 255 (1971).
19. Y. Miura, M. Okazaki, S.-I. Hamada, S.-I. Murakawa, and R. Yugen, *Biotechnol. Bioeng. 19*: 701 (1977).
20. D. F. Gerson and J. E. Zajic, *Dev. Ind. Microbiol. 19*:557 (1978).
21. A. J. Desai, K. M. Patel, and J. D. Desai, *Curr. Sci.* (1987).
22. O. Käppeli and W. R. Finnerty, *Biotechnol. Bioeng. 22*: 495 (1980).
23. P. G. Reddy, H. D. Singh, M. G. Pathak, S. D. Bhagat, and J. N. Baruah, *Biotechnol. Bioeng. 25*: 387 (1983).
24. S. S. Cameotra, H. D. Singh, and J. N. Baruah, *Biotechnol. Bioeng. 26*: 554 (1984).
25. S. S. Cameotra, H. D. Singh, A. K. Hazarika, and j. N. Baruah, *Biotechnol. Bioeng. 25*: 2945 (1983).
26. P. Goswami and H. D. Singh, *Biotechnol. Bioeng. 37*:1 (1991).
27. J. Lapinkas, *Chem. Ind.* (Dec.): 784–789 (1989).
28. N. Pilpel, *Endeavour 100*: 11 (1968).
29. C. F. Gibbs, *Proceedings of the International Biodegradation Symposium* (J. M. Sharpley and A. M. Kaplan, eds.), Applied Science Pub., Essex, England, 1976, p. 127.
30. R. M. Atlas and R. Bartha, *Microbial Ecology: Fundamentals and Applications,* Addison-Wesley, Reading, MA (1981).
31. Anonymous, *Chem. Eng. News 48*: 48 (1970).
32. M. J. Barnhart and J. M. Myers, *Pollution Eng.* (Oct): 110–112 (1989).
33. R. Bewley, B. Ellis, P. Theile, I. Viney, and J. Rees, *Chem. Ind.* (Dec.): 778–783 (1989).
34. R. E. Hinchee and D. C. Downey, *NWWA/API Proceedings of Petroleum Hydrocarbons and Organic Chemicals in Ground Water,* Nov. 9–11, Houston, Texas, 1988.
35. American Petroleum Institute, *Field Study of Enhanced Subsurface Biodegradation of Hydrocarbons Using Hydrogen Peroxide as an Oxygen Source,* Publ. 4448, (1987).

36. E. R. Barenschee, O. Helmling, S. Dahmer, B. Del Grosso, and C. Ludwig, "Kinetic Studies on the Hydrogen Peroxide-Enhanced In Situ Biodegradation of Hydrocarbons in Water-Saturated Ground Zone," in *Contaminated Soil '90* (F. Arendt, M. Hinsenveld and W. J. van den Brink, eds.), Kluwer Academic Publishers, 1990.
37. M. McD. Francis, *Proceedings of the Fifth Annual General Meeting on Biominet,* Calgary, Alberta, 1988, p. 119.
38. G. I. Vecchioli, M. T. Del Panno, and M. T. Painceira, *Environ. Pollution 67*: 249 (1990).
39. B. A. Molnaa and R. B. Grubbs, "Bioremediation of Petroleum Contaminated Soils Using a Microbial Consortia as Inoculum," in *Petroleum Contaminated Soils* (E. J. Calabrese and P. T. Kostecki, eds.), 1989, Vol. 2, p. 219.
40. J. Newton, *Pollution Eng.* (Dec.): 46 (1990).
41. S. Fogel, M. Findlay, and A. Moore, "Enhanced Bioremediation Techniques for In Situ and Onsite Treatment of Petroleum Contaminated Soils and Groundwater," in *Petroleum Contaminated Soils* (E. J. Calabrese and P. T. Kostecki, eds.), 1989, Vol. 2, p. 219.
42. P. E Strup, *Determination of Polynuclear Aromatic Hydrocarbons in Industrial and Municipal Wastewaters, Cincinnati,* U.S. Environmental Protection Agency, Rept. No. EPA-600/54-82-025 (1982).
43. A. Köhler, D. Bryniok, B. Eichler, K. Mackenbock, D. Freier-Schröder, and H. J. Knackmuss, in *Dechema Biotechnology Conferences 4,* VCH; Weinheim, 1990, pp. 585–587.
44. G. J. Mulkins-Phillips and J. E. Stewart, *Appl. Microbiol. 28*: 547–552 (1974).
45. J. M. Foght and D. W. S. Westlake, *Can. J. Microbiol. 28*: 117–122 (1982).
46. K.-I. Hisatsuka, T. B. Nakahara, N. Sano, and K. Yamada, *Agr. Biol. Chem. 35*: 686–692 (1971).
47. S. Ito and S. Inoue, *Appl. Environ. Microbiol. 43*: 1278–1283 (1982).
48. G. Goma, A. Pareilleux, and G. Durand, *J. Ferment. Technol. 51*: 616–618 (1973).
49. A. Oberbremer, R. Müller-Hurtig, and F. Wagner, *Appl. Microbiol. Biotechnol. 32*: 485–489 (1990).
50. R. Müller-Hurtig, A. Oberbremer, R. Meier, and F. Wagner, "Effect of Microbial Surfactants on Hydrocarbon Mineralization in Different Model Systems of Soil," in *Contaminated Soil '90* (F. Arendt, M. Hinsenveld, W. J. van den Brink, eds.), Kluwer Academic Publishers, Dordrecht, 1990, pp. 491–492.
51. R. M. Atlas and R. Bartha, "Effects of Some Commercial Oil Herders, Dispersants and Bacterial Inocula on Biodegradation of Oil in Seawater," in *The Microbial Degradation of Oil Pollutants* (D. G. Ahearn and S. P. Meyers, eds.), Louisiana State University publ. no. LSU-SG-73-01, Baton Rouge, 1973.
52. A. Oberbremer and R. Müller-Hurtig, *Appl. Microbiol. Biotechnol. 31*: 582–586, (1989).
53. I. Bossert and R. Bartha, "The Fate of Petroleum in Soil Ecosystems, in *Petroleum Microbiology* (R. Atlas, ed.), Macmillan, New York, 1984, pp. 355–397.
54. A. Oberbremer, *Einfluss von Biotensiden auf den microbiellen Olabbau in einem Ackerboden: Untersuchungen in Rühr—und Festbettreaktoren,* Dissertation TU Braunschweig (1990).
55. A. Reisfeld, E. Rosenberg, and D. Gutnick, *Appl. Microbiol. 24*: 363–368 (1972).

56. J. M. Foght, D. L. Gutnick, and D. W. S. Westlake, *Appl. Environ. Microbiol. 55*: 36–42 (1989).
57. O. Pines and D. Gutnick, *Appl. Environ. Microbiol. 51*: 661–663 (1986).
58. C. Syldatk and F. Wagner, "Production of Biosurfactants," in *Biosurfactants and Biotechnology* (N. Kosaric,, W. L. Cairns and N. C. C. Gray, eds.), Marcel Dekker, New York, 1987, pp. 89–120.
59. T. V. Koronelli, The Use of Rhodococci for Emulgation and Sorption of Oil Hydrocarbons, in *Proceedings of the Fourth European Congress on Biotechnology* (O. M. Neijssel, R. R. Meer, K. Ch. A. M. van der Luyben, eds.), Vol. 3, Elsevier, Amsterdam, 1987, p. 545.
60. E. Goclik, R. Müller-Hurtig, and F. Wagner, *Appl. Microbiol. Biotechnol. 34*: 122–126 (1991).
61. E. Goclik, R. Müller-Hurtig, and F. Wagner, Effect of Microbial Surfactants, in *Contaminated Soil '90* (F. Arendt, M. Hinsenvald, and W. J. van den Brink, eds.), Kluwer Academic Publishers, Dordrecht, 1990, pp. 489–490.
62. Mikro-Bak Biotechnik, German Patent DE 3909324 (1990).
63. B. E. Rittmann and N. M. Johnson, *Water Sci. Technol. 21*: 209–219 (1989).
64. B. N. Aronstein, Y. A. Calvillo, and M. Alexander, *Environ, Sci. Technol. 25*: 1728 (1991).
65. R. Müller-Hurtig, R. Meier, R. Kindervater, and F. Wagner, Effect of Added Microbial Surfactants on Hydrocarbon Degradation in Fixed Bed Soil Columns, in *Dechema Biotechnology Conferences,* Vol. 3 Part B, Verlag, Chemie, Weinheim, 1989, pp. 969–972.
66. D. Raymond et al. US Patent 4588506 (1986).
67. R. Meier, R. Müller-Hurtig, and F. Wagner, Construction and Operation of a Soil Fixed Bed Reactor for the Examination of Aerobic Bioremediation of Contaminated Soil, *Bioproces Eng.* submitted.
68. R. Meier, R. Müller-Hurtig, and F. Wagner, Einfluss von Biotensiden auf den Abbau eines Modellols in verschiedenen Bodentypen, in *Fachgesprach Umweltsschutz,* Dechema VCH, Weinheim, 1992, Vol. 9.
69. P. H. Pritchard and T. J. Starr, Microbial Degradation of Oil and Hydrocarbons in Continuous Culture, in *The Microbial Degradation of Oil Pollutants* (D. G. Ahearn and S. P. Meyers, eds.). Publication no. LSU-SG73-01, Center for Wetland Resources, Louisiana State University, Baton Rouge, 1973, pp. 39–45.
70. P. H. Pritchard, R. M. Ventullo, and J. M. Sulfita, The Microbial Degradation of Diesel Oil in Multistage Continuous Culture Systems, in *Proceedings of the Third International Biodegradation Symposium* (J. M. Shapley and A. M. Kaplan, eds.), Applied Science Publishing Ltd., London, 1976, pp. 67–78.
71. A. Horowitz and R. M. Atlas, *Appl. Environ. Microbiol. 33*: 647–653 (1977).
72. J. C. Bertrand, E. Rambeloarisoa, J. F. Rontani, G. Giusti, G. Mattei, *Biotechnol. Lett. 5*: 567–572 (1983).
73. G. Mattei and J. C. Bertrand, *Biotechnol. Lett. 7*: 217–222 (1985).
74. G. Mattei, E. Rambeloarisoa, G. Guisti, J. F. Rontani, and J. C. Bertrand, *Appl. Microbiol. Biotechnol. 23*: 302–304 (1986).

75. G. Dumenil, G. Mattei, M. Sergent, J. C. Bertrand, M. Laget, and R. Phan-Tan-Luu, *Appl. Microbiol. Biotechnol. 27*: 405–409 (1988).
76. K. Shailubhai, *Trends Biotechnol. 4*: 200–206 (1986).
77. W. D. Ellis, J. R. Payne, and McNabb, *Treatment of Contaminated Soils with Aqueous Surfactants,* U.S. Environmental Protection Agency/600/2-85/129 (1985).
78. A. S. Abdul, T. L. Gibson, and D. N. Rai, *Ground Water 28*: 920 (1990).
79. H.-W. Hurtig, T. Knacker, H. Schallnass, and G. Arendt, "In Situ Mobilisierung von Restölkonzentrationen in Boden-Entwicklung eines Verfahrens fur Auswahl von ölmobilisierenden Detergentien," in *2. TNO/BMFT Kongress Altlastensanierung, Hamburg 1988* (K. Wolf, W. J. van den Brink, F. J. Colon, eds.), Kluwer Acad. Publ., Dordrecht, 1988, pp. 941–948.
80. K.-H. van Bernem, *Senckenbergiana Marit. 16*: 13–23 (1984).

Index